EARLY IRISH IRONWORKING

B.G. Scott

For My Parents

EARLY IRISH IRONWORKING

B.G. Scott

ULSTER MUSEUM

Publication of this book
has been made possible
through generous grants from the
British Academy,
Council for British Archaeology,
Marc Fitch Fund,
National Museum of Ireland,
Queens University Belfast.

ISBN 0 900761 25 3

Ulster Museum Publication Number 266

Table of Contents

Preface

I was first introduced to the study of early metal-use by Professor Martyn Jope, while an undergraduate at Queens University Belfast. It was he who gently guided me towards post-graduate research into ironworking in Ireland. Shortly after beginning my studies, I was fortunate to meet and learn from the pionneer of archaeometallurgy in Ireland, Professor M.J. (Brian) O'Kelly, who initiated me into some of the mysteries of industrial remains on archaeological sites. I was fortunate also at this time to encounter Professor Michael Duignan who showed me the importance of the early Irish literary and linguistic evidence. To each of these fine scholars I owe a very real debt for their kind encouragement, advice and interest.

In 1972 I was appointed to a post in the Department of Antiquities of the Ulster Museum, and was permitted to continue with some of my research as part of my duties there. Between 1972 and 1984, I was helped greatly by the then Keeper of Antiquities, Laurence Flanagan, and by his deputy Richard Warner, who made study material available and were always ready to discuss problems and ideas. I am most grateful to them, and to the present Director of the Ulster Museum, Mr. J.C. Nolan for permission to use some museum facilities and time in the course of my research and writing.

Over the years, I have been struck by the generosity of colleagues throughout Ireland in making material and information available, often in advance of publication. I am sure that this was due to innate kindness, rather than just amused tolerance of one so blessed of the holy simplicity as to be engrossed by corroded lumps of iron! I remember with gratitude the unfailingly friendly assistance of successive directors of the National Museum of Ireland, Drs. Joseph Raftery, Breandan Ó Ríordáin and Patrick Wallace, and of the present Keeper of Irish Antiquities Dr. Michael Ryan and his colleagues. Had they not made material available for metallographic study, this work would not have been possible. Mr. R. Weatherup, Curator of Armagh County Museum kindly permitted me access to the collections there.

Coming from a state school background in Northern Ireland, I did not have the opportunity to gain a basic grounding in Irish. As a result, what little I do know of the language, modern and ancient, has been learned somewhat haphazardly. Ignorance notwithstanding, I have pursued the subject as far as possible. Without the patient guidance of Professor Gearóid Mac Eoin, my efforts would have come to little. Whatever the merits of the chapter on literary and linguistic evidence, they are due in no small part to him.

A number of excavators have invited me to participate in the uncovering of the remains of early ironworking, and I am most grateful to Professor George Eogan, Dr. Tom Fanning, Dr. Mick Monk, Leo Swann and Richard Warner for these opportunities. I am grateful too to Valerie Keeley, Dr. Chris Lynn, Elizabeth O'Brien and Brian Williams for inviting me to see work in progress on their sites. Other archaeologists have provided information and allowed me access to their recently excavated material, and for this I must thank Vic Buckley, Dr. Richard Ivens, Eamonn Kelly, Drs. Chris Lynn, Barry and Joseph Raftery, Professor Etienne Rynne, Martin Timoney, Professor Bernard Wailes, Richard Warner and Brian Williams. Janet Lang kindly provided samples from the Lisnacrogher ironwork, Kathleen Ryan and Kathleen Moreau provided the Dún Ailinne samples, and Mark Hall collaborated in the examination of the knives described in Chapter 5.3.

In the course of my research, I have been assisted greatly by members of the *Comité pour la Sidérurgie Ancienne de l'UISPP*, and in particular by Drs. Henry Cleere and Radomír Pleiner (Secretary of the CPSA) and Professors Kazimierz Bielenin, Jerzy Piaskowski and Ronald Tylecote, all of whom have made such major contributions to archaeometallurgical studies of European ironworking. I must also record my thanks to Drs. Phillip Andrieux, Peter Crew, Janos Gömöri, Peter Kresten, Gerry McDonnell, Bob Maddin, Gert Magnusson, Elzbieta Nosek, Effie Photos and Gerhard Sperl for many stimulating discussions over the years. I have spent some very fruitful and enjoyable times discussing archaeometallurgy in Prague with Radomír Pleiner, and I am grateful to Dr. M. Richter, Director of the Archaeological Institute ČSAV, Prague and Ms. B. Brabcová for facilitating my study visits there.

Chapters of this book have been read in draft by a number of colleagues who made helpful suggestions and corrections, and for this I thank Drs Henry Cleere, Chris Lynn, Gerry McDonnell, Jim Mallory, Radomír Pleiner, Richard Warner and Professor Gearóid Mac Eoin. Over the years, Richard Warner has always been ready to do battle in his own inimitably gentle way over points of interpretation of the Irish evidence. While we disagree in many areas, his critical evaluations of a number of problems have assisted me greatly in my thinking and writing. Needless to say, none of these scholars bear any responsibility for the demerits of this book. Equally, while the Ulster Museum has participated in its publication, the views expressed throughout are personal, and not to be attributed to that institution.

Dr. Connor Murphy and Ms. Patricia Bonner of the State Laboratories, Ireland, undertook analyses of the iron artefacts, Jim Macrae of the SEM Laboratory, Queens University Belfast, and Dr. Chris Salter of the Department of Metallurgy, Oxford, undertook microprobe analyses. The Department of Mechanical Engineering, Queens University Belfast allowed me access to hardness testing equipment, and Florence Bertin made many of the hardness measurements. Many of the artefact drawings, and all of the schematics were done by Mrs Deirdre Crone of the Department of Antiquities, Ulster Museum, who spent painstaking hours on them. The photographs were printed by Michael McKeown, Department of Design and Exhibition Services, Ulster Museum. A significant number of drawings were taken from the published work of Dr. Barry Raftery, University College, Dublin, who kindly gave permission for wholesale borrowings. Mrs Sandra Neill, Ulster Museum and Mr. Bryan McCabe of W. & G. Baird Ltd. have guided the book through press with their customary skill and efficiency. My secretary Edith McCandless has been of great assistance, and my other staff, past and present, have helped also by tolerating my not infrequent absences of both body and mind. To all of these, I am most grateful.

Publication of this book would not have been possible without the generous grants provided by the British Academy, Council for British Archaeology, The Marc Fitch Fund, the National Museum of Ireland, Queen's University Belfast and the Ulster Museum.

Last, but not least, I can only thank my wife Lesley for enduring with patience and good humour many years of ferrous widowhood.

Readers may find the layout of this book somewhat eccentric. I realise that the most logical order might be seen to have been to place the work on ores, slags and furnaces as Chapter 3, prefacing the metallographic studies and archaeological syntheses. However, I felt that this chapter was sufficiently sparse as to allow its positioning as the last of the chapters dealing with archaeological evidence.

I must also note the reasons for what may be seen as the deviations from what I have tried to define as archaeometallurgical methodology. As I explain in Chapters 1 and 8, archaeometallurgy requires that the evidential net be cast as widely as possible. Were I reviewing this book as written by another I should query the paucity of information on artefact typology, on ore sources, and on the interpretation of slag assemblages. A major problem for me has been the lack of *corpora* of ironwork for the Later Iron Age, and of information on assemblages of industrial debris. My work as Keeper of Conservation in the Ulster Museum has precluded me from making the detailed collection searches and from carrying out the fieldwork necessary. My research in recent years thus has been restricted to that which I have been able to carry out in my laboratory without interfering unduly with the duties of my post, and to reading and writing that could be done in my own time at home. For this reason, I have tried to provide the fullest bibliography possible so as to facilitate the work of others.

In a country whose archaeology were organised differently, I might have expected more people to have been working in the same field, thus providing sources of cross-fertilisation of ideas and approaches. Unfortunately in Ireland, I have been working alone for the most part, and have had to attempt to emulate those earlier polymaths whose era should have now been history. This has forced me to make expeditions into the territory of others, with the risk of antagonising a wider range of interests than is afforded even by a career in politics or the police. To those specialists who may feel that I should have stayed within the boundaries of my training as a prehistorian, I can only observe:

Nach indéuin benar	Any anvil which is struck
for-cain in cách no-da-	teaches he who strikes it;
ben nach sí for-canar	it is not taught.

Brian Scott
1st June 1990

List of Figures

List of Plates

List of Tables

CHAPTER 1

Introduction

1.1 – Background to the study

The subject of this book is iron, a material which has been in use in Europe since the later second millennium BC. The influence of iron and of its production industries on the development of world social and economic formations, especially in the last few centuries, can scarcely be understated. But it was not until the 19th and 20th centuries that metallurgists began to understand properly the complexities of the structure of iron and its alloys, and to apply scientific theory to achieve real consistency of production. For over two millennia, however, such knowledge was not available, and yet ironworkers were able to produce first finely forged tools and weapons and later well-cast artefacts. For the greater part of its history, the shaping of iron and steel was a craft of hand and eye, rather than a problem of engineering, an art rather than a science. It could be made to repay the skill of the smith not only in terms of shape and surface texture, but also by giving properties far surpassing those of any previously available industrial material. Ultimately it was these properties which ensured its adoption by the metalworkers of the Old World.

That Ireland followed the rest of the Old World in adopting iron, in place of copper alloys, as the major industrial metal may be seen from the many native-produced artefacts housed in our national collections. While it proves profoundly difficult to reach consensus definition of an Irish 'Iron Age' in conventional archaeological terms, native development of iron exploitation demonstrates the progress of a technological Iron Age here. Much of the attention of Irish archaeologists in this century has been directed at the period of iron use, with art-styles and habitation sites the subjects of continual detailed study and debate. Relatively little thought, however, has been given in Ireland to its eponymous material, either in terms of its working or of its place in the early economy. Although there has been some study of the surviving debris of ironworking, by the late M.J. O'Kelly (whose pioneering experiments in the early 1950s were among the earliest conducted anywhere: O'Kelly 1961a), as far as I can determine the only technical studies of early iron artefacts found in Ireland, prior to the start of my own work in 1971 (Scott 1971), were some chemical analyses of Viking sword blades from Kilmainham, Co. Dublin, carried out in the mid-19th century (e.g. Mallett 1853), the examination of one object each from Aghnaskea cairns A and B, Co. Louth (Evans, E.E. 1936, 1940), the examination of two split blooms from Co. Fermanagh (Evans, E.E. 1948) and the examination of a bloom from Downpatrick, Co. Down (Tylecote 1962, 263).

The establishment of iron in Europe was part of a complex of major social and economic changes. The adoption of iron is important archaeologically, because through this process we can detect the development of socio-economic structures which could support workers experimenting in a new medium and provide markets for their wares. Its appeal to prehistoric societies is evident from its exploitation as a source of cheap raw material on which the mass production of a wide range of mundane artefacts was based. There had been large-scale production of artefacts before iron was known. There is nothing in pre-Iron Age Europe, however, to compare with the scale of Iron Age organised mass-production as represented by the iron nail jointing of the timber components of a *murus gallicus*. The simple fact of the appearance of iron tells us much about the organisation of industry in early society. It thus seems important to study not only the morphology of iron artefacts, but also the ways their production was organised by early workers, the skills displayed in all stages of their manufacture and even (where possible) how they spoke of their raw materials and techniques.

In order that we may understand better the role of iron and its exploitation in early Ireland (or any other region), four main areas require detailed examination: first, the archaeological evidence for iron-related industrial activity, second, the raw materials – ores, fuels and furnace construction materials – on which ironworking depended and the mechanisms and structures of smelting, third, the levels of iron-working technology employed in converting raw materials to finished artefacts, and fourth the light which early Irish language and literature sheds on the exploitation of iron. These are the fields which make up the study today known (somewhat inelegantly) as *archaeometallurgy*. The present organisation of the study of Ireland's past precludes the mounting of the institutionalised, interdisciplinary research programmes required to carry out such research in depth. Thus, the work presented here is the product of one person's crossing of boundaries between archaeology, linguistics, literary studies, metallurgy, petrology and chemistry, each a far-ranging specialism in its own right. The reader will therefore appreciate that this book could well be subtitled *A Preliminary Survey*, since one person can do little more than define the main problems involved in such a wide field. It is my hope that others will be encouraged to engage in research into the many problems left unresolved here. One useful function of this work may be in defining problems more clearly, even though, during its course, it has seemed that attempting to obtain such definitions has thrown up more problems than solutions! Preliminary conclusions are certainly possible from a preliminary survey, but as we shall see, much more still remains to be done.

* * *

1.2 – The archaeometallurgical study of early iron use and its applications in Ireland

The overall aim of the archaeometallurgist, well described and defined by Cleere (1987), is to identify and define the roles of metals in early society, drawing together and synthesising all of the available sources of information. It is, in effect, a broad study of one aspect of man's exploitation of natural resources (cf. Jope 1966). The first modern works (cf. Scott 1974a) consisted of the compositional analysis and/or metallographic examination of artefacts. As the subject has expanded more of the available evidence has been trawled in, and now, no single line of study dominates. Instead, links between the different bodies of evidence are being established, widening the view of the ways in which our predecessors exploited metals and of their impact on early social, economic and political organisations. While it is iron and iron-use which are the subjects of this book, the work presented here is but part of a broader canvas. A common theme is the question group 'when . . ., how . . ., why . . ., where . . .?'. These may be posed at site level, to define the scope of industrial activity and its degree of technological development, or at regional level as part of the broader analysis of resource exploitation. One obvious focal point is the artefact from a site assemblage, whose scheme of interrogation as defined by Northover (1985: with modifications and additions) may be summarised as:

Metallurgical

what is it made of?
where did the materials come from?
how was it made?
why was it made from that material
how did use affect it
can we match composition and technology elsewhere?
can we reproduce the physical structure experimentally?

Archaeological

what is it?
what was it used for?
why is it here?
why was it made like that?
where did it come from?
how did its use affect users?
why is it in its present state?
what date is it?
what are the comparanda?

Although the two groups can be treated separately, to answer each as completely as possible requires complementary answers from the other. The information obtained can be matched site by site, allowing the broader regional and chronological picture to become clearer.

The second focus of attention obviously is the industrial evidence from a site and again we can readily construct a check-list of technological questions requiring answers:

Metallurgical

what species were used for fuel? what ore was exploited?
what clay was used?
where did the raw materials come from?
how were they brought to the site?
what are the waste materials present?
what process does each structure represent?
what range of processes is represented?
can we duplicate the process(es) experimentally?
what scale of production is represented?

The archaeological questions will be those asked by every competent excavator. Again, it is essential that we treat the two as complementary to obtain maximum effect.

'Traditional' archaeological methodology provides data on the distribution of iron use, on the typological range of artefacts produced and on the chronological ordering of artefact assemblages. However, it tells us little about production technology. Two iron implements may have the same form, but have been manufactured at quite different levels of technology. Finds of ore fragments and 'slags' on a site may indicate smelting, smithing or a combination of both, but without further study, do not take us much further. Finds of iron artefacts tell us that iron was in use, but not how competent were the producers. It is thus obvious that to progress, the scope of investigations must be widened. The two suites of scientific methodology which can be brought to bear are metallography (the study of metal structures and how they were formed) and mineralogy (the study of the structure and composition of minerals). By introducing these, we can see how iron artefacts were produced and thus compare levels of technology spatially and chronologically. From this, we can chart regionally the development of industrial and economic expertise through time. The evidence to be found in early literature is of the utmost importance in helping us to discover not only how artefacts were made, but also (as far as possible) how they were viewed by their makers and users, and how ironworkers fitted into social, economic and political organisation. Archaeometallurgy is not merely concerned with the scientific evaluation of individual artefacts, but rather with developing a detailed picture of the many roles of metal use in society. It is thus neither a subsidiary of, or replacement for traditional archaeology, but rather an augmentation of its aims and methods. In the last thirty years, major contributions to the study of early iron working technology and its impact on society and economy have been made by workers such as Andrieux, Anteins, Bielenin, Cleere, Coghlan, Crew, Gömöri, Kolčin, Kresten, Maddin, Magnusson, Mangin, Martens, Muhly, Nosek, Piaskowski, Pleiner, Sperl, Stech, Thomsen and Tylecote (see *References*).

The greatest single problem to confront the archaeologist studying Irish iron artefacts (and one hardly confined to Ireland) is that the vast majority are typologically nondescript. There is seldom sufficient consistency of detail to permit other than the most general seriation. Indeed, in Ireland, shapes observed on one site may bear little relationship to those of artefacts from a site supposedly of similar period only a few miles away. That ubiquitous general purpose tool the single-edged iron knife shows no immediately-apparent patterns to the variations in its form (Fig. 1.2.1), either within or between assemblages over more than a millennium. Moreover, the sharpening and re-sharpening of knives will have brought about drastic modifications of their original forms (Fig. 1.2.2). The degree to which this affected shape will be unpredictable from superficial examination, since a good steel cutting edge will have required different treatment to one made from soft iron. Equally simple artefacts, such as iron shears are likewise, at present, not amenable to meaningful typological seriation over the same period. Ornaments of iron (for example, pennanular brooches or pins) may be grouped by typological comparison with their copper alloy counterparts. Swords of the period are slightly easier (cf. Rynne 1982; Raftery, B. 1984, 62ff), although there are differences of opinion over their claimed typologies and dates. Similarly, while some spearheads show characteristics which could, outside Ireland, lead to their being firmly attributed to well-defined groupings, the ancestry of others is sufficiently unclear as to lead different scholars to offer datings many centuries apart. Even the shapes of weapons which display an identifiably intrusive ancestry may have had their distinctive features blurred by the manipulations of Irish smiths. The problem of typological simplicity is further exacerbated by the absence of a secure chronology for the majority of Irish Iron Age artefacts, for there is an almost-embarrassing scarcity here of clearly dateable associations for much early ironwork (grave deposits, hoards and stratified deposits on well-excavated sites).

Another major cause of difficulty for scholars is the high degree of susceptibility of iron to corrosion which may reduce the most distinctive shape to a featureless mass. Although simple and cheap, there often is a surprising lack of interest among excavators in the use of X-radiographic screening of assemblages as an aid to identification and interpretation (Scott 1976a). Even where assemblages are well preserved, there are still problems, since the nature of bloomery iron precluded adherence by early smiths to regular patterns, demanding instead that each shape be brought out by hand from the material. Now, the work of the farrier shows just how accurately copies can be reproduced over a period of time, when he forges sets of matched horseshoes using his eyes as sensitive measuring instruments. It is clear that the early smiths here could have anticipated their Anglo-Saxon, Scandinavian and Anglo-Norman counterparts by establishing consistent styles of iron artefact to be reproduced widely throughout the country. They generally tended not to adhere to strict patterns, indicating a strong tendency to work from a general idea for specific artefact types, rather from a specific pattern to reproduce artefacts which now would lend themselves to typological seriation. One reason is certainly the dispersed nature of Irish society in the 1st millennium AD where large industrial groupings were not formed. This precluded the development of organisations geared to the kind of mass-production which gives rise to typologically-amenable uniformity[1]. Thus, chemistry and early polity conspired to ensure that conventional studies based on comparison of consistent features, in most cases can only have limited value. It is thus necessary to study other features of surviving ironwork if the full information which such material holds is to be retrieved.

The problems of poor preservation, lack of useful associations and of sufficient typologically diagnostic features become particularly acute when we consider the question of tool types (sect. 1.5). We have few recognisable smiths' tools of any period before the intrusion of the Vikings. It is possible to infer the use

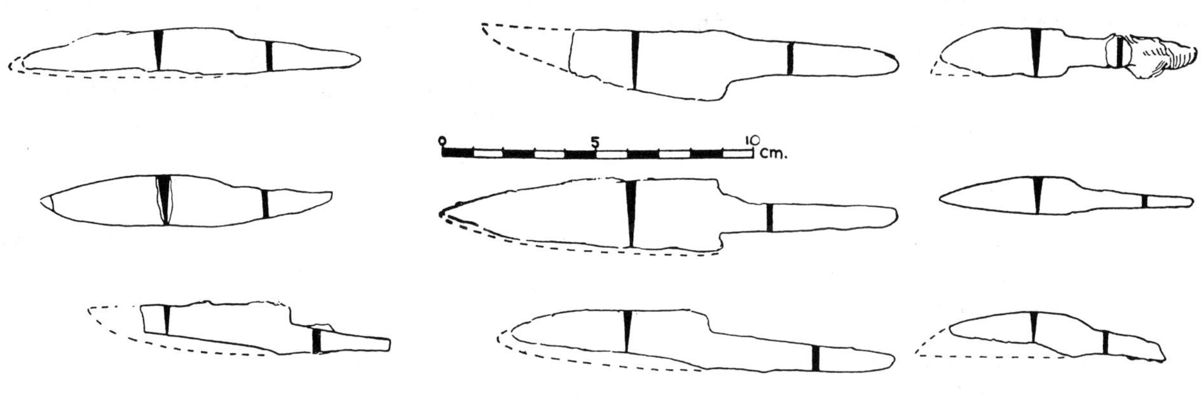

Fig. 1.2.1 Variation in the shapes of single-edged iron knives from Garryduff I, Co. Cork, and one archaeologist's reconstruction of shape (after O'Kelly 1964).

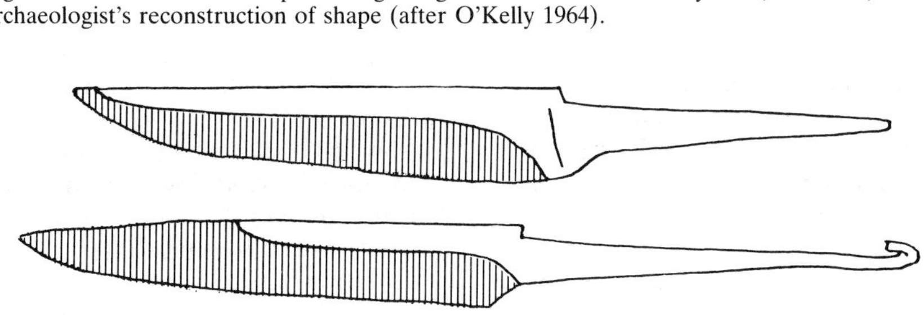

Fig. 1.2.2 Diagrammatic representation of the changes in shape caused by continual sharpening of an iron knife. Shaded areas represent metal removed through sharpening (after Pleiner 1962).

of iron tools both in blacksmithing and other crafts in the Earlier Iron Age, not only from tool marks but also from the properties of the material that would have been required to accomplish particular tasks (cf. Chapter 4.1). It has been argued cogently that the use of a steel chisel would have been required to carve away substrate copper-alloy on the Bann Disc, leaving the fine curvilinear decoration standing out in low relief. But artefacts like these are almost certain to go unrecognised after a long interment, both because of their simple shape and as a result of corrosion damage. If technological progress is indeed to be seen in terms of the search for a better cutting edge, then it is essential that we can identify not only the obvious edged weaponry of different periods, but also the equally important, but far less prominent, contemporary working tools.

The unattractive debris of smelting provides complex data which may be studied to provide a wealth of detail on the conditions during working of the furnace, the efficiency of the process and even in some instances the likely type of ores used. Chemical analyses give us the overall composition of the slag, while the structures and compositions of the various phases present tell us something about conditions during the smelt, about the efficiency of the process and about cooling conditions. Analysis of a single piece of slag will provide information of very limited value. But when a series of slag pieces from one site is analysed and examined, along with other survivals from the smelting operation (such as the remains of furnace linings and fragments of unreduced or partially reduced ores), the results provide an overall picture of the operation of the smelting process. Frequently, 'slag' turns out to result from smithing, either the consolidation of raw blooms after their removal from the furnace, or the forging up of artefacts from stock. Again, the form and quantity of smithing slag gives us an important insight into the level of industrial activity on a site: valuable work in this field includes studies by Sperl (1980), Bachmann (1982), Kresten (1984), McDonnell (1986) and Keesman (1989).

The remains of smelting furnaces themselves are invaluable evidence for production and their numbers and occurrences on sites allows us to make some deductions about output levels. Our understanding of the mechanisms of early iron production has been assisted significantly by experimental studies. Reconstructions of early smelting have attempted to duplicate the process as operated in the past: this includes the work of Gilles (1958), O'Kelly (1961a), Thomsen (1964), Cleere (1972), Pleiner (1969a), Tylecote et al. (1971), Bielenin (1973), Clough (1987), Tholander (1987), Crew (1988), Photos (1988), and Andrieux (1989). Comparison of the products of ancient smelting sites and modern smelting experiments have improved our understanding of how early man managed to make metal. From the data obtained, we have learned much about the technological constraints on early ironworkers and also how they managed to circumvent at least some of them to conduct viable and evidently successful operations. The study of recent bloomery production in India and Africa (Cleere 1963; Tylecote 1965; van der Merwe 1980) should not be overlooked because smiths can still produce iron as a matter of course! Similarly, experiments to study the behaviour of hot iron and steel under the blacksmith's hammer (Pleiner 1973) have played an important role in the interpretations of structures observed under the microscope. Although much has been learned from such studies, they have still the potential to shed further light.

The shape of an early iron artefact conceals an often complex crystal structure whose interpretation is the key to our understanding of the technical competence of individual smiths and the physical properties of their products. Chemical analysis of the metal may, in certain circumstances indicate the type of ore from which it was smelted or, more likely (and of almost equal importance), eliminate particular ore types from consideration (e.g. Salter 1982). Metallographic examination of sections taken from iron artefacts can tell us whether or not steel, or unevenly carburised iron, or even carbon-free iron was used. Weld seams may often show up in polished section as indicators of another important skill of the ironworker, particularly when the weld is between iron and steel, something that indicates a high degree of competence on the part of the smith. Another crucially important technique detectable in this way is the heat treatment of steel which allowed the smith greatly to enhance the physical properties of his tools and weapons. Identification of the structures to which these give rise also allows us to differentiate between mundane ironwork and good working tools, even in assemblages badly damaged by corrosion. Analysis of the slag inclusions which early wrought iron inevitably contains has the potential of offering information on the ores from which they were smelted (Hedges and Salter 1979).

In considering the 'quality' of ancient iron artefacts, it is important to avoid imposing modern metallurgical standards onto past production. There is a subtle point to be made here concerning the application of technological processes. A characteristic of Viking weaponry, for example, is welding which leaves a clean, sharp seam indicating quick, precise work. In contrast, many of the weld seams observed in early Irish artefacts show evidence of working at heat for periods longer than necessary to accomplish a joint. So while one might contrast the skill of native Irish and intrusive Viking smiths in welding technique, the fact remains that in both cases the desired result – the fusing together of two pieces of metal – was achieved. One might be tempted to make the subjective judgement that by working faster, the Viking smith could achieve a higher output than his Irish counterpart. But this would ignore the differences in industrial organisation between the two groups in Ireland. In short, the macrosection of an Irish iron artefact might not look as attractive as that from an equivalent Viking piece, but if the desired effect was achieved in each case, then comparisons of 'quality' made on the basis of microstructure alone may be misleading. In fact, it is in continuous use that 'quality', or lack of it, will have shown. An iron sword or spearhead, made from the poorest of iron, but with sharpened edges and point, would have been quite capable of inflicting appalling injury with one well-directed blow. Equally, the worst quality iron knife, with sharpened edge, would cut effectively for a time. But the majority of Irish swords examined would have stood up only poorly to prolonged combat. Similarly, many knives would have required very frequent resharpening, resulting in metal loss. Better made tools and weapons would have given more reliable and longer service. For these reasons, comments on the effectiveness of tools and weapons, made on the metallographic examinations presented in Chapters 3–5, are intended as a rough guide only as to how they might have performed in regular use.

Although metallographic studies of early ironwork date back to the end of the 19th century (Scott 1974a), the seminal published study for early iron working technology is Kolčin's clear demonstration of technological change in iron knives incorporated into the well-dated stratigraphy observed during the excavations at Novgorod (Kolčin 1959). Here it was found that the method of constructing knife blades changed and developed through time (Fig. 1.2.3) and that within reasonable limits of caution, the technologies of knife blades could be used to position them in a technological/typological classification tied into a chronological framework (cf. McDonnell 1987, 1989; Ottaway 1987). The key feature of this work was the attempt to study statistically representative samples through the stratigraphy in much the same way that the palaeoecologist takes a boring through a peat deposit and then counts the numbers of pollen grains of the various species represented to see how their relative frequencies change along the length of the bore.

When sufficiently large numbers of artefacts have been studied in this way, it is possible to use techniques of numerical analysis to discern recurrent patterns that may set the production technology of one region and/or period apart from others. This is how the development of the technology in the Old World is being charted. At the level of the individual site assemblage, we are able to scan the range of production technology and, at the very least, to assess the likely usefulness of the products for the tasks on which they would have been employed. On a wider scale, metallographic examination and chemical analysis provide the basis for the broad description of the products of regional industries. From these data, it is then possible to go on to examine the structure and organisation of the iron production industries themselves and their roles in wider socio-economic and political frameworks. This is shown particularly well by Piaskowski's long-term research on the assemblage of iron artefacts associated with the major smelting and bloomsmithing complex of the Holy Cross Mountains of southern Poland (Piaskowski 1964, 1985; Bielenin 1974). The fruits of wider surveys still may be seen in syntheses of work conducted for barbarian Europe over the past two decades (e.g. Pleiner 1962, 1980).

It is not yet possible to conduct detailed statistical study of the chemical and metallographic features of early Irish ironwork, since the numbers of artefacts studied so far is too small for such methods to give useful results. There is thus a problem to be faced in the fact that an examination of one artefact may be quite unrepresentative of the general level(s) of skill of smelters and smiths working within a particular region and/or period. Furthermore, by its very nature, the form in which early iron was produced by the bloomery process means that it is frequently very uneven in composition: thus a sample must be sited so as to ensure that the structures observed are as representative of the overall composition as possible. It is not unusual to find that two sections taken from the same artefact show quite different metal structures, even though they are sited but a few millimetres apart. One must also be aware that the non-

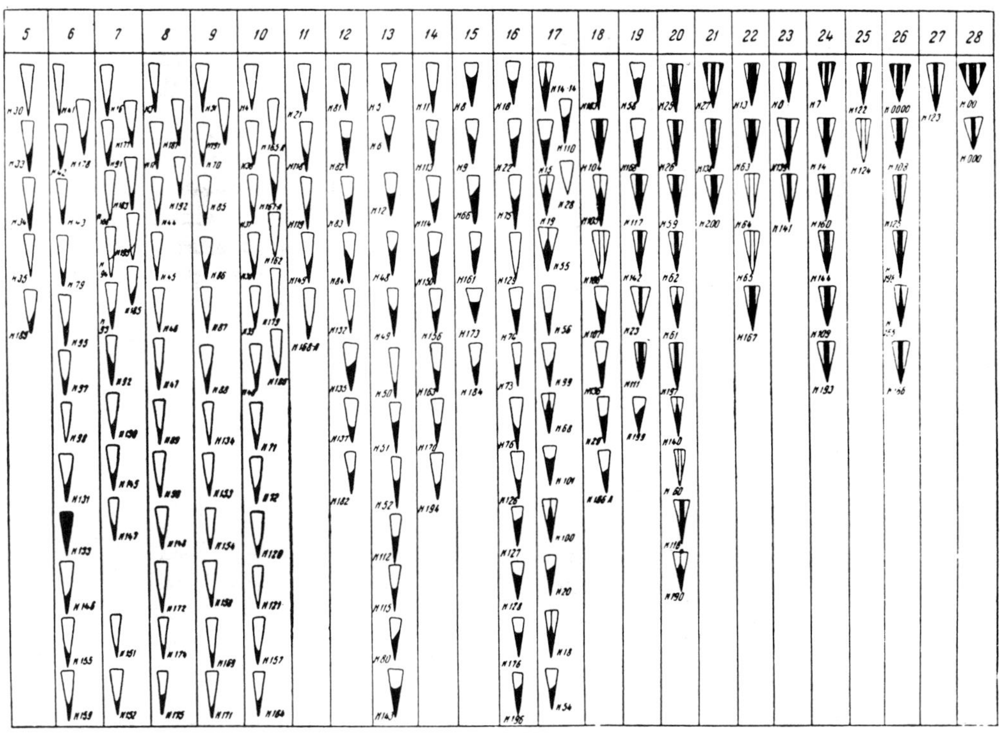

Fig. 1.2.3 The development of knife-blade technology related to the dendrochronologically dated stratigraphy at Novgorod (after Kolčin 1959).

appearance of a particular technique may not be due to any lack of knowledge on the part of the smith, but rather to the fact that its application would not really have been needed in the manufacture of that particular artefact. It is easy to envisage a situation in which metallographic examinations of artefacts from an assemblage revealed no evidence of welding or heat-treatment, simply because they were not particularly appropriate to the manufacture of any of the artefacts examined. Equally, it is possible to cite numerous examples where a technique clearly was not used, resulting in an inferior product. One must beware of giving too wide a significance to the evidence of skill shown by one artefact or one small group of artefacts.

The information provided by excavation, physical examination and analysis, is complemented by the evidence available in the form of early language and literature (Scott 1983a, 1989b). Ireland in particular is endowed with a wealth of written material ranging from law texts to poetry. Although much of the metallurgical information is difficult to extract, it is possible to to find descriptions of early mining and metalworking and to reconstruct a significant part of of the associated vocabulary. Study of early writers may also help to locate early groups of ironworkers, as Piaskowski (1981) has done for the *Kotini*. All of this forms a body of information that is crucial to our understanding of the role of metals in early society, since it provides us with valuable insights into how people regarded and described metals, and into the roles of metals in early society. Thus, although traditionally treated as a separate field, the study of early texts must be regarded as an integral part of archaeometallurgy.

Despite the inadequacy of our present data for Ireland, it is still possible to generalise on the technology employed in the manufacture of various artefact types and from these broadly to trace the technological progress of ironworking here. The archaeological and literary evidence provide some useful cross-comparisons, with regard both to the organisation of production as part of the early economy and the non-technological factors acting to constrain the development of iron industries. In addition, it is possible to make use, albeit tentative, of comparative technology to augment typological seriation. We might draw the analogy here with the art-historical approach adopted to trace the development of major styles and motifs (such as is found on Iron Age decorated metalwork) over a period of time. However, the technological features observed through a particular class of artefact are not sensitive enough for more than very cautious inclusion amongst a range of typological discriminants. Also, by comparing native with intrusive, again, using the art styles analogy, it should be possible to detect outside influences on indigenous technology. In the future, when a great many more artefacts have been examined and analysed, we will be able to describe not only the development of iron-use here in clearer detail, but also to reconstruct more accurately the organisation of early Irish industry and economy as a whole.

* * *

1.3 – The Chronological Framework

Among Irish archaeologists, uncertainty over the major problems of the period from the end of the Bronze Age to the arrival of the Anglo-Normans, is reflected by a proliferation of terminology used to describe its supposed major subdivisions. The terms *Iron Age*, *Hallstatt*, *La Tène*, *sub-Roman*, *sub-Saxon*, *Early Christian*, *Dark Age*, *Early Historic*, *Patrician*, *Celtic*, *Early Mediaeval* and *Viking* are all used in the archaeological literature with an alarming variety of intents and purposes (cf. Champion 1982). The commonly-used term *Early Christian* seeks to use the 4th/5th century AD implantation of Christianity (an acquisition dear to the Irish) to define a period. But the earliest impact of the new creed is seen, at present, not in the archaeological record, but as a stratum of mainly ecclesiastical Latin loanwords in Old Irish. The term *Early Historic* is even more nebulous in conception, and *Early Mediaeval* seems quite meaningless. While the onslaught of the Scandinavians on 9th century Ireland had a major impact on contemporary ecclesiastics and *literati*, (as Ryan – 1982, 45 – succinctly describes it '. . . for concentrated nastiness . . . the early phase of Scandinavian piracy has few parallels in European history . . .'), its major effects were on the Irish seabord. Inland, their artefacts and art styles certainly influenced Irish craftsmen, but the overall way of life was little altered if at all. The particular insistence of many archaeologists both inside and outside Ireland, on the use of the term *Celtic archaeology* is unfortunate for its linguistic connotations. Application of *Celtic* to remains of the period under review is less than helpful since for prehistory at least, we are unable to define when an artefact-type or site was used by speakers of a Celtic, or any other dialect.

The Later Bronze Age of Ireland, subdivided into Eogan's *Bishopsland* (ca 13th–10th centuries BC) and *Dowris* (ca 9th–7th centuries BC) periods, is comparatively well mapped (Eogan 1964). The introduction of iron technology to Ireland undoubtedly came about at some time in the 1st millennium BC. But the mechanisms which brought it about, and the finer dating are the cause of much disagreement (cf. Raftery, J. 1940; Ó Ríordáin 1946; Champion 1971; Scott 1974b and c, 1978a; Warner 1974, 1976, 1983; Raftery, B. 1976a, 1984; Caulfield 1977). Between the end of Eogan's Later Bronze Age Dowris period and the 3rd century BC appearance of finely decorated metalwork, there is a diastem in the archaeological record. What happened to bring this about is quite unclear (see Chapter 3.1), but in any event we are faced with a 'dark age' of at least four centuries for which we have great difficulty in accounting. Even when things become clearer, problems still arise. Indeed, as B. Raftery pointedly remarks, it is well-nigh impossible to obtain consensus agreement even on the fundamental question of what might constitute an 'Irish Iron Age'. He suggests that '. . . the concept of some homogeneous phase of Irish prehistory which can be conveniently labelled "Iron Age" is no longer tenable. . . . In a sense it is almost possible to suggest that several "Iron Ages" existed in the country . . .' (1984, 3ff). While perhaps unpalatable, this view is nevertheless one which may not be dismissed easily. There are major problems with the application of C-14 dating and, in addition, many of the well-understood boundary markers of change in the archaeological record of other regions are, by and large, missing for much of the 1st millennium BC and the earlier part of the 1st millennium AD. It is very difficult to gain a coherent picture of events and changes, for this period. The later 1st millennium AD is, by comparison, much better known.

The breakdown of the usefulness of C-14 dating in the 1st millennium BC is, as Baillie and Pilcher remark (1983, 58f), a catastrophe for the archaeology of the period. Dendrochronological calibration of radiocarbon age has shown that much of the 1st millennium BC is characterised by a flattening of the curve (Baillie and Pilcher 1983, 57, Fig. 1: Baillie 1985, 19ff) which has most unfortunate implications for any attempt to obtain practical C-14 dating in this period. In effect, any C-14 determinations falling into this period are subject to hopelessly confusing effects which render them for all practical purposes useless even in the construction of relative chronological schemes. The most distressing examples of this effect are found with the C-14 dates determined for sites such as Navan, Co. Armagh (Selkirk and Waterman 1970), Dún Ailinne, Co. Kildare (Wailes 1976, esp. 338), Clogher, Co. Tyrone (Warner 1972), and Rathtinaun, Co. Sligo (Raftery, J. 1963, 110). Dates for the earliest phases of activity at all of these sites are in, or near, the zone of confusion and fail totally to refine their chronologies. The main hope for dating events in the Later Bronze Age/Earlier Iron Age transition would now appear to lie in obtaining reliable associations of artefacts with timber which may be dated by dendrochronology[2] and it is to the future examination of crannog sites that we may look for any significant clarification of the absolute chronology of events in this period. For all of these reasons, in this study C-14 dates which fall into this 'disaster area' will be noted where appropriate, but not given any weight.

The nature of society and economy in Ireland during the period ca 600 BC to AD 500, and the chronology of its development, is still not agreed, sometimes seeming to be the subject of a debate of the deaf. In his analyses of decorated sword scabbards and of the Bann disc, Jope (1954; Jope and Wilson 1957) suggested a chronological division, with some geographical bias, of the then identifiable Iron Age material. He saw evidence of Middle La Tène influence of the period ca 3rd to 1st centuries BC, noticeably in north-east Ireland, and influence from the pre-Roman Iron Age of Britain, most evident in central and western Ireland. His perception was not of a cultural duality, but rather of different sources of influence, separated chronologically and spatially. Rynne (1961) sought to reverse Jope's scheme, with influence from pre-Roman Iron Age Britain being felt mainly in the northern part of Ireland, and Continental (including Spanish) influence most heavily in the south. A geographical dualism has been proposed by Caulfield (1977, 1981) who argued for the existence of two main Iron Age strata, one northern and based strongly on contact with, and elements from La Tène period Britain and Europe, the other with a largely southerly distribution and seemingly wholly native. He suggested that this could have given rise to the portrayal of north-south conflict in the (much later recorded) saga *Táin Bó Cúalnge*. Warner (1983) rejected any geographical dualism, suggesting that when all of the material is examined, no meaningful regional variations can be distinguished, and that those variations which do appear are chronological, thus, in this respect, following the original thrust of Jope's arguments.

The question of the level and form of influence of Romanised Britain on Ireland exemplifies some of the difficulties in the archaeology of Ireland in the earlier part of the 1st millennium AD. In other regions major upheavals, such as that caused for much of Britain and Europe by Roman imperialist expansion, provide useful horizons in the archaeological record. But in Roman-free Ireland on the fringe of the *imperium*, no such conveniently sharp boundary horizons are available in archaeological stratigraphy. In AD 70, we find the Roman general Agricola reported by Tacitus as believing that Ireland could be taken and garrisoned by a comparatively small force. This could be interpreted as evidence of the continuing disarray resulting from a socio-economic and political collapse at the end of the Later Bronze Age half a millennium previously[3]. Set against this, in the middle of otherwise scant archaeological evidence for large population groupings, is the fact that large linear earthworks (Lynn 1982a; Walshe 1987) and sophisticated timber trackways (Raftery, B. 1987) were being constructed at the end of the 1st millennium BC. Furthermore, we know that in the early 4th century AD Irish raids on Roman Britain were a major problem there, and that groups of Irish were settling in western Britain. Was Agricola's assessment of the military situation faulty, or did the early centuries AD see a revitalisation of the island?

It can be argued that the Roman conquest of Britain saw the beginning of significant but stealthy influence on Ireland. Artefact types, such as the *gladius*- and *spatha*-derived swords (Chapter 5.1), and styles derived from the Romanised world seeped in and fused into the native assemblage. Given the evidence for interaction across the Irish Sea at this time, it would be surprising to find that Romanised Britain provided no cultural or material component to contemporary or later Ireland. But effects of the interaction were subtle, with none of the evidence pointing to Ireland as having been manoeuvred into informal provincialism. Such influence and interaction as charted by Warner (1976) provided a spread of opportunities to the native Irish in many fields, some of which they chose to take up, some of which they ignored. The nett result is somewhat confusing. Thus, study of the period ca AD 200–500, as with the earlier period, shows La Tène compatible material lingering on in some regions, while in others it is absent. New types which most likely have their origins in this period last on for several centuries. The concept of a patchy spread of regional Iron Age cultures is an attractive, if evasive, explanation for what might otherwise seem to constitute a second 'dark age' (cf. Warner 1987). By the 6th/7th centuries AD, new components in Irish society are detectable. Christianity and literacy, anchoritic monasticism, the codification of elements of early Irish laws alongside those from Roman jurisprudence, educational practices and the strata of Latin loanwords in Old Irish are indirect legacies from the Romanisation of Britain. But as such they do not provide a sharp boundary marker in the Irish archaeological record.

Since the focus of this work is on iron and the native Irish industries which produced, shaped and distributed it, it seems both natural and safe here to adopt a simple, ferrocentric scheme. While conspicuously avoiding some of the wider archaeological issues, it provides a coherent framework for the results of the research. It also has the virtue of underscoring both the importance of iron in the early Irish economy and the continuity of native ironworking traditions revealed through the metallographic research. One must, however, enter the *caveat* that such a simplified sequence, from which many of the interpretative difficulties have been debarred, can lead to an oversimplified image of change and a dramatisation of its most obvious effects. Nevertheless, the difficulties which could distract from the main purpose of the book, leads me to divide the period between the emergence of native produced iron artefacts and the 10th century AD (i.e. before any industrial effects of the Scandinavian intrusion could be felt) into three broad chronological phases, following on from Eogan's Dowris phase of the Later Bronze Age (ca 9th-7th centuries BC).

I will define the Irish Iron Age as the period:

(a) which begins with the first native smelting of iron ores and forging of the metal produced into artefacts, and
(b) during which iron becomes the principal material from which tools and weapons are made.

The appearance of small quantities of iron coincides with the appearance of intrusive Hallstatt elements, and with the sharp decline of the once vigorous copper-alloy industries of the earlier part of Eogan's Dowris phase. This is used here as the first boundary marker, with the transition from copper-alloy to iron technology being taken as marking the beginning of the Irish Iron Age. I term this transition period *Earlier Iron Age A* (*EIA A*), and date it to ca 7th–3rd centuries BC.

The significant increase in the archaeological prominence of iron (though represented still only at a low level by comparison with other regions of the Old World) results from the appearance of types which, while distinctly insular, show foreign middle La Tène origins. Part of a wider resurgence of metal industries in the last quarter of the 1st millennium BC, it acts usefully as a demarcation between the earlier transition phase EIA A, and the emergence of new institutions and industries, typified in north-east Ulster by the distinctive material from Lisnacrogher in Co. Antrim. This appears to result from the re-establishment of firm contact with the outside world, the most prominent and readily identifiable elements being British La Tène. This period, during which native ironworking is becoming established as an industry to supply tools and (particularly) edged weapons, I describe as *Earlier Iron Age B* (*EIA B*), and date to the period ca 3rd century BC–5th century AD (this equates with the EIA 1, 2 and R1 periods of Warner 1976). The 3rd century BC dating of the Lisnacrogher sword scabbards provides a useful lower limit. Difficulties of dating the extinction of La Tène compatible material and of placing in an Irish chronological framework ironwork of seeming Roman connection make the upper limit of AD 500 arbitrary, if convenient (cf. Warner 1986, 1987).

The rest of the period under review, 6th–10th centuries AD, when iron is finally seen to be commonplace, I describe as *Later Iron Age A* (*LIA A*: historically, this has been known as the 'pre-Viking' phase of the 'Early Christian Period'). Metallurgically, the period sees fairly uniform development of expertise in non-ferrous metalworking, with expansion both in production and consumption and the appearance of a small number of identifiably specialised sites. Earlier influences from Roman Britain were replaced by those arising from gradual infiltrations from the Germanic world. There is certainly evidence of significant contacts during this period, reflected in the design of some Irish iron artefacts.

The overall scheme may be summarised as:

Dowris Bronze Age ca 9th–7th centuries BC, the period when contacts with foreign iron users become apparent.

Earlier Iron Age A (**EIA A**) ca 7th–3rd centuries BC, the period when the transition from copper-alloy to iron begins in an essentially Later Bronze Age technological setting.

Earlier Iron Age B (**EIA B**) ca 3rd century BC–5th century AD, the period when iron is becoming fully established as the major industrial metal. A substantial Later Bronze Age technological substrate still survives in the earlier stages, possibly through to the end of the 1st millennium BC. Influences from the Romanised world appear and strengthen, later to be replaced by those from post-Roman and Germanicised Britain.

Later Iron Age A (**LIA A**) ca 6th–10th centuries AD, the period when ironworking and iron artefacts are commonplace.

The intrusion of the Scandinavians did not alter significantly basic patterns of industry here in its early stages. These formidable travellers were nevertheless bringing Ireland, really for the first time, the fruits of mainstream European iron technology to her very doorstep. I would thus add a further stage to the scheme, the period of Scandinavian influence, and term it *Later Iron Age B* (*LIA B*).

Later Iron Age B (**LIA B**) ca 10th–12th centuries AD, the period when the Scandinavian presence leads to exposure of Irish smiths to the mainstream of European metal technology.

The nature of the interface between native and intrusive technology at present is quite unclear, although study of the well-stratified material from Viking Dublin is now under way. LIA B will not be discussed here, since Scandinavian influence on native production and technology constitutes another major area of study.

The next boundary horizon would be the 12th

century Anglo-Norman invasion and conquest. The influences of Scandinavian marauders and land-hungry Anglo-Norman freebooters on patterns of daily life has often been heavily overstated by historians. Viking smiths introduced levels of ironworking and production technology quite foreign to Ireland, apparently exerting significant influence on native technology as less fraught contacts were established between native and intruder. Anglo-Norman ironworking technology was not significantly more advanced than that of the native Irish, while their mass-produced artefacts, which have all of the aesthetic appeal of the quartermaster's store, could not have had an immediate technological impact.

The final phase of this scheme, again outside the scope of this work, would be marked by the introduction of the charcoal-burning blast furnace by Elizabethan speculators at the end of the 16th century, drawn by cheap resources of fuel and labour in Ireland (Scott 1985). This industry was severly disrupted by the great uprising of 1641, in which many English-owned ironworks were seriously damaged or else destroyed. Although production resumed in the aftermath, the industry, foreign owned and without native involvement, was already in decline, not the least reason being the effects of profligate exploitation of forest and woodland which was well on the way to leaving Ireland with the lowest tree cover in Europe in modern times.

* * *

1.4 – The structure of bloomery iron

Iron (chemical symbol Fe) constitutes 4.7% of the earth's crust (second in abundance among the metals only to aluminium). It can be combined with other elements, producing perhaps the most useful group of alloys known to man, one on whose properties virtually every major engineering project of modern times depends. By examining the structures found in ancient iron artefacts, it is possible to tell much about how they were made, and to judge how they would have performed in use. Thus it is useful to look briefly at the nature of iron itself to see how these structures arose.

Metallic iron is found in nature (Buchwald and Mosdal 1985), both as extra-terrestrial meteorites and as crystals of terrestrial origin in rocks such as basalts, including those of the Giant's Causeway in Co. Antrim, or in larger masses such as those found in Greenland. It is, however, far more widely known in the form of chemical compounds (most importantly with oxygen, carbon and sulphur forming iron oxides, carbonates and sulphides) in association with a wide range of minerals, as *iron ores*. The production of metallic iron from its ores is achieved by the process known as *smelting*, carried out at high temperature in a *furnace*. This involves a chemical reaction (*reduction*) between the iron oxide and a *reducing agent*, the gas *carbon monoxide* which in antiquity was obtained by the burning of charcoal. The feature which marks out iron smelting from the smelting of other metals in antiquity, is that it was a solid state process, unlike the smelting of copper, tin and lead which were all produced molten. The end product of smelting was a spongy mass of reduced iron particles and slag known as a *raw bloom*. Since the actual process and its associated structures have been well described elsewhere (e.g. Pleiner 1958; Cleere 1972; Tylecote 1986, 1987; Clough 1987; Tholander 1987), and since Irish furnaces are described in Chapter 6, we shall limit discussion here to the metallic products of early smelting.

(a) The structure and properties of iron and steel: general

Like other crystalline solids, iron has its atoms ordered into a regular three-dimensional arrangement known as a *lattice*. At room temperature, the iron crystal lattice has as its basic building-block the cube. There are atoms at each of the corners, and one at the central intersection of the diagonals, the *body-centred cubic* (bcc) form (Fig. 1.4.1a, b). This stable form is known as α-iron or *ferrite*. Where large numbers of crystals are growing together, they meet at *grain boundaries* and do not get the chance to exhibit their underlying symmetry, showing up instead as a mass of *polyhedral* ferrite grains (Plate 1.4.1a).

An important property of iron is *allotropy*, which means that changes in crystalline structure occur at distinct temperatures of *transformation*. Thus, α-iron is the stable allotrope at all temperatures below 912°C, although losing its magnetic permeability at 770°C. At 912°C, the structure alters from a bcc to a *face-centred* cubic (fcc) lattice (Fig. 1.4.1c, d), the *allotrope* known as γ-iron, *austenite*, which exists up to 1410°C. Above 1410°C the structure reverts to bcc[4] (δ-iron), which it retains until the melting-point at 1540°C. Iron can therefore be said to have two *allotropic modifications*.

Carbon can dissolve in iron to a limited degree, with its atoms fitting into spaces between the iron atoms to form an *interstitial solid solution* (Fig. 1.4.1e). Although carbon atoms are much smaller than those of iron, the spaces between the iron atoms in the bcc lattice are also small, and thus the solubility of carbon in iron at 727°C is only 0.02%, while at room temperature it is negligible. Carbon has a greater solubility in austenite than in ferrite, because the interatomic spaces in the fcc lattice are larger (Fig. 1.4.1f), allowing austenite to hold carbon in solid solution up to a maximum of 2.11% at 1148°C.

When carbon is present in iron at room temperature in amounts above the solubility limit, a second constituent, *pearlite* (Plate 1.4.1b), appears with the ferrite. It consists of a series of alternating laminae of ferrite and iron carbide, *cementite*, an exceedingly hard and brittle phase, in which the atoms of carbon and iron have taken on a new *orthorhombic* lattice arrangement (Fig. 1.4.1g). Thus, pearlite is not itself a phase, but a mixture of two phases. As the content of carbon rises, so the proportion of pearlite to ferrite increases, coupled with an increase in the strength of the metal. At 0.77% C (the *eutectoid* composition[5], usually given as 0.8%), the structure at room temperature consists entirely of pearlite, and the metal is at its strongest. With a carbon

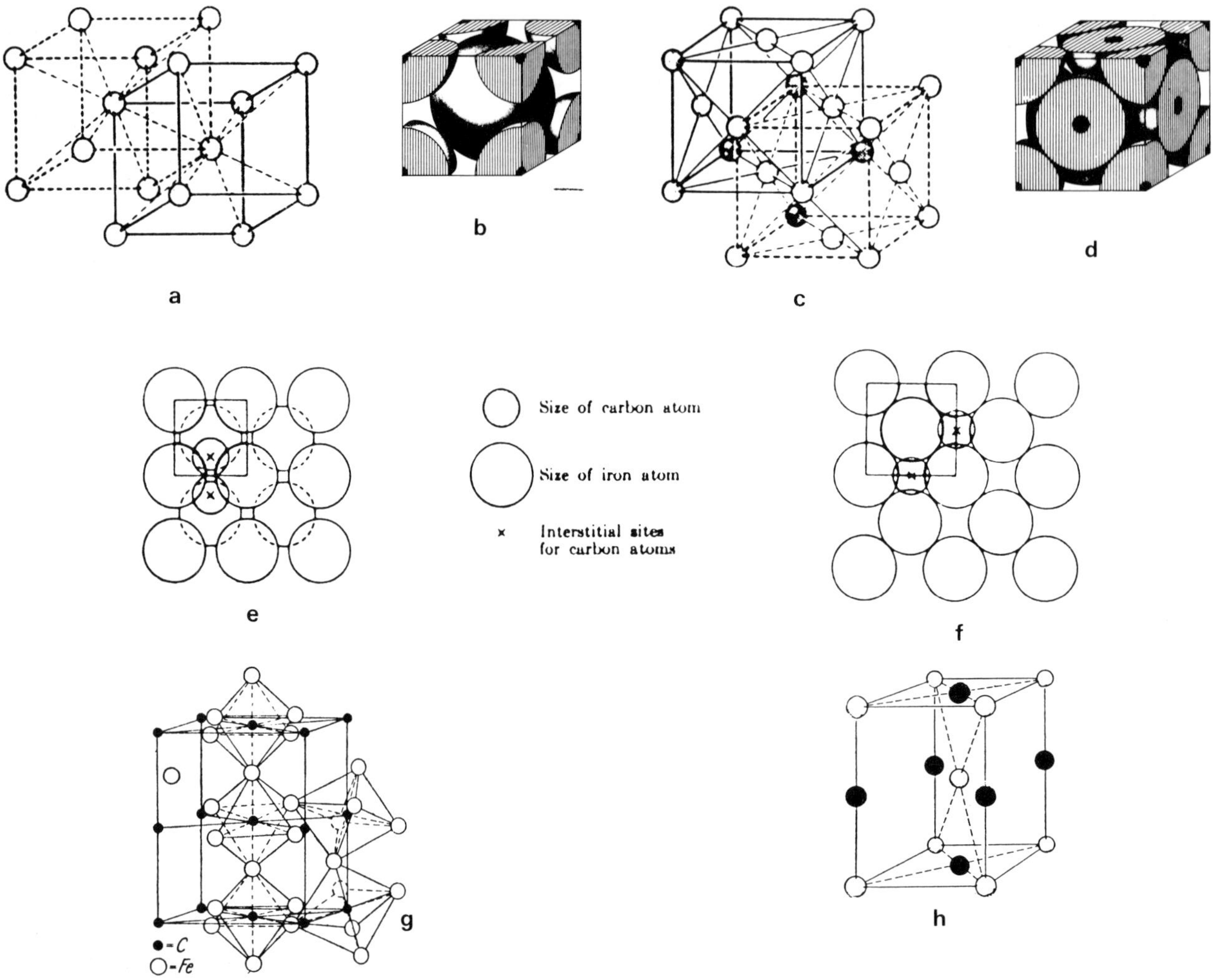

Fig.1.4.1 (a) body-centred cubic lattice of α-iron, (b) packing of one bcc cell, (c) face-centred cubic lattice of γ-iron, (d) packing of one fcc cell, (e) the positioning of carbon atoms in interstitial solid solution in α-iron, (f) the positioning of carbon atoms in interstitial solid solution in γ-iron, (g) orthorhombic arrangement of carbon and iron in cementite, (h) body-centred tetragonal structure of martensite ((a)-(f) after van Vlack 1968: (g) after *Metals Reference Book*: (h) after Hume-Rothery 1966).

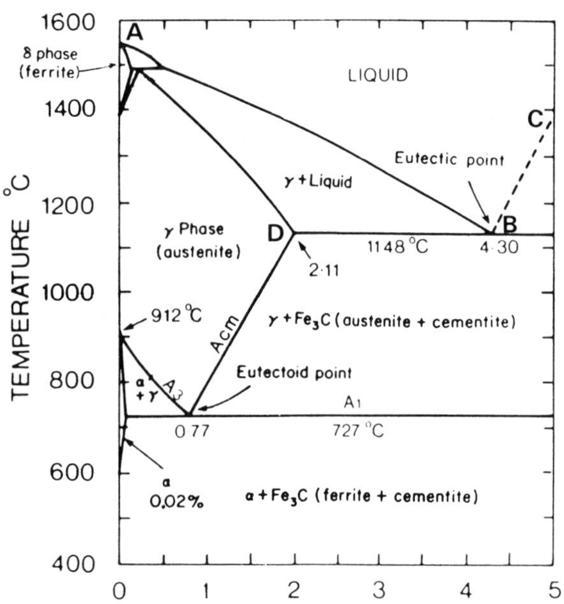

Fig. 1.4.2 The iron-cementite system (after Samuels 1980).

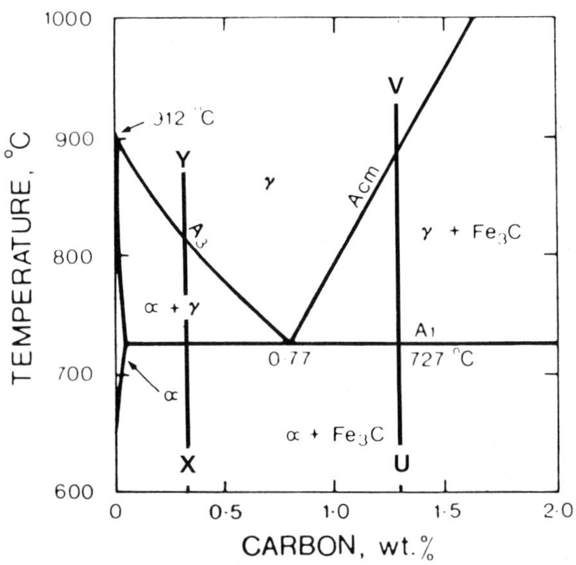

Fig. 1.4.3 Eutectoid region of the iron-cementite diagram (after Samuels 1980).

content greater than ca 0.8%, the structure consists of pearlite and cementite. Free cementite can also occur in low-C iron at ferrite grain boundaries (Plate 1.4.1c). The presence of free cementite in high carbon steels increases hardness but also brittleness. With carbon present above the eutectoid composition, steels are referred to as *hypereutectoid*, while below it, they are known as *hypoeutectoid*. Ferrite and cementite which make up pearlite are referred to as *eutectoid* ferrite and cementite, to distinguish them from *proeutectoid* ferrite and cementite, which form before the eutectoid transformation.

The *constitutional* or *phase* diagram for iron and cementite (Fig. 1.4.2) shows the temperature-related boundaries between the different phases that exist with varying contents of carbon in iron[6]. In fact this is an *equilibrium* diagram, telling us only what structures form *if sufficient time is allowed*. It cannot tell us about the *length of time needed* to reach equilibrium. For this reason, while we can study it to understand behaviour at different temperatures, it must be used with caution and, when cooling rates increase significantly, it is necessary to turn to other data.

Above the line **ABC** on Fig. 1.4.2, everything is liquid. The melting point of pure iron (1540°C) and decreasing melting point caused by progressively higher carbon contents can be seen by following **AB**. Below **ADB**, everything is in the solid state, while between **ADB** and **ABC** we have a mixture of solid and liquid. Since we are dealing with bloomery iron which was produced in the solid state, the area of the phase diagram enlarged as Fig. 1.4.3 is of most interest.

As iron is heated or cooled through the temperatures where transformations occur, the progress is arrested. The arrests are known as *critical points* and usually are denoted by the letter *A* (French *arrêt*) with the subscripts *c* (indicating heating – French *chauffage*) and *r* (cooling – *refroidissment*), followed by a number (0–4 depending on the temperature). For hypereutectoid steels, Ac_3/Ar_3 is often denoted by A_{cm}. As can be seen from Figs. 1.4.2 and 1.4.3, A_1, A_3 and A_{cm} identify the lines that define the critical temperatures associated with the transformation of austenite. For our purposes, we shall treat A_c and A_r as coincident[7] and refer to the transformation temperatures simply by A_{1-3}. Thus, A_3 is the critical point encountered at 912°C, where the bcc/fcc change occurs.

The structures observed in iron and steel depend on the temperature-ranges and the length of time at which the metal was held and worked at heat, the degree of deformation at low temperature, and the rate of cooling. The blacksmith works his metal from around 600°C, up to the top welding temperature over 1200°C, and the structures formed result from the ways in which the critical temperatures are approached during heating and cooling. Using Fig. 1.4.3 we can follow the behaviour of a 0.3% C steel as it is heated and cooled slowly. Following the vertical line **XY** at 0.3% C, we find that the first arrest (A_1) occurs at 727°C. While the proeutectoid ferrite remains unaltered, the constituents of the pearlite dissolve to form the γ-iron solid solution iron containing 0.77% C, austenite. A second arrest point (A_2 – not illustrated) is encountered at 770°C: here, ferrite changes to the non-magnetic state. As the temperature increases, proeutectoid ferrite is absorbed into solid solution, until A_3 is reached at around 805°C where there is a further arrest with remaining proeutectoid ferrite transforming to austenite, so that the whole structure is now austenitic. The austenite formed from the pearlite is richer in carbon than that formed from the proeutectoid ferrite, and if sufficient time is allowed, carbon diffuses into the austenite formed from proeutectoid ferrite to give an even distribution. Following the line **UV** at 1.3% C upwards, the process is the same, except that at this hypereutectoid composition, it is proeutectoid cementite which dissolves into solid solution up to A_3.

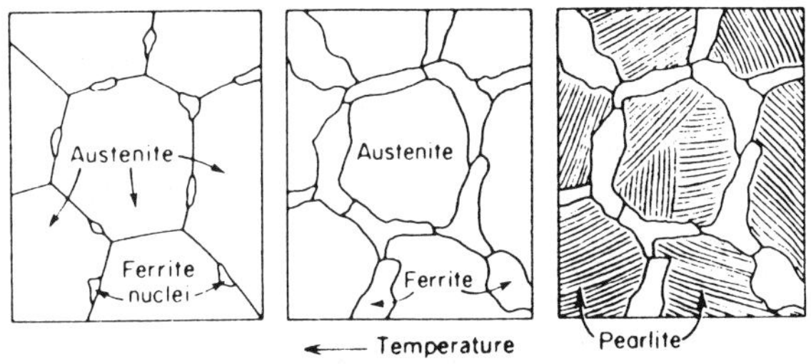

Fig. 1.4.4a Nucleation and growth of pro-eutectoid ferrite at austenite grain boundaries, followed by the growth of pearlite from parent austenite (after Reed-Hill 1973).

If the 0.3% C steel is now allowed to cool slowly back to room temperature, the process is reversed. Following **YX**, the first arrest (A_3) is encountered, and proeutectoid ferrite starts to be precipitated out of solid solution at the austenite grain boundaries (Fig. 1.4.4a). As the temperature falls further, ferrite continues to separate out, and grain growth occurs into the parent austenite grains, while, as a result, the austenite becomes progressively enriched in carbon. At A_2, ferrite regains its magnetic condition. When the temperature falls to A_1, the remaining austenite, which now contains 0.77% C, transforms to pearlite. Cementite grows along crystal planes of the austenite taking up sufficient carbon for its formation. This leaves bands deficient in carbon which thus transform to ferrite (Fig. 1.4.4b). The formation of pearlite involves the diffusion of carbon, and therefore is dependant upon time and temperature. Slow cooling allows time for significant diffusion to occur, resulting in a coarse structure with thick cementite and ferrite lamellae. Faster rates of cooling give progressively finer structures. The size of the pearlite colonies resulting from the austenite-pearlite transformation reflects the size of the parent grains of austenite. Between A_1 and room temperature, there are no

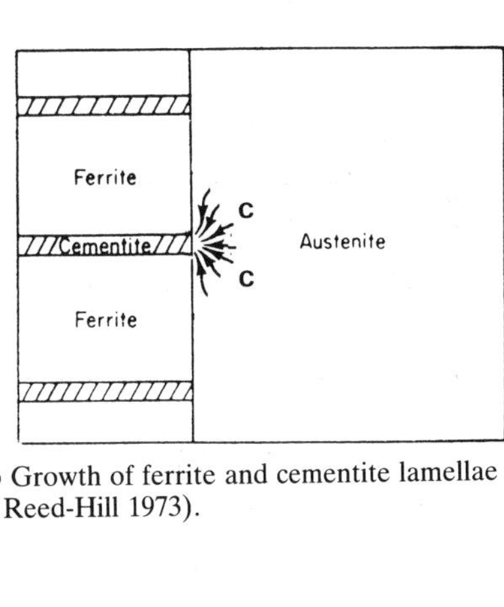

Fig. 1.4.4b Growth of ferrite and cementite lamellae from austenite (after Reed-Hill 1973).

further visible changes. For the hypereutectoid steel (**VU**), the process is very similar, except that it is proeutectoid cementite which is precipitated from the austenite between A_3 and A_1. The austenite becomes progressively poorer in carbon, until at A_1 the eutectoid composition is reached and the austenite transforms to pearlite by the same mechanism.

Hammering metal at room temperature (*cold working*) progressively deforms the crystal lattice, with shearing and slippage of sections of the lattice against one another, increasing the hardness of the metal. As this continues, dislocations pile up[8], and the original crystals are broken up into sub-grains, reducing progressively the possibility for further deformation. Finally the metal becomes so brittle that fractures appear as the force required to produce further deformation exceeds the cohesive strength of the metal. Heating of the metal for a period allows the rearrangement of the lattice and reformation of new, strain-free crystals, with a subsequent reduction in brittleness and restoration of malleability. The degree of cold working has a marked effect on the final grain size of the metal on cooling (Fig. 1.4.5). A low level of deformation often will lead to exaggerated growth, while severe cold work will produce a fine grain structure. Where some areas are more deformed than others, there may be marked contrasts in grain size on final cooling (Plate 1.4.1d). The presence of phosphorous must be taken into account, since it also promotes ferritic grain growth, leading again to contrasts in grain size. If metal is forged above A_3, then the stress introduced by deformation will be balanced at least in part by recrystallisation. At the end of work, the metal is still at heat and there is the opportunity for a complete mitigation of the effects of deformation. This can result in the same structure as would be obtained by cold working followed by heating.

The process of heating metal to soften it is known generally as *annealing*. In modern practice, heating to soften and to modify the structure of ferrous metals is divided into a number of processes of differing effect

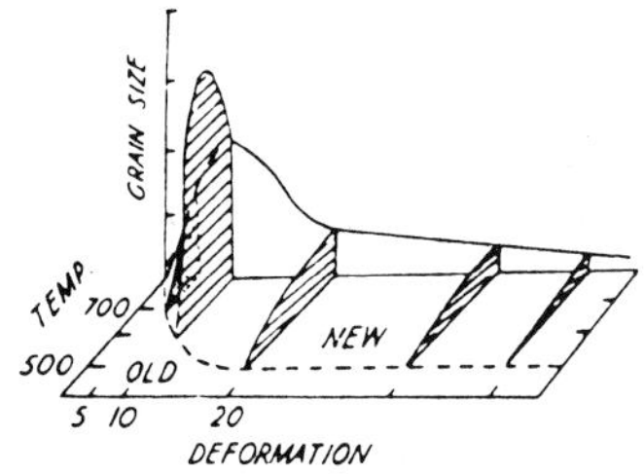

Fig. 1.4.5a The relationship between grain size, degree of deformation and annealing temperature for a 0.18% carbon steel, time constant (after Rollason 1973).

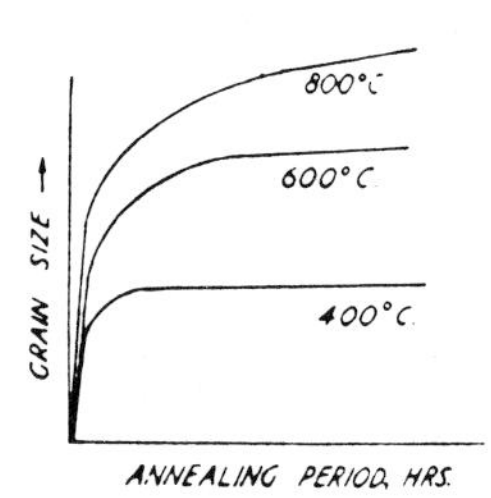

Fig. 1.4.5b Effect of annealing time and temperature on grain growth (after Rollason 1973).

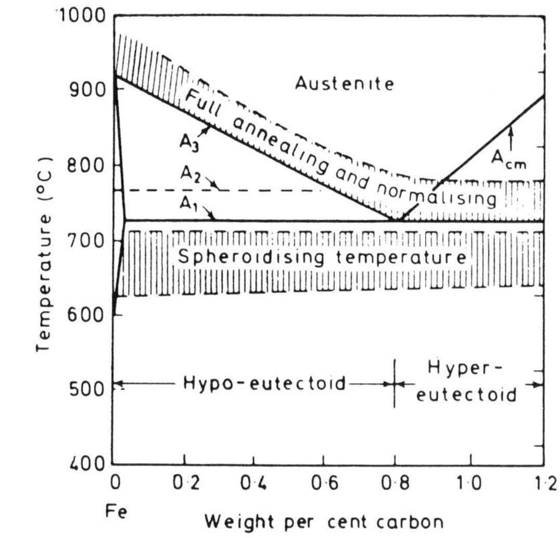

Fig. 1.4.6 Eutectoid region of the iron-cementite diagram showing temperature bands for full annealing, normalising and spheroidising (after Chadwick 1972).

(Fig. 1.4.6). Heating metal to above A_1 followed by cooling will affect the final grain size (Fig. 1.4.7) and hence the structure and properties of the metal. The strain induced by cold working can be relieved by heating at around 550°C, when deformed grains recrystallise. Heating just below A_1 causes both proeutectoid and eutectoid cementite to coalesce and form *spheroidised cementite* (Plate 1.4.1e), leading to an increase in ductility. With comparatively short periods, the pearlite colonies retain their outlines, but the cementite lamellae begin to break up into chains of carbide particles. In modern practice, this treatment, referred to as *spheroidising*, is prolonged and gives a structure which consists of coalesced carbide particles in a ferrite groundmass, with the progressive spheroidisation of the carbide as larger and larger masses are formed. If the metal is held above A_3 for a time and then cooled slowly, *full annealing*, the growth of large austenite grains and the cooling rate lead to the growth of large ferrite grains and of coarse pearlite. If instead, the metal is allowed to cool in air after being held above A_3, *normalising*, finer grained ferrite and fine pearlite results.

An interesting structure indicative of the treatment which a metal has received is the *Widmanstätten*[9] structure (Plate 1.4.1f). This can be formed when an iron-carbon alloy is cooled after heating in the auste-

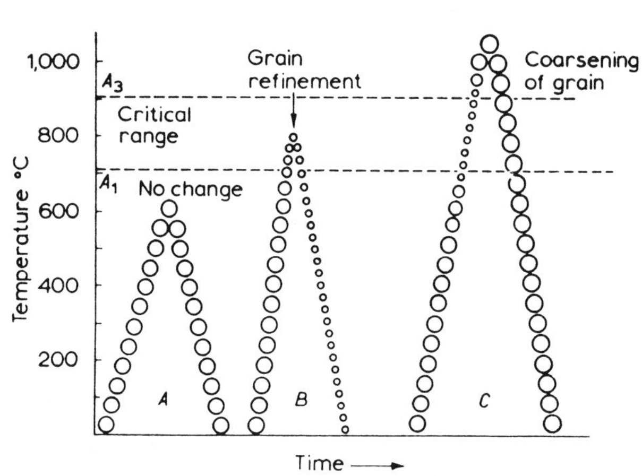

Fig. 1.4.7 Grain growth pattern at different temperatures.

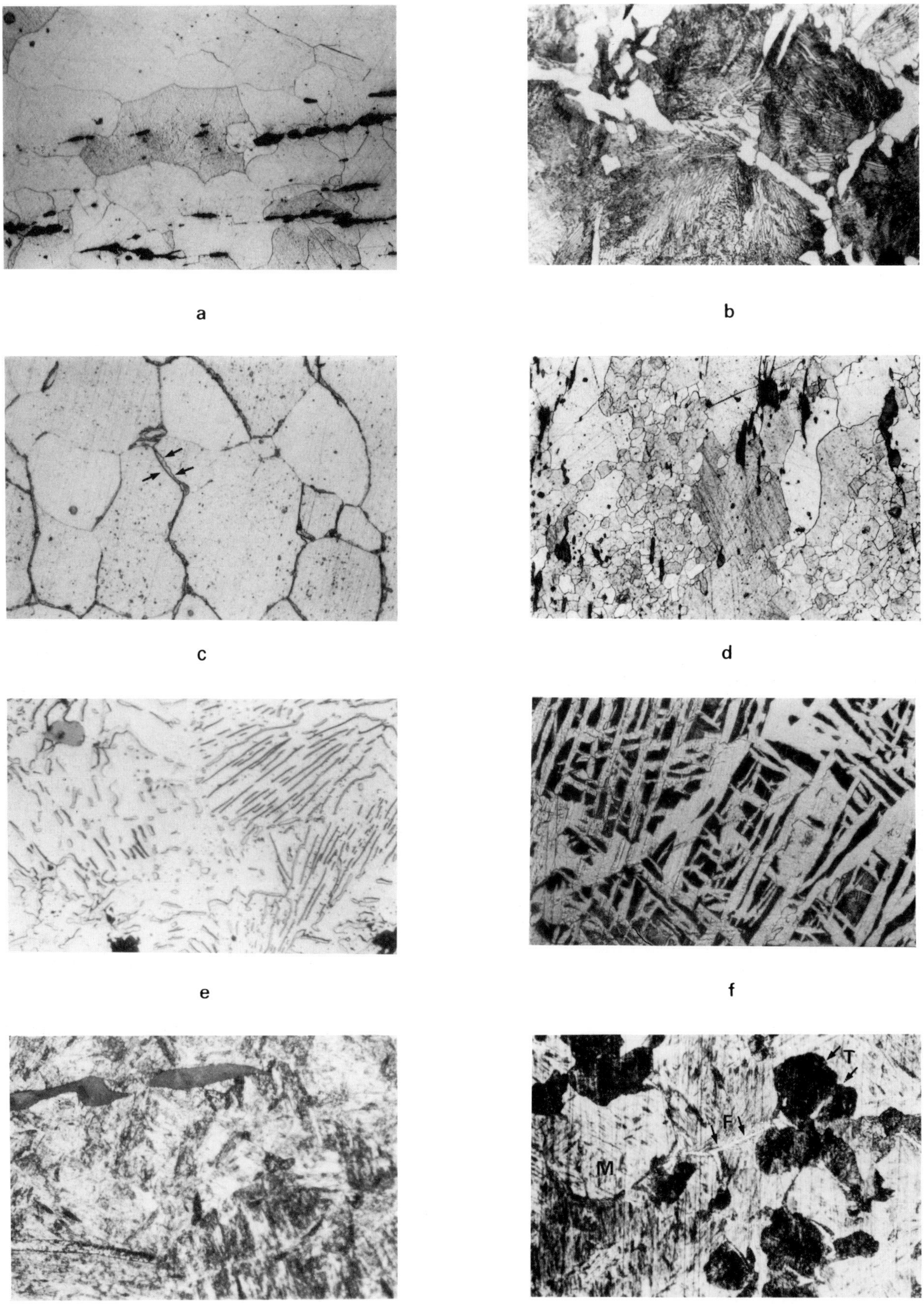

Plate 1.4.1 (a) ferrite with slag inclusions x50, (b) pearlite with ferrite x250, (c) cementite (arrowed) at ferrite grain boundaries x 250, (d) uneven ferritic grain growth x100, (e) spheroidised cementite x250, (f) Widmanstätten ferrite (light) and pearlite (dark) x100, (g) martensite x250, (h) troostite (T-->) with martensite (M) and remanent ferrite (F-->) x500. (a)-(d) and (f) etched 4% Nital, (e) etched Picral, (g)-(h) etched Bisulphite.

nite temperature range (i.e. above A_3) for long enough to allow the growth of large austenite grains. Alternatively, the structure can be formed by heating in the austenite temperature range followed by air cooling. In each case, complete separation of proeutectoid ferrite or cementite at the austenite grain boundaries to form a network is not possible and instead there is ejection along preferred planes of the austenite crystals, rather than at austenitic grain boundaries (Fig. 1.4.8). This gives the proeutectoid phase a jagged appearance. Whichever route has been followed, this structure is considered most undesirable in modern steelmaking, as it leads to a lowering of ductility and a low resistance to shock.

If steel is heated to a temperature above A_3, held there until the α to γ transformation is complete, and then plunged into water, *quenched*, a very different structure results from that produced by slower cooling. There is insufficient time for the γ-α transformation to occur by diffusion of the carbon, and austenite instead transforms to a new structure, *martensite* (Plate 1.4.1g). This transformation always goes to completion, or near completion in plain carbon steels (i.e. those without any significant content of other alloying elements). The transformation to *body-centred tetragonal* (bct) martensite (Fig. 1.4.1h) begins in the range 400–200°C depending on carbon content. The combination of a non-cubic structure and the retention of carbon puts the crystal lattice under considerable strain, making martensite very much more difficult to deform than any other phase except cementite, the degree of hardness depending on carbon content (Fig. 1.4.9). The change from fcc to bct structure arising from quenching causes an expansion in volume, but one which is superimposed on the normal cooling contraction. This produces severe internal stresses which can cause physical cracks, *quench cracks*. Thus, one skill of the early smith in hardening his metal was the avoidance of cracking during heat-treatment.

The effects of cooling rates on the structures formed by rapid cooling from above A_3 can be illustrated from a *continuous cooling transformation* diagram. Fig. 1.4.10 shows the CCT diagram for a plain eutectoid steel, where the X axis is the logarithm of time, and the Y axis temperature. Two curves represent the start of the transformation and completion: the dotted line joining the two curves represents the end of the zone of pearlite transformation. The shaded area enclosed by the start and completion curves represents the transformation process. The horizontal line (M_s) represents the temperature at which the martensite transformation begins.

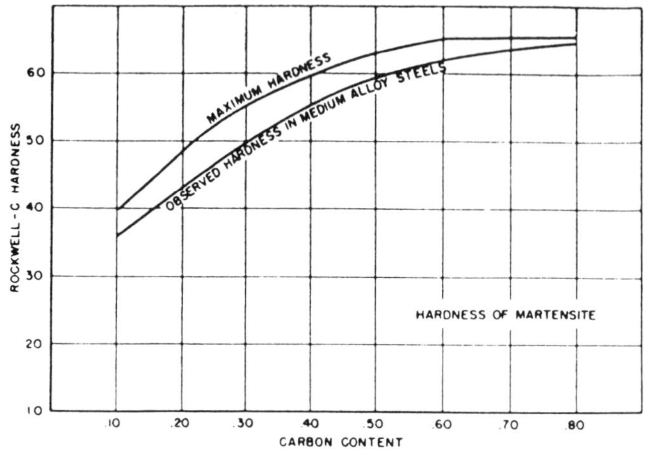

Fig. 1.4.9 The influence of carbon content on the hardness of martensite (after Maddin 1987).

By superimposing lines representing the time taken to cool the steel from above A_3 (temperature v the log of time after the start of cooling), we may get an indication of the structures likely to be formed at different rates. Taking the slowest rate of cooling first, the line **SA** cuts the start curve at **B** and the end curve at **C**. This cooling rate is very slow, and the product is coarse pearlite, and once the transformation has been completed, no further change can take place. At a faster cooling rate, as in air-cooling, represented by the line **SD**, the start curve is intersected at **E**, and the transformation zone traversed until the same curve is intersected at **F**. This represents a cooling time of several minutes only and the end product is again pearlite, but much finer in form. The line **SG** represents the sort of cooling rate achieved by quenching in oil, where the time spent in the zone of pearlite transformation is too short for all of the austenite to transform to pearlite. The remaining

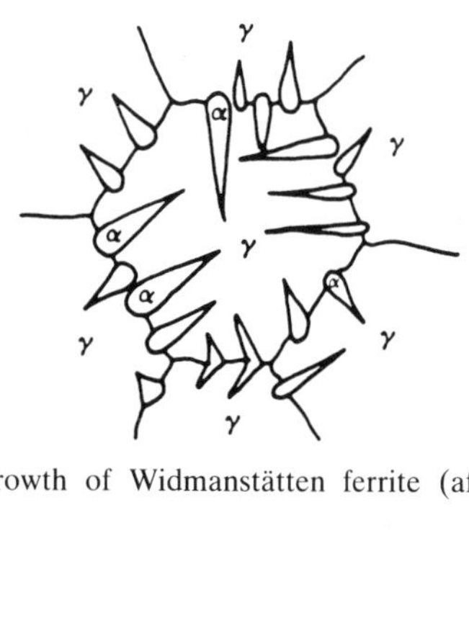

Fig. 1.4.8 Growth of Widmanstätten ferrite (after Chadwick 1972).

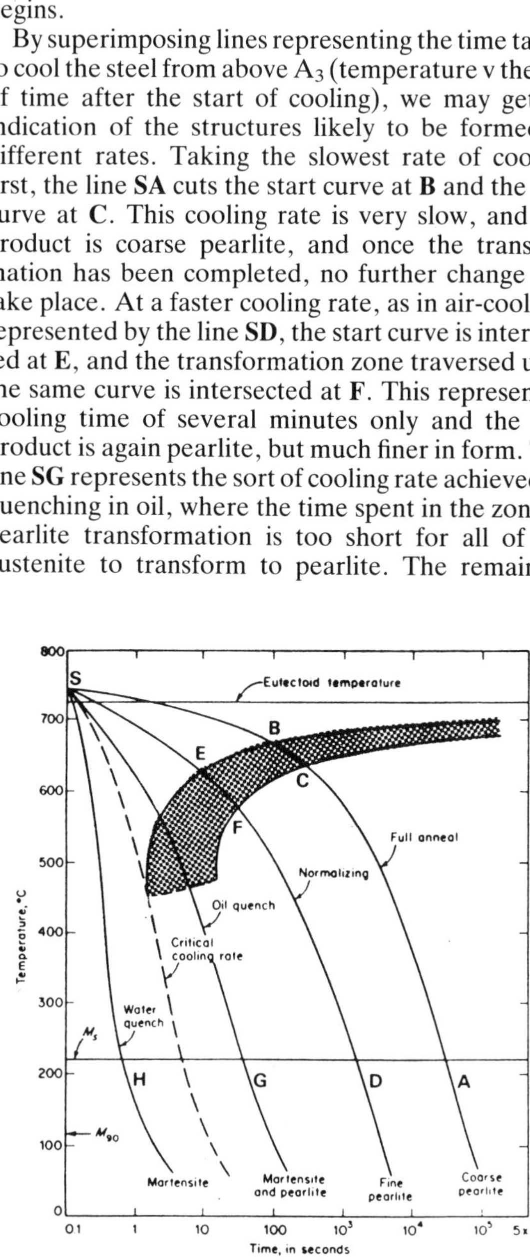

Fig. 1.4.10 Schematic diagram to show the effects of different cooling rates on the microstructure of a eutectoid steel (after Reed-Hill 1973).

austenite begins to transform to martensite when the curve intersects the martensite start line (M_s). The final structure will be martensite, with very fine pearlite whose structure cannot be resolved under the light microscope, and fine pearlite (resolvable under the microscope). For a faster cooling rate still, resulting from a quench in cold water, the cooling curve **SH** does not enter the transformation zone at all and so the structure remains as austenite until the M_s line is intersected, when transformation to martensite begins: the final structure will be fully martensitic.

In quenching of large pieces, the cooling rate at the surface is sufficiently fast for complete transformation to martensite. However, further into the body of the metal, it is slower because the path along which heat is lost is longer. Thus, the interior of a quenched piece not infrequently shows a range of structures resulting from slower cooling. The most prominent is black-etching, and with a structure which cannot be resolved at the magnifications on a light microscope. It has been commonly known as *troostite*, and treated as a distinct phase in its own right[10], although it is now known to be exceedingly fine pearlite. In quenched steel, it is found in increasing amounts with martensite towards the interior of a quenched piece (Plate 1.4.1h), where the slower cooling still produces a very finely lamellar pearlite. Such structures can be produced deliberately in thinner pieces by quenching in a medium of lower thermal capacity (i.e. which conduct heat away from the metal more slowly), such as oil: this is known as *slack quenching*. Thus, quenching plain carbon steels can produce a continuum of structures from exterior to centre starting with martensite, grading into martensite and fine, unresolvable pearlite and ending with finely lamellar pearlite.

Martensite is *metastable* at room temperature, in other words it is altered with comparative ease. It is glass-brittle and, for a steel to provide its optimum properties, say in a cutting edge, this must be modified if fractures are to be avoided. Reheating a quenched piece towards A_1 will gradually alter the martensitic structure, reducing hardness and increasing toughness[11], the process known as *tempering*. The changes occurring on tempering are not readily determined under the light microscope until the later stages. However, their effects are noticeable from progressive modification of physical properties. In the temperature range 50–200°C, internal stresses are relieved, some carbon is ejected as ε-carbide, the lattice becoming less tetragonal, tending towards bcc. This may be detectable indirectly by the darker etching appearance of tempered martensite over untempered martensite. In the range 200–350°C, cementite is formed from the ε-carbide and from the decomposition of remaining martensite, leaving finely divided cementite in a ferrite groundmass. Here the cementite is in the form of minute globules (as opposed to the lamellae in pearlite). Tempering in the range ca 350–650°C induces a progressive coarsening of the cementite globules. This formation of spheroidised cementite develops much more quickly from martensite than from pearlite.

Tempering may occur automatically during the quenching process if the metal is not cooled rapidly enough towards the end. For example, a piece might be quenched by plunging it into water, but withdrawn before it reached room temperature. In such a case, martensite formed by the initial cooling would then be *autotempered* by the heat still remaining in the core of the piece. Thus, the appearance of tempered martensite is not necessarily conclusive proof that a steel has been both quenched and tempered deliberately. Properly executed, however, the temper will provide the ideal medium for creating a cutting-edge. Smiths judge temperature by the colours produced by oxidation of the surface, and when the correct colour appears, 'fix' the temper by re-quenching the piece.

If the metal is only held for a short period above A_3 prior to quenching, not all of the proeutectoid ferrite or cementite will have dissolved, with the result that the quenched structure will show both martensite and ferrite together (cf. Plate 1.4.1h). Since tempering takes place below A_3 (in fact, below A_1), network ferrite and cementite cannot be formed during this process and if observed, must have survived from the pre-heat treatment structure.

(b) *The structure and properties of early bloomery iron and steel*

All of the structures illustrated in this chapter are taken from early iron objects from Ireland, showing how archaeological material can correspond to its modern counterparts. However, although many ancient metallurgical structures can be defined quite readily, there are a number of difficulties resulting from the peculiarities of bloomery iron. This has given rise to much study and debate amongst specialists, and the following are among the reading essential for anyone wishing to go into greater detail: Chilton and Evans 1955; Maddin et al. 1977; Nosek and Mazur 1987; Piaskowski 1965, 1971, 1974, 1976, 1982a and b, 1984; Pleiner 1964, 1967, 1971a, 1973, 1974, 1982: Rehder 1989; Thålin 1967; Tholander 1978; Tylecote and Thomsen 1973; Tylecote and Gilmour 1986. A very accessible and readable modern textbook is that by Samuels (1980).

Iron can be reduced as solid metal from its oxides at temperatures around 800°C (well below the melting point of pure iron). But iron ores consist of more than just compounds of iron and the knack of smelting consisted of being able to separate the iron formed from the other unwanted materials[12], principally silicates, the *gangue minerals* as *slags*. In combination with some of the iron oxides, the gangue minerals fuse and melt at temperatures in the range 1150–1350°C (depending on the ore) to form slags.

The simple way to achieve separation between iron and slag would have been to smelt above the melting point of iron thus allowing the molten metal to separate off from the molten slag. In fact, during the smelting process, reduced iron can take up carbon by diffusion from carbon monoxide. This lowers the melting point of the metal, until at a carbon content of 4.3% (as can be seen from Fig. 1.4.2) it falls to around 1150°C. Molten iron, *cast iron*, could be produced, although its brittleness[13] made it unsuitable for use in this form. It could be converted into useable iron and steel by allowing some or all of the carbon to burn off in oxidising conditions and this process formed the basis of mediaeval and later iron and steel production. Because two stages were involved in making useable iron, this is known as the *indirect process*.

The solution to the problem of separating metal from unwanted gangue minerals was to reduce iron at temperatures *below* its melting point, but where semi-molten or molten slags were formed. Smelting was carried out around 1200°C, and particles of iron reduced from iron oxide sintered together to produce the *bloom*, a spongy mass of metallic iron intimately intermixed with slag (Plate 1.4.2a). This gives its name commonly to the smelting (*bloomery*) process, the (*bloomery*) furnace, and to the (*bloomery*) iron produced. Because only one step was involved in producing iron, this solid state smelting is referred to also as the *direct process*. In Ireland, in the period under review, only the direct process appears to have been used and so the remainder of the discussion will be restricted to bloomery iron.

To produce useable stock for the making of artefacts, the ironworker first had to heat the bloom and hammer it both to weld together the particles of iron into a solid mass, and to expel as much of the slag as possible. It was crucial that the bloom should be heated to temperatures where the slag was at least viscous and at which the iron particles would weld together. Very careful hammering was required, particularly in the early stages of consolidation, so that the bloom simply did not break up. Once this operation was complete, the smith had a billet of iron from which he could begin to make artefacts. Not all of the smelting slag particles could be expelled during the initial consolidation into stock and a characteristic feature of wrought iron is the presence of *slag inclusions*. In addition to iron oxides, these inclusions contain other phases deriving from the gangue minerals and are often quite complex. Elements such as manganese (Mn), vanadium (V), titanium (Ti) and aluminium (Al) which are found in bulk metal analyses, but which could never have been reduced to the metallic state in the bloomery smelting furnace, result from the slag particles included in metal taken for analysis, although Mn, Ti and V could, in theory, form carbides. While strings of slag in wrought iron are considered beneficial to its mechanical properties, too much entrapped slag can be a source of weakness (as well as providing convenient entry points for the penetration of corrosion into the interior). The amount of entrapped slag present is therefore a measure of the skill (or lack of it) of the early iron producer in expelling slag during consolidation of the bloom (Plate 1.4.2b).

In modern iron- and steel-making, metal is produced in the molten state, so that the elements present in minor and trace quantities and any inclusions, by and large are distributed evenly. In the bloomery process, this liquid mixing could not take place, and so, in addition to entrapped slag, it is by no means unusual to find not only carbon, derived from the charcoal fuel, but also small amounts of other elements such as nitrogen (N) derived from the air blast and phosphorous (P) and arsenic (As) from the ore, locally concentrated. Unlike iron and steel made today, the product of the early smelting furnace was a metal of highly variable composition, although relatively 'pure' in comparison with modern alloy steels. Homogeneous iron and steel is a hallmark of modern production, while heterogeneity is the mark of antiquity. It is important to remember that when we analyse pieces of either the blooms produced in antiquity, or of the artefacts forged from them, the contents of elements such as C, N, P and As (as well as other elements) determined are *average* values. Microprobe analyses do provide accurate compositions for particular points or in specific structures and are a valuable complement to bulk analyses.

The various alloying elements found in blooms entered into solid solution by *diffusion* at high temperature (above A_3 in the austenite field) during the smelting process. The uptake of these elements was determined by conditions within the furnace and the length of time of residence. For example, the more reducing the conditions at temperatures above A_3, the greater was the degree of penetration of carbon, deriving from carbon monoxide, into the austenite lattice. This diffusion of carbon into iron is known as *carburisation*, the rate of diffusion (which obviously is very much slower than when metal is molten) being governed by temperature (Fig. 1.4.11). When carburisation is identified as being the result of the smelting process, it is referred to as *primary carburisation* to distinguish it from *secondary carburisation*, a treatment applied outside the smelting furnace, and used on finished, or near finished artefacts (see below).

It is common to observe in blooms areas of almost pure iron lying beside steel of high carbon content (Plate 1.4.2c). This results from the uneven exposure of iron particles to carburising and to oxidising conditions during the smelting process. The distribution of carbon in the bloom could be affected significantly also by the presence of phosphorous which acted as a barrier to its diffusion. A heterogeneous bloom yielded heterogeneous stock, and when this in turn was forged up into artefacts, they too inherited the uneven distribution of elements, albeit further modified by the forging process. Plate 1.4.2c shows a variability in composition that is absolutely typical of so many artefacts made from bloomery wrought iron. Such heterogeneity obviously has a major bearing on the structures and properties of finished artefacts, and for this reason, it is essential that samples for examination are large enough to contain representatives of all of the possible variations in structure. The smaller the sample, the greater the uncertainty over interpretation. Equally, there are instances where it may prove necessary to take more than one sample from a specimen. The taking of minute samples from bloomery metal whose locations are unrelated to the type and form of artefacts is, by and large, a fruitless exercise. Where damage to appearance is invoked as a reason against sampling, it should be remembered

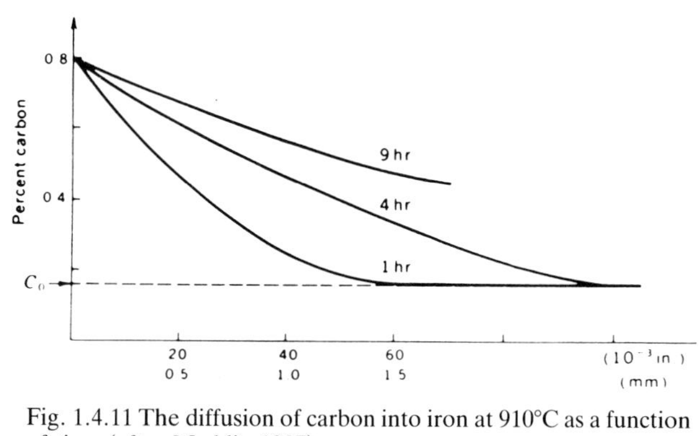

Fig. 1.4.11 The diffusion of carbon into iron at 910°C as a function of time (after Maddin 1987).

that by judicious use of synthetic resins and inert fillers, it is possible to disguise any sampling site to the point that it is only detectable through X-radiography.

A significant proportion of early artefacts were made from carbon-free iron which, in itself, would have been if anything inferior to the average bronze. As with bronze, however, the hardness of iron can be increased by cold working (Fig. 1.4.12), the greater the degree of cold working (expressed as reduction in the thickness of the piece being worked), the higher the hardness. Cold working can be seen both from its effect on final grain form and also sometimes where the lack of further work has left grains distorted (Plate 1.4.2d). From metallographic studies of early iron artefacts carried out in recent years, it has become clear that phosphoric iron played a significant role in some production industries. Minor amounts of phosphorous in solid solution increase the hardness of iron more rapidly than any other element (Fig. 1.4.13), and give a brilliant sheen to the polished metal. Thus, phosphoric iron was sometimes introduced deliberately to produce a patterned effect. Arsenical iron, an oddity to modern metallurgy, also appears to have been deliberately exploited in antiquity, and arsenic too increases the hardness of iron, although detailed figures are not available.

In modern practice, a local concentration of carbon is produced deliberately at the surface of pieces to toughen this region. The process, known as *case hardening*, consists of heating the metal, under reducing conditions, in contact with carbon to give a 'skin' of steel. In antiquity, ironworkers realised that they could convert parts of iron artefacts to steel by heating it in the reducing zone of the forge fire[14]. Although they would not have known how the process worked, what in fact happened was that, just as iron could absorb carbon during the smelting process, so iron in the fire could also absorb carbon when kept away from the oxidising air blast. This process, here termed *secondary carburisation* (to distinguish it from carburisation in the smelting furnace), converted the surface region of the artefact to steel, and as a consequence, there is a gradient in concentration of carbon from high at the metal surface, to low in the interior.

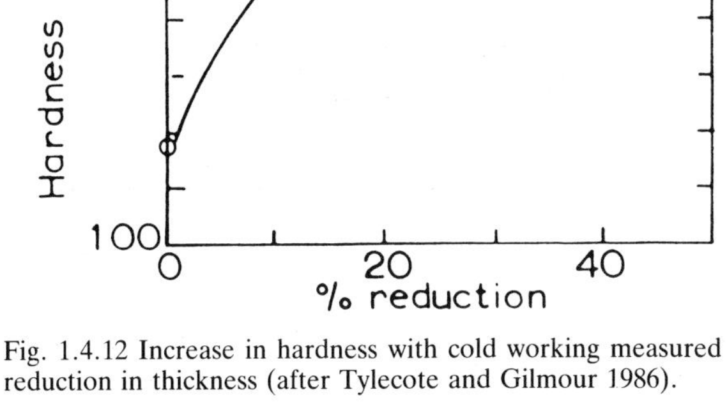

Fig. 1.4.12 Increase in hardness with cold working measured as reduction in thickness (after Tylecote and Gilmour 1986).

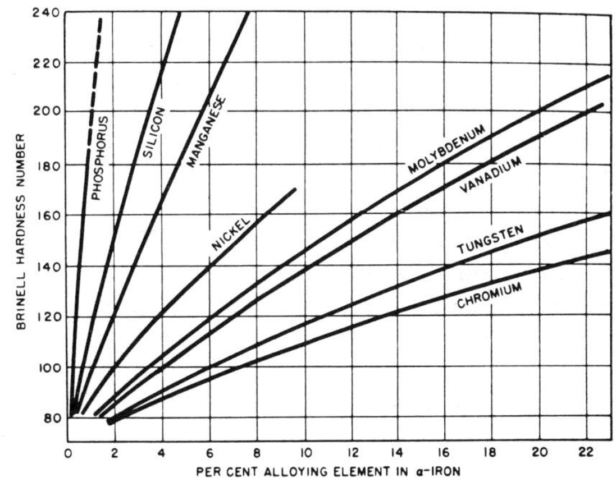

Fig. 1.4.13 Increase in hardness resulting from solid solution of elements in iron (after Maddin 1987).

The longer the process was continued, and the higher the temperature, the deeper the penetration of carbon into the austenite lattice.

It will be obvious that there are instances in which it can be very difficult to distinguish between deliberate carburisation of an artefact in the forge fire, and the product of forging out a billet of iron unevenly carburised during the smelting process. Equally, it is not impossible that occasionally, some uptake of carbon also occurred unintentionally. A primary criterion for the identification of secondary carburisation is a distinct gradient in carbon content from high at the surface to low in the interior. Such treatment was not infrequently given to the cutting edge of an artefact to improve its quality. And in identifying deliberate secondary carburisation, it is important to remember that it is most likely to be present on a working surface or edge. It will be appreciated also that carbon will not stay fixed in the distribution arising from primary or secondary carburisation if heating operations are prolonged. During forging, carbon will diffuse from areas of high content into those of low content and this again can make interpretation difficult.

A further complication to interpreting structure is the fact that not only can carbon diffuse into iron, but can also be lost by oxidation. This process, known as *decarburisation*, occurs when steel is heated in in oxidising conditions. The carbon combines with oxygen to form carbon dioxide and the surface is consequentially depleted of carbon. With a piece which has suffered decarburisation, we will find a carbon gradient this time from low at the surface (where a coarse ferritic structure may occur) to higher in the interior. However, again such negative gradients can also result from uneven primary carburisation, and the positioning of the carbon-depleted area must be taken into consideration. Recent experimental work by Crew on the forging of blooms (Crew pers. comm.) and by Pleiner on the forging of consolidated stock into artefacts (Pleiner pers. comm.) has shown that there are significant metal losses, in the form of *hammer-scale* produced by the oxidation of the hot metal. It was not only in forging that decarburisation occurred, but also when artefacts, usually weapons, were placed on cremation pyres. Again, the removal of metal by oxidation will influence the final microstructures that we observe, and may affect significantly the appearance of the distribution of elements, particularly at the surviving surface of an

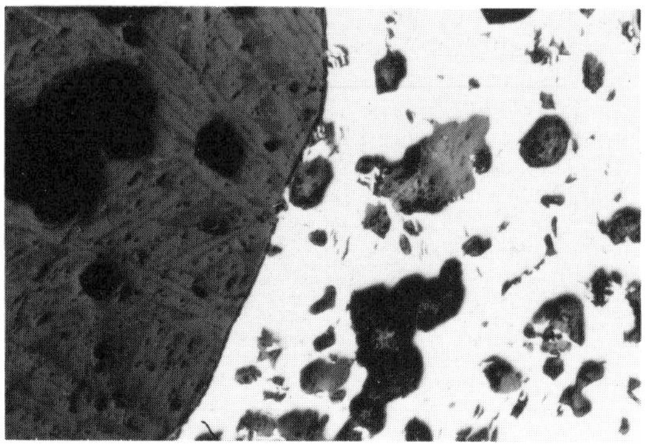

a

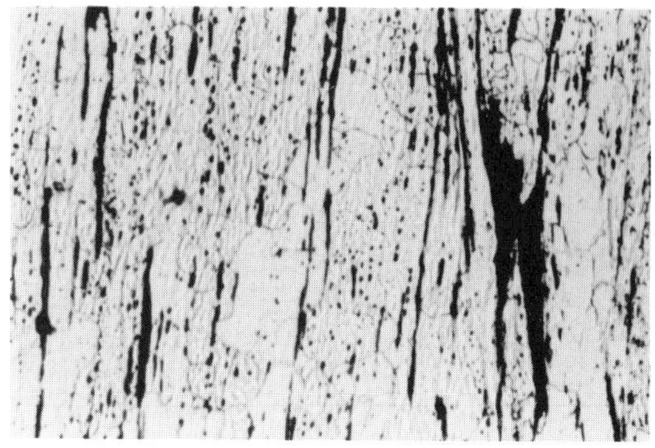

b

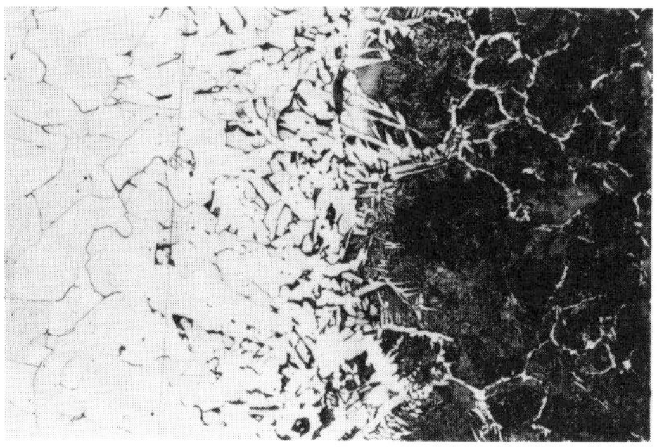

c

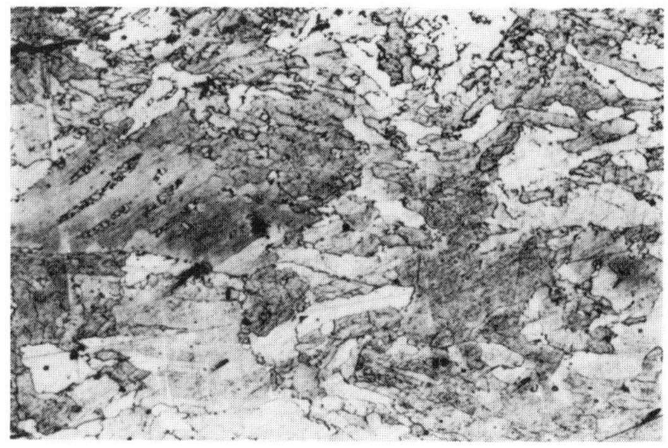

d

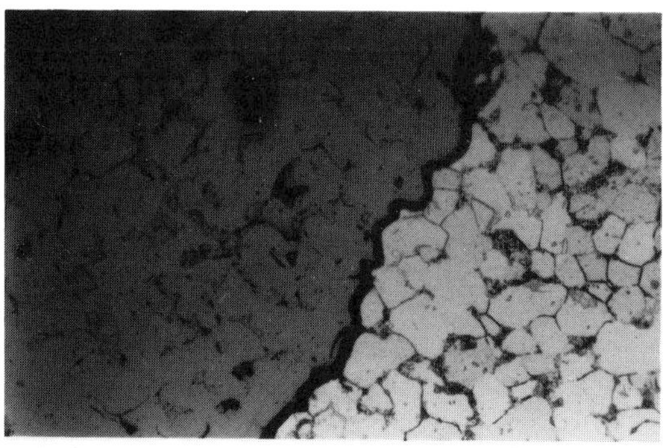

e

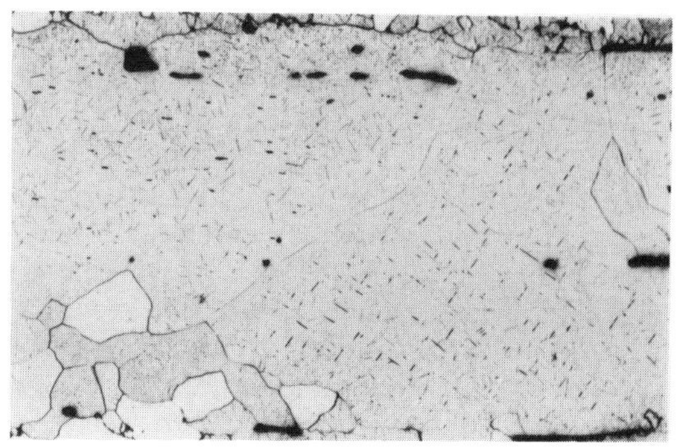

f

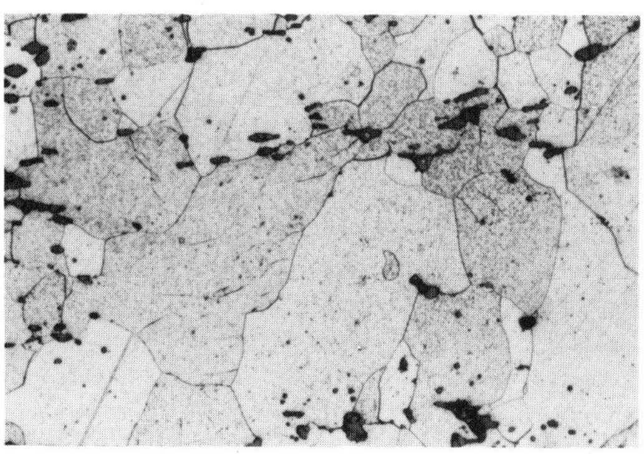

g

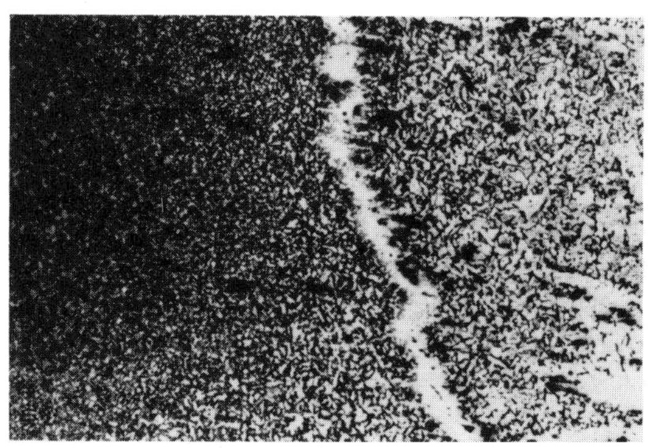

h

Plate 1.4.2 (a) structure of a partially consolidated bloom showing ferrite and slag, x25, (b) slag inclusions and ferrite, x50, (c) local carbon gradient in heterogeneous stock x100, (d) cold worked ferrite x100, (e) grain structure of original metal (light) outlined in corrosion products (grey) by the remains of interstitial pearlite x100, (f) acicular precipitates in ferrite x250, (g) ferrite with texture resulting possibly from age hardening x250, (h) diffusion of carbon across a weld seam x100 (Etched 4% Nital).

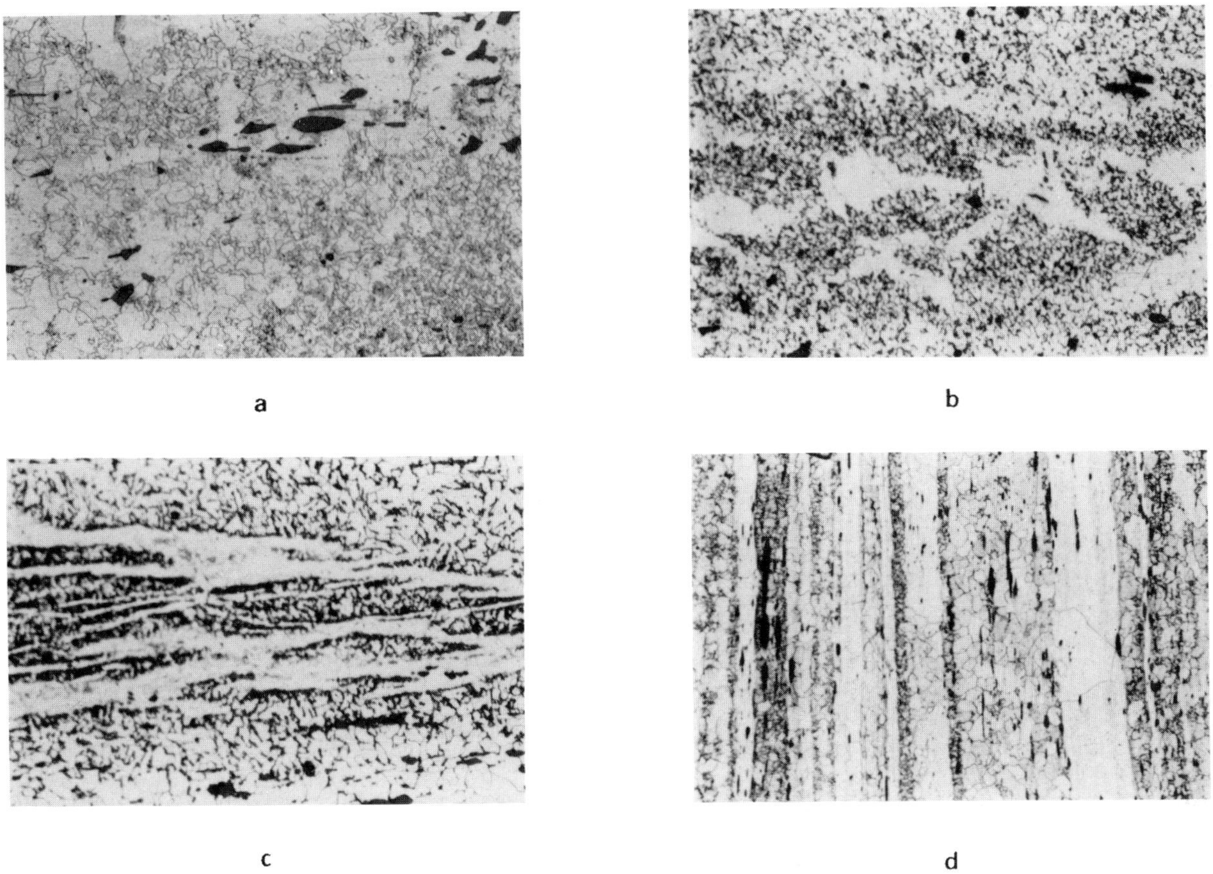

Plate 1.4.3 (a) weld seams left over from the consolidation of the bloom x100, (b) local segregations of P/As x100, (c) ferrite banding resulting from segregation in stock x50, (d) laminar structure resulting from fold-welding x50 (Etched 4% Nital).

artefact. Finally, we must be conscious always of the loss of metal resulting from the corrosion of an artefact. While remanent structures can be identified in the corrosion envelope (Plate 1.4.2e), thus giving some guide to the overall structure at and close to the original metal surface, their interpretation is fraught with difficulty.

In conjunction with the structures resulting from carburisation, it is interesting also to note the needle-like precipitates sometimes observed inside ferrite grains (Plate 1.4.2f). These needles are usually identified as nitrides such as γ'-Fe_4N, although some, in fact, may be carbides. The nitrides result from the diffusion of nitrogen into the iron. The sources of this are the nitrogen in the air blown into the furnace, and the heating of iron in contact with nitrogenous materials. The process of *nitriding*, the parallel of carburisation, is part of modern industrial practice. The story of *Weland the Smith* feeding iron filings from sword blades to 'tame birds' to improve the quality of the metal has been interpreted as suggesting nitriding (Chapter 7, note 28), and an increase in the nitrogen content of metal can be achieved by heating it in contact with chicken droppings. It is noticeable that in at least some cases, acicular precipitates occur in larger grains showing elevated concentrations of phosphorous. Levels of phosphorous could also be raised by the analogous process of *phosphiding*, heating iron in contact with a material such as bone ash which is rich in phosphorous. Some of the high phosphorous iron used in pattern welding could have been produced in this way.

Ferrite containing carbon above the level of solid solubility, and which has been cooled rapidly from above A_3 will be supersaturated with carbon at room temperature. Subsequent low temperature treatment will result in the precipitation of very fine particles of ε-carbide, possibly also of nitrides. While these will be too fine to view under the light microscope, they are detectable from the pits which form on etching (Plate 1.4.2g). There is a subsequent increase in the hardness of the ferrite – *age hardening*. The lower the temperature of the secondary treatment, the slower the rate of precipitation. If there is no subsequent heat treatment, then the precipitation process is exceedingly slow, although on the archaeological time scale[15] it may be of importance as an indicator that quenching of an artefact was attempted even if the starting material was quite unsuitable for it.

From the discussion of heat treatment above, it is obvious that a quenched and tempered steel was greatly superior to even the best of copper alloys. It was this property that would have made the early products of bloomery smelting so significantly more attractive than even the best non-ferrous artefacts. We have seen that early ironworkers could produce steel either by primary or secondary carburisation. In the case of primary carburisation, one simple way to obtain steel from a consolidated but heterogeneous bloom would have been to quench it and, using the brittleness of martensite, break off the steel areas from the more malleable low-carbon areas.

Ferrous metals can be fused together at high temperatures (well above A_3) and under the pressure of forging – *hammer welding*. For the two surfaces being welded to bond together, they had to be clean, although entrapped slag (carried over from the smelting process) would have acted as a *flux* to remove scale: alternatively, smiths could have used a flux such as sand. In either case, some of this material remained at the junction between the two surfaces and shows up as the chain of small inclusions following the line of the seam. Different carbon contents require different welding temperatures: wrought iron is welded at a white heat, while carbon steel requires a lower temperature. The welding together of pieces of iron and steel is thus no mean feat, since it requires a careful judgement of temperature. Under the microscope, weld seams show as discontinuities between areas, often marked by an abrupt change in carbon content and in the nature and orientation of slag inclusions. Weld seams may also be marked by enrichment of such elements as arsenic, nickel and cobalt through preferential oxidation at the hot metal surfaces.

The identification of seams resulting from the deliberate welding together of two or more pieces can often be difficult for a variety of reasons, one of the commonest being the diffusion of carbon across the them. Carbon will diffuse from areas of high concentration into those where carbon is either present in low concentrations, or completely absent, the governing factors including temperature, time, and the presence or absence of P and As. Thus, an abrupt change in carbon content may not mark a weld seam, but rather an interface between areas of high and low phosphorous content. If the artefact has been held at welding heat for any length of time, the carbon content of the low-C side of a weld seam will have been significantly increased by diffusion across it and the resulting gradient may obscure the presence of the weld-seam (Plate 1.4.2h), suggesting instead either primary or secondary carburisation. A further complication is that welding seams also occur randomly distributed as a result of the original consolidation of the bloom where the particles were welded together (Plate 1.4.3a). Their survival can be a cause of some confusion in interpreting the metal structure, and thus the technological history of an artefact.

Two other structures which can cause a certain degree of confusion are the banding resulting from carbon segregation (*Zeilengefüge*, *structure en bandes*), and the white lines associated with local enrichment of arsenic (also elements such as nickel and cobalt). The precise mechanism of production of such structures has been the subject of intense debate. Structures where bands of ferrite and pearlite alternate can arise from a variety of mechanisms in modern practice. The main cause in artefacts made from bloomery stock was the heterogeneity of primarily carburised metal. With heterogeneous stock (Plate 1.4.3b), heavy forging flattened out segregated areas in the direction of working, giving the section a striped appearance (Plate 1.4.3c). After forging, alternate bands of ferrite and ferrite with pearlite can appear to result from multiple weld seams. And in the absence of carbon, uneven distribution of phosphorous could promote differences in ferritic grain size, so that the forging out of such heterogeneous material resulted in bands of large and small ferrite grains. White lines, which on microanalysis show elevated contents of arsenic (and also in some cases nickel, cobalt and copper) arise from the enrichment of metal surfaces when they are being heated during forging operations. Iron oxidises at the heated surface to form scale, thus leaving behind preferentially elevated contents of elements like arsenic and phosphorous in these regions (cf. Figs. 4.2.2–4.2.4). The welding of a piece which has undergone such an enrichment will leave a prominent white line coincident with the weld seam. Segregations of arsenic in the original stock also seem to leave white stripes in the final structure (cf. Plate 4.2.21c). And although it has been assumed that the distributions of phosphorous and arsenic are coincident, there is some evidence to suggest a degree of mutual exclusion (cf. Figs. 4.2.2–4.2.4). It is thus necessary (as we shall see in Chapters 3–5) to view such structures with some suspicion!

A most interesting feature observed in some early artefacts is the appearance of numerous laminae in the cross-section (Plate 1.4.3d). The traditional view has been that laminated structures were produced by welding together large numbers of thin plates (the process usually described as *piling* or *faggotting*). In fact, these pieces were made by welding together a relatively small number of pieces of metal and then folding over and welding the composite stock a number of times: even carrying this out using a single piece of metal would have produced similar results. This technique, which we shall refer to as *fold-welding*, brought to a peak of perfection by mediaeval Japanese sword-smiths (well documented in Chikashige 1936), would rapidly multiply the number of weld seams and laminae observed in the section. The number of layers so produced is equal to 2^n where n is the number of times a simple two-piece block is folded over. If the starting stock contained a number of different pieces of metal welded together, then theoretically, the number of laminae that could be observed in section will be Z x 2^n where Z is the number of components in the original stock. But as the welding process proceeded, thinner layers were probably totally lost through oxidation. Furthermore, excessive time at a high temperature would undoubtedly have led to diffusion of carbon across weld seams, thus obscuring them and thus further confusing interpretation. Enrichment by oxidation of the substrate will have contributed to the formation of distinctive banded structures. In addition to this, use of inhomogeneously carburised stock as starting material will have introduced banding into the section. Thus, while these pieces can have the distinct technological advantage over simple forgings from bloomery iron by giving a more even distribution of the content of carbon, phosphorous, arsenic and other elements, their interpretation is far from simple.

This description of structures observed in early bloomery iron is of necessity brief, since the variety is almost endless. It is intended only to be a basic guide to assist in the understanding of the metallographic evidence discussed in the later chapters. The reader interested in pursuing the subject further is referred to the articles cited at the start of the section, and to the other metallographic studies listed under *References*. Techniques of sample preparation and of the metallographic examination of iron are described in most basic works on metallurgy (see Appendix 2).

* * *

1.5 – Blacksmithing tools and techniques

Just as we can reconstruct at least a part of the metallurgical history of a piece by examining its microstructure, so from its shape we can gain a good impression of the types of tools used to shape it. Although early literature names a number of smithing implements (Chapter 7.5), only a very limited number of those which would have been used by blacksmiths in Ireland during the period under review have survived or (perhaps more accurately) have been identified. The shaping techniques used on Irish iron artefacts attest to the use of a variety of tools which do not yet appear in the archaeological record. Unfortunately, we do not have anything to compare with hoards of blacksmithing tools such as those from Mästermyr in Sweden (Thålin-Bergmann 1979, 109f) or Sadovec in Bulgaria (Pleiner 1962, 119), or indeed the Viking tools from Kilmainham, Co. Dublin (Coffey and Armstrong 1910, 118, Fig. 7).

As with his modern counterpart, the toolkit of the early blacksmith was based on the anvil, tongs and hammer. Despite this, the inside of a modern blacksmith's forge often presents a bewildering array of implements in racks round the walls, with the anvil and the hearth as its foci. In particular, there will be a wide variety of tongs used to grip different shaped and sized pieces of metal, and different sizes of hammers and cutting tools. Furthermore, all blacksmiths will have favourite tools which they have made themselves for particular pieces of work and this adds further to the layman's confusion. But the variety is deceptive, disguising the fact that a small number of tool types is being used to carry out a large number of operations.

One of the principal problems in the identification of smithing tools is that corrosion will have reduced the great majority to featureless scraps and fragments. Identification from the unpromising mass of material that constitutes a significant proportion of most excavated assemblages of ironwork will require close examination, based on X-radiography. Were such a program to be carried out, it is likely that we could add a number of potential tool types to the otherwise meagre list. But this will not be enough, since another major problem is that some simple forms could have been used in other crafts as well as ironworking. The chisel, for example, is basically a bar of iron with a cutting edge at one end and a striking surface at the other. The same shape of tool could have been used in stone carving, woodworking and blacksmithing. Similarly, a small pointed rod could have doubled as a bradawl for woodworking, or a punch for making holes in leather, non-ferrous and ferrous metals. Metallographic study can indicate, for example, whether an edged tool could be expected to have performed reasonably in ironworking. Thus, it seems unlikely that the chisel from Feerwore, Co. Galway (Plate 4.2.18; examination no. 26) could have been used to cut iron, hot or cold, because it is of such poor quality soft iron. But even where we do find a chisel with a good steel edge showing heat-treatment, this does not necessarily mean that it was used for metalworking – it could be a high quality woodworker's tool. In the absence of a survey of potential smiths' tools, and of metallographic data, I will restrict the discussion to clearly identifiable types and to illustrating the evidence for the use of others yet to be identified. Other evidence is discussed in Chapters 3–5.

Fig. 1.5.1 illustrates the use of some tools whose early counterparts can be inferred from Irish EIA A-LIA A iron artefacts and techniques which we can observe:

Tongs and *hammers*: used to grip and hammer hot metal. The jaws of the tongs will vary greatly, depending on the shape of the piece to be gripped. The hammer varies in shape and weight according to use, from sledges for heavy hammering to light ball-peened hammers. Hammers were used not only for hammering metal directly, but also for applying force to other tools to cut and pierce. The tongs were used to grip and twist metal in the production of twisted rod and bar work, a technique which was practised throughout the period under review.

Setts and *chisels*: these are tools for cutting metal, hot or cold. Chisels are hammered hand held (Fig. 1.5.1a), while setts are held with a rod (Fig. 1.5.1b). In modern practice, the chisels and setts used for cold work are thicker and more massive than those for hot cutting, and hot chisels are longer to protect the hand from the heat of the stock being cut.

Fullers: these are blunt-ended tools similar in shape in shape to setts and chisels which are used to thin down metal (Fig. 1.5.1c), and to make shoulders before drawing-down (Fig. 1.5.1d).

Swages: these are two-part tools (Fig. 1.5.1e) with grooved faces. They are used to reduce thickness and to give a good rounded finish. In modern practice, the lower half is set into a hole in the anvil, while the upper is hand-held.

Mandrels: while the anvil would have provided the surface for most work, some curving and bending of metal, as in the forging of small rings and of spear-head sockets would have been done round mandrels, which are simply tapered, round-sectioned bars (Fig. 1.5.1f).

Punches and *drifts*: the simplest way of forming a hole in a piece of metal was to pierce it, hot, using a pointed bar or punch (Fig. 1.5.1g). If required, the hole could then be expanded by the use of a wider tapered bar or drift (Fig. 1.5.1h).

Files: as in modern practice, these would have been used in the finishing of work, removing rough surfaces, making fine adjustments to shape, and in

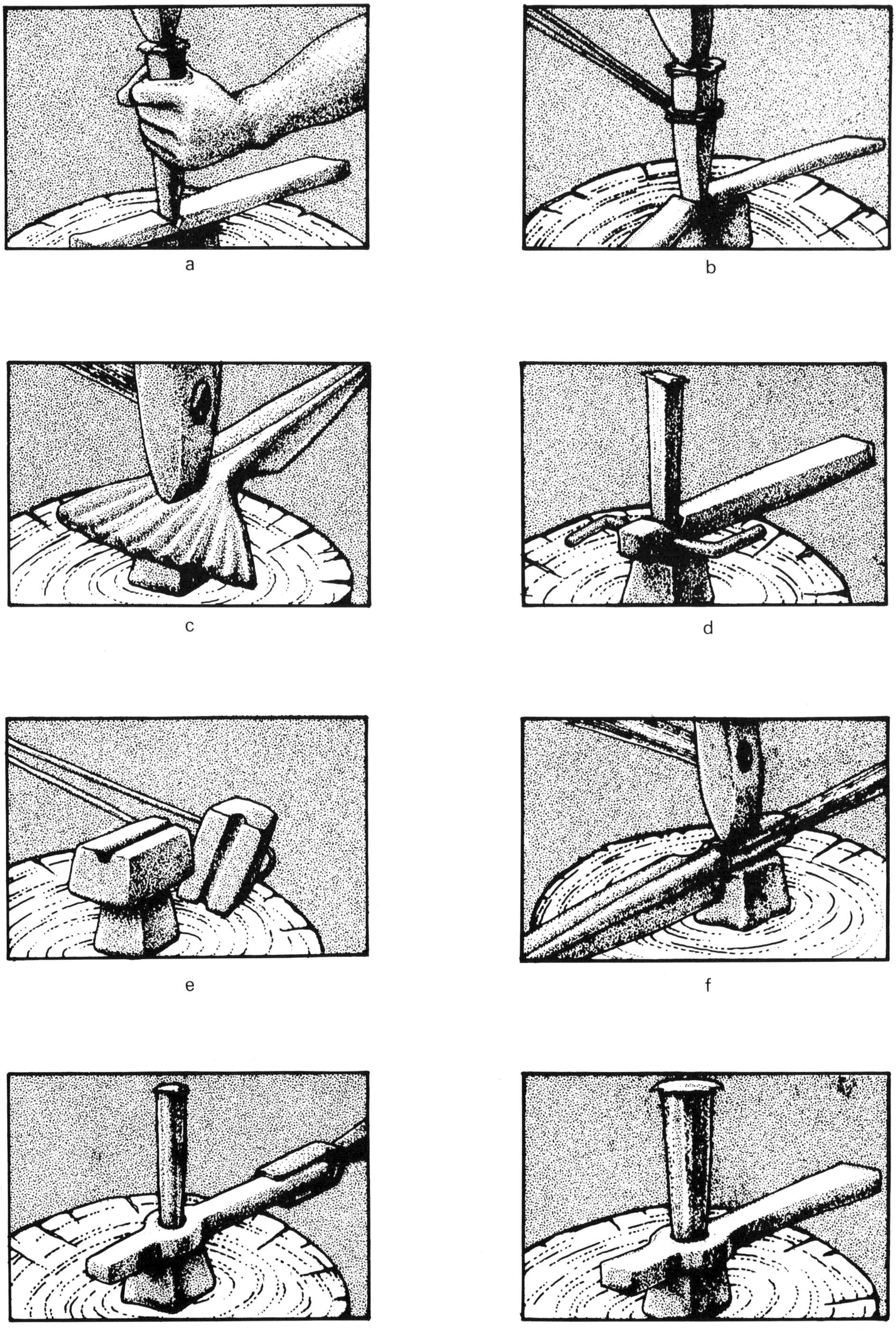

Fig. 1.5.1 Basic smithing tools and techniques: (a) cutting stock with a chisel, (b) cutting stock with a sett, (c) drawing down by increasing width, (d) making a shoulder using a fuller, (e) swages, (f) use of a mandrel, (g) use of a punch to make holes, (h) use of a drift to enlarge holes.

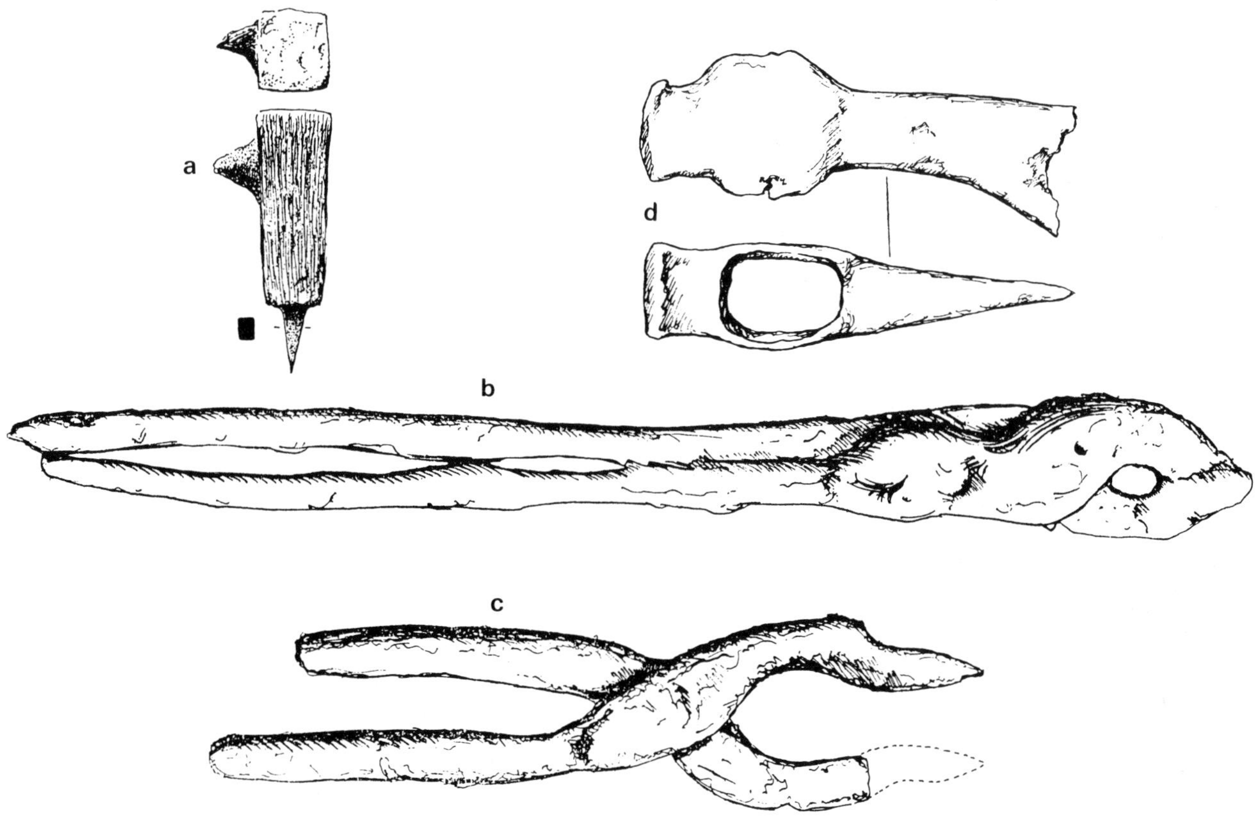

Fig. 1.5.2 Smithing tools from Irish sites, all x½: (a) anvil from Garryduff I, Co. Cork (after O'Kelly 1964), (b) and (c) tongs from Garannes, Co. Cork (after Ó Ríordáin 1942), (d) sett(?) from Cahercommaun, Co. Clare (after Hencken 1938).

sharpening. There is also evidence that whetstones and grindstones were used as well.

The only anvils securely dated to the period under review[16] are the stone block found in the smelting and smithing area at Clogher, Co. Tyrone (Chapter 6.5: Plate 6.5.7), and the small iron anvil from Garryduff I, Co. Cork (Fig. 1.5.2a). This small specimen, with a working surface 3 cm x 3 cm, and a tiny bick iron, has a spike at the base where it would have been driven into a block of wood. Apart from the excellent textual evidence for this arrangement of anvil and block in Ireland (Chapter 7.4), such anvils are widely known also from outside the country (e.g. for Europe see Pleiner 1962).

Examples of tongs come from Garannes, Co. Cork (Fig. 1.5.2b, c). Twisted bar work, produced by gripping the two ends of a hot bar and twisting in opposite directions, is seen, for example, on the ring handles of the EIA A Drumlane, Co. Cavan cauldron (Fig. 3.2.3), on LIA A bucket handles from Lagore, Co. Westmeath (Hencken 1950, 114, Fig. 47), and from the herringbone panels of the Lagore collar (Fig. 1.5.4 below): it requires much practice to be able to produce an even twist No blacksmiths' hammers are attributable to the period under review, but likely examples of setts, which might have doubled as heavy hammers, come from Cahercommaun, Co. Clare, Lough Faughan, Co. Down and Lagore, Co. Westmeath (Fig. 1.5.2d). A file, now lost, was found at Carbury, Co. Kildare (*CIIAA* 589), and this might have been a metalworking tool.

We can best see how tools were used by looking at the ways in which some early Irish iron artefacts were formed (Fig. 1.5.3). The attractive horned animal from Clogher, Co. Tyrone has its horns and a scroll under the chin formed by the cutting-back of strips of metal using a chisel. Similarly, the bifurcation of bifid pins such as those from Garryduff show the use of a chisel to cut and separate metal components.

Reducing metal stock to a required thickness, *drawing down*, can be carried out using a fuller, either by increasing length or width. The range of applications is wide – from the production of sheet metal such as that used in the cauldron from Drumlane, Co. Cavan (Fig. 3.2.3), and the spreading out of metal in the forging of axes such as those from Lough Mourne, Co. Antrim (Fig. 3.3.1) and Rathtinaun, Co. Sligo (Fig. 3.3.2), to the shaping of a nail point. The ribbing on EIA B sword blades such as that from Lisnacrogher (Plate 4.2.3a) and the lunate depressions on the 'Co. Fermanagh' and Drumshanbo, Co. Leitrim spearheads (Plates 5.3.18 and 5.3.19) show the use of the fuller, this time to provide a decorative feature. Sudden reduction in thickness is seen clearly on such artefacts as the tanged, single-edged knife or a sword, where a fuller would have been used to form the tang by drawing down the metal at one end of the blade. When smiths made spearheads, the sockets were formed by spreading out a flap of metal which was then bent round a mandrel.

The opposite to drawing down is *upsetting* which produces local thickening of metal, as in the forging of the terminals of the iron pennanular brooch from Lagore, Co. Westmeath. The snout of the Clogher animal (Fig. 1.5.3e) has been shaped in this way. Upsetting was also important in the formation of

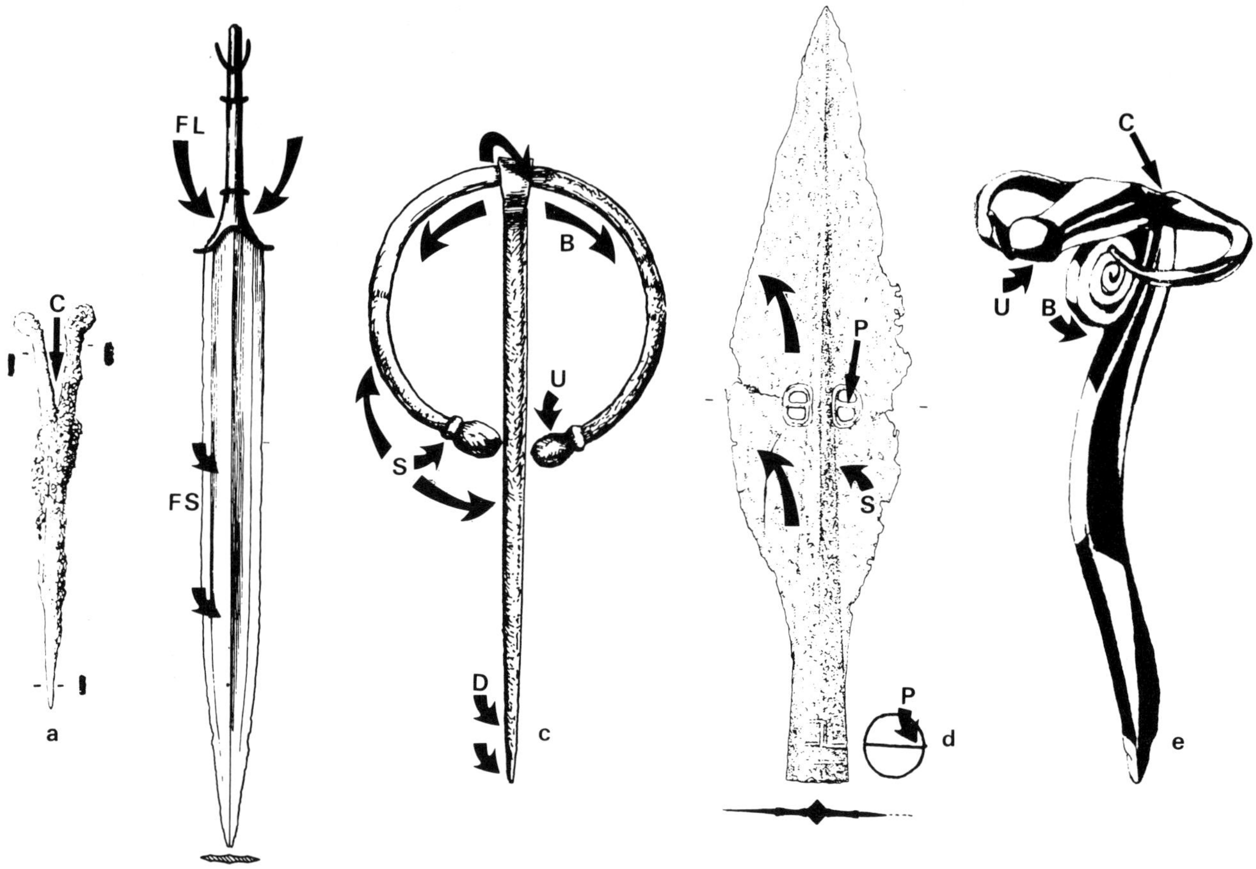

Fig. 1.5.3 Smithing techniques observed on early iron artefacts: (a) bifid pin from Garryduff I, Co. Cork, (after O'Kelly 1964), (b) triple-ribbed blade from Lisnacrogher, Co. Antrim (after Raftery, B. 1984), (c) iron pennannular brooch from Lagore, Co. Westmeath (after Hencken 1950), (d) spearhead from Corrofin, Co. Clare (after Raftery, B. 1984), (e) horned beast from Clogher, Co. Tyrone (Courtesy of R.B. Warner). B=bending: C=cutting: D=drawing down to a point: FL=fullering to draw down by increasing length: FS=fullering to decrease thickness and increase width: P=punching: S=swaging: U=upsetting.

sharp bends, to counteract the natural thinning out of metal bent towards a right-angle. The formation of a good right-angle is a skilled task, requiring the metal to be upset in the region of deformation (Fig. 1.5.1a), as in the forging up of the socketed axe from the Rath of Feerwore, Co. Galway (Fig. 4.2.1). Even the forming of gentle bends requires skill to avoid local thinning of metal.

Swages would have been used, amongst other things, to produce thin, round-sectioned rods which could be made up into pins and brooches. Thicker stock would also have been produced as the starting point for a variety of artefacts, including rings, chain-links and vessel attachments. Punches and drifts were used to produce, amongst other things, rivet holes, sockets in axes, and the perforations in the blades of the spearheads from Roodstown, Co. Louth and Corrofin, Co. Clare (cf. Fig. 4.1.1i). An interesting use of the punch is seen from a spearhead of probable EIA B date from Moyfin, Co. Meath (Fig. 4.1.1h) which has small spots of red enamel set in small depressions which would have been made with a special shaped punch. Use of punches with shaped ends is known elsewhere in Iron Age Europe (e.g. Pleiner 1962, Taf. III; Lang 1987, 63).

Although bloomery iron and steel were much easier to weld by heating and hammering than their modern counterparts, the technique required skill and a good judgement of heat. Welding was used not only to join separate pieces of metal together, such as adding steel cutting edges to iron tools and weapons, but also to close faces together, as in the forging of the socket of an axe or spearhead (Fig. 1.5.1f), and the closing of chain links and rings. For operations such as the forming of a chain link or the forging of a ring-head, the two ends of the bar often would have been flattened slightly, *scarfed*, and then overlapped for closure.

One piece which draws many of these techniques together is the so-called 'slave collar' from Lagore, Co. Westmeath, which dates to no later than the mid-7th century AD. In its construction we can see the use of the hammer, tongs and anvil (for gripping and hammering), of chisels and setts to cut, of fullers to draw down, of the tongs to produce twisted bar work, of swages to produce the stock for the chain links and of the technique of hammer welding is the assembly of the components.

The smith[17] took as his starting point two billets of iron and ten square-sectioned rods, one longer than the rest. Eight of the rods were twisted for most of their length, one was left plain and the longest was bent to form an eye to which the chain would be attached (Fig. 1.5.4a). The ends of the twisted rods were then welded together in pairs so that their twists opposed, and two pairs were each welded together, separated by one of the plain rods, to form two panels (Fig. 1.5.4b). To form the backing for these panels,

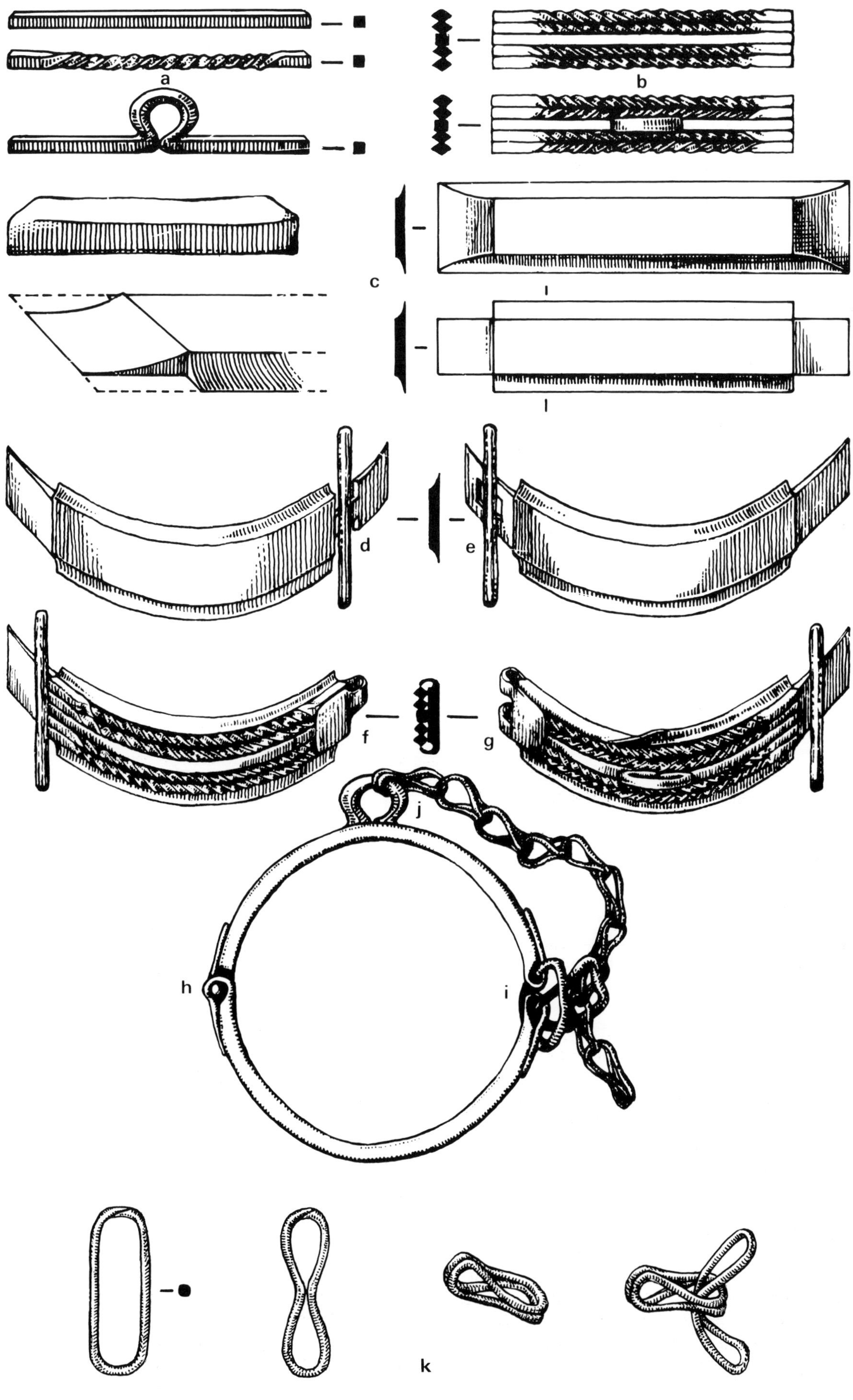

Fig. 1.5.4 Schematic diagram (not to scale) of the construction of the Lagore, Co. Westmeath collar and attached chain.

the billets were forged out into two thick plates with tapered edges and ends cut parallel with the edges (Fig. 1.5.4c). At the end of one of the plates, metal was cut out so as to leave a flap attached by a thin strip (Fig. 1.5.4d). A square was cut out of the equivalent end of the other plate (Fig. 1.5.4e) and both plates were bent to the required curvature. The two panels were welded at their ends against the backing plates and the end-flaps of the backing plates were bent back over the ends of the panels and welded to them (Fig. 1.5.4f, g). The tapered edges of the backing plates were bent round, to cover part of the outermost twisted rods. When the two cut flaps at the end of the plates were bent round they formed one and two eyes respectively through which a rivet was inserted to form the hinge for the collar (Fig. 1.5.4h). The loops formed at the other ends of the plates held the two links which formed the closure mechanism (Fig. 1.5.4i). Finally, the first link of the attachment chain was inserted and closed over (Fig. 1.5.4j,k).

As with the survey of the metallurgy of bloomery iron, this survey of smithing tools and techniques is, of necessity, brief. It is interesting to note that only a tiny proportion of the sites which have produced evidence of ironworking (Appendix 1) have produced certain blacksmiths' tools. While it is most likely that implements such as tongs and anvils would have had long use, probably being passed from one generation to the next, it is difficult to believe that those so far identified represent the sum total of survivals. This is an area requiring further study.

* * *

Notes to Chapter 1

1 Against this hypothesis it could be argued that the earlier 1st millennium AD in Ireland shows remarkable uniformity in such aspects as the various schools producing fine non-ferrous metalwork. However, by its very nature, fine metalwork is a specialist product, not to be ascribed a basic industrial role.

2 A list of C-14 dates for sites of all periods in Ireland, including calibrations made using the latest data, has recently been published (O'Kelly 1989, 347–352).

The breakdown of the usefulness of C-14 dating is particularly unfortunate since otherwise, it would now be possible to date carbon extracted from artefacts. The application was originally proposed by van de Merwe (1969), although twenty years ago, dating required the virtual destruction of large objects. With improved technology, it is now possible to obtain reliable dates from less than 100 mg of carbon and this would be obtainable from a metal section. Equally, carbon can readily be extracted from slag for dating and has the advantage of having been sealed against organic contamination in a silicate envelope. The dating of carbon from slag and artefacts may prove useful in sorting out dating sequences for the latter part of EIA B and LIA A, but is unlikely to help for earlier artefacts. One technique which just might conceivably help in relative seriation is fission track dating which, if suitable slag inclusions (i.e. with uranium content) could be identified and their track counts compared against standards (cf. Scott 1974d)

3 However, it should be remembered that the Anglo-Norman conquest of Ireland in the period AD 1169–1175 was accomplished by extremely meagre forces.

4 δ-iron only exists above 1410°C and is of no interest to our discussion.

5 The lowest temperature at which a solution is completely liquid is known as as the *eutectic* temperature, and the composition of the solution at this point is the eutectic. The eutectic reaction involves the separation of solid crystals (S) from a liquid (L). The *eutectoid* ('eutectic-like') reaction is so called on analogy with the eutectic reaction, and involves the separation of solids from another solid phase (S_a). The two can be summarised as:

Eutectic reaction $$L \underset{\text{cooling}}{\overset{\text{heating}}{\longleftrightarrow}} S_1 + S_2$$

Eutectoid reaction $$S_a \underset{\text{cooling}}{\overset{\text{heating}}{\longleftrightarrow}} S_b + S_c$$

6 The diagram used here is of the iron-cementite system. This is not a true equilibrium diagram since cementite is not an equilibrium phase, decomposing under the right conditions to give graphite. However, this is exceedingly rare in ordinary steels, since once formed, cementite (which nucleates much more readily than graphite) is very stable and thus can be treated as if it were an equilibrium phase.

7 The critical points for cooling occur at slightly lower temperatures than those encountered during heating.

8 A structure resulting from cold-working is the *Neumann band* seen in deformed ferrite. It has not been observed so far in Irish specimens. Good examples are illustrated by Samuels (1980, 175 and 203, Fig. 45).

9 The structure is named after Alois von Widmänstatten who in 1810 first published an account of the structure of a meteorite which fell near Zagreb in Yugoslavia in 1751.

10 The term *troostite* originally referred to a dark etching, nodular phase, unresolvable under the light microscope. Electron microscope studies have shown it to be extremely fine pearlite. However, for the sake of ease of reference, the term is retained here. Another term found in archaeometallurgical literature is *sorbite*, used to describe the finely divided carbide in a ferrite matrix resulting from tempering, while *sorbitic pearlite* was used to describe finely lamellar pearlite resulting from rapid cooling rates. They are much less easy to define than troostite, and so will not be used here.

The terminology of the transformation products obtained on progressively faster cooling rates from the austenitic range in plain carbon steels is quite confusing in British and American metallurgical literature from the late 1930's onwards. Two good accounts of the differences between the older and newer terminologies in terms of actual structures are given by Hume-Rothery (1966, 219ff) and Kennon (1980).

11 The rather vague terms 'toughness' and 'hardness' refer here to resistance to deformation.

12 Some iron ores, however, can comprise almost pure iron compounds, and could be reduced without the production of any significant quantity of slag (Clough 1987), but these do not concern us here. Similarly, the smelting of hammer-scale, virtually pure iron oxide, would be a slagless process. Again, we have no evidence for this, unless this is the meaning behind the Old Irish term *íarn aithlegtha* (see Chapter 7.3).

13 Of all of the ancient civilisations, only the Chinese came to grips with the problems posed by cast iron, basing industrial production on this metal. The interested reader is referred to the works of Needham (1970) and Tylecote (1986).

14 Rehder (1989) argues that secondary carburisation would have been exceedingly difficult to achieve in a forge fire because of the problems of maintaining a sufficiently high temperature in conjunction with an adequate concentration of carbon monoxide, a proposition strongly challenged by Wagner (1990). Carburisation of cutting-edges was common practice in antiquity. This question requires further investigation.

15 Hardening may also be solely a function of age and occur irrespective of the cooling regime. Thus, its interpretation in ancient artefacts must be treated with caution.

16 Wood-Martin (1886, 166ff) reported 'anvils' from crannogs at Loughtamand and Randalstown, Co. Antrim, as well as tongs and hammers. Unfortunately their contexts are so poor that they are excluded from consideration.

17 Fig. 1.5.4 is intended as a summary of the main stages in the construction of the collar. For the sake of simplicity it is laid out as if the whole collar was made at the same time. In fact, the collar as we have it now represents the work of two smiths, one who made the original, and a second who carried out extensive repairs (Scott 1978b).

CHAPTER 2

The Foreign Background to Irish Iron Metallurgy

2.1 – Metallurgical transitions: some general comments

If there is any certainty in our understanding of the establishment of iron as the primary industrial material of the Old World, a process whose effects are felt to the present day, it is that it was not aimless. From this starting point, we may seek to identify some general features of metallurgical transition processes common to the chronologically and geographically disparate Old World cultures in which they occurred. The rise of copper alloys, then of iron, to focal positions in early industry and economy could not have happened without the diligent application of human curiosity and ingenuity. But curiosity which led to considerable investment of time in learning to go from making simple trinkets of native metal, through to prospecting, mining, smelting and shaping, would not have been enough unless the producers of metals were able to eat! Significant benefits must have accrued, not only to consumers but also to producers. It is clear that when considering these transitions, we are dealing not with theoretical metallurgical investigations, but with practical applications of hard-won knowledge.

While there must also have been economic and social structures whose needs and capacities both stimulated and facilitated experimentation, metals not only had to be of use, but also must have been widely perceived as being of use. Whatever the cultural milieu, innovative Semitic, classical Mediterranean or barbarian European, *demand*, both potential and actual, was a prerequisite for supply to become a practical proposition. These critical steps in the development of man's ability to dominate his environment were not taken by theoreticians, but by practical craftsmen adopting a market-orientated stance in response to market forces. One can only support Muhly's robust dismissal as absurd of the denial of the profit motive in early trade and industry (1982, 253), even if the problem is more complex than he suggests. The disposal of the products of metallurgical labour must have been to the benefit of the disposer, since inheritance of the gene of altruism is not conspicuous as a successful evolutionary trait!

The capacity of prehistoric metallurgists to develop working theory was extremely limited, yet it is evident that they possessed a deep sympathy for, and empirically, at least, a practical understanding of metals[1]. Even if, in regions such as barbarian Europe on the threshold of the Iron Age, the key to extraction of iron – that the unmetallic-looking bloom concealed useable metal – was given to them by others, professional metalworkers there soon unlocked the secrets of manipulating at least some of the mechanical properties of iron. They were intensely practical, always with an eye to the marketplace in which the sale of their skills guaranteed their survival in societies where craftsmen were faced with the stark alternatives of usefulness and redundancy. It was the ability to 'deliver saleable goods' that kept the metallurgical transitions on course.

Globally, prehistoric metallurgical transitions involved the *addition*, not substitution, of new ideas (Scott 1978a). They were cumulative processes. As Piggott notes '. . . production of iron may be seen . . . as an innovative stage within a pyrotechnical continuum . . .' (1982, 19). Iron smelting, as the process of converting friable, coloured stone into lustrous, malleable metal is the direct conceptual descendent of the earlier smelting of copper. The skills of percussive working of stone were modified to the needs of metalworking, the basic skill of precisely applying force to a specific point remaining central[2]. However, earliest Bronze Age groups generally retained their full range of stoneworking skills. Initially, it would seem, stone and metals fulfilled industrially similar roles. The transition to iron again carried over the skill of percussive shaping, as well as those of locating and extracting ores, furnace construction and operation. Here, however, we can see a difference from the stone/metal transition, since once iron becomes established, there is a distinct shift in the role of copper alloys. Iron became the primary industrial material, while copper alloys are used principally in non-utilitarian contexts. Like our modern technocracy, the metalworkers of the Late Bronze Age, from Ireland to the Balkans, must have formed a distinct professional class. That they had built extensively and continuously on their technological inheritance is abundantly clear from those products of their experience which have found their way into the archaeological record.

Both the stone/metal and non-ferrous metal/iron technological transitions include phases in which old and new are employed side by side, with gradual role changes becoming apparent. Thus, for the first technological transition, we find as a general sequence:

Stone dependence .... Stone dependence plus use of native metals	STONE AGE
Stone dependence plus smelted metal artefacts .... Skeuomorphic interaction	TRANSITION
Metallurgy providing the bulk of tools and weapons	METAL AGE

Similarly, for the non-ferrous metal/iron transition, the general sequence is:

Non-ferrous metals only .... Non-ferrous metals plus meteoric iron and 'accidental' production of iron in copper smelting	BRONZE AGE
Non-ferrous metal plus iron smelted deliberately .... Decorative bimetallism based on non-ferrous metals .... Functional bimetallism based on ferrous metals .... Skeuomorphic interaction Prestige iron artefacts	TRANSITION
Iron artefacts predominating, non-ferrous metals in decorative roles, including decorative bimetallism	IRON AGE

These steps were taken as the economic and industrial climates allowed, at rates determined by speed of learning and the availability of raw materials.

In earlier publications (Scott 1977a, 1978a; cf. Scott and Francis 1981), I have suggested that the transition from stone to copper alloy took place *despite* the extra labour involved in metal tool production. Two rather basic flow diagrams (Figs. 2.1.1 and 2.1.2) compare the efforts involved in stone- and copper-tool production. There is a significant distinction to be made between the aesthetic and status values placed on metals in a non-industrial context and the decision that metals offered a greater return than stone for effort invested. The differences in availability and accessibility of artefact-grade stone and metal ores, while varying from region to region, do not immediately point to the desirability of abandoning effective technology for one quite different and more complex. Obviously, metals rapidly attained high status values, a fine illustration being

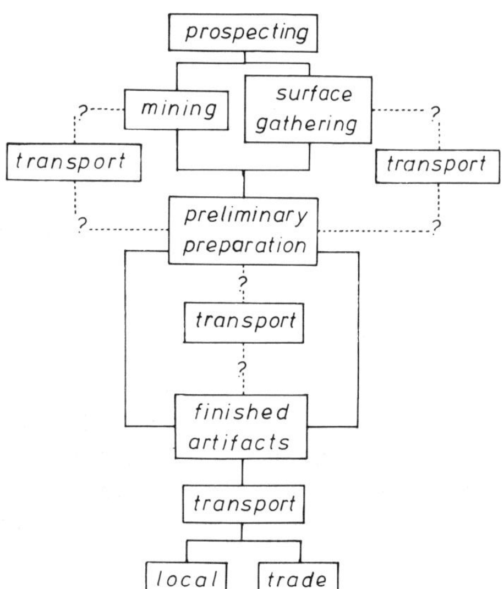

Fig. 2.1.1 Simplified flow diagram for the production of stone artefacts.

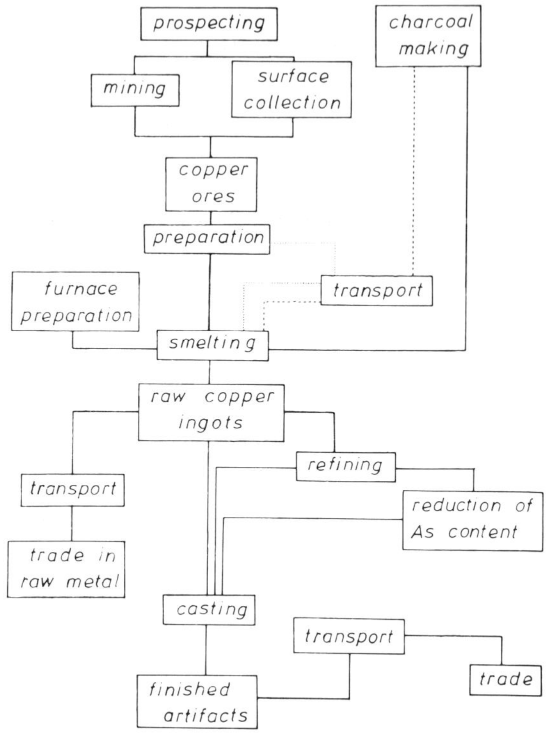

Fig. 2.1.2 Simplified flow diagram for the production of copper artefacts.

the cast cuprous metal copies of *C. antiqua Poli* shells among late Únětice peoples (Píč 1899; cf. Scott 1978a, 194 n. 12). We may see further evidence of this from the reproduction on polished stone of such features as casting seams copied from copper axes, themselves ultimately skeuomorphic of polished stone axes (e.g. Childe 1950, 107ff; cf. the skeuomorphism displayed by copper and stone axes from the Chalcolithic necropolis at Varna, Bulgaria – Ivanov 1978). But to make prestige ornaments is one thing, to change the contents of one's toolkit to new, initially less easily produced types, is quite another.

The shift from stone- to non-ferrous metal-dependence throws up the conceptual difficulty that one is trying to reconstruct remote prehistoric perceptions of the properties of unlikes. Since we know that a transition from stone- to non-ferrous metal dependence took place in the Old World, it is obvious that copper and its alloys gained higher values than lithics both in terms of status and of achievement against effort involved in production. Given that a major thrust of early developments in tool- and weapon-making was directed towards the production of a better cutting edge, can we say that the edge of a well-worked copper artefact bestowed greater benefits on its user (i.e. not only was it sharper, but also easier to obtain and maintain for daily work) than that of a collection of freshly-struck flint flakes or shaped tools? The durability of the metal cutting edge is greater than that of single lithic tools, but how significant a factor was this in securing the industrial ascendency of metals, particularly in regions with abundant flint and stone resources and comparatively poor in copper ores? If farmers of the Irish Late Neolithic in flint-rich north-east Ulster had anything of the nature of their modern counterparts, the selling of metals, even via trinket loss-leaders, would have required some highly competent marketing simply to convince them that some at least of their regular essential requirements could be more easily obtained with the assistance of metals. One factor which would have obviously have had an effect, albeit not readily detectable, is peer pressure – 'keeping up with the Jones'. In any event, the choice rested with small groups or even individuals and they chose to move from stone to metal tools over a period of time. In general terms, this marks a change from an era of disposable tools to one of conservation and recycling. The degree to which earliest Bronze Age recognition of the potential benefits offered by recycling metal was a significant factor in the stone/metal transition must be uncertain, although recycling certainly became a major part of Bronze Age metallurgical organisation.

As with the questions surrounding the establishment of industrial non-ferrous metallurgy, the reason why there was a need to find and adopt wholesale a substitute for non-ferrous metals has not been convincingly explained. We can divide the question 'why adopt iron?' into two parts; first 'why was iron considered of more use than copper-alloy?', second, 'why was a substitute needed at all?'. In answer to the latter, one set of factors offered as the spur to change is commodity shortage, brought about by exhaustion of regional resources, embargoes on the dispersal of raw materials, or disruption of supplies of copper and tin by widespread unrest and hostilities. This has been cogently argued for the Mediterranean region in the period beginning sometime around the 13th century BC (e.g. Snodgrass 1980). While plausible for this area, such explanation does not cover, with any comfort, the situation in barbarian Europe. The metalworkers of the Hallstatt C period, for example, continued to produce non-ferrous metalwork of the highest quality at the same time that they were manufacturing finely-wrought iron artefacts. Major socio-political developments undoubtedly were taking place, but evidence for the disruption of resource exploitation and distribution in this region is not easy to find.

Human curiosity and ingenuity stimulated and fuelled the copper-alloy/iron transition not only in the Near East but also in Barbarian Europe. Alone, they could not have brought about wholesale industrial change, even among the most innovative of societies. As with the transitional relationship between metals and stone, iron must have been able to compete with and be considered superior to copper-alloys (cf. Spratt 1982, 82f). In this transition, we *can* compare physical properties. Thus, Maddin et al. (1977, 124f; cf. Maddin 1982) convincingly demonstrate that the metal which we most commonly encounter as the product of early bloomery iron smelting by no means compares favourably with competently-worked bronze (Fig. 2.1.3). However, the discovery of the iron-carbon alloys, the potential properties of which were significantly better than those of bronze, was an obvious incentive to the early metal exploiters to persevere.

In view of the metallurgical sophistication of the Mediterranean lands, the role of meteoric iron as a conceptual link between salvaged prills and the 'black metal of heaven' should not be underplayed. Evidence from the Aegean alone of the use of meteoric iron in prestige trinkets from Minoan and Mycenean

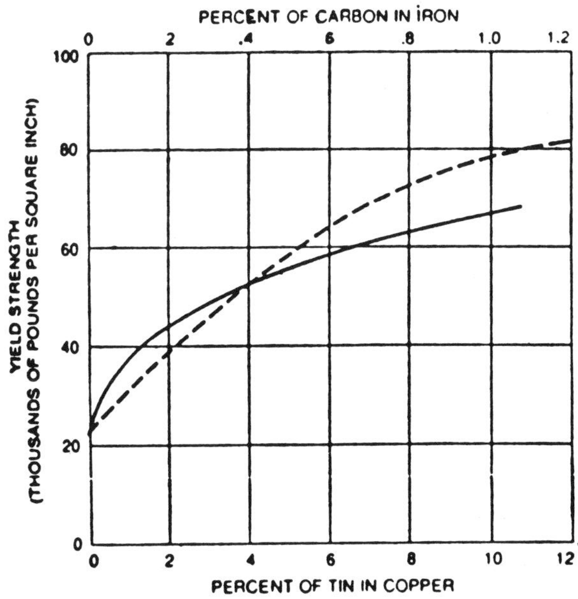

Fig. 2.1.3 Comparison of the strengths of iron-carbon (----) and copper tin (—) alloys (after Maddin 1987).

contexts is impressive (Varoufakis 1981). As with the relationship between native copper and the smelted metal (cf. Scott and Francis 1981), so there would be nothing in the nature of meteoric or telluric iron to instruct the early metallurgist in its production: use in the cold of native metals is not metallurgy. However, the importance lies in its being an important conceptual link between the first *smelted* metal and an already familiar and prized material. Meteoric iron would have informed a metal-using audience of the existence of another metal and would have ensured that at least some were able to identify it when it was produced in non-ferrous metallurgical operations.

In some respects, the emergence of iron (as opposed to its developed technology) might be seen to resemble the first use of aluminium in Europe following Deville's discovery of a practical way of producing it in 1855. The initial prohibitive production costs (e.g. Tylecote 1976, 149f) gave it great prestige value resulting, amongst other things, in its use for fine table cutlery, and for medals struck in France by Napoleon III. When the cheaper and more efficient technique of production by the electrolysis of molten cryolite was perfected, the price plummetted. But the intervening period was one in which simple rarity gave the metal a highly inflated value. The combination of the prestige value of possessing a metal imbued with heavenly properties, and the useful physical properties of at least some of the earliest metal, would have been a powerful driving force towards ensuring the continuation of experimentation.

The use of iron ore fluxes in copper smelting in Late Bronze Age industrial environments (Pleiner 1986) led to the regular production of pieces of iron, some highly carburised to *steel*, as a by-product. There is a high incidence of high-C metal at the beginnings of the European transition (Pleiner 1980, 388f; updated in Pleiner 1987, 82), and to this list is to be added the growing body of late second millennium BC steels from the eastern Mediterranean region (cf. Waldbaum 1978, 24ff) at sites such as Idalion on Cyprus

(Tholander 1971), Ta'anach in Palestine (Stech-Wheeler et al. 1981), the Baq'ah Valley in Jordan (Piggott and McGovern 1982), and Pella in Jordan (Smith, R.H. et al. 1984). Pleiner (1988, 36) notes the occurrence of an iron rod embedded in a cuprous cake of Ha B_1 period from Suchdol in Bohemia, suggesting that it may have been used to assist in the removal of iron particles from a molten copper bath. An analogous find, though later in date, is the iron rod embedded in a block of copper slag from Poggio Civitate, an Etruscan site of the 7th/6th centuries BC in central Italy (Warden et al. 1982, 29f). The use of iron ore as a flux in copper smelting as discussed by Tylecote (1974), of iron[3] in copper (Cooke and Aschbrenner 1975; Craddock and Meeks 1987), and the implications of these in combination as considered by Wertime (1980, 12f) and Charles (1980, 166), all indicate the possibility that iron was a regular enough by-product of copper smelting in some regions to stimulate its collection.

The production of what could not unreasonably be termed 'steel', and the discovery of hardening by rapid cooling were the keys to the acceptability of iron as an industrial and martial substitute for copper alloys. And as Pleiner (1969a) has shown, the product of some three hours operation of a Želechovice-type bloomery furnace was a eutectoid steel, with only a small area near the tuyère zone having been decarburised (the conditions leading to the production of high-carbon blooms are summarised in Guyan et al. 1973, 84–85). The development of the bloomery process, presumably by experiment, producing blooms with carburised areas by use of high fuel to ore ratios (cf. Tylecote et al. 1971), would have consolidated belief that a viable substitute for even the best weapons-grade copper alloy had been found (cf. Fig. 2.1.4). One must note, however, that the Želechovice furnaces used by Pleiner in his experiments have what has been interpreted as a special chamber where the bloom could escape decarburisation. Also, high fuel to ore ratios can, if not handled carefully, lead to reoxidation of the bloom. And when relatively carbon-free iron started to become a common product, the recognition of *carburisation* as a way of producing the same desirable physical properties would have sustained the momentum. But recognition of how carbon-free iron could be steeled may not be a primary discovery, and one must take issue with the hypothesis that secondary carburisation (Chapter 1.4) of artefacts was the principal spur to the spread of ferrous techology (cf. Stech-Wheeler and Maddin 1980, 116f; Stech-Wheeler et al. 1981). In a recent contribution to the debate, Rehder (1989) argues that phyical conditions in a forge fire would have made secondary carburisation a much more difficult prospect than is commonly implied. Although his appreciation of the European archaeometallurgical evidence is open to criticism, Rehder's analysis of the thermodynamics of carburisation, and his conclusions, may prove to have implications for our understanding of the course of development of the earliest ironworking (cf. Chapter 1, note 14).

If the model furnished by Pleiner for the introduction of Bronze Age metallurgists to the new metal is accepted, then we see that the discovery of secondary carburisation (i.e. outside the smelting furnace) would have been unnecessary to spur on the emer-

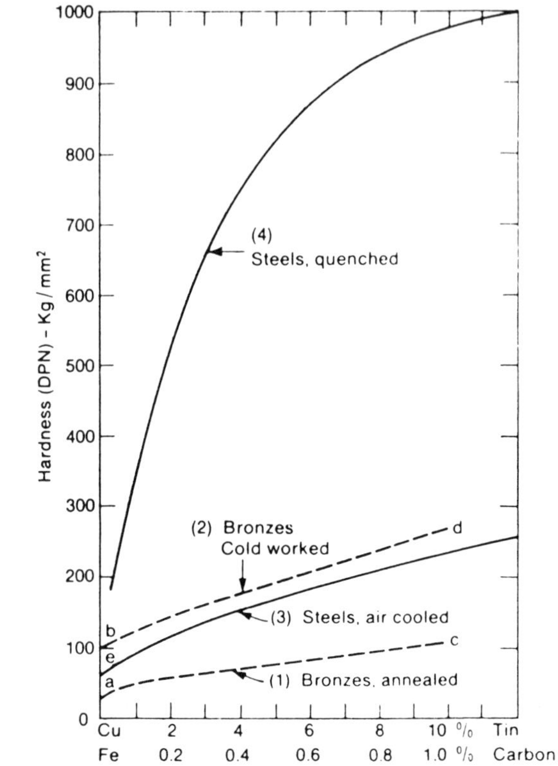

Fig. 2.1.4 Comparison of hardness of processed iron-carbon and copper-tin alloys (after Smith, C.S. 1967).

gence of iron, by and large, in a phase in which highly carburised iron was first being recovered from the debris of copper smelting and then produced in the smelting furnace. The subsequent development of iron smelting proper would have produced blooms which, if not completely carburised, would have had areas of high-C content. Given the metallographic evidence for the early knowledge of quenching, it seems highly probable that steels were separated off from consolidated blooms by the simple technique of quenching them and then separating out the hard, brittle areas by hammering. Obviously critical to the debate is the difficult question of criteria used for distinguishing between carburisation of the bloom during the smelting process, and carburisation of stock or of finished artefacts in the forge fire (cf. Piaskowski 1971; Pleiner 1971a).

There seem to be strong arguments for secondary carburisation to be a later development. Thus, artefacts such as those from the Baq'ah Valley discussed by Piggott and McGovern (1982), and pick from Mount Adir in northern Gallilee examined by Davis et al. (1985) would have been made of stock forged from primarily carburised blooms and not low carbon iron secondarily carburised in the forge fire. The assumption that the reheating of a bloom during forging to expel entrapped slag and to weld together the particles of iron would lead to significant enrichment seems highly questionable: oxidative decarburisation would seem much more likely. Many blooms examined metallographically show uneven volumetric carbon distributions which must result from the smelting regimes. Furthermore, even if significant surface penetration of carbon occurred during consolidation of the bloom, the carburised surface regions would have been prone to loss as oxidised scale. As Maddin et al. (1977, 124) illustrate, the rate of solid state diffusion of carbon is slow (cf. Fig. 1.4.11), and even allowing for control of oxidative surface decarburisation, it is difficult to envisage the Mount Adir pick (though not the Idalion knife) being carburised thoroughly in the forge fire. Mediterranean artefacts for which bloom-stage carburisation is suggested to explain high-C contents may instead be interpreted either as the products of carburisation of *artefacts*, or else (and more likely) of the use of primarily carburised blooms. An alternative explanation is that it was not carburisation but *quenching* that was early recognised and that the first generations of iron producers simply quenched inhomogeneous consolidated blooms, broke off the brittle martensitic areas and reworked this material. Traces of welding seams would tend to be obliterated by long forging, thus obscuring the multi-piece construction of such artefacts (cf. Pleiner 1973).

The silvery, arsenic-rich surface of a votive bull from a Hattian grave at Horoztepe in Anatolia is argued by C.S. Smith (1973) to have been produced by a cementation process allowing inwards diffusion of arsenic. Although presumably a purely decorative technique, this third millennium BC piece is evidence of sophisticated knowledge of a complex interaction between substances well before the appearance of iron. Substituting iron as the substrate and carbon as the intrusive element, would lead to carburisation. Given the very long history of prehistoric progress by observation and analogy (smelting here being a prime example), it seems not unreasonable to envisage the principle of decorative cementation being applied to the products of ironworking. Thus, although one must view the idea that carburisation outside the smelting furnace, rather than the early production of primary high carbon steel, ensured the establishment of iron as the major utility metal, with some unease, it would be unwise at this time to rule out entirely the possibility that this was an early discovery, based on earlier technology.

An alternative explanation for the discovery of iron production is prompted, in part, by early literary references to 'molten iron' (discussed in greater detail in Chapter 7.3), in part by dissatisfaction with the conventional view that the transition to iron must involve the immediate development of solid-state smelting. There is no theoretical reason why the earliest ferrous metal produced could not have been *cast iron* smelted under highly reducing conditions. White cast iron could be produced not only in a shaft furnace but also in a simple enclosed bowl furnace. On analogy with the oxidative refining of raw copper and manipulation of arsenic content, as well as the cupel refining of gold, cast iron could have been converted to steel by oxidative decarburisation, leaving no immediately recognisable evidence[4]. Again, experimental work shows that it is comparatively simple to decarburise white cast iron in a bowl furnace. Grey cast iron too can be decarburised, although this is more difficult to accomplish. Since metalworkers in Bronze Age industrial settings were thoroughly acquainted with liquid-state smelting, it is easier to postulate a conceptual link with molten iron, than with solid-state smelting and the singularly unmetallic appearance of a bloom. The scarcity of cast iron in the archaeological record was an objection raised by Pleiner (1969b, 26) to this suggestion, originating from Krawczuk and Piaskowski's reading (1958; cf. Halleux 1974) of the famous (or infamous) 'cast iron' passage in the *Meteorologica* of Aristotle (Chapter 7.3). However, the idea that any occurrence of cast iron in pre-Mediaeval times results from accident has gained acceptance by default. There has never been any detailed study either of the overall distribution of cast iron in prehistory (cf. Tylecote 1986, 139), or of the possibilities of distinguishing between bloomery metal and that produced by decarburisation of cast iron in structures other than the Mediaeval finery.

Experimental work by Tylecote and Boydell (1978) to reconstruct the copper smelting processes employed at Timna in Israel, used haematite as a flux, and produced coppers of high iron content. In one attempt at crucible refining, a layer of cast iron was produced (Tylecote and Boydell 1978, 46). A recent study by Photos (1989a) has shown that cast iron could be decarburised in furnaces of the type depicted on Greek black figure ceramics. Indeed, Photos' study demonstrates that Aristotle's στόμωμα was indeed steel produced by decarburisation. Aristotle's description, coupled with the 6th/5th century depictions on ceramics, suggests strongly that this steel production technique was restricted to special use. The wider exploitation of iron and of primary and secondary carburised metal would have been based on the bloomery process.

A phenomenon that largely has escaped comment, namely that amongst the earliest iron artefacts examined metallographically, there is a very high proportion of good steel, has been noted by Pleiner (1980, 388f; 1987, 82; 1988) and Piaskowski (1987a, 160, Table 1). As iron technology became more widespread, the proportion of homogeneous steel fell sharply, with typically heterogeneous carbon distribution becoming the norm. Although much more research needs to be done to provide a final answer, this situation accords closely with the idea that the earliest iron was cast iron, decarburised to steel. This allows us to suggest that it was bloomery smelting which was, in the first instance, accidental, with blooms being produced when the fuel to ore ratio was too low. Thus, the transition to iron would have involved two discoveries: first, cast iron and decarburisation (a difficult process), second, solid-state bloomery smelting. Such early acquaintance with molten iron, easily explains the strong semantic overtones of 'melting' in Indo-European words used in connection with the smelting of iron (see Chapter 7.1). Although this can be explained within conventional wisdom as semantic carry-over, the possibility of iron being first produced in the cast state and then refined by methods already proven for copper and gold, should not be dismissed. This would appear to be a critical area requiring further research.

For barbarian Europe, including the Atlantic margins, we may view earliest encounters with iron as provoking the same reaction to its rarity as postulated for earliest encounters there with the non-ferrous metals. Its association in the Mediterranean lands with the heavens could have accompanied the first artefacts to reach Europe, stimulating and sustaining interest. But the technological transition took place in the context of vigorous and thriving non-ferrous metal industries, and one must ultimately give most weight to the explanation that it was the properties of the material which aroused practical interest, and thus led professional workers in long-established production industries to experiment and to innovate. It seems likely also that the reputation afforded to the metal in general by the best early steels, overcame any resistance that might have been raised by uncarburised iron being produced by less skilled smelters. The versatility of copper alloys was still fully exploited and indeed when we move outside the field of weaponry, iron was used in all types of 'menial' roles whereas copper alloys tended to follow precious metals into 'prestige' roles. The only clear argument against the case for recognition of superior properties as the principal spur to change would be a demonstration of a significant diminution of production of non-utilitarian artefacts coincident with the first experiments with iron. Lower overall production of copper alloys would not be valid support for the 'scarcity forcing change' theory, since the adoption of iron would, perforce, have reduced overall production of copper and its alloying constituents due to the fall in demand for copper alloy tools and weapons. Only significant impoverishment of the assemblage of non-ferrous non-utilitaria would demonstrate the existence of a resource crisis forcing major industrial change. This does not appear to be the case.

* * *

2.2 – From copper-alloy to iron in barbarian Europe

Knowledge of iron spread to Europe out of the innovative Near East in the earlier centuries of the second millennium BC (Fig. 2.2.1). From this focus it was carried and transmitted through the Balkans and along the Mediterranean sea-routes. In the period ca. 1500–1000 BC, the Mycenean world became acquainted with iron, some smelted, some meteoric in origin, but in small amounts, while central and northern Europe (still firmly in the earlier stages of Bronze Age technological developments) also received products of the new technology (e.g. Pleiner 1980, 1981, 1986). The arrival of iron in Europe (but not necessarily, *pace* Powell 1976, 3, of knowledge of *smelting* technology) is signalled by occasional very early pieces in Hallstatt A/B and Montelius III/IV horizons (cf. Randsborg 1972, 41–42 and 42 *n*121). The earliest European iron finds are the dagger hilt from Gánovce in northern Slovakia, dating to perhaps the 15th century BC (Pleiner 1981, 115, abb. 1) and a small punch or chisel found beneath a log-built trackway at Baageroostervelde in Holland whose timbers have been dendro-dated to around 1350 BC (Charles 1984). Although still comparatively rare, iron is known from Hallstatt $B_{2\text{-}3}$/Montelius V contexts from northern Italy to southern Scandinavia (Fig. 2.2.2).

In Barbarian Europe, acquaintance with iron seems initially to have come through a movement of small artefacts from regions where production, and the reputation of properties, had become firmly established. It is clear that by the 8th century BC, many metalworkers were actually beginning to come to terms with the particular difficulties which the smelting and working of iron posed. It is not yet clear how long after the first circulation of pieces of imported iron it was before European metallurgists began to *smelt* iron. But they certainly were experimenting with the *hammering* of iron and with combining it with their traditional non-ferrous stock.

A central feature of the transition from copper alloy to iron was the interaction between workers in established technology and those developing the new. Certainly the earliest European *shaping* of iron would have been at the hands of those previously working in copper alloy, but once the technology was properly implanted, some metalworkers, adventurous, inquisitive and sensing the great potential of iron, must have begun to specialise, though still working in a strictly Late Bronze Age industrial setting. Of particular interest in this respect are those artefacts which show the new (and, presumably, mainly imported) metal used in conjunction with copper alloys – *decorative bimetallism*. This may have been more than just a desire to display possession of a prestige metal and perhaps we might consider also the possibility that the iron had magical properties ascribed to it. Pieces include the iron-inlaid copper alloy knives from Arnitslund and Kjøldbymagle in Denmark (Montelius IV: Stjernquist 1961, 79 and Plate LVI), the 10th/9th century iron-rivetted copper-alloy sword from grave 75 at Selciatello in Etruria (Villanovan IA: Hencken 1968, I, 45, and 46, fig. 34.i), iron-inlaid copper alloy pins and rings from Ha B contexts (Reith 1942, 9ff and abb. 4.5; cf. Kimmig 1964, 274ff; Podborský 1970, 173ff), swords of the Mörigen type (Sprockhoff 1931, 47ff; Pleiner 1981, 118 abb. 5), and a bronze votive bull from Býčí Skála in Czechoslovakia with iron inlays on its body (Šorm and Schwarz 1969, 85–86 and Tab. 10–12). Such pieces are the analogues of the Mycenean rings which have meteoric iron in their bezels (Varoufakis 1981) ostentatiously displaying the possession of a new, precious and possibly thaumaturgical material.

Following these, and showing an increasing supply of iron coupled with greater confidence on the part of metalworkers, are those artefacts where iron was used functionally in conjunction with copper alloy (cf. Waldbaum 1982 for a survey of such artefacts in the Near East and Mediterranean). Typical of this *functional bimetallism* are the iron-bladed daggers of Golovjatino and Gamów types with *kreuzformig* copper alloy hilts (Podborský 1970, 153ff and abb. 25–26), the *Lappenäxte* from Hallstatt cemetery with iron blades and run-on copper alloy sockets (Plate 2.2.1: Kromer 1959, II, pl. 142. Grab 702:3–4), and the fibula from grave 178 at Selciatello Sopra in Etruria with copper-alloy pin and intricate openwork body (Villanovan IC: Hencken 1968, I, 109, and 108, fig. 95). Of great interest here are two small Late Bronze Age copper-alloy bulls from Vestby in Norway, which show the use of *iron* chaplets in the casting process (Rosenqvist 1954, 128ff), and the bronze-coated iron socketed axe also of Late Bronze Age date from Hemse in Sweden (Serning 1984, 7–8). In addition, detailed observations of how tool-marks were made on copper alloy artefacts (e.g. Lowry et al. 1971) indicate the likelihood of the not-infrequent use of quality ferrous tools in Bronze Age industrial settings. Although the earlier pieces may contain

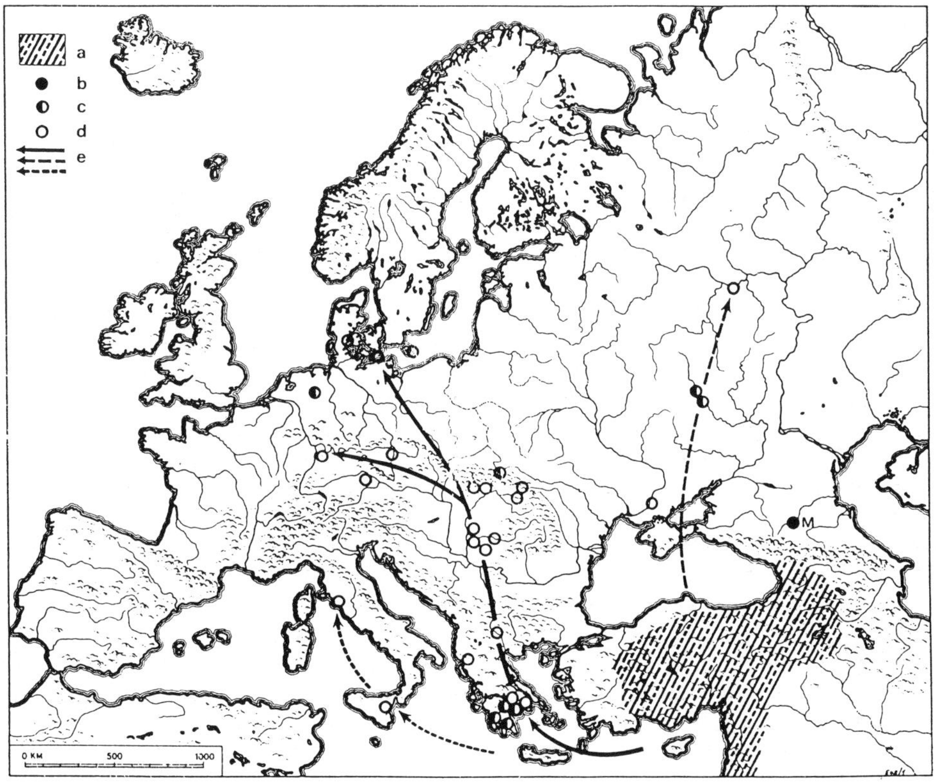

Fig. 2.2.1 The earliest routes of iron into Europe: (a) homeland of iron technology in the 15th-11th centuries BC: (b) objects of meteoric iron: (c) iron finds of the period 15th-13th centuries BC: (d) iron finds in Late Bronze Age contexts: (e) suggested routes for the spread of the earliest iron objects into Europe (after Pleiner 1981 with addition).

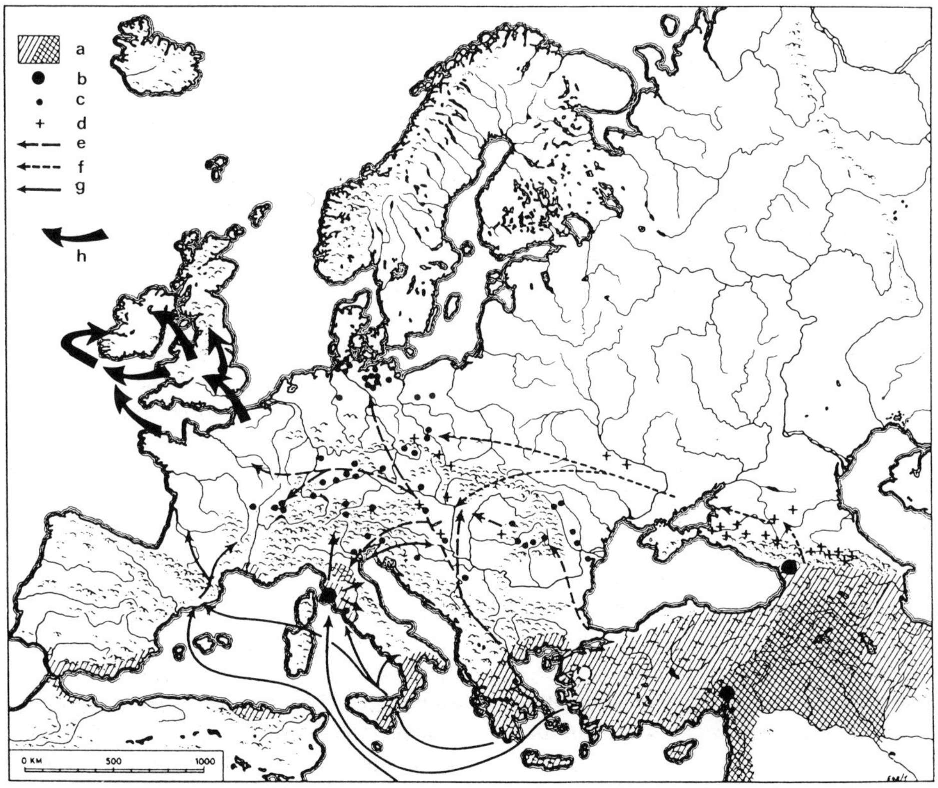

Fig. 2.2.2 Distribution and spread of iron and of the knowledge of iron technology in Europe during the Late Bronze Age (*ca* 8th century BC) on the basis of associations: (a) regions with full iron-based industries: (b) firmly-established centres of iron production: (c) iron finds of the Late Bronze Age (ca Ha B_2): (d) iron artefacts of eastern Late Bronze Age type: (e) suggested Thraco-Veneto-Illyric route: (f) Cimmerian route: (g) Greek, Phoenician and Etruscan route: (h) Atlantic route (after Pleiner 1981 with additions).

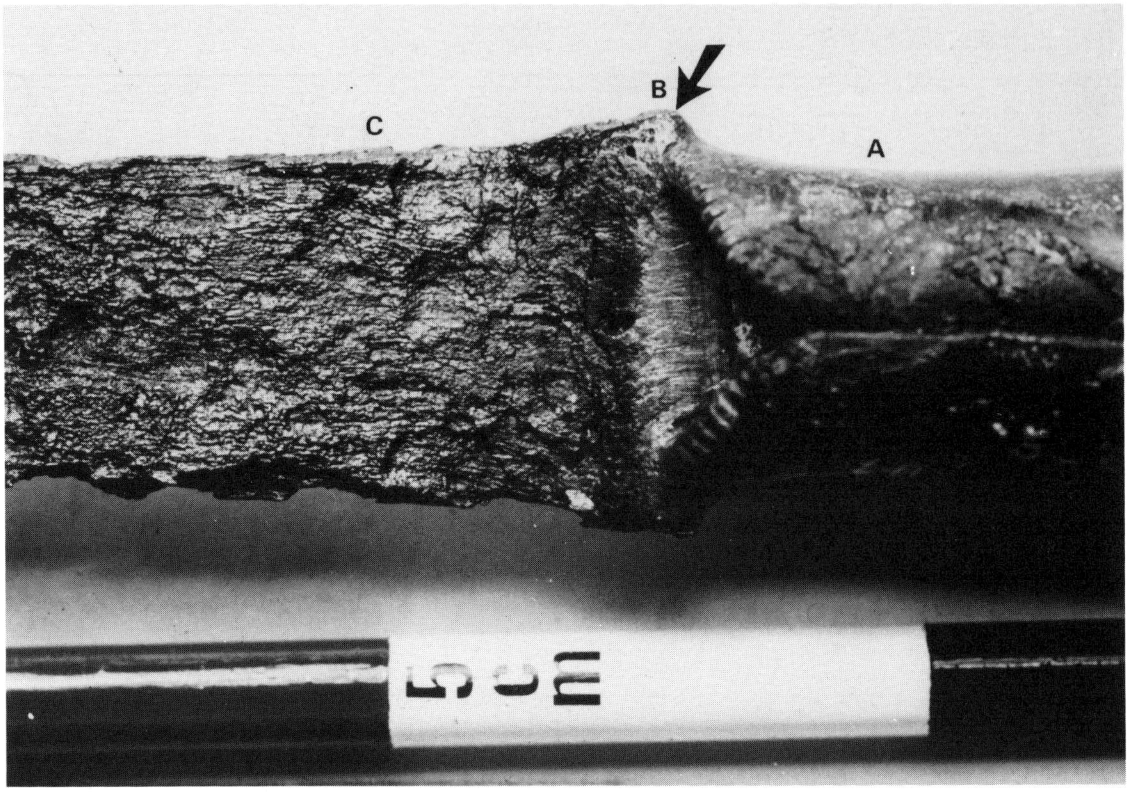

Plate 2.2.1 Lappenaxt from grave 702 at Hallstatt, Austria, showing a copper-alloy socket (A) run onto an iron blade (C) and the junction (B) between them *(Courtesy of the Naturhistorisches Museum, Wien)*.

nothing but imported iron, the combination of old and new is not only technologically significant, but also is symbolic of the initial grafting of iron technology onto the mature metallurgical stock, to be followed later by its separate planting after a period of maturation.

The 8th century BC (Hallstatt B/Montelius V), saw the arrival on the Continent of substantial, and in some cases at least, well-made iron artefacts, including the rapier-like blade from Singen am Hohentweil in Germany (Kimmig 1981; cf. Pleiner 1980, 380f) apparently produced by fold-welding (Boll et al. 1981). In the succeeding Hallstatt C/Montelius VI phase the availability of iron began to increase, although it was still very much secondary to copper alloy. What does seem evident, however, is that the *smelting* of iron was a skill being mastered by European metalworkers, although as Pleiner notes (1980, 384f), evidence for the earliest smelting operations is at best sketchy. It could be argued that this lack of evidence for local production indicates importation not just of finished artefacts but also of stock from a foreign source – but from where? In the Hallstatt C assemblage we see not only the copying in iron of artefact shapes previously made in copper alloy, but also the copying *in copper alloy* of features of iron products. This horizon of *skeuomorphic interaction* is a further characteristic of the transition process. While a good point could be established on a copper alloy sword by suitable configuration of the mould, it was most easily produced on the iron counterpart first by forging and then by cutting and/or grinding. A number of copper alloy swords show this feature (Plate 2.2.2) in imitation of the form of the points of the iron swords which were originally modelled on the copper alloy Mindelheim type (cf. Cowen 1967, 384f). Another group to show the interaction between ironworking and the old traditions are looped and socketed axes which, while taking on forms somewhat different from their copper alloy models, still show skeuomorphism in forged iron of such previously cast features as socket-mouth mouldings and vertical ribbing on the body (cf. Kromer 1959, I, 111, 157–158, 184; II, Taf 82.1, 82.2, 161.9, 190 - graves 469, 789, 1000). Perhaps we might consider the wheel to have come full circle with the appearance among the late Hallstatt assemblage of iron artefacts inlaid with copper alloy pieces (e.g. Eichhorn et al. 1974).

The 7th and 6th centuries BC in Britain seem to be the time when native craftsmen there were making their first aquaintance with ironworking. The first artefacts are either imports of Continental Hallstatt C/D, or show unmistakable signs of having been produced locally in a Late Bronze Age industrial context. Different regions of Britain appear to have received the new technology in different ways, as underlined by Alexander's (1981) model[5]. The best known pieces come from the Llyn Fawr (Wales) hoard, where an iron spearhead, and a sickle of native manufacture, were found together with part of an imported Hallstatt C iron sword, and two cauldrons and an assortment of other typically Late Bronze Age copper alloy types (Fox 1939a). While the spearhead is typologically nondescript, the sickle (*pace* Alexander 1981, 66 note 1) bears close comparison to the Killaloe type of the Irish Late Bronze Age (Fox 1939b, 225f). The British series also includes the unpublished iron swan's-necked pins from Moel Hiraddug, iron ring-headed pins from sites at Merthyr Mawr Warren, Dinorben, Penmaen and Bosherton (Savory 1976, 252), the iron vase-headed

Plate 2.2.2 Copper-alloy and iron swords from Hallstatt cemetery showing skeuomorphism of point from iron to copper-alloy (*Courtesy of the Naturhistorisches Museum, Wien*).

pin from Ffrid Faldwyn (O'Neill, B.H. 1942) and the iron swan's neck pins from All Cannings Cross and Swallowcliffe Down, Wiltshire (Dunning 1934). Of particular note in this context is an iron pin from Fengate with its iron stem and copper-alloy disc head (Eogan 1974b), demonstrating bimetallism in the British transition period also. Of less easily demonstrable date, but certainly attributable to Late Bronze Age experimentation, are the iron looped and socketed axes (Rainbow 1928; Manning and Saunders 1972). Directly comparable in concept are a spearhead from Kempsford, Glos. (Tylecote and Gilmour 1986, 112, Fig. 46, and 115–117), and a sword of 'uncertain provenance' from Reading Museum (Tylecote and Gilmour 1986, 153, Fig. 63(3), and 150f), both of which are strongly skeuomorphic of British Late Bronze Age types. At Sheepen Hill, Essex, a small corroded iron stud was found in association with a cauldron[6] (Hawkes and Smith 1957, 161).

The establishment of iron production industries brought to an end the apprenticeship of European and British metalworkers in the blacksmith's profession. It also marks the start of the process of replacement of copper alloy by iron as the primary utility material, especially for weapons, cutting and percussive tools, and the process of redeployment of copper-alloy to a primarily non-utilitarian role. It is no accident that the working of iron was so closely bound up with copper alloy working, since percussive shaping, which is the crucial skill underpinning early exploitation of iron, was developed to a very high degree indeed among the metalworkers of the Late Bronze Age. It is surely no exaggeration to say that serious iron exploitation could not have begun had the skills of hot hammering of copper alloy not been long established. While smelting technology presented its own particular difficulties, a raw bloom of iron would have been of little value without the skill required to consolidate it and then apply the hammer to form a useable tool or weapon. The beautiful lines of the iron Mindelheim swords, imitating and indeed improving their cast copper alloy models show the mature skills of the percussive shaper of copper alloy applied in a new medium. The Later Bronze Age of Europe, Britain and Ireland is a period characterised by metallurgical excellence, one in which the professional delight of the cratfsmen in their mastery of gold and copper alloy is clear both from the aesthetic and technological qualities of their products. Their skill, displayed through the beating of metal and casting of complex three-dimensional shapes, has not been surpassed. These metalworkers created well-designed and effective tools, utensils and weapons. In one sense, it is strange that, as their control over non-ferrous metals reached a peak, experimentation with a new, more difficult material should begin and so rapidly become established. However, it was surely this excellence and energy which were the foundations of innovation.

That the earliest ironworkers of Europe and Britain could shape iron superbly is not in doubt. In terms of the ability of the earliest European ironworkers to manipulate the mechanical properties of the new material, our knowledge is less complete, although a significant number of metallographic examinations of central European Hallstatt period artefacts has been built up, particularly through the researches of Piaskowski and of Pleiner. While many Hallstatt period iron artefacts subjected to metallographic examination show relatively simple technology (i.e. working up of unevenly carburised stock into artefacts without deliberate manipulation of properties: Pleiner 1980, 388–389), a significant number do show both the production of steel and a more developed understanding of the material (e.g. Piaskowski 1969). The production of sheet metal, such an important skill in the European Later Bronze age, is also seen from pieces such as the tubular ring from Býčí Skála (which has been shown not to be of cast iron as previously claimed: Stranský 1984). Knowledge of the occurrence and distribution of heat-treated artefacts is sketchy, although martensitic structures are by no means unknown. One group of artefacts which shows just how far some smiths of the period had reached in their experimentation comprises Polish Hallstatt period socketed axes, including those from from Wietrzno-Bóbrka (Piaskowski 1960), Dzierdionów, Jezierzyce and Goplano (Piaskowski 1969). These pieces between them show the main features of advanced iron technology, including carburisation, the welding together of iron and steel, the use of high-nickel metal, fold-welding and heat treatment. Fold-welding (cf. Chapter 1.4) seems a very early discovery, being observed not only in central European artefacts, but elsewhere, as in the axe from Eskilstuna in Sweden (Tholander 1978).

Among the developments in European Later Bronze Age socio-political and economic organisation is the development of larger and larger groupings, with increasing sophistication of social stratification and economic organisation (cf. Cunliffe 1976; Pleiner 1979). The final establishment, in the latter part of the Bronze Age, of the craft-professional as a member of society who could be fully supported by society in return for the products of his specialisation also created the climate under which such workers gained a degree of freedom to devote time to experimentation and research. We see this clearly in the non-ferrous metalwork of the period and, by the start of Hallstatt C/Montelius VI, also in their involvement with the new technology. However, it would be difficult to postulate an autonomous development of ironworking for any European region, and it would not seem feasible for the origins of the industry to be grounded in travellers' and traders' accounts of how the artefacts they were bringing into a region were made.

At present, we may see only the arrival of actual pieces of iron in Later Bronze Age Europe, but it would be surprising if some at least had not been produced by travelling ironworkers. While it cannot, at present, be proven, it is much easier to envisage European non-ferrous metalworkers learning *at first hand* from foreign experts (presumably with a suitable rate of return!) and then going on to build on this knowledge and pass it to their fellow craft specialists. Equally, one can envisage the buying-in of the new expertise by the status-conscious and affluent upper tiers of European Late Bronze Age society. In either case, the often disappointing performance of iron relative to good copper alloy could have been offset by comparison with the better products that an experienced worker would have been more likely to achieve. Certainly, the superficially poor comparison between iron and copper alloy, in terms both of ease of production and shaping, and consistent mechanical performance, would have created a marketing resistance that second-hand tuition would have not readily overcome. Once, however, the eccentricities of the iron smelting process had been basically sorted out and the merits of heat-treated carburised iron demonstrated, then the skilled metallurgists of the European Late Bronze Age clearly lost no time in exploiting its potential.

As iron, for whatever reasons, became the basis of an economically viable production industry, so the blacksmith finally emerged as a craft specialist – a professional in his own right. Since the transition took place among European peoples using Indo-European languages, it is to this time also that we must look for the roots of metallurgical vocabulary relating specifically to ironworking among the members of the Indo-European family of languages. Undoubtedly, most of the ideas relating to metalworking, mining and smelting already existed, and it would have only been in those areas unique to ferrous metallurgy that new specialist terminology would have been required (Scott 1983a, 1989b).

It should also be at this time that the ideas of magico-religious powers having been granted to the smith started to develop. There are certainly grounds for believing that at least one strand of thinking would have linked iron in its meteoric incarnation with the heavens (e.g. Forbes 1964, 199; Eliade 1978, 19f). Also, since a steel cutting edge would have significantly superior properties to those even of the best copper alloys of the time, the triumph of weapons forged from the new material over those made in traditional ways could have elevated the blacksmith to the position of 'armourer of the gods'. And it would have been good commercial sense for smiths to cloak their skills in magic. Innate respect for the supernatural and for the power of nature would have made no little contribution to the establishment of iron techology as an integral part of the social and economic fabric. If we accept that the European Later Bronze Age was a time when greater emphasis than ever before was placed on the martial arts, then the provider of superior weapons could not have helped but enjoy special status, particularly if they fostered and encouraged belief in the supernatural aspects of the gift being bestowed, through the smith as intermediary, on a warrior class. It is interesting to speculate whether the foundations of the ultimate success of iron in superseding copper alloy was based not so much on its superior properties for use in tools as on its value in weaponry coupled with some astute salesmanship at both sacred and secular levels.

* * *

Notes to Chapter 2

1 This is not to suggest in any way that prehistoric 'science', as the term has been used recently, was involved in early metallurgy, even if its development involved at least one element of 'scientific method' – *the ability to reproduce results consistently*.

2 Flint-knapping on the one hand, and sheet bronzeworking and iron forging on the other, might not seem immediately to be related. But similarities of technique are obscured by the nature of the end result. The knapper aimed at revealing the desired shape by removing the material surrounding it. The

Bronze Age sheet metalworker and Iron Age smith achieved their goals by plastic deformation. Flint-knapping relied on fracture patterns, while sheet metalworking and blacksmithing were based on the malleability of hot metal. But in each case, misplaced blows are at best a hindrance, and it is the long tradition of working by the precise application of percussive force to a specific point that is crucial here.

3 One is reminded here of Pliny's tale of the sculptor Aristonidas' depiction of the madness of Athamas, of which he writes:

> *Aristonidas artifex, cum exprimere vellet Athamantis furorem Learcho filio praecipitato residentem paenitentia, aes ferrumque miscuit, ut robigine eius per nitorem aeris relucente exprimeretur verecundiaea rubor.*
>
> When the artist Aristonidas desired to represent the madness of Athamas subsiding in repentance after he had hurled his son Learchus from the rock, he made a blend of copper and iron, in order that the blush of shame should be represented by the rust of the iron shining through the brilliant surface of the copper.
> Book XXXIV.xl: Rackham 1952, 230.

This can be compared with the results reported by Craddock and Meeks (1987) on coppers of high iron content.

4 It is worth restating here that throughout the late M. J. O'Kelly's experiments with smelting iron in a bowl furnace, he frequently produced white *cast* iron rather than a bloom.

5 While Alexander's proposals for the regional sub-division of Britain into metallurgical transition zones (1981) may or may not find favour with British archaeologists, his comments on Ireland should be discounted.

6 Curwen's report (1948) of iron in the hoard from Sompting, Sussex, has been called into question by Briggs (1987, 175), and should probably be discounted.

CHAPTER 3

The Earliest Iron in Ireland – Earlier Iron Age A

3.1 – The Later Bronze Age background

Although Ireland is the farthest offshore island of Europe, metallurgy here in the latter part of the Bronze Age, spanning approximately the centuries between 1100–600 BC kept reasonably closely in step with contemporary European technology. The Irish Later Bronze Age is the age of the specialist professional metalworker as shown by a major expansion in the output of artefacts. Some of these are distinctively Irish in concept, others demonstrating strong contacts with and influence from Britain initially and later the wider European theatre (cf. Eogan 1964; Herity and Eogan 1977, 167ff). It is not, however, the level of output that is as striking as the sheer quality of craftsmanship. This was an effervescent industry, receptive of new ideas, but always innovative. Multi-component sheet copper-alloy vessels (Hawkes and Smith 1957; Briggs 1987), delicately ornamented sheet goldwork (e.g. Eogan 1981), and the production of complex castings typified by side- and end-blow horns (Holmes 1978), all offer eloquent testimony to Irish mastery of three-dimensional shaping of metal. These skills, along with those of prospecting, mining and smelting, should have provided the same foundations as in other regions for the establishment and development of thriving iron production industries. After ca 600 BC however, the pattern deviated from that of the rest of barbarian Europe to such an extent that we cannot talk with any confidence of an Irish Iron Age *per se* beginning much before the start of the 3rd century BC.

By far the greater part of our knowledge of the Later Bronze Age in Ireland is based on metal artefact types, although there is a small number of occupation sites whose origins may be attributable to this period. In addition, there is palaeoecological evidence for different forms of land use and of animal husbandry. A restricted range of imported Hallstatt artefacts, appropriate to the ca 7th/6th centuries BC, appear in significant numbers. Although we do not have as complete a picture as would be desirable of life in Ireland in the earlier 1st millennium BC, the sum of the evidence allows us to infer that the form of economy which existed was broadly comparable with that more fully seen elsewhere in barbarian Europe.

There is evidence of a falling-off of external contacts and influences towards the end of the 7th century BC, and this coincides with the virtual disappearance of the metal industries which for several centuries had been so productive. The period ca 600–200 BC is best described as a 'Dark Age', a pessimistic title which it seems to fully merit (Warner 1974: cf. Scott 1974b). While, for example, Jope (1961, 307) was able to describe the British series of iron daggers as '. . . the products of a coherent and complex workshop tradition operating continuously from the 6th through the 5th and 4th centuries BC . . .', Ireland can produce no such continuity for this period, let alone coherent tradition. From the scant surviving archaeological evidence, there appears to be a major discontinuity between later Dowris and the emergent Iron Age (exemplified by the fine iron weapons from Lisnacrogher, Co. Antrim – Chapter 4), and a gap in the archaeological record that cannot easily be explained (Scott 1974b). There are none of the characteristic indigenous Later Bronze Age artefact types independently dated after ca 600 BC. Indeed, one is hard pressed to detect *any* activity, industrial or otherwise, after this time, and it is as if the depopulations of Ireland described in the fairytale world of *Lebor Gabála* were reality and not the pseudo-history of well over a millennium later. While one is cautious of using a term like 'collapse', it is nevertheless difficult to describe the disappearance of the highly-organised Dowris metal industries, and thus of the structures which must have supported them, in any less dramatic a fashion.

As Eogan has demonstrated clearly, one significant feature of the rise of the Later Bronze Age metal industries is the evidence for the emergence of major regional schools of metalworking (Eogan 1974). This could only have been possible with a widespread economic organisation and society in which there was a high level of communication. This strongly implies continuing evolution of socio-political and economic organisations that created the markets whose servicing supported specialist craft-professionals and provided the industrial climate necessary for experimentation and innovation. If industrial production diminished so dramatically, we can only deduce that it was because the social, economic and political structures on which it depended had passed away. Had the organisation capable of promoting widespread and (by prehistoric Irish standards) large-scale production ability survived in any coherent form, we might reasonably expect the ongoing production of the major metal types, albeit in an impoverished industrial setting. Experimentation with iron, to be expected in this context, is likely to have continued at a low level, but would surely have continued to ensure the emergence of iron-use in a technological Iron Age at a much earlier date.

It is possible that the concentration of evidence as metalwork has deceived us into believing that the earlier part of the Dowris phase was one of uniform progress in all areas. Indeed, in view of the possibility that the very few settlement sites of the period may be the fragmentary survival of a record of major socio-economic and political change initiated in the earlier 1st millennium BC, it seems reasonable to regard the Dowris metal industries as the last major achievement of an already-decaying system. When the old order finally passed, there would no longer have been the structures to support production at previous levels. This would give the appearance in the archaeological record of a collapse, when in fact this seemingly sudden 'event' was but the culmination of a much longer process.

The demise of organised Later Bronze Age metal industrial output in Ireland coincided with the arrival of artefact designs from other regions on the threshold of full-scale industrial transition from copper-alloy to iron use. It is an attractive hypothesis which attributes the causes to the arrival of invaders from the world of the iron-users (cf. Waddell 1978; Champion 1982), even more so if responsibility for the implantation of the Goidelic dialect could also be laid at their door. This proposition is difficult in view of the archaeological evidence for the state of the country in the latter part of the 1st millennium BC, since there is no trace among the scant survivals of these centuries of any major influx of people. Thus, B. Raftery's suggestion that the stimulus to build defended hilltop settlements might have resulted from the insecurity caused by small groups of iron-using intruders (1976b, 352) seems unlikely. Nevertheless, amongst the remnants of a collapsed society, small groups of intrusive settlers would be virtually undetectable in the archaeological record.

One might envisage the arrival of Goidelic-speaking intruders who infiltrated, in small numbers, a Dowris-period Ireland where social and economic decay was advanced. If such people had brought with them a knowledge of ironworking, it is possible to hypothesise both the implantation of iron technology and the introduction of the indigenous population to the Goidelic dialect. As Mallory (1984, 68) notes, a new language cannot be imposed on a population by intruders, but is rather learned because the indigenes see its acquisition as opening access to benefits such as desirable innovations that newcomers have brought with them. Iron metallurgy, drawing on seemingly extensive local resources in north-east Ulster at least (Chapter 6.2), would seem one such motivation. While this hypothesis finds little favour amongst Irish archaeologists, not least on the dubious grounds that there is not a significant body of identifiably early ironwork (Mallory 1984, 67), it cannot be dismissed readily. Some early pieces undoubtedly do exist, suggesting that originally the number was significantly larger until the chemistry of iron, oxygen and water conspired to reduce it. The debate inevitably degenerates into trial by decibel, since all of the interrelated problems admit of no easy solution.

That the Irish Later Bronze Age metal industries were receptive to outside ideas is shown by foreign influences on artefact design. If these industries lost contact with the outside world, or the societies which they served became stagnant so that contemporary developments elsewhere simply passed them by, this would surely not have halted the work of metal smiths. The suggestion that industry declined so rapidly because there were no longer Bronze Age societies outside Ireland with which to trade surely places too great a limitation on indigenous abilities and an exaggerated emphasis on dependence on foreign influences and resources. It is not impossible that a sudden shortage of either of the stock metals, copper and tin, led to the collapse, but this is undemonstrable at present[1]. It is worth considering the problems that could have been posed for Dowris period metalworkers by the necessity of relying on external tin sources (selling tin in exchange for, amongst other things, Irish gold). Had this source ceased to be available, then manufacturing would have shifted to total reliance on scrap, a resource that was apparently depleted regularly by the deposition of large votive hoards of bronze artefacts. Resource shortages may well have been contributory to the Dowris collapse, but were hardly the sole cause.

It has been proposed that miniature copper-alloy artefact types are evidence either of a serious shortage of metals in Ireland at the end of the Later Bronze Age or of the impoverishment of the 'Dark Age' (most recently, Lynn 1983a, 55). However, the miniature socketed axe seems a perfectly viable tool type at the bottom end of the size range, while the miniature spearhead, competently used, would have been just as lethal as its larger relatives, albeit not inflicting such spectacular wounds. In fact, examination of hoard evidence shows that 'miniatures' occur before and during the Dowris period (Fig. 3.1.1). The hoards from Lahardan, Co. Clare and Newport, Co. Mayo which are dated to the earlier part of Dowris, contain socketed axes whose lengths go from 5.0 cm down to 3.2 cm (Eogan 1983, 68 and 109). The eponymous Dowris, Co. Offaly hoard contains spearheads ranging in length from 34.2 cm down to 10.0 cm, and axeheads ranging in length from 10.7 cm down to 4.7 cm (Eogan 1983, 117f). Two 'miniatures', a socketed axe 4.3 cm in length, and a spearhead 8.0 cm in length, come from the hoard from Kilmurry, Co. Kerry, which also includes a continental import in the form of a knobbed bracelet of Hallstatt C date. If one further considers pieces like the kite-shaped spearhead of length 9.0 cm found at Tattenamona, Co. Fermanagh along with two other such spearheads of lengths 14.2 cm and 17.2 cm, it is easy to see that what has been singled out as 'miniaturisation' is, in fact, a misidentification of the lower end of a size-range for particular artefacts[2]. It makes as little sense to single out the largest artefacts from the range as evidence of a sudden upsurge in the availability of metal, as it does to take the smallest members as evidence of a dying industry painfully eking out the last dregs of its resources.

While Ireland is reasonably generously endowed with copper sources (Briggs 1976; cf. Scott 1977a, 10, Fig. 1), metalworkers here from earliest times, have always faced a scarcity of tin resources. Both Flanagan, L. (1978, 158f) and Jackson, J. (1978, 121f), however, point out that during the earlier part of the Irish Bronze Age, certainly, the amounts of tin required to sustain output would have been well within the capacity even of the meagre Irish deposits. With the increase in production of bronze in the Later Bronze Age, there is still no need to envisage a sudden (or even gradual) exhaustion of resources. Interpretation of the results of a comprehensive series of analyses of Bronze Age artefacts has shown that recycling of metal must have been commonplace from the start of the Bronze Age in the British Isles at least (Northover 1982; cf. Scott 1977a). For every copper-alloy artefact which has survived, at least one more is bound to have ended its life in the crucible. It is quite in order to extrapolate from Northover's work and conclude that recycling would have played an important a role *throughout* the Bronze Age in Ireland. In the Irish Later Bronze Age, the use of

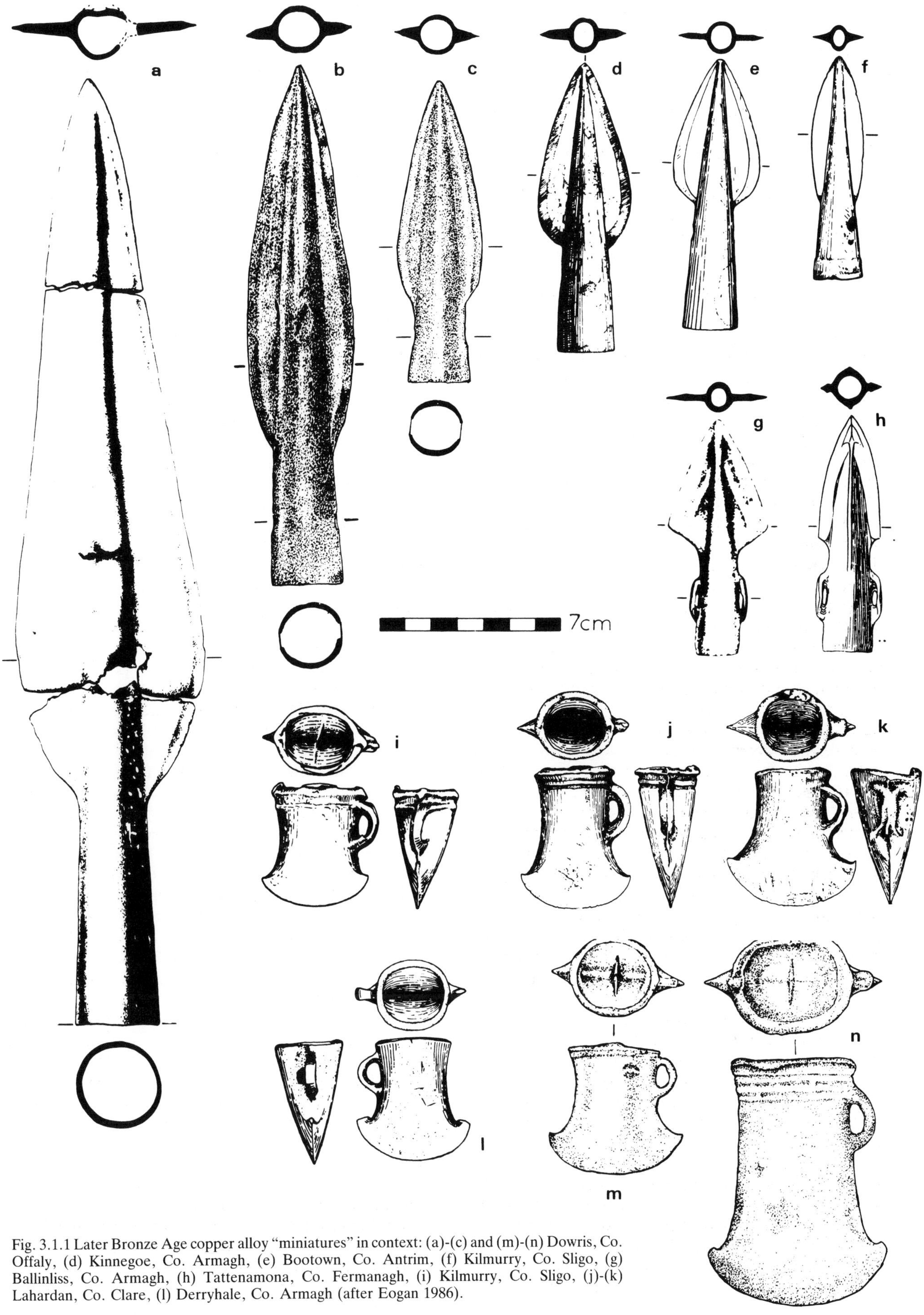

Fig. 3.1.1 Later Bronze Age copper alloy "miniatures" in context: (a)-(c) and (m)-(n) Dowris, Co. Offaly, (d) Kinnegoe, Co. Armagh, (e) Bootown, Co. Antrim, (f) Kilmurry, Co. Sligo, (g) Ballinliss, Co. Armagh, (h) Tattenamona, Co. Fermanagh, (i) Kilmurry, Co. Sligo, (j)-(k) Lahardan, Co. Clare, (l) Derryhale, Co. Armagh (after Eogan 1986).

scrap as a major raw material cannot be simplistically interpreted as evidence that accessible ore deposits were running out. Recycling of damaged artefacts on this scale, with the assemblage of founders' hoards, is rather to be taken as a sign of the increasing sophistication of Later Bronze Age industrial organisation. Those who made the artefacts were not only reducing the effort involved in the chain of manufacture, but also broadening their economic dependence to include more than one source of raw material. As Northover (1982, 50f) notes, the exploitation of scrap is indeed to be expected as part of the movement patterns of artefacts in metallurgically active regions. Given the obvious indigenous expertise in locating and exploiting ore resources, coupled with the extensive use of scrap as a material resource in the Later Bronze Age, the later paucity of artefacts demonstrates a lack of the industrial organisation and markets necessary to sustain regular and continuing production on the scale previously attained. Since 'invasion' and 'resource crisis' are unhelpful, we must look to potential internal socio-economic and political change as a primary explanation.

The archaeological subdivision of the European transition period rests on a wealth of artefactal and monumental evidence. Through this, the process is seen as part of a suite of changes in the ordering of society and industry. For Ireland, however, there is only a handful of enigmatic artefacts and sites on which to attempt to base reconstructions. An area to which comparatively little consideration has been given is the suite of reasons for the decline and fall of the society and economy which we see represented by the products of the Dowris phase industries. I have suggested elsewhere (Scott 1973, 1978a) that the process began well before its effects become apparent. The increase in the size of socio-political units suggested by the metalwork groupings parallells the mainstream trend in Europe and Britain towards urbanisation. Evidence for Dowris period beginnings for the hilltop settlement at Rathgall, Co. Wicklow (Raftery, B. 1976b, 345f), similar, though less well defined evidence from Clogher, Co. Tyrone (Warner 1972), and newly-emerging evidence from Haughey's Fort, Co. Armagh (Mallory 1987) indicates that up until then, developments in the potentially defensive siting of settlements were in step with the rest of the barbarian world[3]. The emphasis on weaponry, in particular the sword, suggests a period of uncertainty, of martial ostentation, and of violence. However, at the same time, we also find evidence for the establishment of what was to become the dominant socio-economic and political pattern for well over 1500 years, represented in later times by the small, self-supporting farmstead unit.

The trend in the barbarian world towards the increasing size of community groupings which permitted construction of large, strongly fortified and eminently defensive settlements (cf. Cunliffe 1976; Härke 1979, 21ff) has produced monuments which dominate the archaeological landscape. Now seen as a development of the later Bronze Age rather primarily an Iron Age phenomenon, the large defended hilltop settlement is evidence of communal anxiety, of the coming together of larger numbers of people than in earlier times, and of greater community effort directed to a common goal. Although the hillfort also appears in Ireland, its numbers are small in comparison with the many thousands of small enclosed settlements, *ringforts*, and water-girt residences, *crannogs*. A Later Bronze Age date is indicated by the finds from a ringfort-like structure at Aughinish, Co. Limerick, where an iron horsebit was found in association with Dowris-style artefacts (Kelly, E.P. 1974). However, while the typical Irish ringfort of the 1st millennium AD would have never posed any obstacle to a serious attacker, and was primarily an obstacle to be put in the way of livestock and fellow-competitors in what one writer has described as the bloody Irish sport of cattle-raiding, the water-girt site may be seen to have a more seriously defensive purpose.

While there is still debate over the definition of the term *crannog* in terms of construction (e.g. Lynn 1983a), the view is taken here (*contra* Lynn) that it may usefully include both artificial islands and natural water-girt sites where advantage has been taken of shallows. The important point is the imposition of a water barrier between occupants and potential enemies. There are a number of such sites for which Later Bronze Age associations may reasonably be claimed: Knocknalappa, Co. Clare (Raftery, J. 1943), Moynagh Lough, Co. Meath (Bradley 1986), Rathtinaun, Co. Sligo (Raftery, J. 1972, 3), Lough Eskragh (Collins and Seaby 1960; Williams 1978) and Island McHugh (Davies 1950; Simpson pers. comm.), both in Co. Tyrone, and the earliest stratum at Ballinderry crannog no. 2, Co. Offaly (Hencken 1942). A Later Bronze Age spearhead was recovered from a crannog in Lough Enagh, Co. Derry (Davies 1941). Later Bronze Age artefacts are recorded (without detail of context) from crannogs at Loughnaglack and Monalty Lough in Co. Monaghan, while part of a sword mould is reported as having come from the '. . . surface of an ancient crannog . . . at a depth of 21 feet (6.40 m) underneath peat . . .' in Co. Fermanagh (Herity and Eogan 1977, 193). As with this last report, the destroyed crannogs at Ballykinler and Tyrella North, Co. Down (*ASCD* 183) which supposedly produced copper-alloy artefact types associated with iron, are too poorly reported for any positive attributions to be made. A copper-alloy swan's neck pin from Ballyhoe, Co. Westmeath is a possible crannog find (*CIIAA* 437). An iron looped and socketed axe (sect. 3.2) is reported as having come from a crannog in Lough Mourne, Co. Antrim (Wakeman 1898, 214f). The contexts of these last two artefacts are dubious, but cannot be set aside lightly.

Hillforts are prominent in Irish archaeology (e.g. Raftery, B. 1972, 1976b). They may well prove, however, to have obscured, in academic thinking, the emergence of a fundamental opposite, and ultimately much more significant trend, away from urbanisation and towards fragmentation represented by the move (by extended family units?) to defensible, but anonymous small water-girt settlement sites. The dispersal of population away from large centres would have diminished greatly, if not removed, potential for large-scale industrial organisation. The long-term implications of this for industrial development can scarcely be overstated.

I have argued elsewhere (Scott 1973) that the disruption of the proto-historic Irish *cóiceda* (the major political subdivisions in the 1st millennium

AD) and events which followed, was due (*pace* Byrne 1971) primarily to the inherent instability of large political blocs based on the unit known in historic times as the *tuath*. Binchy (1970, 7) demonstrates the widespread occurrence of derivatives of Indo-European **teutā* (>Irish *tuath*), a distribution indicating the similar paths of expansion of population groupings. In later prehistoric Ireland, the formation of large groupings could have been accomplished through the banding together of a number of what we might term 'proto-*tuath*s' whose members acknowledged an overall ruler (an arrangement seen from the later 1st millennium AD onwards). Internal dynastic strife (cf. Ó Corráin 1971), and (since a state of tension, latent or actual is, on historic precedent, the norm between co-existing power blocs), competition and pressure from neighbouring groups, not to mention epidemic and famine, would have made for a more or less constant state of political flux, leaving such federations prone to frequent fission and realignment. This fragmentation potential, expressing itself as the drawing-in of small communities on themselves in times of stress, continued to be a dominant feature of Irish prehistory and history long after the end of the Later Bronze Age.

We can suggest that an innate instability in the socio-political and economic infrastructure of Dowris period must have contributed to its downfall. The precise combination of circumstances will probably always be hidden from us, but single-factor explanations, such as loss of resources, major climatic deterioration (which there was undoubtedly – Mitchell 1976, 158f), or the intrusion of foreign elements are too simplistic. What is of particular significance to this study is the organisational trend that emerges in the Later Bronze Age, away from the type of society that could support large regional schools of metalworking as defined by Eogan (1974a), and towards a form of dispersed industry based on small settlement units. While elsewhere in Europe iron industries grew in size and co-ordination to the point where they were capable of providing mass-produced nails for timber-laced ramparts, Irish socio-political and economic organisation simply precluded such development.

The lack of communication here between the skilled and innovative which this must have engendered, prevented early Irish craftsmen from participating in a number of important technological innovations, including the use of the potter's wheel and of the shaft-furnace for iron smelting. Thus, in the later 1st millennium AD, we find the first urban settlements in Ireland founded by intrusive Scandinavians in stark contrast to native sites, not only in terms of the organisation of society, but also in the degree of commercial and industrial contact and interaction with other groupings. While after the christianisation of Ireland, crafts associated with the church flourished (undoubtedly gaining much from the widespread contacts among clerics), and fine non-ferrous metalwork was produced to the highest standards, the technology of materials not directly exploitable to the glory of God failed to keep in touch with the mainstream of European innovation.

The Later Bronze Age 'Dark Age' may thus perhaps be interpreted as a period which saw the beginnings of a radical restructuring of society, and which took perhaps the whole of the later part of the 1st millennium BC and the first centuries of the 1st millennium AD to complete. As we shall see below, from the mid-1st millennium AD onwards, we may at least speak with confidence about levels of industrial output, albeit on almost a cottage-industry basis in most instances. During the earlier period we see perhaps two main periods of experimentation with iron, the first virtually stillborn as a result of ongoing change, the second significantly retarded as a result of the effects of that change.

* * *

3.2 – The transition from copper-alloy to iron in Ireland: Earlier Iron Age A

The Dowris phase sees the arrival in Ireland of Hallstatt C period artefacts coming from a background where the transition from copper alloy to iron was firmly under way. Swords of Eogan's Class 5 (Eogan 1965), albeit late and insular versions of the classic Hallstatt C Gündlingen variety, along with characteristic chapes (Eogan 1965, 170) and such artefacts as the bracelet from Kilmurry, Co. Kerry, the 'flesh fork' from Dunaverny, Co. Antrim (Scott, J. G. and Powell 1969, 125), and the as yet unpublished finds from Navan, Co. Armagh (Selkirk and Waterman 1970; Lynn et al. forthcoming), attest to contact with a variety of emergent iron-using societies. The 7th/6th centuries BC are the time when small-scale experimentation with iron technology was occurring at about the same time as in Britain.

In a recent paper, Briggs (1987, 172f) drew attention to two Dowris period cauldrons which seemed to incorporate iron components[4], and to a bucket with patches secured by iron rivets. In fact, one of these cauldrons, a Leeds type A from Tulnacross, Co. Tyrone contains no iron. In the case of the type B cauldron from Lisdrumturk, Co. Monaghan, Briggs identifies correctly a total of thirteen iron rivets, of which five are probably in secondary positions (Briggs 1987, 174, Fig. 10). A well-preserved bucket from Derrymacash, Co. Armagh shows secondary patches secured by iron rivets, a fact noted by the original reporter of the find (Dugan 1897) within a short period of its discovery and there seems no doubt as to the antiquity of the iron components. The survival of such bimetallic couples is by no means uncommon, despite the obvious tendency towards preferential corrosion of the iron components. Briggs suggests that these iron components strengthen arguments for the long survival of such vessels, ouside the chronological constraints proposed by, amongst others, Hawkes and Smith (1957). The precise dating must rest on some resolution of the problems of determining the time-span when these vessels were in use. But even allowing for long life, it seems inescapable that they indicate experimental use of the new metal well before the start of EIA B.

At Aughinish, Co. Limerick, the excavator (Kelly, E.P. 1974) reported the association of a very badly corroded iron horsebit with a Dowris period style

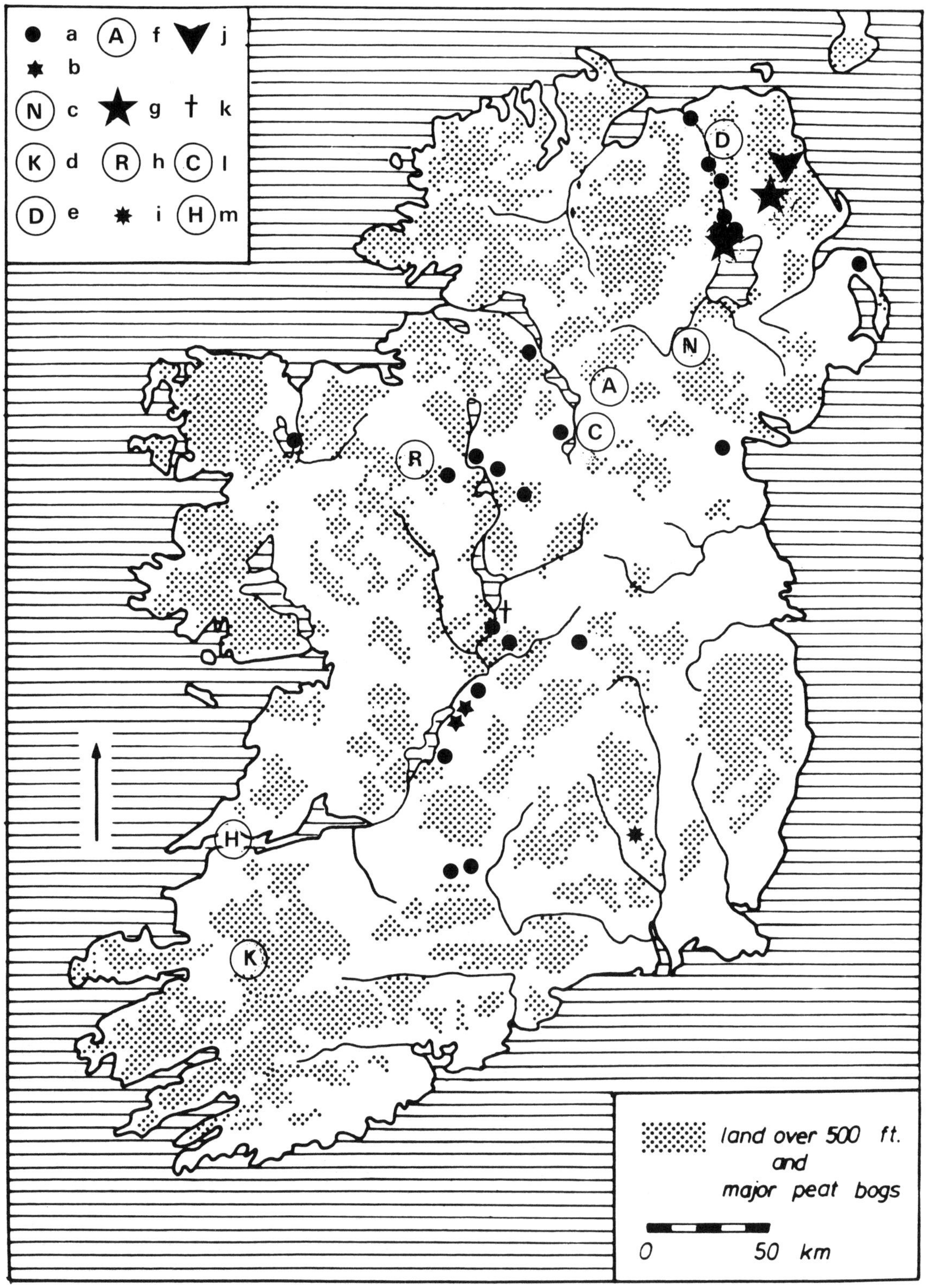

Fig. 3.2.1 Distribution of provenanced intrusive Hallstatt C and D period material in Ireland, and of Irish EIA A iron artefacts: (a) Eogan Class 5 swords, (b) Eogan Class 4 chapes, (c) Navan, (d) Kilmurry, (e) Dunaverny, (f) Altartate cauldron, (g) iron looped and socketed axes, (h) Rathtinaun axe (and other iron artefacts), (i) Dunbell swan's neck pin, (j) Glenariff axe, (k) Athlone sword, (l) Drumlane cauldron, (m) Aughinish.

tanged chisel and a copper alloy knob-headed pin of Hallstatt C period type. Excavations at Rathtinaun crannog, Co. Sligo produced an iron shaft-hole axe, iron three-pronged fork and an iron swan's neck pin in association with Late Bronze Age types (Raftery, J. 1972, 2–3), while an iron swan's neck pin is recorded as having been recovered during the 19th century destruction of one of a group of ringforts at Dunbell, Co. Kilkenny (*CIIAA* 436). From the river Shannon comes a spearhead (unfortunately not amenable to metallographic examination due to its recent use as a poker!) which Rynne (1979) would see as a Hallstatt C period import from Iberia. In addition to these, a small group of indigenous iron artefacts in the tradition of the European iron skeuomorphs of copper alloy types is also prominent.

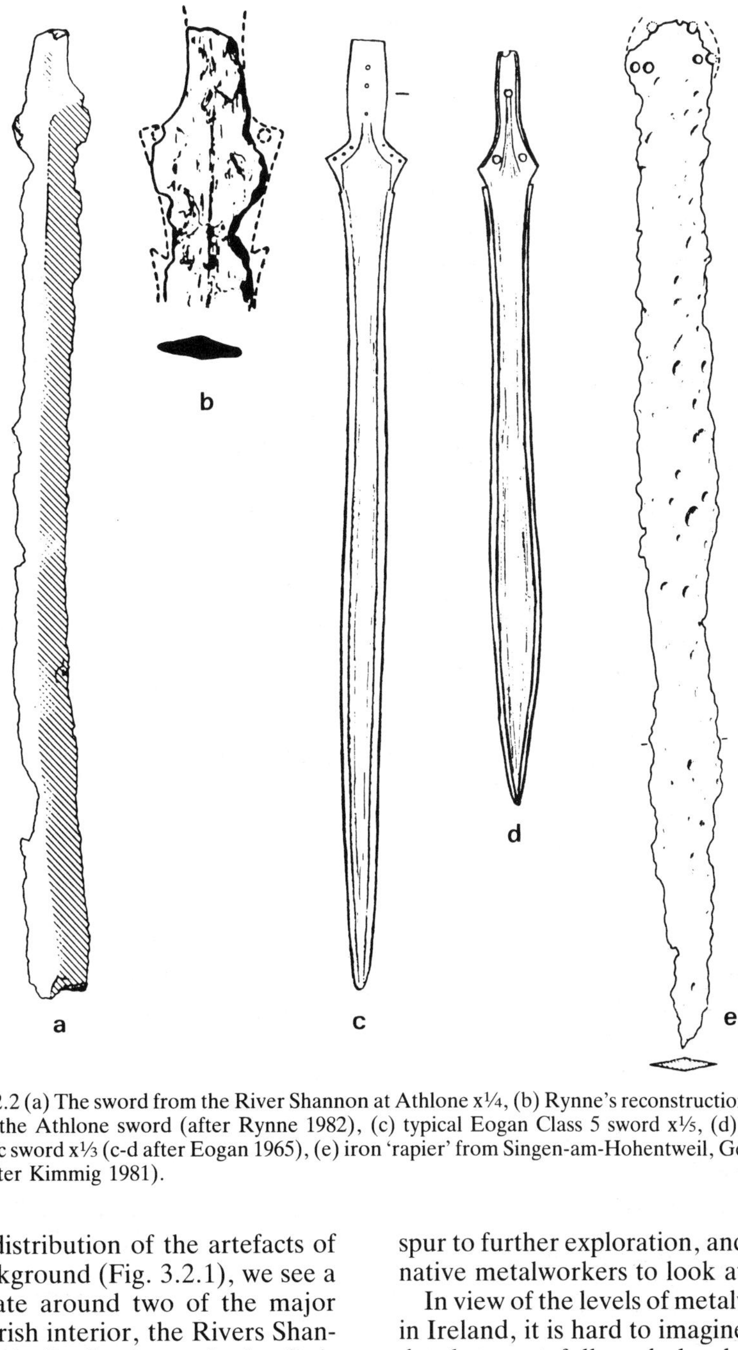

Fig. 3.2.2 (a) The sword from the River Shannon at Athlone x¼, (b) Rynne's reconstruction of the hilt of the Athlone sword (after Rynne 1982), (c) typical Eogan Class 5 sword x⅕, (d) Eogan Class 4c sword x⅓ (c-d after Eogan 1965), (e) iron 'rapier' from Singen-am-Hohentweil, Germany x¼ (after Kimmig 1981).

If we examine the distribution of the artefacts of Hallstatt C period background (Fig. 3.2.1), we see a tendency to concentrate around two of the major entry routes into the Irish interior, the Rivers Shannon and Bann. While this distribution undoubtedly is biassed by, amongst other things, dredging of the two major rivers and the activities of collectors, it is nevertheless noteworthy that both of these routes lead to major and readily accessible iron ore deposits (cf. Fig. 6.2.1). The predominantly martial nature of the material accords more with the intrusion of adventurers (cf. Burgess 1974, 213) than any communal movement of settlers. Such groups would have passed by outcrops of the laterites of Co. Antrim when entering by the Bann, and by *Sliabh an Iarann* ('Mountain of Iron') by Lough Allen where rich sideritic ore nodules are conspicuous in stream beds. Had they numbered metalworkers with experience of iron, such potential wealth would have hardly passed unnoticed. While most emphasis is usually put on the mining and smelting side of iron production in antiquity, an equally critical skill was that of prospecting, recognising where ores most easily worked by contemporary technology could be obtained. Identification of readily available Irish ores could have been a spur to further exploration, and as such, a stimulus to native metalworkers to look at these raw materials.

In view of the levels of metalworking skill available in Ireland, it is hard to imagine that, had indigenous developments followed a level course, the progress of ironworking here would not have parallelled other regions of the barbarian world, with the creation of blocs whose power was based on a monopoly of readily-accessible iron ore (cf. Härke 1979). We have seen that a major feature of the technological transition in Europe is the interaction between the working of copper alloy and iron. Although Ireland produces her own examples, they are few in number. Furthermore, since we do not see the earliest ironwork in proper association with the copper alloy types, it is impossible at present to trace the immediate influence of ironworking on established artefact types, but it can be traced in later times (see Chapter 4.1).

I have suggested (Scott 1974b, 1978a) that the iron types which can quite clearly be seen as resulting from the first native experimentation with iron by workers from the later stages of the Dowris period industries are two looped and socketed axes from Co. Antrim (Wakeman 1898, 214–242; *CIIAA* 576, 577: Plates 3.3.1a, 3.3.2a). Arguing away from a transition

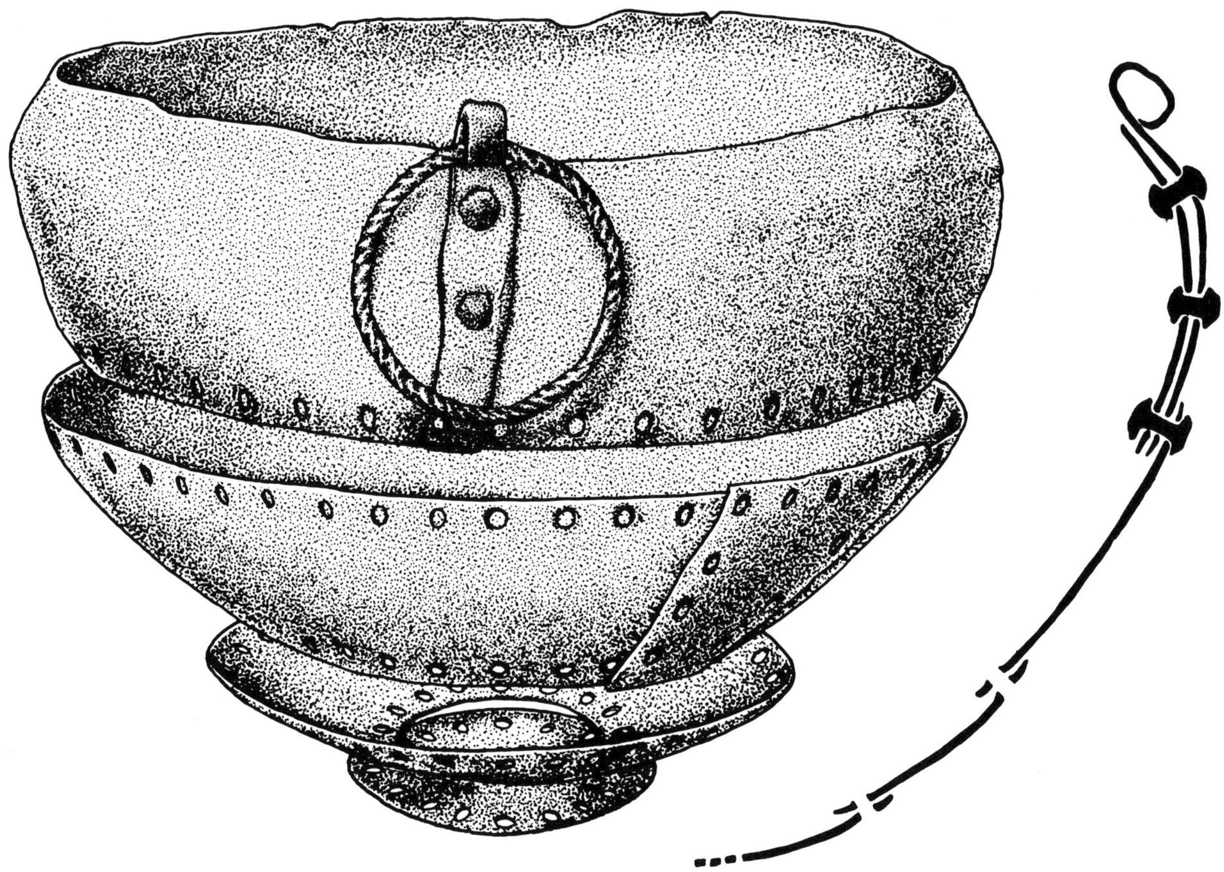

Fig. 3.2.3 Schematic diagram (not to scale) of the construction of the Drumlane cauldron (after Scott 1978a).

period origin for this type, Manning and Saunders averred that '. . . the forging of iron types originally designed to be cast in bronze is a considerable feat indicating a developed iron technology, rather than one in its fumbling infancy.' (1972, 280). It is difficult, however, to view them as other than skeuomorphs demonstrating the skill of the Later Bronze Age metalworkers being applied quite successfully to the new metal. Both Irish examples are close in form to the British series and are reasonable copies of the copper alloy model, including the thickened moulding round the socket mouth. The slightly 'banana-shaped' appearance is the result of the way in which the forging was approached. And while the welding of the Lough Mourne axe left something to be desired (see examination 1), these pieces show a sound knowledge of a wide range of smithing techniques. While copper alloy was still the dominant utility metal, the finely-forged iron Gündlingen swords (Cowen 1967), hardly products of 'fumbling infants' are contemporaries of the cast copper alloy forms which they imitate. The looped and socketed axes are the Irish analogues of the British series and might be considered as imports, were it not that the analytical evidence indicates a possible local manufacture at least for the Lough Mourne axe.

Of possibly similar date-range is a badly disfigured artefact from the River Shannon at Athlone, Co. Westmeath (Coffey 1907) which Rynne (1982, 93f) has identified tentatively as a Hallstatt C period sword (Fig. 3.2.2a). (I now accept this interpretation and do not reject the piece as in an earlier publication: Scott 1974b, 10). The Shannon 'sword' presents greater difficulty since the piece certainly is not, as Rynne notes directly comparable either with classic continental iron Hallstatt period weapons, or with the intrusive Llyn Fawr piece. He suggests a reconstruction (Fig. 3.2.2b) resembling a Later Bronze Age weapon. The problem of reconstructing detail on corroded edges of iron artefacts is to curb the natural temptation to take artistic and archaeological license. The depth of ricasso and positioning of rivets suggested by Rynne would turn this piece into a potential skeuomorph of an Eogan class 5 weapon (the Hallstatt C type: Fig. 3.2.2c) or class 4c (Fig. 3.2.2d) an indigenous weapon showing the influence of class 5 (Eogan 1965, 13 and figs. 63–64). As a potential skeuomorph, the piece invites close comparison with the iron sword of uncertain provenance from Reading Museum (Tylecote and Gilmour 1986, 112), and those from Skocjan, Banat and Tirol in Rumania (Pleiner 1981, Abb. 2.3–5), which may also be seen as skeuomorphic of local Later Bronze Age forms from the region, as is the remarkable iron 'rapier' (Fig. 3.2.2e) from the Hallstatt B_3 grave at Singen am Hohentweil (Kimmig 1981).

Another piece, certainly skeuomorphic of a Hallstatt D period type cauldron, is the rivetted sheet-iron cauldron from Drumlane, Co. Cavan (Milligan 1886, 149; Scott 1974b, 18, Plate II; *CIIA* 555; Raftery, B. 1984, 228f). The piece was originally compared by Mahr (1934, 22) with types of the 'early Middle Ages', but this is quite unconvincing. The construction diagram (Fig. 3.2.3) shows that the cauldron was put together from concentric sheets jointed together in the classic Later Bronze Age style with roughly domed-headed rivets. Pottery whose forms are skeuomorphic of Hallstatt D cauldrons is known from

Clogher, Co. Tyrone (information courtesy of R.B. Warner), while the famous wooden cauldron from Altartate, Co. Monaghan (possibly also two other examples from Gartagowan Bog, Co. Tyrone and Clogh Bog, Co. Fermanagh: Mahr 1934, 20) shows local copying in another medium. The decoration on the Altartate cauldron is close to that on pottery with a possible 6th/5th century BC dating from Blewburton Hill, Berks. (Harding 1974, 147, fig. 40F). With the Drumlane cauldron so close in concept to the Later Bronze Age types, and the apparent productivity of the Hallstatt D form in different media, it is difficult to interpret this piece, like the axes, as other than a copy in a new material of a well-established form. It is also possible that an escutcheon, attached by rivets to a fragment of iron sheet and still containing a suspension ring of twisted, square-section bar from Crane Island, Co. Leitrim[5] represents another vessel similar in form and date-range to the Drumlane cauldron.

From the excavation of a crannog at Rathtinaun, Co. Sligo came a curious multi-piece iron shaft-hole axe (Plate 3.3.3a) in association with an iron three-pronged fork and an iron swan's neck pin (Raftery, B. 1984, 8). While this object, if found in a context of which the excavator was less certain, would probably be offered a mediaeval date, its small size and three-piece construction are of interest. Like the axes of the Kilbeg hoard discussed in Chapter 4, it could be interpreted as having been made from the products of several smelts, thus potentially indicating the sort of size of bloom being produced during the transition period. In view of the technique, I include very tentatively as of possibly similar context the simple axe from Glenariff, Co. Antrim, whose design is very similar to the basics of the Rathtinaun axe, while the quenched structure of the blade is not out of keeping with either the Rathtinaun or Toome axes (but cf. Chapter 1.2 for the difficulties of offering dates on purely technological grounds).

* * *

3.3 – Metallographic examinations of EIA A iron artefacts (Tables 3.3.1 and 3.3.2)

(1) UM 614–1930 socketed axe, Lough Mourne, Co. Antrim (*CIIA* 577: Plate 3.3.1)

Two sections were taken, one from the cutting edge (*section 1*), the other (*section 2*) from the socket (Plate 3.3.1a).

Section 1: *Non-metallic inclusions*: Numerous elongated two- and three-phase slag inclusions run parallel to the long axis of the section.

Macrostructure: Etching with Nital indicated an uneven distribution of carbon content (Plate 3.3.1b), this appearance being accentuated by Oberhoffer.

Microstructure: Structures range from pure ferrite, through ferrite with interstitial pearlite, to near eutectoid composition with large pearlite colonies (GS 2–4) enclosed by a network of Widmanstätten ferrite. Ferritic grain-size ranges from 2–4 towards the back of the specimen, and from 5–7 near the cutting edge. Grains are regular and there is no trace of any deformation resulting from cold-working. Traces of at least two weld seams, marked by slag inclusions for parts of their lengths, are visible (cf. Plate 3.3.1b). One of the weld seams continues up to the cutting edge, while the other appears to run out several millimetres from the cutting-edge. The central area of high carbon content is seen to concentrate on one side of the main weld seam, with some diffusion across it (Plate 3.3.1c). There is a strong inverse carbon gradient from centre to outside, indicating that this area was carburised before the weld was made.

Hardness: mH_v ferrite 127–174, pearlite 216–259: H_v 110–234.

Section 2: *Non-metallic inclusions*: Numerous two- and three phase inclusions occur following the line of curve of the section. Corrosion has penetrated deeply along the main line of slag, and where the weld has failed, inclusions are large.

Macrostructure: Nital etching shows a highly uneven carbon distribution (Plate 3.3.1d), while etching with Oberhoffer indicates only very minor segregation.

Microstructure: One major weld seam is visible (Plate 3.3.1e), indicating the region in which the edges of the socket were folded one over the other. The weld between the two was not completely successful as shown by voids containing large inclusions near the start of the seam. Corrosion has penetrated following the slag inclusions (cf. Plate 3.3.1e). On one side of the seam, areas of ferrite with interstitial pearlite grade into areas of almost eutectoid composition showing moderately coarse pearlite enclosed by Widmanstätten ferrite. On the other side of the seam the structure is almost entirely ferritic, although there has been some diffusion of carbon across the seam. Ferrite grain size varies (GS 3–5 to 7–8) markedly. The size of the pearlite enclosed by Widmanstätten ferrite ranges from 4–5.

Hardness: mH_v ferrite 157–257, pearlite 297–305: H_v 133–182.

Interpretation: Fig. 3.3.1 is a diagrammatic reconstruction (not to scale) of the forging process, based on this metallographic examination, examination of the surface, and X-radiography. The artefact was worked up from unevenly carburised stock which was first forged out into a thick sheet. The central strip of high-carbon metal was produced by deliberate secondary carburisation of part of the face of the sheet which later was the surface of a weld with a low-C section. A thickened area was left along one edge and a tongue of metal at the other. The piece was folded into a cylinder, one end of which was forged down, welding the two faces of the sheet together, the other end being left open as the socket. Although virtually obliterated on the outer surface, the seam is clearly visible on the inside the of socket and in the metallographic section from this region. This suggests that the two edges of the folded-over sheet were overlapped considerably, although the final weld was not of the best quality. The loop was formed by pinching out a lobe of metal from the completed socket and making a hole through it with a rectangular-sectioned drift.

The small tongue at the other end of the sheet was

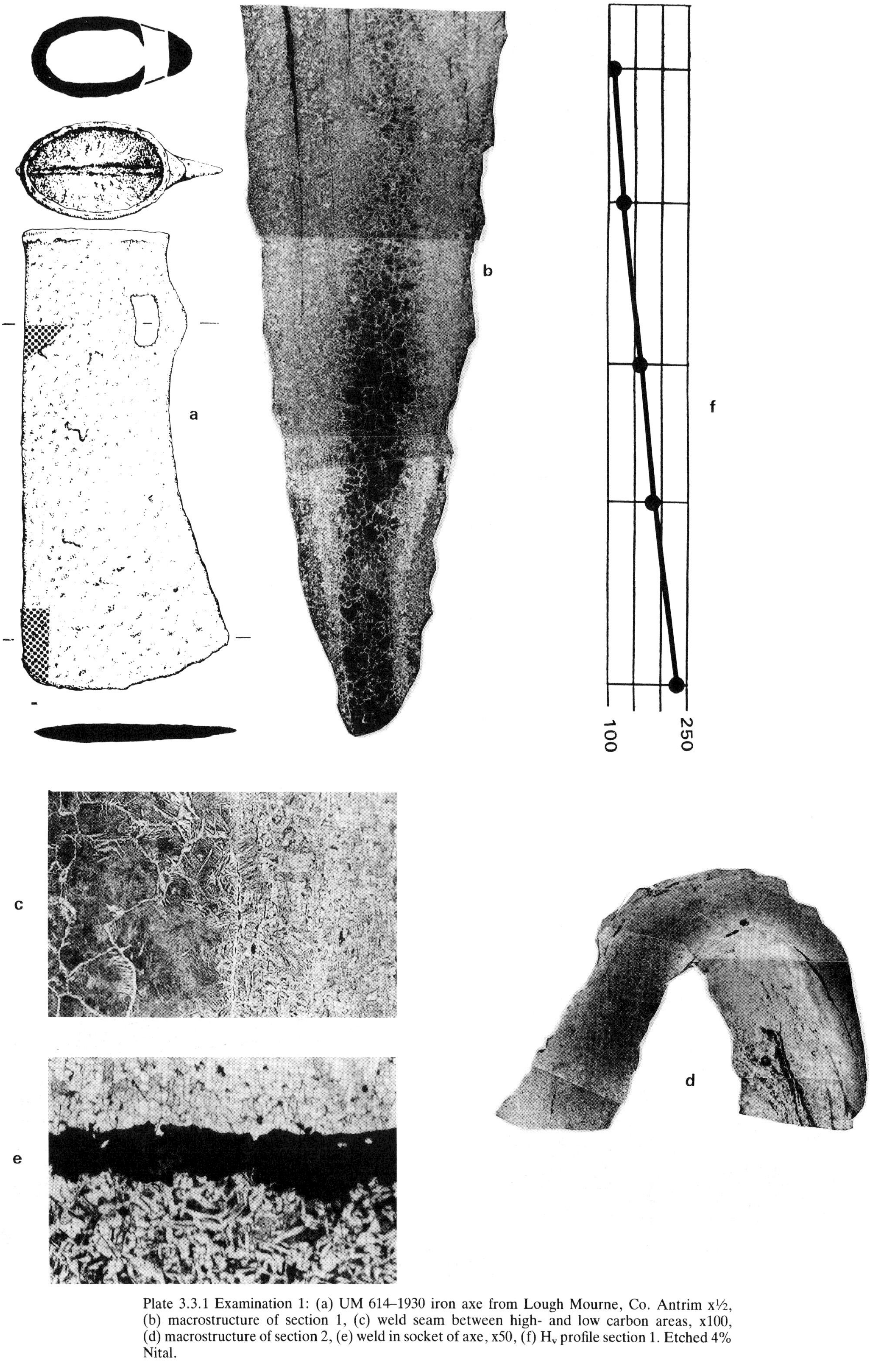

Plate 3.3.1 Examination 1: (a) UM 614–1930 iron axe from Lough Mourne, Co. Antrim x½, (b) macrostructure of section 1, (c) weld seam between high- and low carbon areas, x100, (d) macrostructure of section 2, (e) weld in socket of axe, x50, (f) H_v profile section 1. Etched 4% Nital.

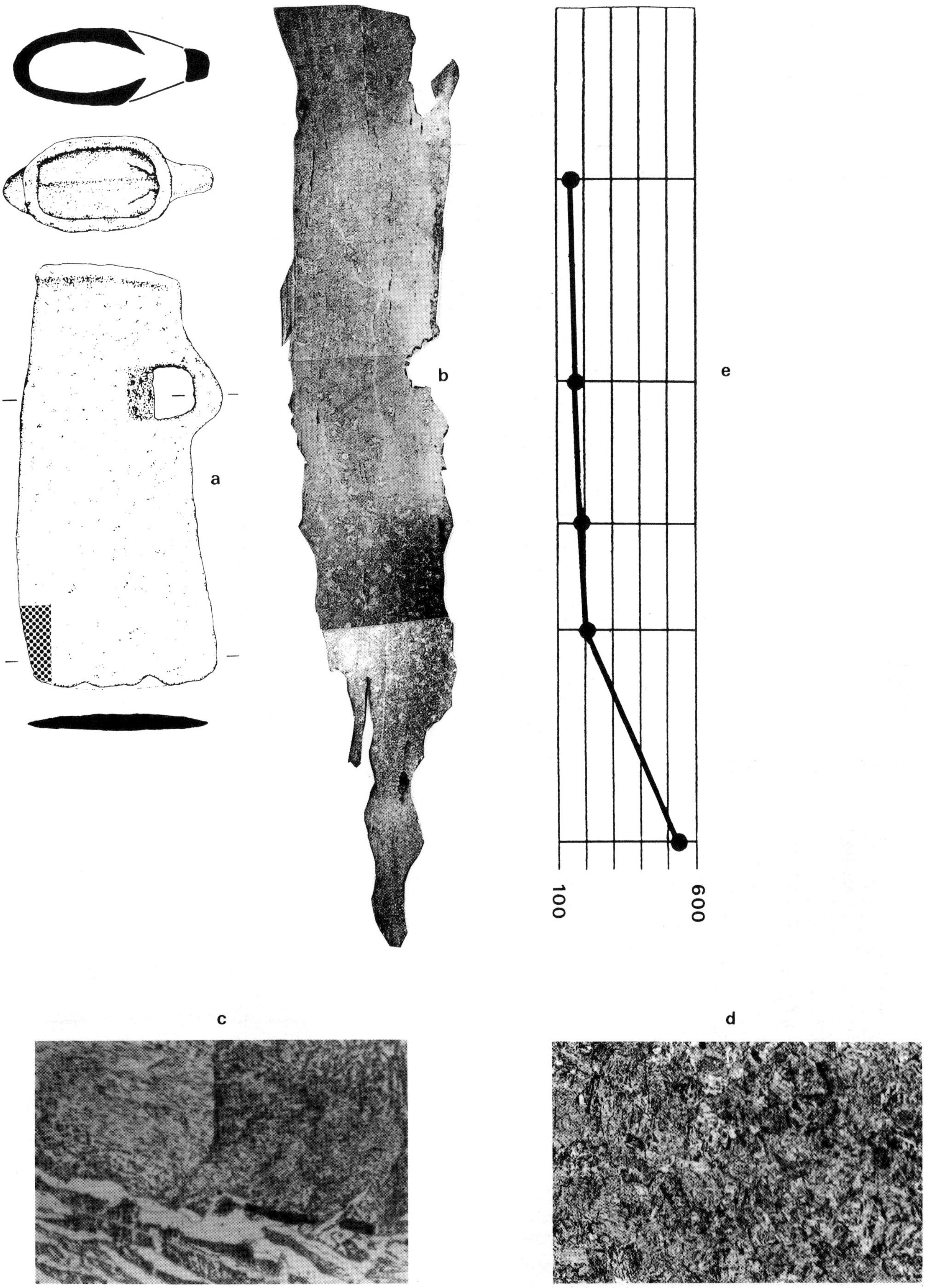

Plate 3.3.2 Examination 2: (a) UM 615–1930 iron axe from Toome Bar, Co. Antrim x½, (b) macrostructure, (c) pearlite showing carbide spheroidisation, x500, (d) martensite at cutting edge, x100, (e) H_v profile. (b) and (d) etched 4% Nital, (c) etched Picral.

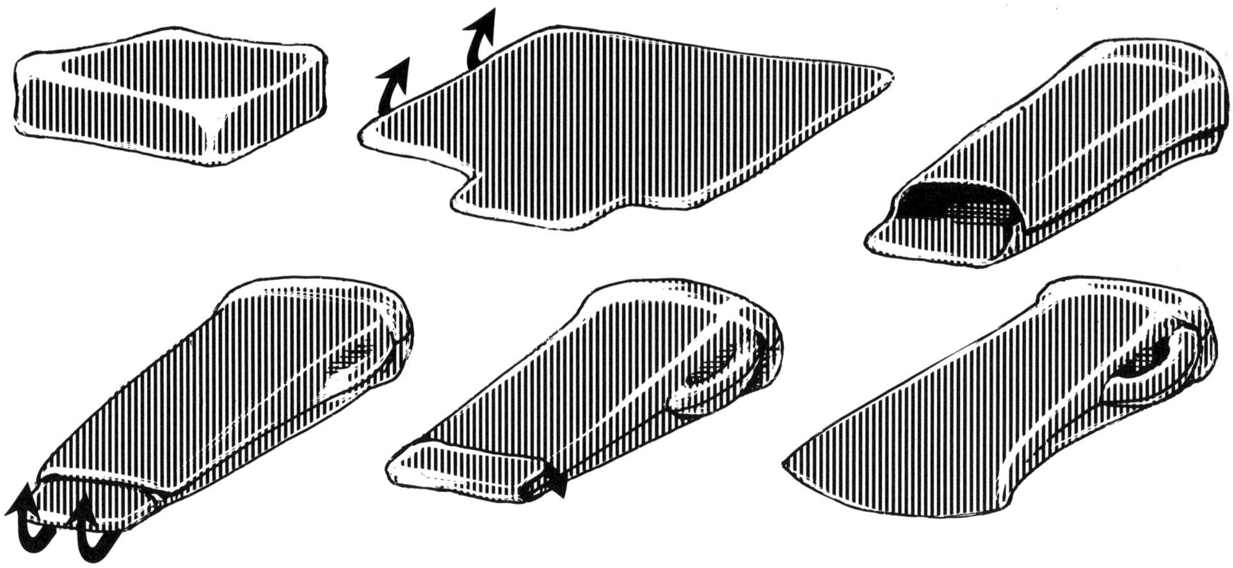

Fig. 3.3.1 The forging of the Lough Mourne axe (not to scale).

deliberately carburised, then folded back against one face of the blade and welded to it. This gave a steel centre to the cutting edge between two areas of much lower carbon content, and a section similar in appearance to those resulting from the mediaeval sandwich-welded technique of blade construction (e.g. Piaskowski 1964; Tylecote 1986, 197f). The steel centre would have been protected by the softer iron surround and exposed only as far as necessary by sharpening.

The difference in carbon content on either side of the seam in the socket suggests that the reason for only partial success was the difficulty of welding together iron and steel, although the same weld was accomplished in the forging of the cutting edge region. The whole axe probably underwent prolonged forging above A_3, followed by fairly rapid cooling in air. Chemical analysis (Table 3.3.2) revealed low P, Cu and Ni, but high Al and Ti (from the slag inclusions) indicating the possibility that the metal could have been smelted from the lateritic iron ores of Co. Antrim (see Chapter 6). SEM examination of a small section, showed high Al and Ti in the slag inclusions.

(2) UM 615–1930 socketed axe, Toome Bar, Co. Antrim (*CIIAA* 576: Plate 3.3.2)

A section was taken from the cutting edge, parallel to the long axis (Plate 3.3.2a).
Non-metallic inclusions: Single and two-phase slag inclusions are present in lines running parallel to the specimen surface.
Macrostructure: Etching with Nital shows variations in C-content (Plate 3.3.2b). Etching with Oberhoffer indicates no other significant segregations.
Microstructure: At the top of the section, away from the cutting edge, there occurs a ferrite network enclosing large pearlite colonies (GS 1–2) whose carbide shows spheroidisation (Plate 3.3.2c). Towards the cutting-edge, the carbon content decreases giving Widmanstätten ferrite with interstitial coarse pearlite showing spheroidisation. Approaching the cutting-edge, martensite with remanent ferrite and first fine pearlite, grading into troostite appears towards the section surfaces. The structure of the cutting-edge is almost entirely martensitic (Plate 3.3.2d). There is a distinct carbon gradient running back towards the body, showing deliberate secondary carburisation of this area. Traces of one major weld seam are visible, being difficult but not impossible to detect. Similarly, a number of possible bloom welds may be observed.
Hardness: mH_v ferrite 180, pearlite 254, troostite 436, martensite 730: H_v 153–551.
Interpretation: Like the Lough Mourne axe, this implement was made from one piece of metal (cf. Fig. 3.3.1), of comparatively high carbon content overall, but also with areas free from carbon. This was forged out and folded round and welded, although the cutting-edge was formed by welding together the two faces without a fold-over and without carburising the inner faces of the sheet. The loop was drifted out through a lobe pinched out from the body and a thickening around the mouth of the socket was forged out. The area of the cutting-edge appears to have had its carbon content supplemented by deliberate carburisation. Away from the cutting-edge, where there are patches of high-C metal with coarse pearlite near the surface, martensite does not occur, indicating that only the cutting-edge was quenched, following a deliberate plan to improve its quality. The survival of network ferrite with martensite indicates that the metal was not held for long enough above A_3 before quenching took place. Spheroidisation of the pearlite away from the cutting edge indicates prolonged heating during forging, which would account also for the blurring of the main weld seam.

(3) NMI E21:1786 shaft-hole axe, Rathtinaun, Co. Sligo (*CIIAA* 583: Plate 3.3.3)

Surface examination of this piece shows a complex construction[6]. A section was removed from the cutting edge (*section 1*), and a second section (*section 2*) was taken from the top of the blade (Plate 3.3.3a).
Section 1: *Non-metallic inclusions*: A line of very large slag inclusions runs down the centre of the

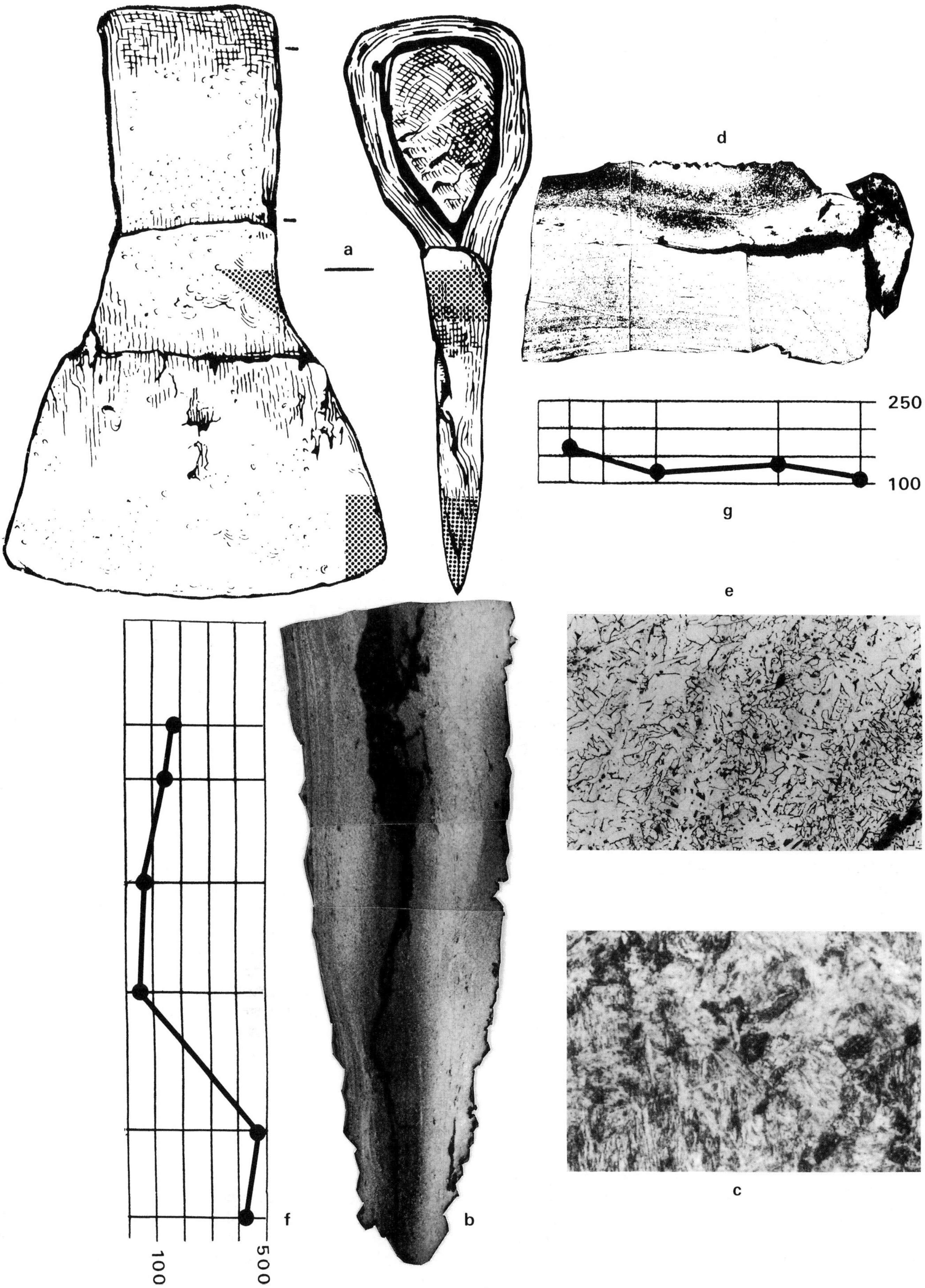

Plate 3.3.3 Examination 3: (a) E21–1786 iron axe from Rathtinaun, Co. Sligo x⅞, (b) macrostructure of section 1, (c) martensite, troostite and remanent ferrite near cutting edge, x500, (d) macrostructure of section 2, (e) banding in section 2, x50, (f) H_v profile of section 1, (g) H_v profile of section 2. Etched 4% Nital.

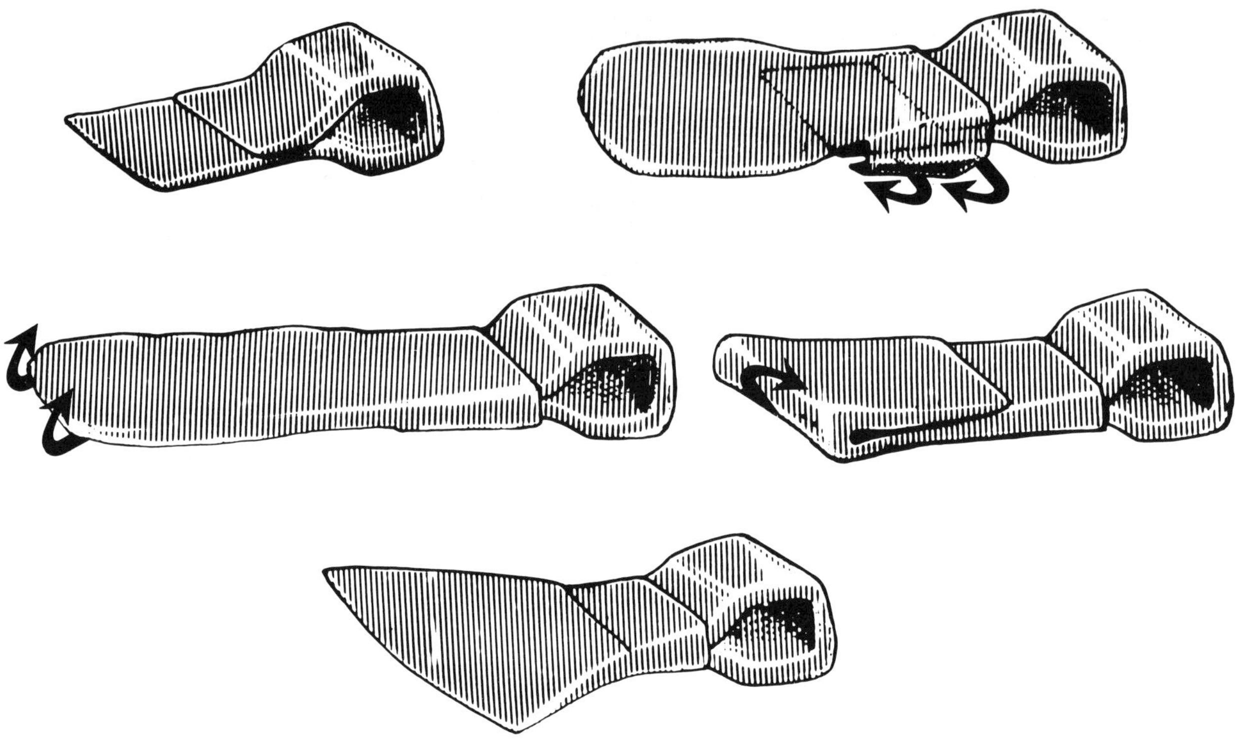

Fig. 3.3.2 The forging of the Rathtinaun axe (not to scale).

section and corrosion has penetrated along its entire length.

Macrostructure: Etching with Oberhoffer reveals no significant segregations. Nital shows an uneven distribution of carbon, with a central high carbon region (Plate 3.3.3b).

Microstructure: There is a very wide variation in distribution of grain size and of carbon content. The central feature is a seam resulting from a weld which was not successfully closed at any point in this section. Towards the cutting edge, there is a concentration of carbon on either side of the junction between the two pieces in the centre of the section and an inverse gradient towards the surface. The high carbon region shows patches of martensite with remanent ferrite and troostite (Plate 3.3.3c), grading into a region of very fine grained ferrite (GS <8) with interstitial spheroidised pearlite. Towards the top of the section, a Widmanstätten structure with fine pearlite occurs in patches, but with ferrite (GS 3–7) predominating.

Hardness: mH_v ferrite 228–279, pearlite 370, troostite 814, martensite 1444: H_v 182–498.

Section 2: *Non-metallic inclusions*: There are numerous and often quite large two- and three-phase inclusions, some marking the lines of weld seams. Corrosion has penetrated along the central weld seam between the two largest pieces of metal, following massive inclusions in areas where the seam was not closed.

Macrostructure: Etching with Oberhoffer showed evidence of banding which was pronounced when a Nital etch was applied. A major weld seam, unclosed for most of its length, runs parallel to the long axis of the section, separating the two components of the blade face (Plate 3.3.3d). At the top of the section is a piece of metal at right-angles to the other two pieces. There is much penetration of corrosion here.

Microstructure: Away from the central weld seam, the right-hand piece of metal can be seen to have a low carbon content, but with distinct banding of ferrite (GS 4–6) with interstitial pearlite (Plate 3.3.3e). The left-hand area shows an inner, almost purely ferritic (GS 2–4), piece welded to an outer piece of composite stock which shows pronounced banding of Widmanstätten ferrite enclosing coarse pearlite (GS 1–2), and finer ferrite (GS 4–5) with pearlite. The high carbon area shows evidence of bloom welds. The seam between the two largest pieces has been closed near the bottom of the section, with some carbon diffusion across. Towards the top of the section, fusion of metal was not achieved between any of the components. The banded structure of the right-hand component, resulting from fold-welding, splays out near the top of the section. This can be attributed to the attempt to weld the top piece over it, obviously with much force, but no success.

Hardness: mH_v ferrite 144–187, pearlite 287–367: H_v 115–219.

Interpretation: Fig. 3.3.2 is a diagrammatic representation (not to scale) of the construction of the axe. One end of a piece of fold-welded stock was bent round and welded back on itself to form the shaft-hole. The area of this weld was covered by a flat plate which was welded over it, the ends being hammered over the top and bottom edges of the blade. The end of the blade was then drawn down to a rough spade shape which was folded back on itself at the widest part, and an attempt was made to weld it back over the second piece of metal. The fold was made at the point where the cutting edge now is.

That the drawing down was not carried out with much skill can be seen from several cracks and splits showing at the surface. There was a slight overlap of metal at one side of the blade and this was folded over to cover the edge and welded down. The weld between the two surfaces was not closed properly and this failure has been accentuated by the ingress of

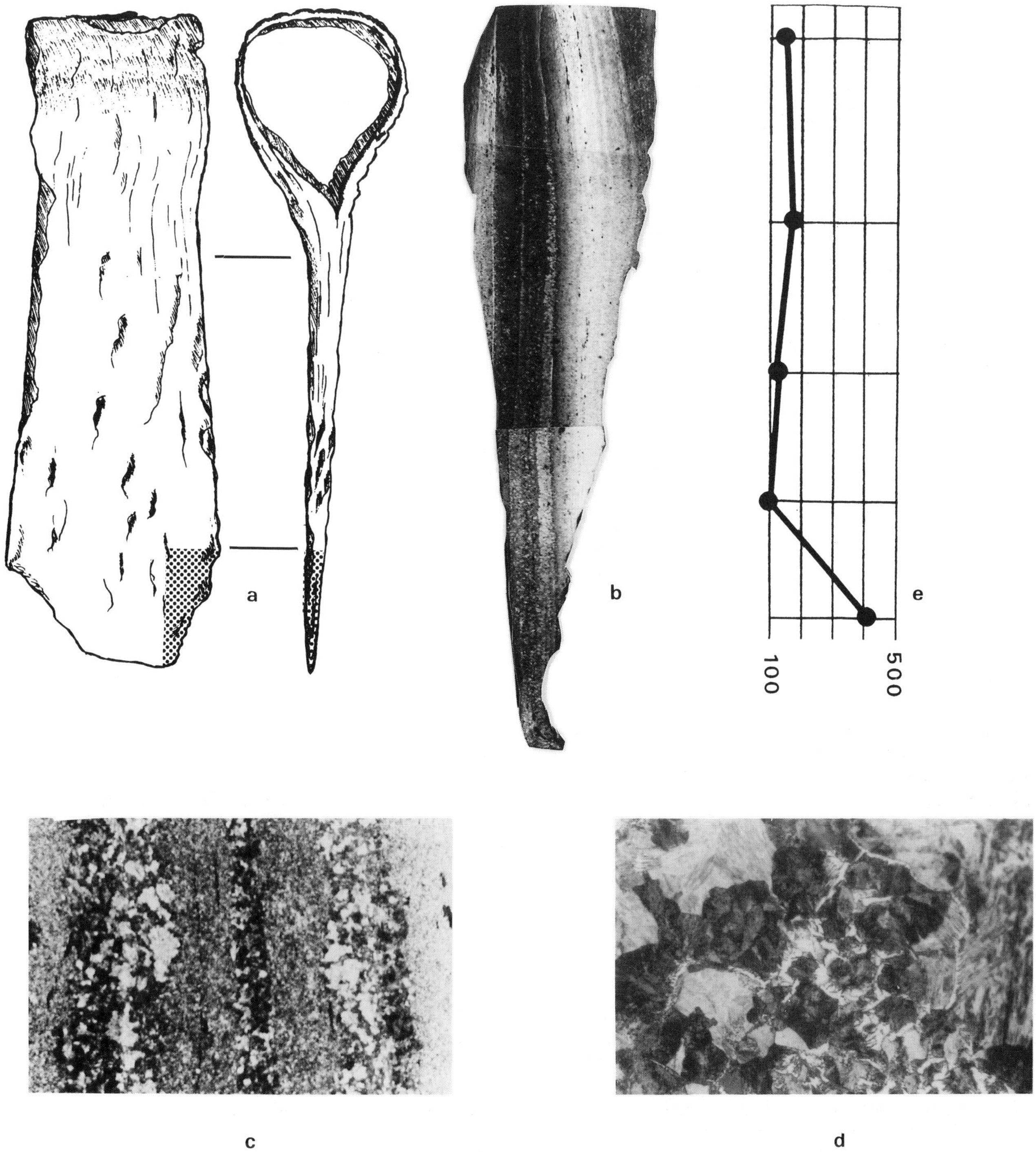

Plate 3.3.4 Examination 4: (a) UM 616–1930 iron axe from Glenariff, Co. Antrim x⅔, (b) macrostructure, (c) banded structure, x50, (d) martensite, troostite and remanent Widmanstätten ferrite, x500, (e) H_v profile. (b)-(c) etched 2% Nital, (d) etched Bisulphite.

corrosion. The area around the point at which the metal was folded over was carburised, thus giving the concentration of carbon towards the centre of the cutting edge. The cutting edge region was then heated above A_3 for long enough to allow most, but not all of the high carbon region to undergo the α-γ transformation and then quenched. The very fine grain of the ferrite towards the outer surfaces in section 1 is indicative of a stay above A_1 long enough for grain refinement, followed by rapid cooling. Tempered martensite coupled with troostite indicates that the rate of cooling in the central region was sufficiently fast for martensite to form, but that the rate was not fast enough for complete transformation, and that the heat of the metal further up the blade (where a Widmanstätten structure is observed) was sufficient to cause some tempering of that martensite which did form. In summary, this piece shows that its maker was aware of a wide range of techniques, but was not best practised in their execution.

(4) UM 616–1930 shaft-hole axe, Glenariff, Co. Antrim (Plate 3.3.4)

A section was taken from the cutting edge running parallel to the long axis of the blade (Plate 3.3.4a). *Non-metallic inclusions:* A scatter of one- and two-

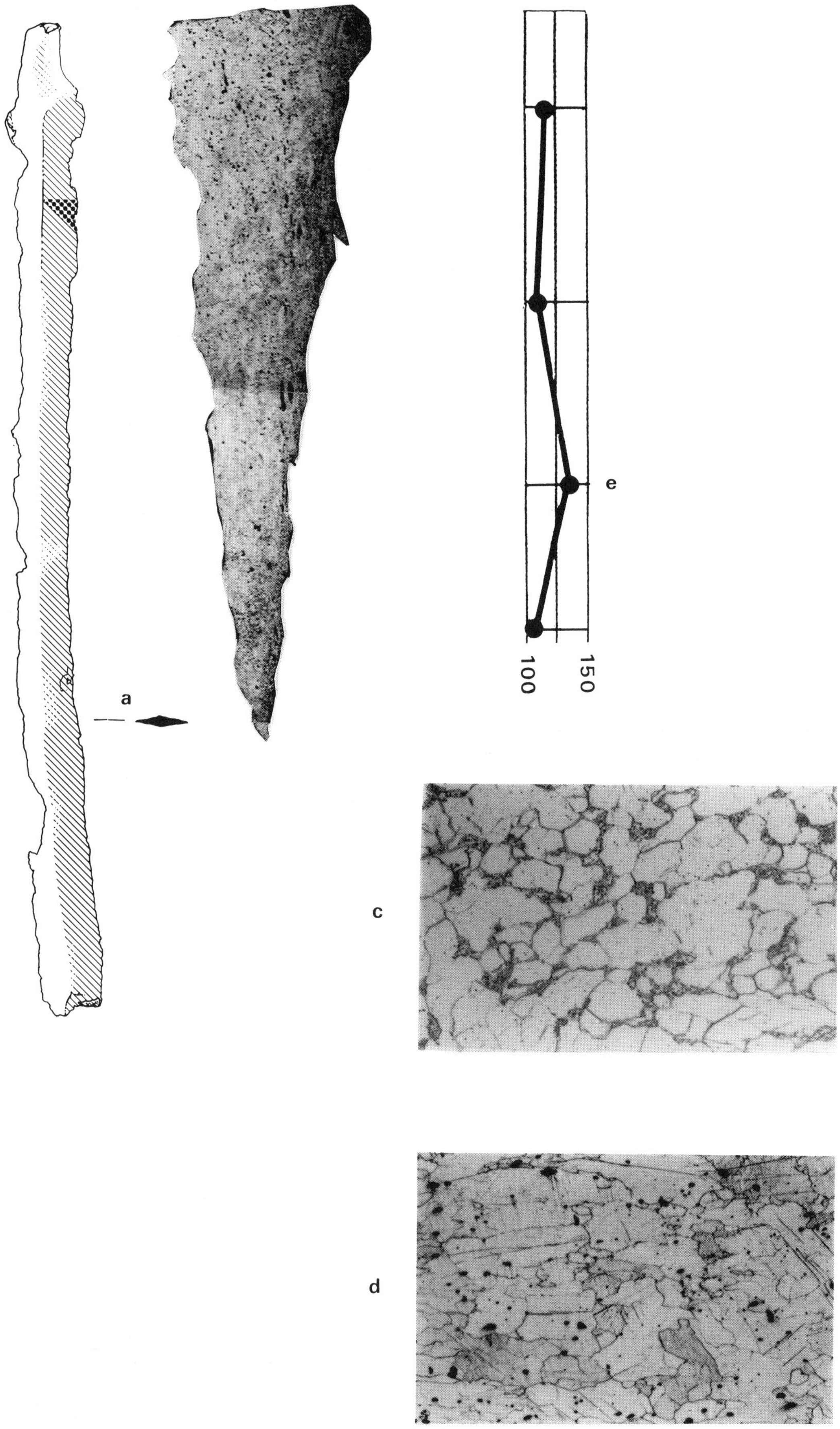

Plate 3.3.5 Examination 5: (a) NMI WK23–A124 iron sword from the River Shannon at Athlone, Co. Westmeath x⅓, (b) macrostructure, (c) ferrite with interstitial pearlite, x250, (d) distorted ferrite near cutting edge, x100, (e) H_v profile. Etched 2% Nital.

phase inclusions run parallel to the long axis of the section.

Macrostructure: Nital etching revealed pronounced banding (Plate 3.3.4b), while etching with Oberhoffer showed no other significant segregations.

Microstructure: The pronounced banding visible in macro-section resolves into areas of very fine ferrite (GS 8–<8) beside high-C areas (Plate 3.3.4c). Two weld seams run along the long axis of the section, parallel to the banding. At the cutting edge the structure is complex with martensite and troostite, as well as the remnants of a Widmanstätten ferrite network (Plate 3.3.4d). This structure continues back in the high-C bands. Although most ferrite grains are tiny, there is one area of larger grain growth (GS 5–6).

Hardness: mH_v ferrite 164–219, pearlite 231–291, troostite 522, martensite 944: H_v 110–413.

Interpretation: This implement was made from heterogeneous stock, with areas greatly different in carbon content. Forging up led to pronounced banding. The blade was not held long enough above A_3 for the α-γ transformation to go to completion. Coupled with the banding, quenching gives a heterogeneous distribution of structures. Nevertheless, the operation did result in increased hardness, and this would have been quite a serviceable tool.

(5) NMI WK23:A124 sword, River Shannon at Athlone, Co. Westmeath (Plate 3.3.5)

A V-shaped specimen representing half of the cross-section was cut from the blade (Plate 3.3.5a).

Non-metallic inclusions: The polished specimen shows a relatively small number of two-phase slag inclusions, some of which are spherical.

Macrostructure: Oberhoffer indicated no significant segregations, while Nital showed little variation in carbon content (Plate 3.3.5b).

Microstructure: The structure is entirely of ferrite (GS 4–5) with coarse interstitial pearlite showing spheroidisation of carbides. (Plate 3.3.5c); some excessive ferritic grain growth can be observed (GS 1–2). Grain boundary cementite is visible in association with ferrite grains. The principal feature is the deformation of the ferrite grains indicating a degree of cold working (Plate 3.3.5d). There is a very slight increase in carbon content from one face of the blade to the other.

Hardness: mH_v ferrite 163–184: H_v 109–127.

Interpretation: The blade was forged from a single piece of metal very low in carbon. There seems to have been an attempt was made to increase the hardness of the edges by cold-hammering, although this was not really successful. It could never have been a particularly serviceable weapon.

Table 3.3.1

Summary of Metallographic results

Number	Artefact	Sample site	structural features										Carbon content%	Hv 1000gm	Structural components	Grain size	mHv 50gm
			Wid	F/W	B/W	SEG	C/B	F/B	A	CAR	H/T	C/W					
1 Plate 3.3.1	Socketed axe Lough Mourne Co. Antrim UM 614–1930	*Section 1* cutting edge	+	–	–	–	–	–	–	+	–	–	0.00 – 0.70	110 – 234	ferrite pearlite	2 – 4 5 – 7 2 – I	127– 174 216– 259
		Section 2 socket	+	–	–	–	–	–	–	–	–	–	0.00 – 0.55	133 – 182	ferrite pearlite	3 – 5 7 – 8 4 – I	157– 257 297– 305
2 Plate 3.3.2	Socketed axe Toome Bar Co. Antrim UM 615–1930	cutting edge	+	–	?	–	–	–	–	+	+	–	0.10 – 0.70	153 – 551	ferrite pearlite troostite martensite	5 – 6 2 – I	180 254 436 730
3 Plate 3.3.3	Shaft-hole axe Rathtinaun Co. Sligo NMI E21:1786	*Section 1* cutting edge	+	–	–	–	–	–	–	+	+	–	0.00 – 0.50	182 – 498	ferrite pearlite troostite martensite	3 – 7 <8 5 – I	228– 279 370 714 1444
		Section 2 top of blade	+	+	–	–	+	–	–	–	–	–	0.00 – 0.50	115 – 219	ferrite pearlite	2 – 4 4 – 6 I – 2	144– 187 287– 367
4 Plate 3.3.4	Shaft-hole axe Glenariff Co. Antrim UM 616–1930	cutting edge	+	–	–	–	+	–	–	–	+	–	0.00 – 0.50	110 – 413	ferrite pearlite troostite martensite	5 – 6 8 –<8 I	164– 219 231– 291 522 944
5 Plate 3.3.5	Sword River Shannon at Athlone Co. Westmeath NMI WK23:A124	cutting edge	–	–	–	–	–	–	–	–	–	+	0.00 – 0.01	109 – 127	ferrite pearlite cementite	1 – 2 4 – 5 I GB	163– 184

I = interstitial GB = grain boundary

Wid = Widmanstätten structure F/W = fold-welding B/W = bloomery welds F/B = ferrite banding SEG = P/As segregation C/B = carbon banding

A = acicular precipitates H/T = heat treatments C/W = cold working CAR = secondary carburisation

+ = present ? = uncertain occurrence – = absent or not detected

NMI = National Museum of Ireland (Dublin) UM = Ulster Museum (Belfast)

Table 3.3.2

Chemical analyses

Reg. no.	Mg	Al	Si	P	S	Ca	Ti	V	Cr	Mn	Co	Ni	Cu	Zn	As	Zr	W	Au	Pb
1 UM 614–1930/1	tr	0.511	0.044	tr	nd	0.044	0.121	0.008	tr	0.129	0.033	0.010	0.052	tr	0.066	tr	tr	tr	tr
/2	tr	0.320	0.049	tr	nd	0.067	0.091	0.005	tr	0.089	0.012	0.044	0.019	tr	0.057	tr	tr	tr	tr
2 UM 615–1930	0.007	0.025	0.126	0.107	0.154	0.172	0.001	tr	0.035	0.121	0.022	0.020	0.037	0.028	nd	0.005	nd	tr	0.034
3 NMI E21:1786/1	0.006	0.011	0.102	0.065	tr	0.030	tr	nd	nd	0.022	0.008	0.045	0.009	tr	0.141	nd	nd	nd	nd
/2	0.003	0.014	0.091	0.044	tr	0.038	nd	tr	tr	0.014	0.014	0.025	0.004	0.007	0.265	0.006	tr	nd	0.033
4 UM 616–1930	0.003	0.026	0.131	0.020	tr	0.071	tr	nd	tr	0.009	tr	0.008	0.100	nd	0.017	nd	nd	nd	nd
5 NMI WK23:A124	0.001	0.046	0.009	0.046	nd	0.002	nd	nd	0.066	0.004	tr	0.018	0.025	nd	tr	nd	nd	nd	nd

tr = trace (P, Pb <0.02%, As <.01%, Au <0.005, others <0.001) nd = not detected
NMI = National Museum of Ireland UM = Ulster Museum (Belfast)

* * *

3.4 – Summary and Conclusions

The five metallographic examinations, the structure of the Drumlane cauldron, and identification of the iron components in the copper alloy cauldron and bucket, show that Irish metalworkers of the transition period learned quickly to forge and joint iron fairly competently. Despite the uncertainties of dating, we may see these artefacts as evidence of the first indigenous experimentation with the new technology. From the point of view of the conversion of pieces of stock into recognisable artefacts, the forging techniques displayed include drawing down and the production of sheet, upsetting, punching and drifting, rivetting and twisting. All of these were long-established in the repertoire of Bronze Age metalworkers and their direct transfer to the forging of iron artefacts presents no problems. In the Lisdrumturk cauldron and Derrymacash bucket, iron is used in conjunction with copper alloy, functional bimetallism, but again drawing on established tradition.

It is in the handling of the production of iron, and of the properties of the material that we see uncertainty alongside genuine innovation. Smelting operations probably produced comparatively small quantities of iron, with some uneven primary carburisation: success in removing entrapped slag varied. There is also clear evidence of welding, secondary carburisation and quenching, crucial new techniques that would have had to have been acquired in addition to those carried on from the Late Bronze Age.

The unclosed weld in the socket of the Lough Mourne axe and those in the Rathtinaun axe do show that their smiths had not come to grips entirely with the difficulties of hammer welding, although obviously aware of the basic practice[7]. We can detect a further lack of experience in controlling the quality of the stock, in terms of heterogeneous carbon distributions. The occurrence of well-developed Widmanstätten structures in three of the artefacts indicates prolonged heating above A_3. This shows, perhaps, a lack of confidence in working, with the metal being held at elevated temperatures for longer than necessary to complete shaping operations. The coarseness of pearlite and the degree of carbide spheroidisation also indicates that operations at heat were prolonged. Nevertheless, the general level of technology displayed by these artefacts is consistent with that shown by contemporary metalworkers in other regions in transition who were also experimenting in an unfamiliar medium. By and large, bloomery iron would respond favourably to existing skills, even while its subtler properties and their exploitation were still being explored.

The River Shannon sword shows how successful was its smelter/smith in expelling entrapped slag from the bloom. It is tempting to see the evidence of cold working from this particular piece as an attempt to increase the hardness of the cutting edge using a technique that would have been familiar to the worker in copper alloy (e.g. Allen et al. 1970, 23 and 226f). Cold working was evident from the Baageroostervelde chisel (Charles 1984), and may be another feature to be sought in transition technology. If so, then this would be a good example of technological transfer from copper-alloy to ironworking.

The discovery of martensitic structures in the Toome, Rathtinaun and Glenariff axes is of prime importance, since they demonstrate knowledge of the most important property of iron – its potential for very high hardness when alloyed with carbon. It is worth recording that, while the sample is exceedingly small, and while the dating of the Glenariff axe is questionable, three out of five edged tools showing both carburisation and quenching potentially is significant. As Pleiner (1980) and Piaskowski (1987b, 160, Table 1) have demonstrated, a feature of the transition period across barbarian Europe, is the high proportion of steels and heat treated artefacts. It is not unreasonable to suggest that this shows Ireland following other regions technologically.

The Toome and Lough Mourne socketed axes and the Rathtinaun shaft-hole axe show also a startling innovation, one difficult to parallel in contemporary transition technologies[8]. In each case, it is clear that secondary carburisation has been used, in the penultimate stage of forming, to raise the carbon content in the region of the cutting edge. These carburised surfaces have then been positioned by folding and welding so that they lie in the centre of the section. In later periods, pieces of high carbon metal were sandwiched between pieces of lower carbon content, the idea being that the more ductile metal would protect the steeled areas, leaving them exposed at the very cutting edge only. The makers of these early Irish pieces achieved the same result, but by a different route. The later technique required the separate production of pieces of steel that could be welded to iron. This folding-in of a carburised surface is the obvious alternative. It can only be described as a very sophisticated design which indicates awareness of the ability of iron to take up carbon to produce a superior metal, of the length of time required for significant penetration, of the need to protect a brittle steeled area and of how effectively to protect it.

The artefacts examined represent the first appearance of good quality ironworking in a Later Bronze Age industrial setting. The range of techniques and abilities which they display is quite consistent with transition ironworking in other regions. The formation of steel-centred cutting edges, a striking achievement, shows just how keenly Irish metalworkers could compete for quality with their Old World contemporaries. Although we do not have evidence as yet of of decorative bimetallism from Irish finds, we can observe functional bimetallism and skeuomorphism, both general features of the experimental phase of transformation. It is not surprising to find such workmanship produced among metalworkers and industries as innovative and energetic as those of Dowris period Ireland. However, just as contacts with other regions which were coming to grips with iron technology were being exploited, and iron was making its first appearance, the period of vigorous industrial and economic activity in which they were produced came to an end.

The identity and origins of those from whom metalworkers of Later Bronze Age Ireland learned of the manufacture of iron and the forging of iron artefacts are unclear. Technologically, some features of the artefacts examined are common to contemporary pieces in barbarian Europe and the Near East. The steel-centred edges, in contrast, from their design and execution, at present seem to be unique. However, we may state with some confidence that there seems to be no question of an independent discovery of iron metallurgy here.

As B. Raftery (1984, 7–8) notes, the principal component of the assemblage of intrusive material of Hallstatt C period is the class 5 sword, for whose immediate origins we should look to south-east Britain, thence to a Continental ancestry. We must remember also the broad associations of the material in the Llyn Fawr, Glams. hoard, in particular the possible Irish parallels of the iron sickle, and look to western Britain as another possible source of transmission. Pieces like the Kilmurry bracelet indicate a more direct French route, while some of the pins draw our attention to Scandinavia. The Irish iron looped and socketed axes stand within the distinctly British series as discussed by Manning and Saunders (1972): while identified here as native products, they are nevertheless indicators of broader British contacts. Their background, like that of the swords, is one of a region where the technological transition was beginning under the stimulus of continental iron-users. It must be said that evidence for the earliest ironworking in southern Britain, now at least, is no better than for Ireland. Direct links between Ireland and iron-users in Atlantic Europe are in evidence, although we cannot gauge yet the relative strengths of influence from Britain and Europe.

Contacts between Ireland and the outside world in the Later Bronze Age were not one way. It is certain that products of Irish metal workshops travelled widely, along with traders and craftsmen. The introduction of iron technology to Ireland would not have been possible without opportunities to see iron being smelted and worked and to learn from those who were familiar with the technology. In the light of present knowledge, it is not possible to identify the source or sources from which the earliest technology here derives. The fact that heat-treatment was observed in three of the artefacts examined, with carburisation, shows a high proportion of 'quality' workmanship in such a small assemblage. This contrasts with the results of work on later artefacts (Chapters 4 and 5), where the proportion of good quality metal is much lower. There is thus broad agreement with the trend observed in other regions for a primary horizon of advanced technology which was not sustained as iron became more common after the transition was complete. The best conclusion that can be drawn in the light of present evidence is that knowledge of iron technology arrived here through foreign visitors and Irish metalworkers returning from southern and western Britain and France. Irish ore resources were available to whoever had the expertise to exploit them.

Wherever their origins, it seems clear that the numbers involved in the introduction of the new technology were small and their impact limited. It does appear that the upheaval in socio-economic and political spheres which may be inferred were instrumental in aborting first experiments with iron in a truly Later Bronze Age industrial setting. We might otherwise have expected to have seen vigorous ironworking centres being established in areas with abundant and accessible iron ores. The failure to establish monopolies (as opposed to small industries), particularly in north-east Ulster which is a principal focus of the developed Earlier Iron Age, bespeaks lack of real opportunity, not indifference. All that was required in terms of material resources and expertise was available, but quite clearly, the structures on which earlier metalworking was based had passed away.

* * *

Notes to Chapter 3

1 cf. problems of interpreting metallurgical developments of the analogous period in the E. Mediterranean as summarised by Waldbaum 1978, esp. 67f, Waldbaum 1982).

2 It might be noted in this context that a significant number of the tanged, single-edged 'knives' from Viking Dublin have blade lengths under 6 cm. Nevertheless, their metallography shows them to be of high quality manufacture, capable of taking and holding exceedingly effective cutting edges (cf. Tomtlund 1973). Simply because function cannot be immediately identified from shape is not a reason for writing an artefact off as of little practical use.

3 Although Proudfoot (1954) claimed Later Bronze Age origins for the defences around Cathedral Hill, Downpatrick, recent excavations (Hamlin and Lynn 1988, 61f) demonstrate clearly that the earliest bank and ditch were constructed in the mid- to later-1st millennium AD.

4 The Tulnacross, Co. Tyrone cauldron has what appeared to be the badly corroded remains of an iron strengthening band running around the inside of the rim and *under* the cast-on straps for the ring handles. The preliminary examination by the author, on which Briggs (1987) based his interpretation was incorrect. Detailed study has shown that this deposit is almost certainly organic in origin and that, despite its appearance and physical properties, has a negligible iron content.

The reference by Eogan (1983, 47) to 'an iron concretion' on a class 3 sword from the (dubious) Youghal, Co. Cork hoard seems incorrect. Examination shows this piece to have a very well preserved patina of the form typical for copper alloys from archaeological contexts, with no signs at all of identifiable iron corrosion products. Weatherup (1982, 62) draws attention to a Later Bronze Age sword showing repairs effected with iron rivets. In fact the joint is modern, with iron rivets being used in addition to soft solder, to cobble together a hilt and blade which certainly do not match.

In an earlier publication (Scott 1974b, 17), I drew attention to the report by Evans (Evans, E.E. 1941) of the occurrence of 'a remarkable mound of iron slag several feet high, containing burnt stones, sherds, fragments of fused clay and sand and bits of bog iron ore' on a sandhills site at Magheragallen, Co. Donegall. An Iron Age date for similar pottery to that found at Magheragallen had been claimed. However, Dr. J. Waddell, to whom I am grateful for the information, suggests that in fact the pottery belongs mainly to the Earlier Bronze Age ceramic tradition, and that as with other such sandhill sites, the stratigraphy at Magheragallen is hopelessly confused.

5 Unpublished and in private hands - information courtesy of the National Museum of Ireland.

6 I am most grateful to Dr. J. Raftery for access to this piece in advance of his own definitive publication.

7 Hammer welding of gold was practised in Later Bronze Age Ireland (e.g. Tylecote 1986, 109), and so even the basics of the technique of jointing metal through exertion of pressure were known before the advent of iron.

8 At the 1989 conference of the *Comité pour la Sidérurgie Ancienne*, held at Ameliówka in Poland, Janet Lang reported on her metallographic examination of an early iron Cypriot dagger or dirk, dating perhaps to the 11th century BC. She noted an equally sophisticated constructional scheme, coupled with heat-treatment. I am most grateful to Mrs. Lang for permission to refer to this work in advance of her own publication.

CHAPTER 4

The Establishment of Iron Technology in Ireland – Earlier Iron Age B

4.1 – The establishment of ironworking

The second phase of the Irish Earlier Iron Age is characterised by a resurgence of fine non-ferrous metalworking skills, and a level of iron production which, although iron was still comparatively scarce, indicates that the technology was firmly established here. A much greater range and number of artefact types were being made in iron than in the transitional EIA A, and it is clear that the takeover by iron as the primary industrial material, coupled with the redeployment of non-ferrous metals towards a non-industrial use, was virtually complete. Raftery's *Corpus* of the material of the period (*CIIAA*) illustrates well the splendid quality of the decorated non-ferrous metalwork of the time. It shows also the re-emergence (albeit on a smaller scale) of the skills of mining, smelting, alloying, casting and shaping of metals, so prominent by their absence after the demise of the Dowris period high-output copper alloy and gold industries. Although we have sound reasons for believing that iron was beginning to be produced in increasing quantities throughout EIA B, we have, as yet only a handful of sites which have produced clearly identifiable remains of smelting and forging (Chapter 6.5). As with the preceding period, we are thus forced to look at artefacts largely in isolation from the social and economic background in which they were produced. Furthermore, our dating evidence is still largely uncertain for the relatively few ferrous survivals of the period.

The broad assemblage identified with EIA B comprises mainly fine non-ferrous metalwork whose artistic congeners are to be found in La Tène Europe. While the likelihood of some technological contribution from outside must be taken into account, overall the production technology employed (as opposed to the artistic repertoire) indicates survival of much of the expertise of the Later Bronze Age. And survival it must be, unless one desires to see a re-establishment of technology stemming from some 'invasion' from outside (cf. Waddell 1978), bringing with it craftsmen responsible for a re-education of the native population. Although EIA A is a period of obscurity, there is evidence nevertheless of some threads of continuity which account better for the skills evident in EIA B than does foreign intrusion. It is particularly noticeable in terms of the use of sheet metalwork – pieces such as the fine trumpet from Loughnashade, Co. Armagh (*CIIAA* 781) and the *repoussée* discs typified by the specimen from Monasterevin, Co. Kildare (*CIIAA* 785, 786). Casting technology too is seen again at its former level of expertise, and in purely technological terms, it is as if the 'dark age' had never intervened, but rather that the expertise of the Dowris industries had gone on developing and responding to stimuli from the outside world.

The appearance of the highly distinctive iron swords and their ornamented copper alloy scabbards, deriving their form from the La Tène world, marks the establishment in Ireland of iron production as a significant industrial activity. The type site, but indifferently examined in the last century, is the supposed crannog at Lisnacrogher, Co. Antrim[1], which produced elegant scabbards for the earliest of which Jope (1954, 90f), has suggested a third century BC date. It is a well-recognised fact that the Irish weapons, identifiable as the counterparts of British and European La Tène II swords, are all appreciably smaller in size than the mean of their foreign contemporaries. Since their scabbards, most of which bear distinctly insular decoration, were made to measure, there seems little doubt of the swords having been forged in Ireland. The Lisnacrogher weapons form part of a series of very finely *shaped* iron pieces, including the two spearheads from the same site, the Corrofin, Co. Clare; Mullaghmore, Co. Roscommon; Moyfin, Co. Meath and 'River Shannon' spearheads (the last three not admitted into *CIIAA*[2]), and the swords from Edenderry, Co. Offaly and Ballinderry, Co. Westmeath (Fig. 4.1.1). Although as we shall see below, the metals from which the Lisnacrogher blades were forged were of indifferent quality, the expertise shown in the forging of these and the other weapons bespeaks a tradition of working with iron that surely pre-dates the 3rd century BC. The dating of the decoration on the copper-alloy scabbards, provides us with a link, albeit tenuous, with the experimental phase of EIA A. It must be remembered, however, that the surviving products of the forges which produced the Lisnacrogher material and its contemporaries need not represent more than a couple of years work in total, and that we are thus seeing only a fraction of the potential output.

Although iron is poorly represented in the assemblage of the period, we may yet detect its use not only from those artefacts which have survived, but also from its interaction with non-ferrous metalworking and other crafts. Thus, in the iron quillon of the Dún Ailinne sword, we have a good example of a skeuomorph of copper alloy forms. During the haphazard 'excavations' at Lough Crew, Co. Meath, what was claimed as a fragment of an iron compass (now lost) was recovered, and if the identification was accurate, it was very likely of the sort of tool used to incise the compass-drawn decoration on the enigmatic bone slips found at the site (Raftery, B. 1984, 251f). Both Jope (Jope and Wilson 1957) and O'Kelly (1961b) argued convincingly that the exceedingly fine raised curvilinear decoration on the Bann disc, Petrie Crown, and the Cork Horns (*CIIAA* 793, 820, 821:

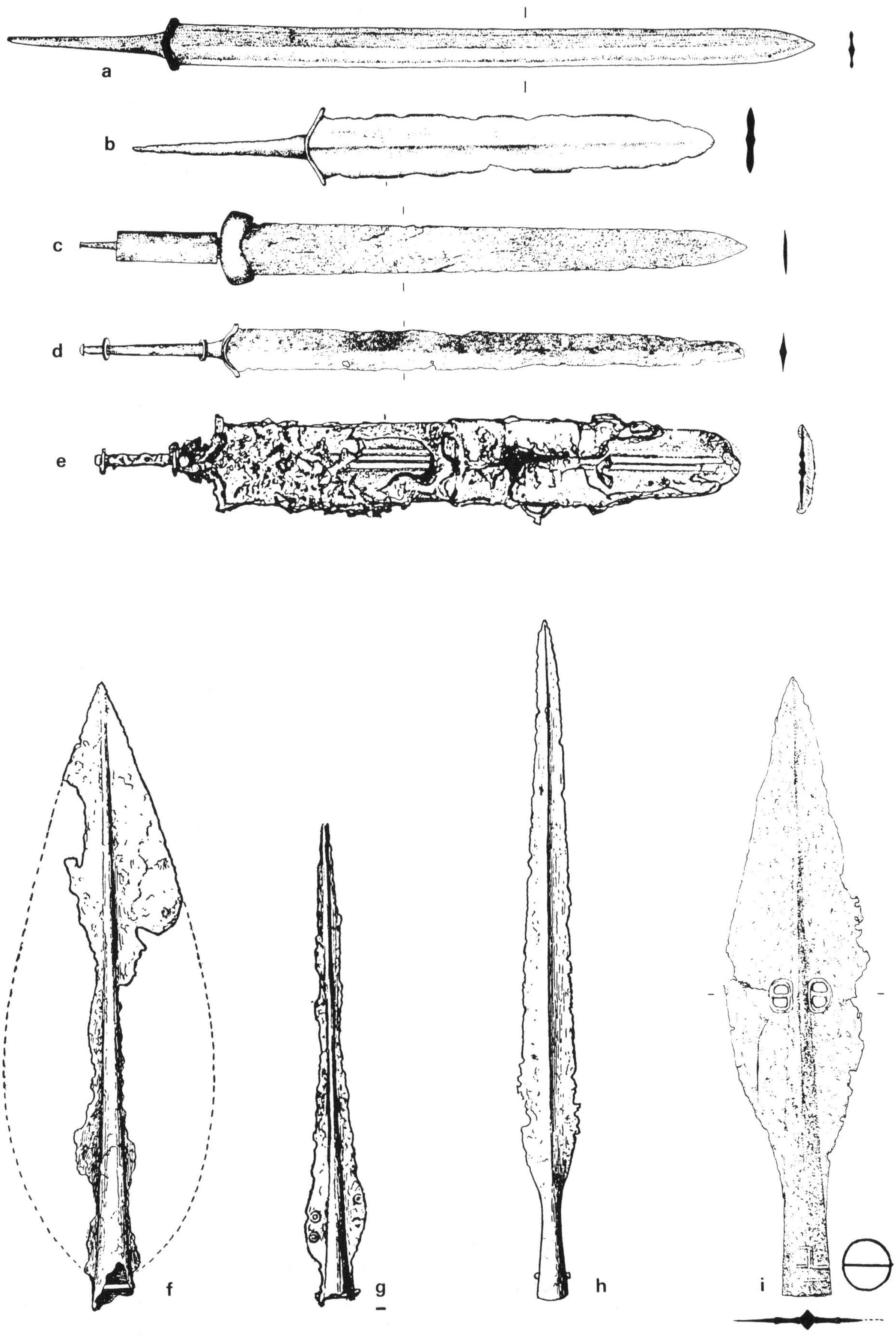

Fig. 4.1.1 Swords and spearheads of EIA B: (a) Ballinderry, Co. Westmeath, (b) Edenderry, Co. Offaly, (c) 'Ireland' (no locality), (d) Cashel, Co. Sligo, (e) Dangan Lower, Co. Galway (f) 'River Shannon', (g) Moyfin, Co. Meath, (h) Mullaghmore, Co. Roscommon, (i) Corrofin, Co. Clare (a-d x¼ and i x½ after Raftery, B. 1984, e x¼ after Rynne 1985, f-g x⅓, h x¼ after Coffey 1910.)

Raftery, B. 1984, Plates 84–88) was executed by cutting back the substrate to leave it standing out in low relief. While we have, as yet, no metallurgical confirmation of this interpretation, it seems safe to suggest that this could only really have been carried out using steel cutting edges (cf. Lowery et al. 1971, 170). The incised decoration on an unlocalised Monasterevin-style disc (*CIIAA* 792; Raftery, B. 1984, Plate 95), the finely incised decoration of the Irish sword scabbards (Raftery, B. 1984, Plates 26–31), the carving of the curvilinear decoration both on rotary quernstones (e.g. *CIIAA* figs. 183–184), and the fine granite monolith at Turoe, Co. Galway (Duignan 1976) all stand as candidates for the use of iron tools. Intriguingly, a small chisel was recovered from the Rath of Feerwore, a site occupied during EIA B, and by which the Turoe stone may have been sited originally. The metallographic examination of this piece (sect. 4.2) however shows that the tool was poorly made, and really would not have been especially suited for such a task as the cutting back of the void areas of the decorative scheme. An iron file (now lost) came from an EIA B burial at Carbury Hill, Co. Kildare[3] (*CIIAA* 589), and traces of file marks[4] may be observed on various copper-alloy pieces, including an unprovenanced Irish Type E horsebit (*CIIAA* 116), also on the iron side links of an unprovenanced type B horsebit (*CIIAA* 29).

Hot and cold setts and chisels would have been needed for the cutting of metal, although the chisel from the Rath of Feerwore (Plate 4.2.18) would not have been up to this task. Drawing down of metal would have required a range of fullers, while the production of triple-ribbing on weapons such as the swords from Lisnacrogher and Ballinderry would have required both top and bottom fullers. Some form of swage would have been required to form the fine midribs of the spearheads described, with the sockets being closed round a mandrel. The perforations for rivets would have been made with fine punches, and punches were used also to create the larger perforations in the blades of the spearheads from Corrofin, Co. Clare and from Roodstown, Co. Louth (*CIIAA* 285). The small spots of red enamel on the blade of the Moyfin, Co. Meath spearhead (Fig. 4.1.1g) are surrounded by raised ridges, and their form and uniformity of size suggest strongly the use of a specially constructed punch or die[5]. The rather aimless rocker tracing on one of the Lisnacrogher spearheads (BM 80.8–2.117: Plate 4.2.9) must have been done with an iron edged tool.

Punches and drifts may have been required to make the perforations in the side-rings of copper-alloy horsebits to take stop-studs (Fig. 4.1.2), the more normal practice being to cast them as an integral part of the ring. The bit from Clongill, Co. Meath illustrated in Fig. 4.1.2 is of special interest because the stop studs on one side are inserted into holes, while on the other side they are cast on! We may again infer the use of tools made of iron rather than copper alloy for this task. The question also then arises as to the reason for this, since casting of stop-studs as a part of the ring seems a much easier way to achieve the desired result. It might be suggested that it arose out of the manufacture of rings with an iron core covered by a copper-alloy flashing (claimed for a bit from Killucan, Co. Westmeath: Howarth 1971, Cat. no. D5; *CIIAA* 82). If this is the case, then it must also be noted that in Ireland, the majority of separately-added stop studs are found on bits which appear to have solid cast copper-alloy rings. The making of iron cores for side-rings is, however, quite common amongst the contemporary British bits, whose stop-studs are in the main separate additions (e.g. at Llyn Cerrig Bach nos. 47–49; Fox 1946, 80, and Plates XXI–XXII). It is presumably to Britain that we should look for the origin of the Irish practice.

A second feature of the bits is the V-moulding found frequently on the side-links (Fig. 4.1.3). Warner (1983, 174f) described this as a 'duck head' in its later form, a motif central to his EIA 2 art style (Fig. 4.1.4). Its origins lie certainly in the shape produced by the forging of an 'eye' in iron: the same effect is produced by the weld-seam where the bar has been welded back on itself. Given the vision of metalworkers of the period in taking and remoulding shapes, it is unwise to be dogmatic. Nevertheless, the combination of stop-studs and the V-moulding seems strongly indicative of stylistic interaction between workers in copper-alloy and iron. The argument will be clinched when, as seems likely sooner rather than later, the find of one or more bits of this form in iron is announced.

As we saw in Chapter 2, a prominent feature of the transition to iron in barbarian Europe was the interaction between workers which resulted in the making of artefacts combining the established metal with the new. This grew out of the phase in which experimentation centered on making iron artefacts, and in turn gave way to the making of iron artefacts with copper-alloy embellishments. In Ireland we have seen the evidence of Later Bronze Age experimentation with iron. Given uninterrupted technological progression, we might have expected to find developed a similar distinct horizon of bi-metallic artefacts here. In EIA A and B, what bimetallism there is, is almost entirely functional. Only with the Ballymoney mirror (*CIIAA* 537) and the Ballyshannon anthropoid sword-hilt (*CIIAA* 244), the latter an import, do we see decorative bimetallism.

There are a number of Irish horsebits where repairs have been effected in antiquity, using iron components. A particularly interesting use of iron here is the repair to the pivot of the mouthpiece of a bit localised to Co. Antrim (*CIIAA* 100) where the broken copper-alloy pivot has been replaced by an iron rod held in place by the casting on of copper-alloy to infill the lost metal of the centre-link. An iron rod has also been used in the side-link of a Type D bit from Streamstown, Co. Westmeath (*CIIAA* 84). While these combinations are secondary to the original design, they nevertheless demonstrate technological interaction between workers in the different metals. We may also note in this context the copper alloy La Tène style fibula from Dundrum, Co. Down (Jope 1962, 30; *CIIAA* 367) which has a spring coiled around a thin iron rod (cf. the use of an iron core in the spring of a La Tène 1b fibula from Ravensburgh Castle, Herts., dated to the mid 4th century BC – Dyer 1976, 157 and 423, Fig. 3). The scarcity of functional bimetallism in artefact design seems indicative of the effects of the events at the end of the Later Bronze Age whose principal legacy was the re-ordering of industrial activity on a scattered and

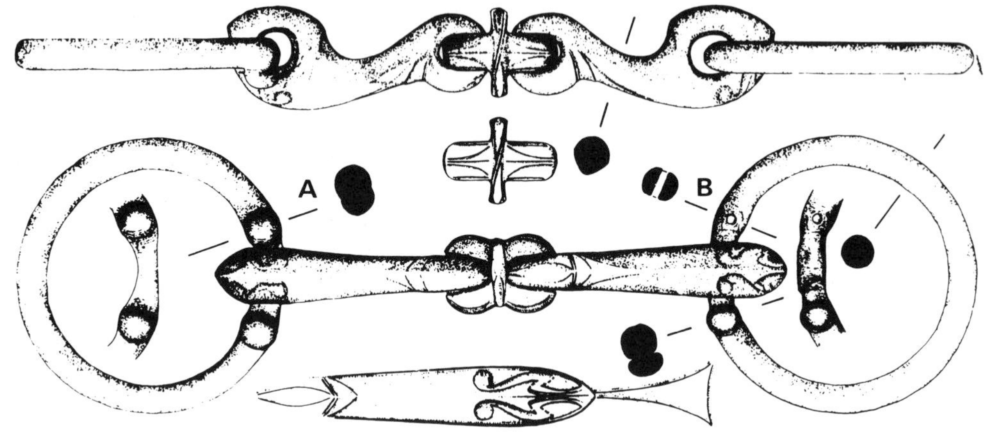

Fig. 4.1.2 Horsebit from Clongill, Co. Meath (*CIIAA* 78) showing cast-on (A) and pegged (B) stop studs.

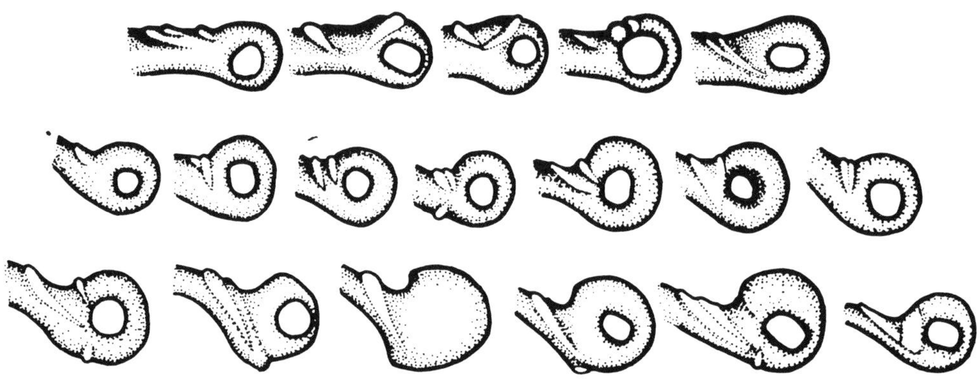

Fig. 4.1.3 V-mouldings on horsebit side links (after Raftery, B. 1984)

small scale where the opportunities for practical exchange and interplay of ideas between professional metalworkers were exceedingly limited. Thus, we see only vestiges of the initial transition experimentation and of interaction: the final stage in which copper-alloy adorns iron is really only represented by the Ballymoney mirror. The loose placing of non-ferrous hilt furniture on sword-blades is obviously not bimetallism. While the evidence is almost entirely negative, it nonetheless helps to clarify our understanding of the development of the earliest ironworking here.

Allowing for the ravages of corrosion and the subsequent bias to which this has surely given rise, it is still instructive to note that of all of the artefacts of all materials admitted by Raftery into his *Corpus* as belonging to the indigenous Irish Iron Age which he dates to the period ca 300 BC–300 AD, only a very small proportion is of iron. Of particular interest is the virtual absence of iron in personal ornament. Of the 26 fibulae listed there, only one (from Feerwore, Co. Galway; *CIIAA* 369) is made of iron (the Dundrum example cited above includes an iron component): since *CIIAA* was published one other iron example has been recovered from excavations at Kiltierney, Co. Fermanagh (Hamlin and Lynn 1988, 26, Fig. 33). Excluding the EIA A iron swan's neck pins, then of the 40 metal pins listed by B. Raftery, only one (uncertain) example from Carbury Hill, Co. Kildare (*CIIAA* 440) is of iron. Of the 139 horse-bits and parts thereof, only one (unprovenanced) bit (*CIIAA* 134), and the rings of Type B bits, one from

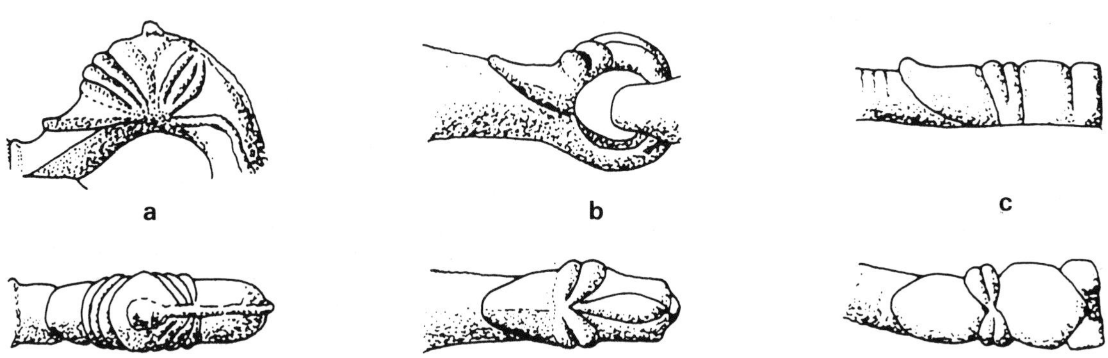

Fig. 4.1.4 The 'duck head' in EIA B art: (a) mirror-handle, (b) horsebit, (c) early zoomorphic penannular brooch (after Warner 1983).

Ballynahinch, Co. Down (*CIIAA* 16), and one unprovenanced (*CIIAA* 29), are of iron: of the 94 Y-shaped pieces, only one is of iron. If we exclude such groups as the Lough Crew bone slips and figure representations, then of some 478 artefacts listed in *CIIAA* which could (in theory at least) be candidates for manufacture in iron, only 47 were, in fact, made of iron – slightly less than 10%. Assemblages from the pre-Roman Iron Age of southern Britain show a much greater and varied use of iron, as, obviously, do those from elsewhere in barbarian Europe (to take but one example, we might consider the assemblage of metalwork excavated from Manching in Bavaria – Jacobi 1974). At first sight, there is stark contrast between the finely cast and wrought copper-alloy and gold products of the period and the few iron pieces other than the swords and spearheads. Nevertheless, the survivals which have come down to us show that skills in ironworking had survived.

It is an immediately attractive proposition that the artefacts most likely to bring together the best skills of early smiths will be weapons. As noted earlier, much technological progress in the exploitation of materials is seen in terms of the improvement of the cutting edge, and early texts, with their descriptions of razor sharp weapons capable of major amputations, provide an additional prompt to a search for high quality blades. The logical starting point of metallographic study is thus the swords and spearheads of native origin (also, happily, by far the largest body of surviving ironwork from EIA B). We begin with the weapons from Lisnacrogher, Co. Antrim, whose ornamented components provide some dating pedigree for the ironwork. The swords from Lisnacrogher would seem to be comparable with continental Middle La Tène period weapons, although as earlier remarked, they stand with the rest of the Irish swords in being significantly smaller than their counterparts elsewhere. The reason for this is not clear, although we can dismiss summarily any suggestion that it was the result of a shortage of metal (cf. Warner 1974), if for no other reason than the spearheads from the same site (standing close comparison with Middle La Tène period spearheads outside Ireland) would pass for size without remark if found among a continental Iron Age assemblage. With their splendidly decorated scabbards, the swords invite interpretation as parade weapons, the real fighting being done with the spear[6]. As Mallory (1982, 108, Fig. 2: Fig. 5.2.2) illustrates graphically, Irish EIA B swords are hardly the weapons used by heroic swordsmen: they were capable of damage as any sharply pointed implement, but not of great use in other than a pub brawl. *CIIAA* lists 71 spearbutts of various types (*CIIAA* 292–363), each of which presumably may be taken as representing at least one spearhead. In addition, it would seem more than likely that the admission into the EIA B assemblage of a number of spearheads passed over by B. Raftery (cf. note 2) gives a more reasonable estimate of the availability of iron than one based on his narrower criteria.

The typological and chronological orderings of Irish EIA B artefacts present many problems, chief among which is the virtual absence of associations useful for dating purposes. Thus, while the starting dates for La Tène style swords is provided by stylistic dating of scabbard decoration and chape style, there is no real way at present of establishing the demise of this style of blade. Both Rynne (1982, 1983) and B. Raftery (1984, 62ff) offer typologies based primarily on hilt furniture rather than blade form. It is only at Dún Ailinne, Co. Kildare that a sword of any recognisable EIA type has been found during a modern scientific excavation (Wailes 1976, 327), and the excavator offers (uncalibrated) C-14 dates which centre on the 1st century AD (Wailes 1976, 338; Raftery, B. 1984, 70). The C-14 dates illustrate the problems of dating, since the form of the Dún Ailinne weapon would put it into Rynne's (1983) putative early grouping.

Identification of the precise positioning of those EIA B swords which lack hilt furniture or scabbards is also problematic. Thus, B. Raftery (following Rynne 1983) would include swords from Edenderry, Co. Offaly and Cashel, Co. Sligo into his Type 1 on the basis of their surviving bell-shaped quillons, but excludes entirely from the EIA B assemblage those from Kildrinagh, Co. Laois whose iron blades are identical in form and (reconstructed) size, but have lost any fittings they might formerly have possessed. The two Kildrinagh swords are included here as of EIA B on the basis of their form. The first is a slender, tapering weapon with sloping shoulders (Plate 4.2.4a) very similar in form and dimensions to the Dún Ailinne, Co. Kildare blade (seemingly the only example of an iron quillon) and to another two swords from Edenderry, Co. Offaly (*CIIAA* 249, 250). The second (Plate 4.2.4b), represented only by the tang and upper section of the blade conforms to the general characteristics (followed by B. Raftery) attributed by Rynne (1982) to his 'La Tène' group.

Both Rynne and B. Raftery take the non-ferrous accessories as typologically diagnostic, not the blades themselves. There are a number of swords and sword fragments which have organic hilt components skeuomorphic of metal fittings. These are assigned by Rynne (1982, 1983) to the later part of EIA B as part of his 'Ultimate La Tène' group. The question of subdivision on the basis of metal fittings as opposed to organic ones (cf. Rynne 1983), is difficult, since it is quite clear that it is the medium and the design of the hilt furniture, not really the shape of the blade, which alters significantly. The only definite difference which can be identified between the two groups is the lack of organic hilt furniture on a ribbed blade (such as those from Lisnacrogher and Edenderry), but the significance of this is unclear. While appreciating the caution underlying the approaches of Rynne and B. Raftery, it is apparent that all of the blades have such similar form and dimensions as to allow those lacking hilt furniture (metal or organic) to be typed with those on which it still survives. Although agreeing with both Rynne and Raftery on the broad characteristics of EIA B swords, the view is taken here that the blade form based on La Tène II design persisted throughout EIA B (and perhaps beyond), while it is another group of weapons which show the potential influence of late- and post-Roman Britain on weaponry and fighting styles[7].

A quite improbable type is that represented by two finds, one from Toome Bar, Co. Antrim, the other from the excavations at Altanagh, Co. Tyrone (Williams 1986). In each case, only the upper part of an obviously longer 'sword' has survived, and in the case of the Altanagh specimen, no coherent metal core survives. The principal features (Plates 4.2.7 and 4.2.8) are a thick, round-sectioned tang and a thick blade of rhomboidal section. The Altanagh specimen came from the second phase of occupation of the site, beside a pit for which the excavator offers a calibrated C-14 date of 250–610 AD (UB-2564 – 1585 ± 70 BP: Williams 1986, 58). This would place the type in later EIA B/early LIA A. It is possible that these are early versions of expanded-ended swords (Chapter 5.1). In fact, unless the blades originally expanded out, it is difficult to see how such a weapon could have been particularly effective, or even if the type in fact was a weapon at all. Thus, while including these two specimens among the assemblage of EIA B weaponry, one must enter the *caveat* that both identification and attribution can only be tentative.

If the La Tène swords at least, can be ordered, in part, on the basis of embellishment, the spearheads almost entirely lack even this basic aid. Thus, although B. Raftery (1984, 108) admits the possibility that more examples than he allows into the EIA B assemblage might be identified, he is unwilling to make comparison with foreign assemblages, even those from amongst whose makers the bearers of La Tène artefact design to Ireland might have come. Again, while recognising the difficulties inherent in typological comparisons between iron artefacts, the view is here taken that several more weapons can be broadly dated to EIA B with at least as much confidence as one has over the undecorated Lisnacrogher spearhead (*CIIAA* 282) and the perforated specimen from Roodstown, Co. Louth (*CIIAA* 283). Thus, the corrosion-damaged weapon from the River Shannon (Plate 4.2.10a) has good parallels at La Tène itself (Vouga 1923, e.g. Plate XIII.1), and in La Tène I Champagne (e.g. Bretz-Mahler 1971, Planche 92.8). We do know that the small spearhead from Dún Ailinne, Co. Kilkenny (Plate 4.2.11a) was recovered from a context which the excavator would date to EIA B (Wailes pers. comm.). It goes without saying that detailed study and typological assessment of iron spearheads from Ireland is overdue!

More mundane, but no less interesting ironwork is represented by a small group of socketed and shaft-hole axes, three of which come from a hoard at Kilbeg, Co. Westmeath (*CIIAA* 584a–c: Plates 4.2.14–16), associated with two Y-shaped pieces, one of copper-alloy the other of iron (*CIIAA* nos. 170 and 232). The square-mouthed socketed axe and chisel from Feerwore, Co. Galway (Plates 4.2.13 and 4.2.18), come from the excavation of a site which also produced one of only two iron fibula known from Ireland (Raftery, J, 1944; *CIIAA* 369). The axe itself is of a form well known from La Tène contexts across barbarian Europe[8]. The three axes of the Kilbeg hoard, along with a stray find from Castlereban, Co. Kildare (Rynne 1958; *CIIAA* 582: Plate 4.2.17), are of particular interest on account of their method of construction in which the shaft is formed separately from the blade and welded to it: this recalls the earlier use of such construction methods in the axe from Rathtinaun (Chapter 3.3: cf. Fig. 3.3.2). What can only be described as two iron 'flat axes', one from among old finds at Lagore, Co. Westmeath, the other from '. . . a crannog in Co. Antrim' (Wakeman 1898, 241), may be interpreted as blades intended for similar axes.

As with the weapons, all of the other iron specimens attributed to EIA B are dated on the basis of association. The shaft-hole axes from Kilbeg, Co. Westmeath and Castlereban, Co. Kildare are given a broad date-range centred on 0 AD from the association at Kilbeg with two Y-shaped pieces (one of iron) in the Kilbeg hoard, being unique in their design (although comparable in concept to the Rathtinaun axe discussed in Chapter 3.2). The Feerwore chisel (Plate 4.2.18), the iron shears from Carbury Hill, Co. Kildare (Plate 4.2.19), and the Lisnacrogher, Co. Antrim sickle (Plate 4.2.20) would, on shape, fit equally well into assemblages of a millennium or more later. This last group consists of artefacts which have long currency in Ireland, and it may well be that at least some stray finds of these types have simply been wrongly dated, but with their being no real hope of future correction.

* * *

4.2 – Metallographic examinations of EIA B iron artefacts

(6) UM B2 sword, Lisnacrogher, Co. Antrim (*CIIAA* 238: Plate 4.2.1)

X-radiography confirmed the survival of metal beneath the corrosion products, and sections were taken (Plate 4.2.1a) from the cutting edge (*section 1*) and from the midrib area (*section 2*).

Section 1: *Non-metallic inclusions*: One- and two-phase inclusions are numerous.

Macrostructure: Etching with Nital showed slight variation in C distribution (Plate 4.2.1b), while Oberhoffer revealed some minor segregations.

Microstructure: The structure is primarily fine grained ferrite (GS 6–8) with interstitial pearlite, but with areas of larger (GS 3–5) ferrite grains (Plate 4.2.1c). There is a tendency for the interstitial pearlite to be more common towards the centre of the section, but examination of the corrosion capsule revealed remanent structures similar to those of the surviving metal, indicating that any differences in carbon distribution may be attributed to inhomogeneity of the stock.

Hardness: mH_v ferrite 269.

Section 2: *Non-metallic inclusions*: One-, two- and three-phase inclusions are numerous.

Macrostructure: Nital indicated low carbon metal, while etching with Oberhoffer indicated no significant segregations (Plate 4.2.1d).

Microstructure: The structure is almost entirely ferritic, of variable grain size (GS 2–7), with interstitial pearlite (Plate 4.2.1e) and some grain boundary cementite, with patches of excessive ferrite grain growth (GS >1). Some acicular precipitates can be observed in the larger ferrite grains.

Hardness: mH_v ferrite 233.

Interpretation: The structures of the two sections graphically illustrate the degree of inhomogeneity

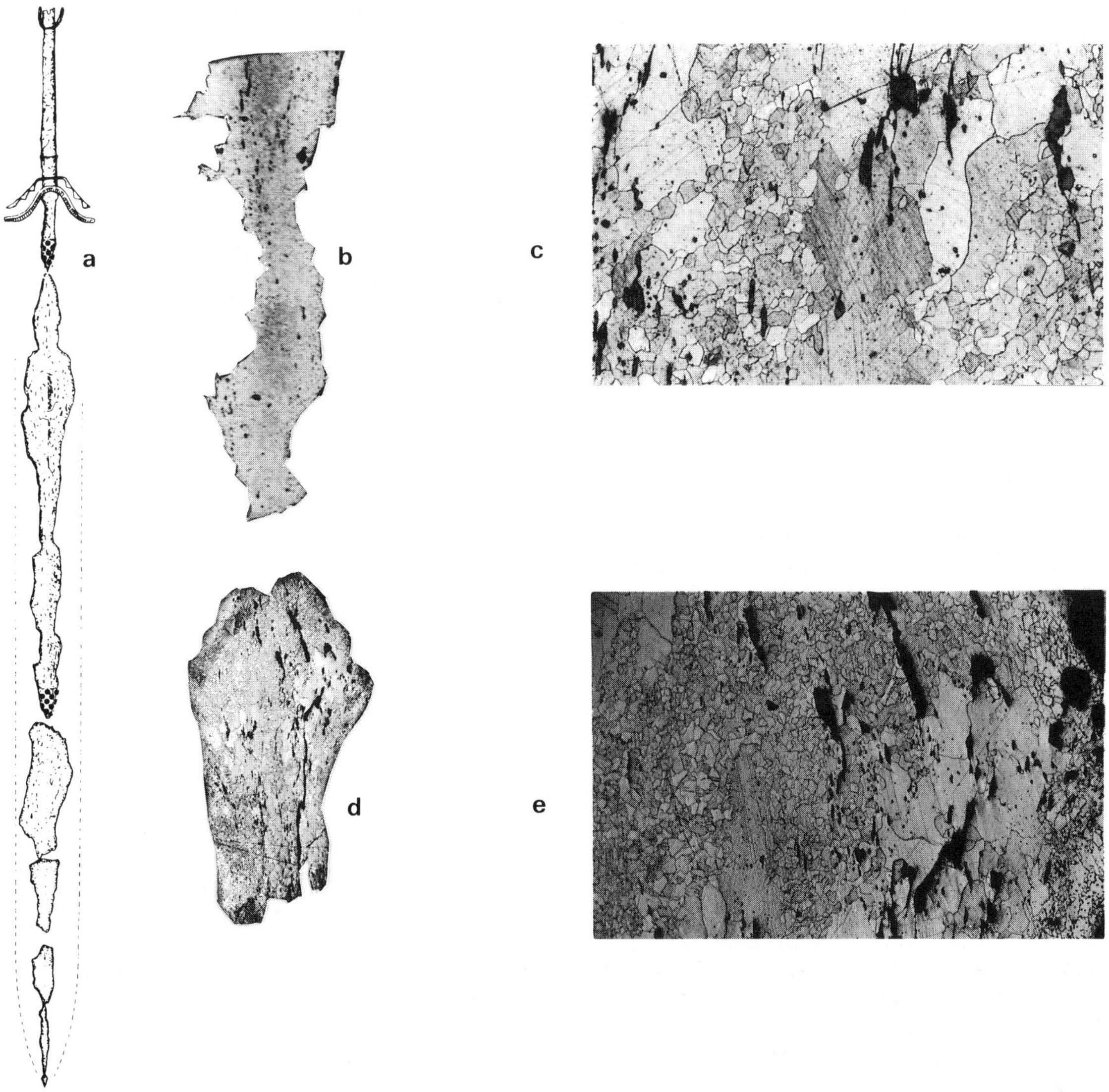

Plate 4.2.1 Examination 6: (a) UM B4 iron sword from Lisnacrogher, Co. Antrim x¼, (b) macrostructure of section 1, (c) ferrite and slag, x100, (d) macrostructure of section 2, (e) ferrite and slag. Etched 4% Nital.

which can be encountered in early bloomery iron over very short distances. The blade was forged from a piece of inhomogeneous bloomery iron in which much slag had been left during consolidation of the bloom. Final forging was followed by air cooling and no attempt at heat-treatment was made. Whatever its original appearance, this would not have been a particularly serviceable weapon.

(7) UM B5 sword, Lisnacrogher, Co. Antrim (*CIIAA* 241: Plate 4.2.2)

Three large fragments of corroded metal (Plate 4.2.2a) were recovered from the inside of the scabbard in which they had been trapped since it was found. Despite their fragmentary nature, a substantial core of metal survived in one piece, giving a virtually complete cross-section of the original blade.
Non-metallic inclusions: Only a few single-phase inclusions occur, although several are quite large.
Macrostructure: Etching with Nital revealed a variable distribution of carbon, with pronounced banding. A large weld seam, along which corrosion has penetrated deeply, is visible in the right-hand area of the section (Plate 4.2.2b). Etching with Oberhoffer indicates no other significant segregations.
Microstructure: The carbon content is highly variable and the banding seen in the macrosection is seen more clearly. Ferrite grain size is small throughout (GS 7–8). In the central area, there is a band of high carbon metal, with pronounced Widmanstätten ferrite and coarse pearlite (GS 4–5: Plate 4.2.2c), some of which shows spheroidisation of the carbide (Plate 4.2.2d). In addition to the corrosion-penetrated seam seen in macrosection, at least three other weld seams are visible, possibly bloom welds.
Hardness: mH_v ferrite 315, pearlite 448: H_v 182–234.
Interpretation: The piece was forged from an unevenly carburised piece of metal. Banding was accentuated by fold-welding, in conjunction with welds left over from the consolidation of the original stock. This would have been a better weapon than the other two from this site.

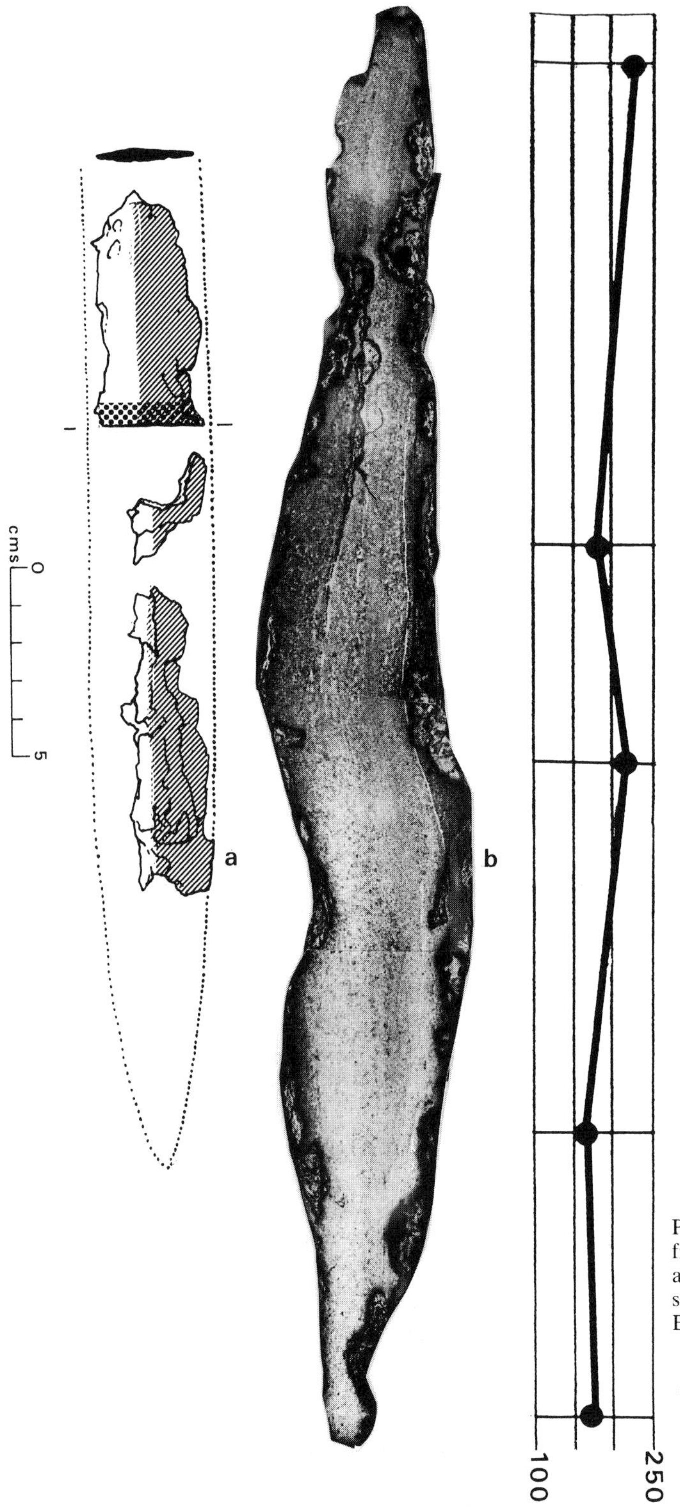

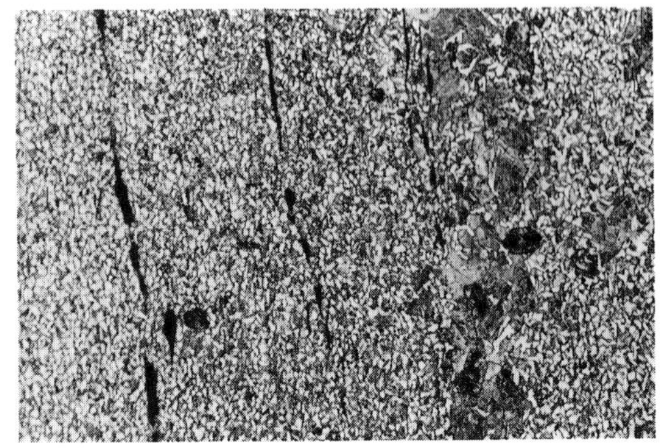

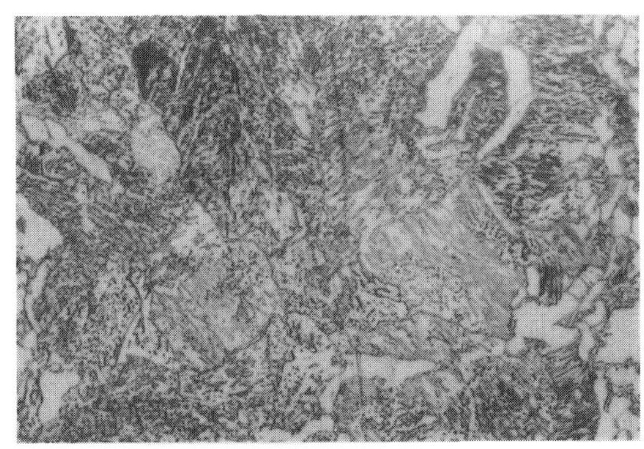

Plate 4.2.2 Examination 7: (a) UM B5 fragments of an iron sword from Lisnacrogher, Co. Antrim, (b) macrostructure, (c) fine ferrite and interstitial pearlite with a high-C band x100, (d) pearlite showing spheroidisation of carbide lamellae x500, (e) H_v profile. Etched 4% Nital.

(8) BM 80.8–2.116 sword, Lisnacrogher, Co. Antrim (*CIIAA* 239: Plate 4.2.3)

A section was taken from the cutting edge of the blade through to the midrib[9] (Plate 4.2.3a).
Non-metallic inclusions: A number of elongated inclusions run in chains parallel to the long axis of the blade. SEM examination showed some inclusions with significant quantities of Mn in conjunction with Fe and Si, while others contained significant quantities of Al (6–7%) and Ti. One, two- and three-phase structures were observed.
Macrostructure: Etching with Nital and Oberhoffer revealed no significant segregations (Plate 4.2.3b).
Microstructure: The structure is mainly ferrite of uniform grain size (GS 6–7), with some interstitial pearlite (Plate 4.2.3c), and traces of grain boundary cementite. In some areas the ferrite shows some elongation, while the pearlite shows distinct spheroidisation.
Hardness: mH_v ferrite 100–183: H_v 129–140.
Interpretation: This finely shaped piece was made from bloomery iron with virtually no carbon content. Forging seems to have been prolonged at around A_1 resulting in spheroidisation of the pearlite, followed by cooling at a normal rate, possibly also with some forging on falling heat. There was no attempt at hardening.

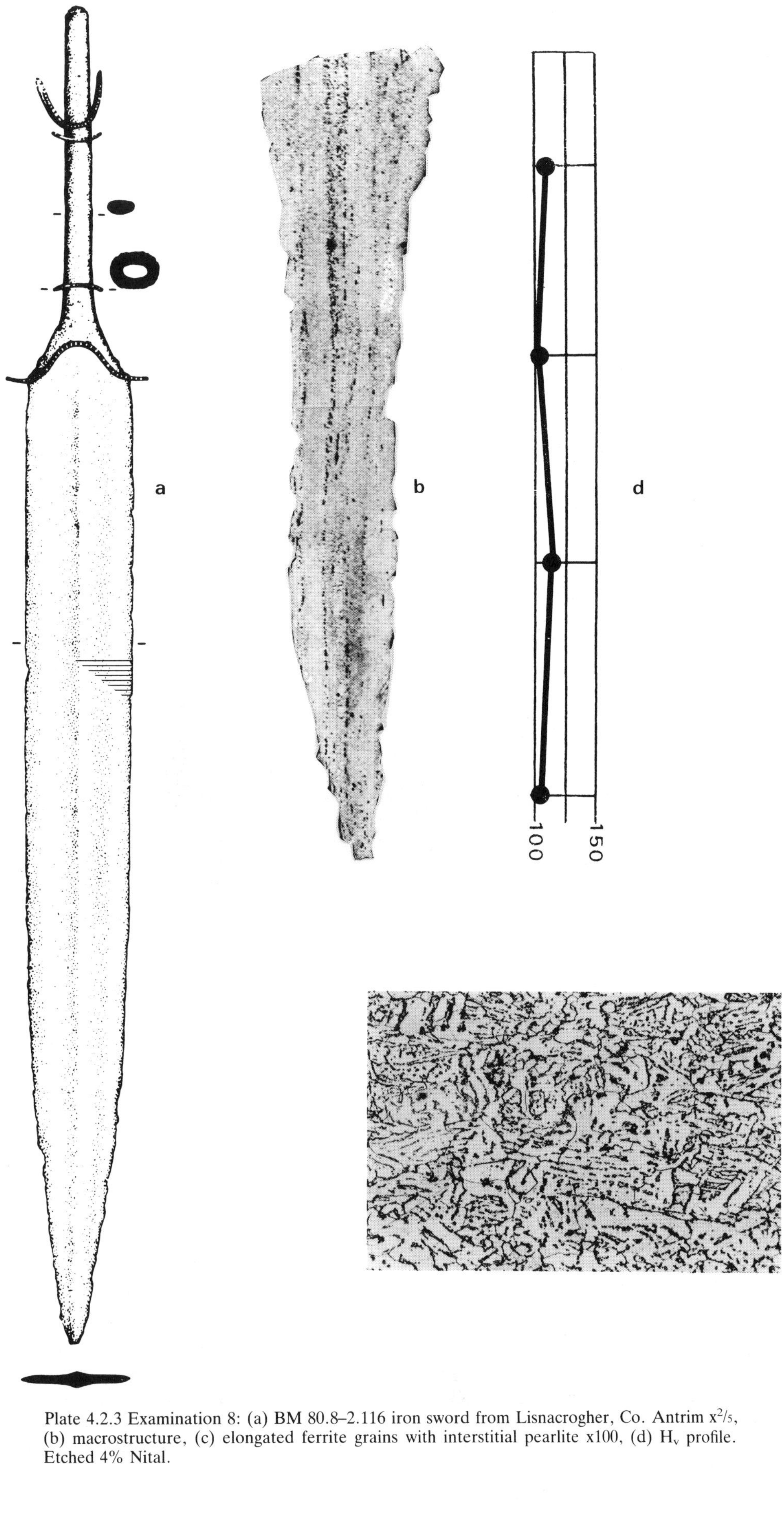

Plate 4.2.3 Examination 8: (a) BM 80.8–2.116 iron sword from Lisnacrogher, Co. Antrim $x^{2}/_{5}$, (b) macrostructure, (c) elongated ferrite grains with interstitial pearlite x100, (d) H_v profile. Etched 4% Nital.

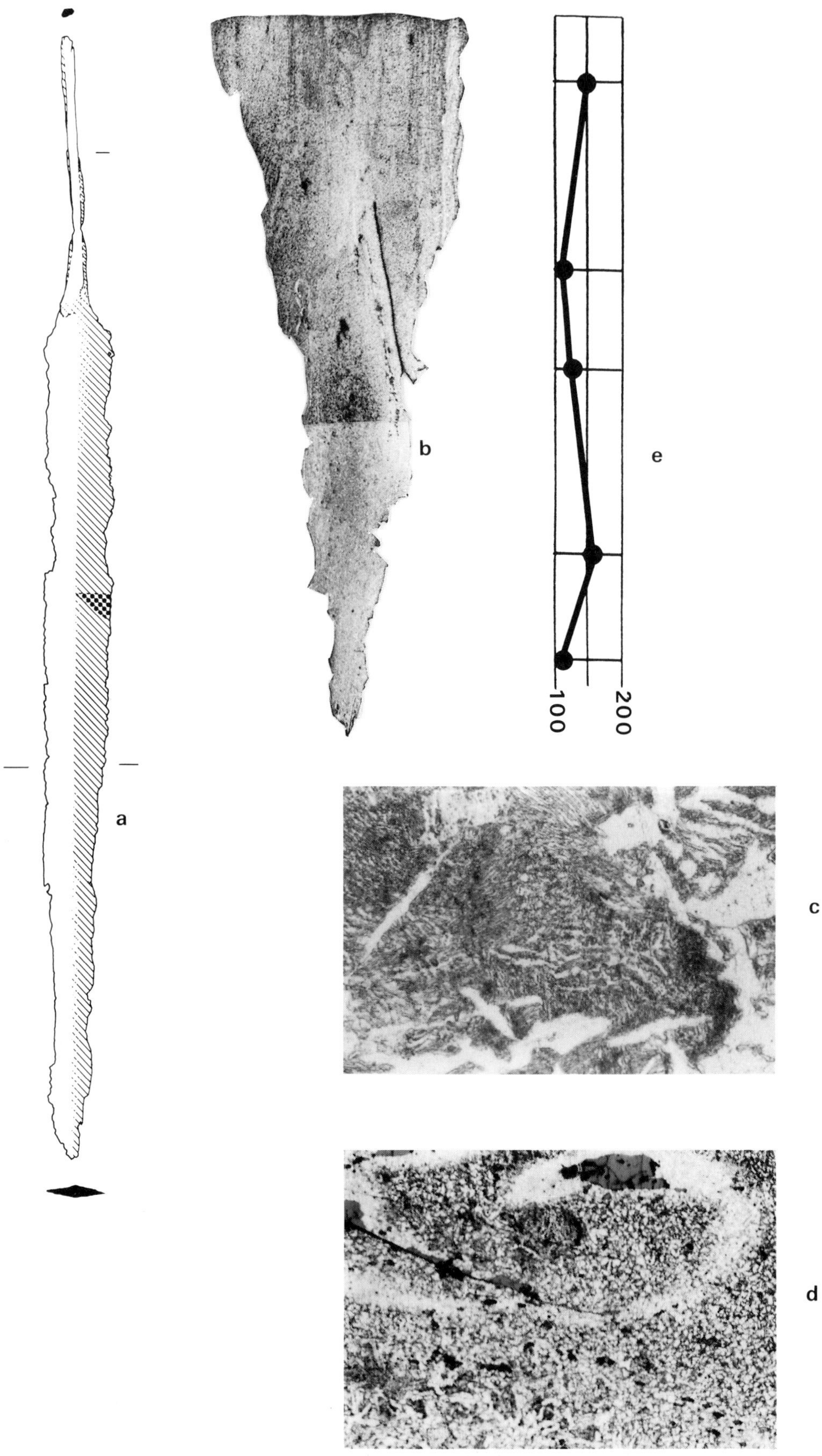

Plate 4.2.4 Examination 9: (a) NMI WK 10:W135:F468 sword from Kildrinagh, Co. Laois x2/5, (b) macrostructure, (c) pearlite showing spheroidisation of carbides x500, (d) remains of bloom welds x50, (e) H_v profile. (b) and (d) etched 4% Nital, (c) etched Picral.

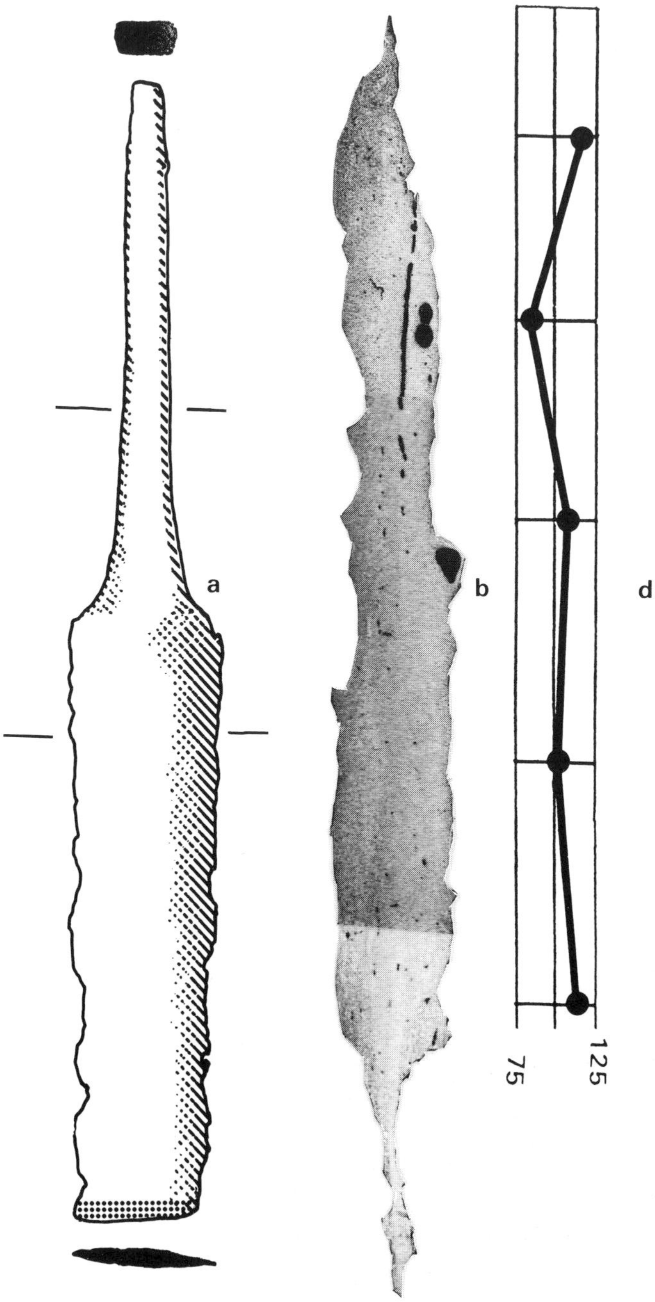

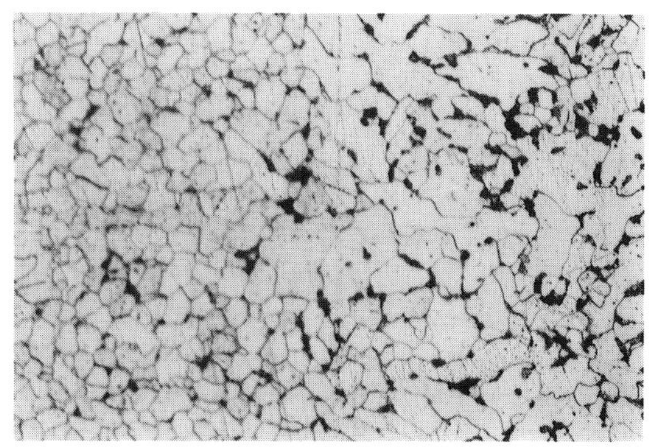

Plate 4.2.5 Examination 10: (a) NMI WK24:W137:P470 sword from Kildrinagh, Co. Laois x⅕, (b) macrostructure, (c) ferrite with interstitial pearlite x100, (d) H_v profile. Etched 4% Nital.

(9) NMI WK10:W135:F468 sword Kildrinagh, Co. Laois (Plate 4.2.4)

A V-shaped section was taken from the cutting edge to the midrib (Plate 4.2.4a).
Non-metallic inclusions: Numerous one- and two-phase inclusions run parallel to the long axis of the specimen.
Macrostructure: Etching with Nital revealed a highly inhomogenous C distribution, and traces of bloom welds, although Oberhoffer showed no other significant segregations (Plate 4.2.4b).
Microstructure: Towards the midrib area, ferrite (GS 7–<8) and ferrite with coarse pearlite (GS 6–7) predominate. Towards the cutting edge, the structure is pearlite and ferrite with significant spheroidisation of carbides (Plate 4.2.4c). Several weld seams, including likely bloom welds, marked by lines of small inclusions are visible in the body of the section (Plate 4.2.4d). There is a significant increase in carbon content towards the cutting edge, indicating that it underwent secondary carburisation.
Hardness: mH_v ferrite 183, pearlite 263–369: H_v 115–162.
Interpretation: The distribution of carbon indicates that the sword was forged from fairly heterogeneous stock, and that the edge underwent secondary carburisation. The degree of spheroidisation of carbides throughout results from the metal being held around A_1 for some time. It is not impossible that this degree of spheroidisation results from overzealous tempering of martensite resulting from quenching, since spheroidisation of carbides proceeds much more rapidly by this route than from pearlite. Given that the edge was carburised, this explanation may not be ruled out lightly. The weld seams suggest that, in addition to survivals from consolidation of the original bloom, the stock had been fold-welded. The structure of the slag inclusions is consistent throughout the section, noticeably so on either side of the weld seams. The final heating led to grain refinement. This would have been a quite serviceable weapon.

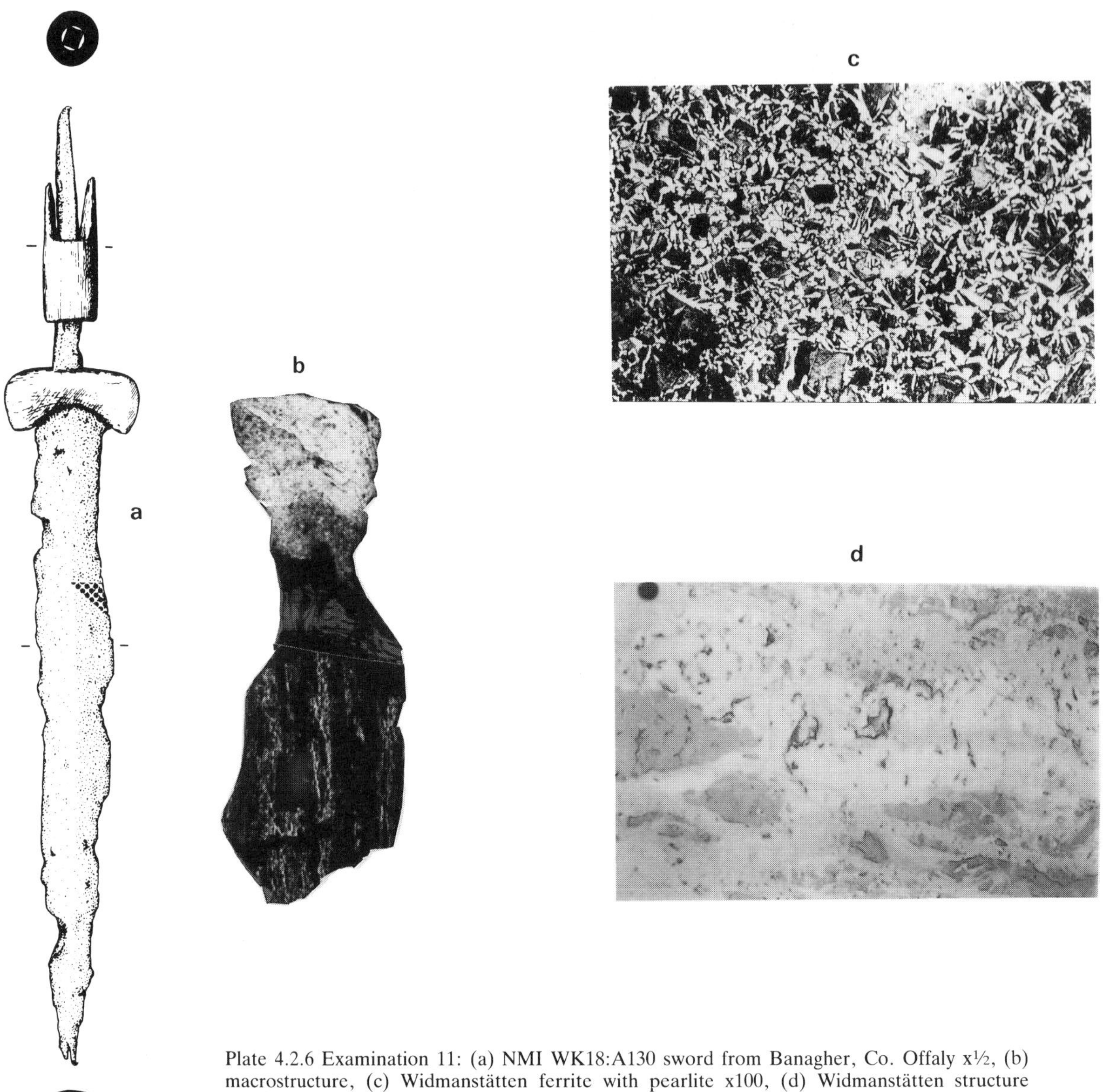

Plate 4.2.6 Examination 11: (a) NMI WK18:A130 sword from Banagher, Co. Offaly x½, (b) macrostructure, (c) Widmanstätten ferrite with pearlite x100, (d) Widmanstätten structure fossilised in corrosion products x100. Etched 4% Nital.

(10) NMI WK24:W137:P470 sword Kildrinagh, Co. Laois (Plate 4.2.5)

A section was taken across the broken end of the surviving blade (Plate 4.2.5a).
Non-metallic inclusions: These are relatively few and are mainly one-phase.
Macrostructure: Etching with Nital and Oberhoffer shows no real heterogeneity (Plate 4.2.5b).
Microstructure: The structure is almost entirely ferritic, with some interstitial pearlite (Plate 4.2.5c), and of fairly regular grain size (GS 4–6), but with a patch of fine ferrite (GS 8) at one edge. The pearlite lamellae show some spheroidisation of carbide.
Hardness: mH_v ferrite 135–178: H_v 99–113.
Interpretation: The blade was forged from a piece of virtually carbon-free stock and could not have been a particularly effective weapon.

(11) NMI WK18:A130 sword, Banagher, Co. Offaly (*CIIAA* 254: Plate 4.2.6)

The blade has been very badly damaged by corrosion, so much so that metal only survives in the thickest part of the section, at the midrib, with all of the cutting edge region having been completely converted to corrosion products. A section containing metal was obtained for examination along with contiguous corrosion (Plate 4.2.6a).
Non-metallic inclusions: These are scarce, but where they occur they are large, single-phase and light grey in colour.
Macrostructure: Etching with Nital showed a very uneven distribution of carbon, but Oberhoffer revealed no other significant segregations (Plate 4.2.6b).
Microstructure: Carbon is very unevenly distributed, with structures ranging from ferrite (GS 7–<8), through ferrite with interstitial pearlite (GS 3–4) to the eutectoid. There is a pronounced Widmanstätten structure in places (Plate 4.2.6c), with coarse pearlite. Several 'ghost' lines occur. Towards the cutting edge, the massive corrosion retains metal islands as well as remanents of similar structures to those observed in the main body of surviving metal (Plate 4.2.6d), suggesting that originally the metal was completely heterogeneous throughout the section.
Hardness: mH_v ferrite 255, pearlite 404.
Interpretation: The blade was made from highly

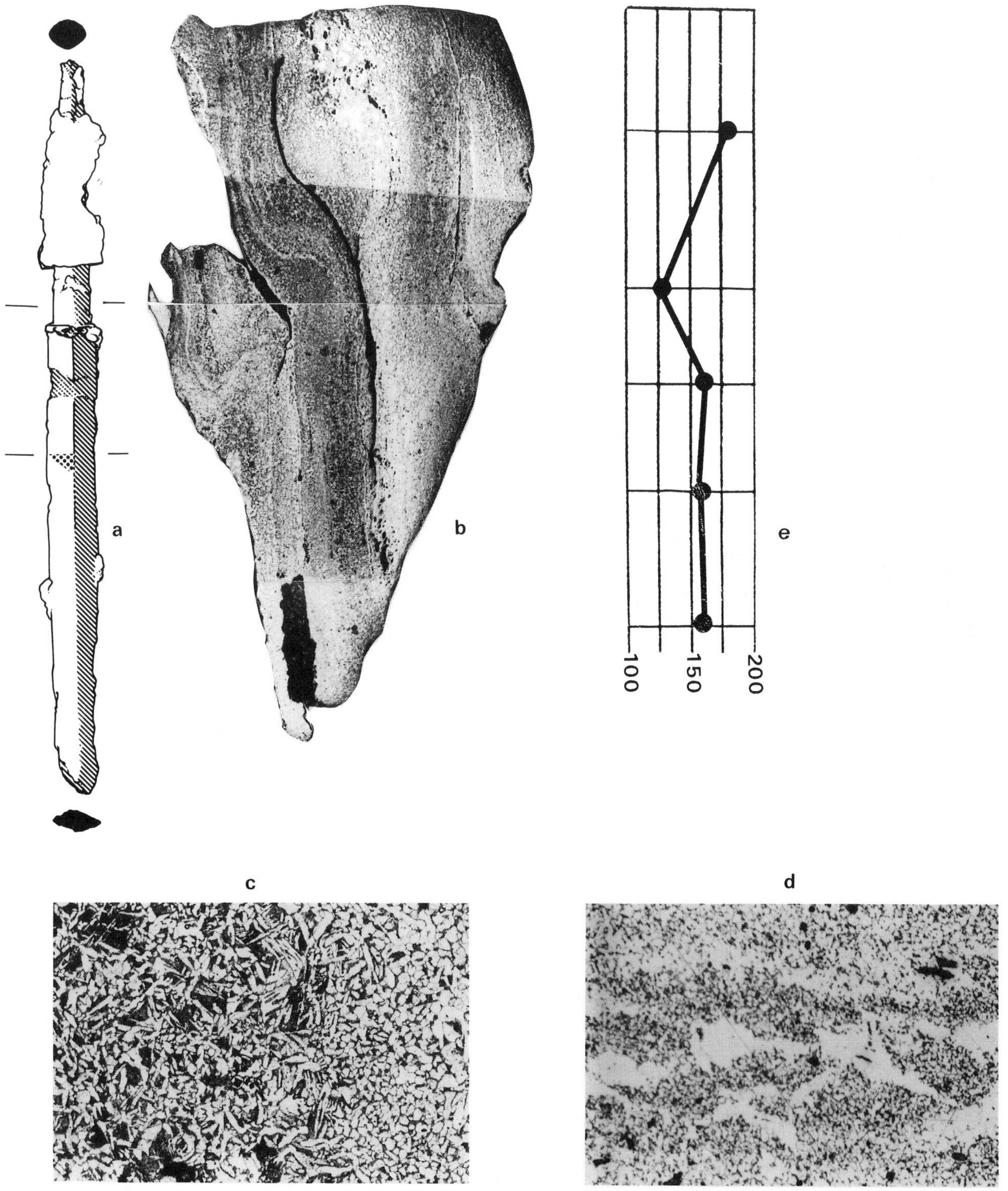

Plate 4.2.7 Examination 12: (a) UM 640–30 sword(?) from Toome Bar, Co. Antrim x½, (b) macrostructure, (c) ferrite and pearlite x100, (d) 'ghosts' x100, (e) H_v profile. Etched 4% Nital.

heterogeneous stock which was allowed to cool normally after forging. There is no evidence of any form of heat treatment.

(12) UM 640–30, sword (?) from Toome Bar, Co. Antrim (Plate 4.2.7)

A section was taken from the cutting edge through to the midrib (Plate 4.2.7a).
Non-metallic inclusions: Two or three areas with large single-phase inclusions occur in conjunction with lines of penetration by corrosion.
Macrostructure: Nital indictes an uneven distribution of C: Oberhoffer indicates no other significant segregations. Two major weld seams, penetrated now by corrosion, can be observed (Plate 4.2.7b).
Microstructure: The structure ranges from patches of large grained ferrite (GS 1–2), through fine-grained ferrite (GS 7–<8), ferrite with interstitial pearlite (GS 5–6), to Widmanstätten ferrite with pearlite (GS 5–7: Plate 4.2.7c). There is pronounced banding

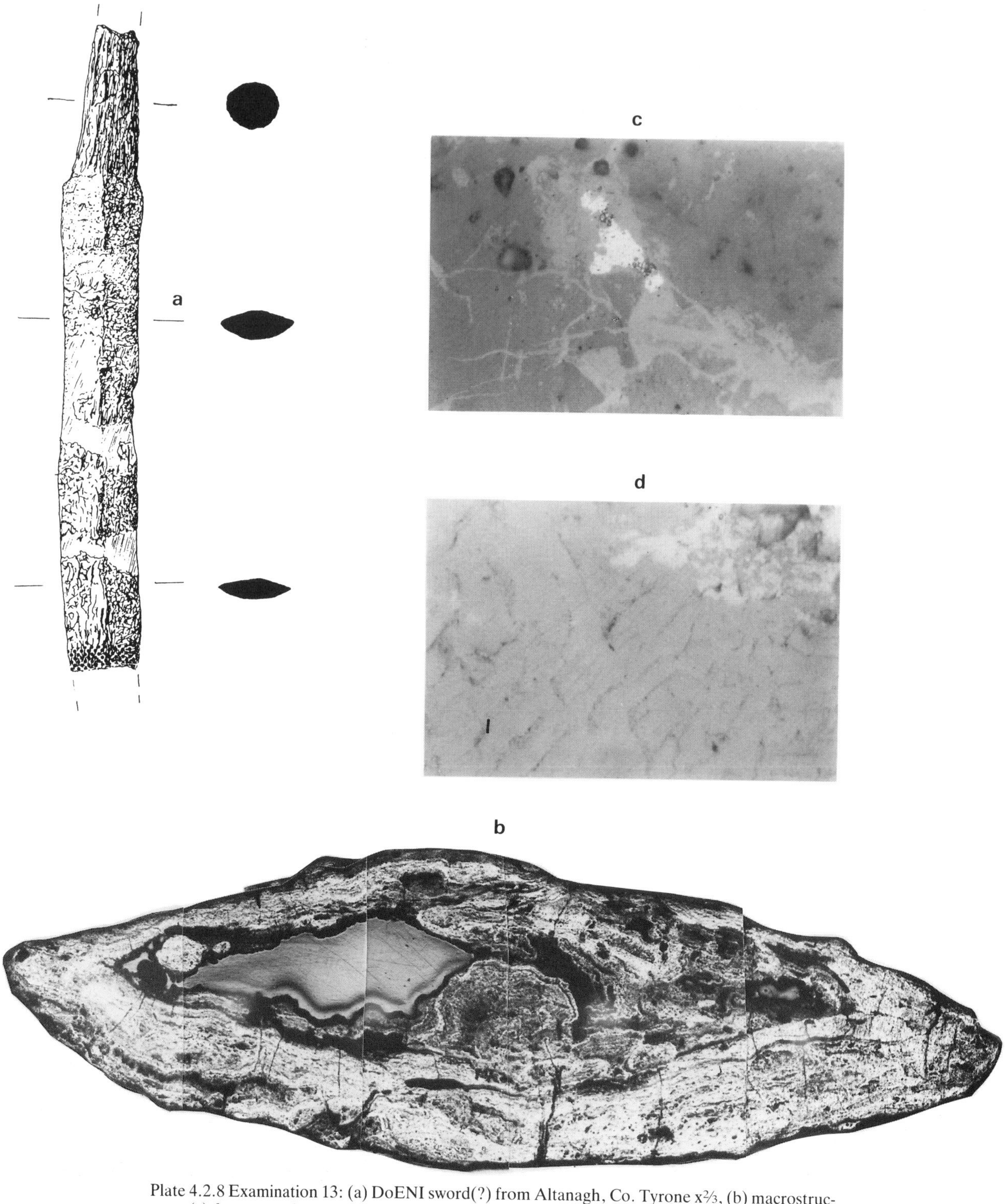

Plate 4.2.8 Examination 13: (a) DoENI sword(?) from Altanagh, Co. Tyrone x⅔, (b) macrostructure, (c) fragment of remanent ferrite x250, (d) Widmanstätten structure, outlined by remanent carbides fossilised in corrosion products x250.

throughout (cf. Plate 4.2.7b). The pearlite throughout is coarse, showing some spheroidisation of carbide. As well as the two major welds, there are numerous traces of bloom welds (Plate 4.2.7d).
Hardness: mH_v ferrite 194–242, pearlite 255–295: H_v 125–180.
Interpretation: The original stock was of very variable carbon content. The two major welds would appear to the result either of very crude fold welding, or else to have survived along with the other bloom welds from the consolidation of the original stock.

(13) DoENI Sword (?) from Altanagh, Co. Tyrone (Plate 4.2.8)

This piece was wholly corroded, but a sample was taken from across the broken end of the blade (Plate 4.2.8a).
Non-metallic inclusions: There are a number of one-phase inclusions embedded in the corrosion ground mass. These are orientated roughly parallel to the central axis of the section.
Macrostructure: The entire cross section comprises

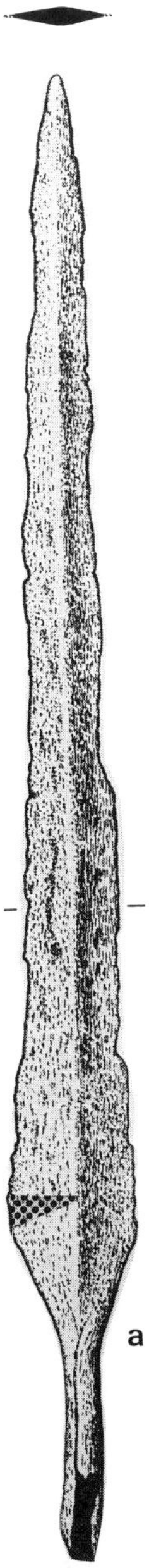

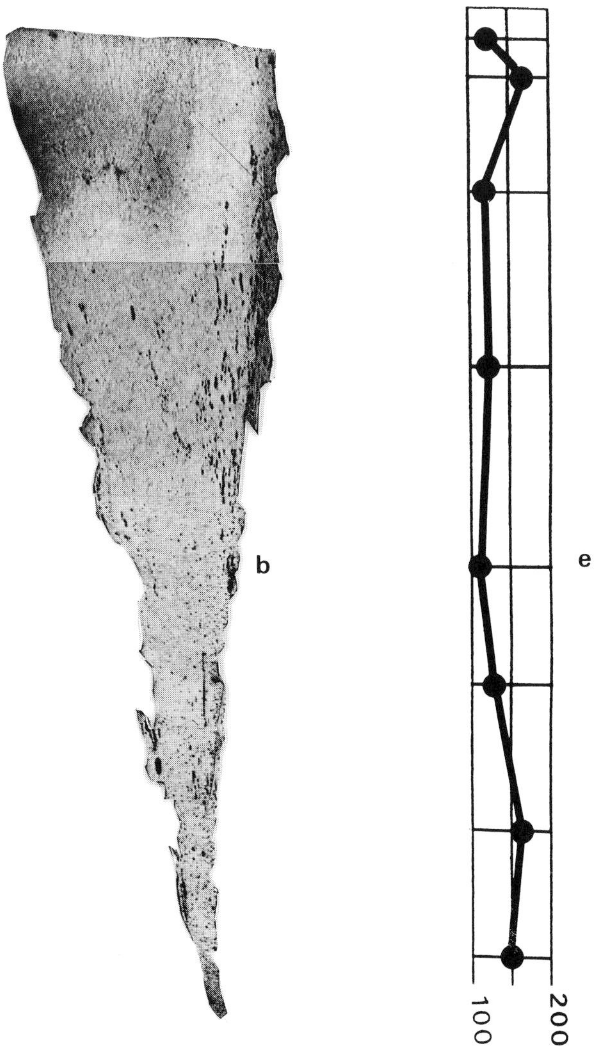

Plate 4.2.9 Examination 14: (a) UM B3 spearhead from Lisnacrogher, Co. Antrim x⅓, (b) macrostructure, (c) ferrite and pearlite near midrib x100, (d) large ferrite grains near cutting edge x250, (e) H_v profile. Etched 4% Nital.

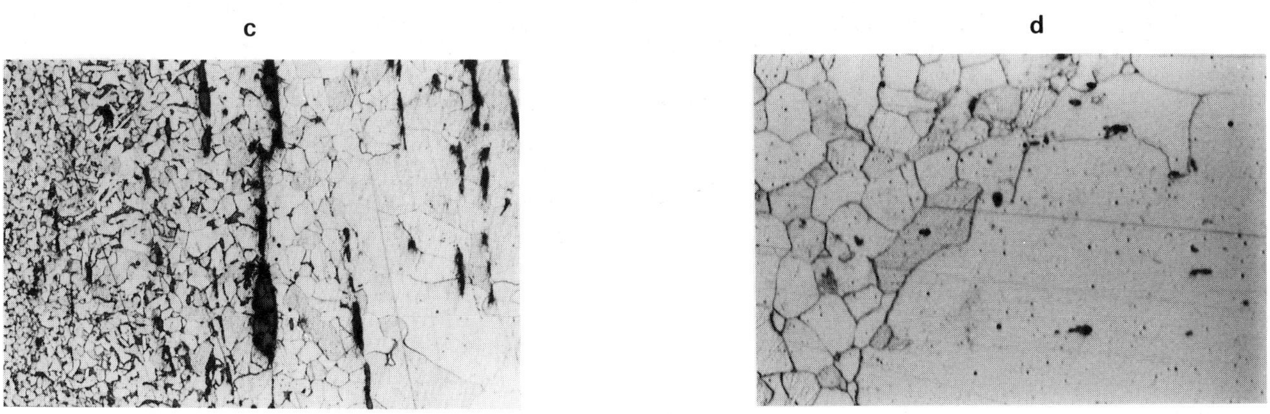

corrosion products from which no idea of original macrostructure could be gained (Plate 4.2.8b).
Microstructure: Within the corrosion groundmass, it was possible to detect flecks of remanent metal, including ferrite (Plate 4.2.8c), ferrite with fine pearlite, and remanent carbides, in one area outlining where there had been an area of Widmanstätten ferrite (Plate 4.2.8d). Although in most sections like this, slag inclusions are the last features to disappear, none could be traced in this specimen. Interpretation of the morphology of corrosion products is fraught with uncertainty, but occasionally remanent structures can be observed (Scott 1976b, 1989a). There are several small patches which retain the relief of a Widmanstätten structure.
Interpretation: From the meagre survivals, it seems

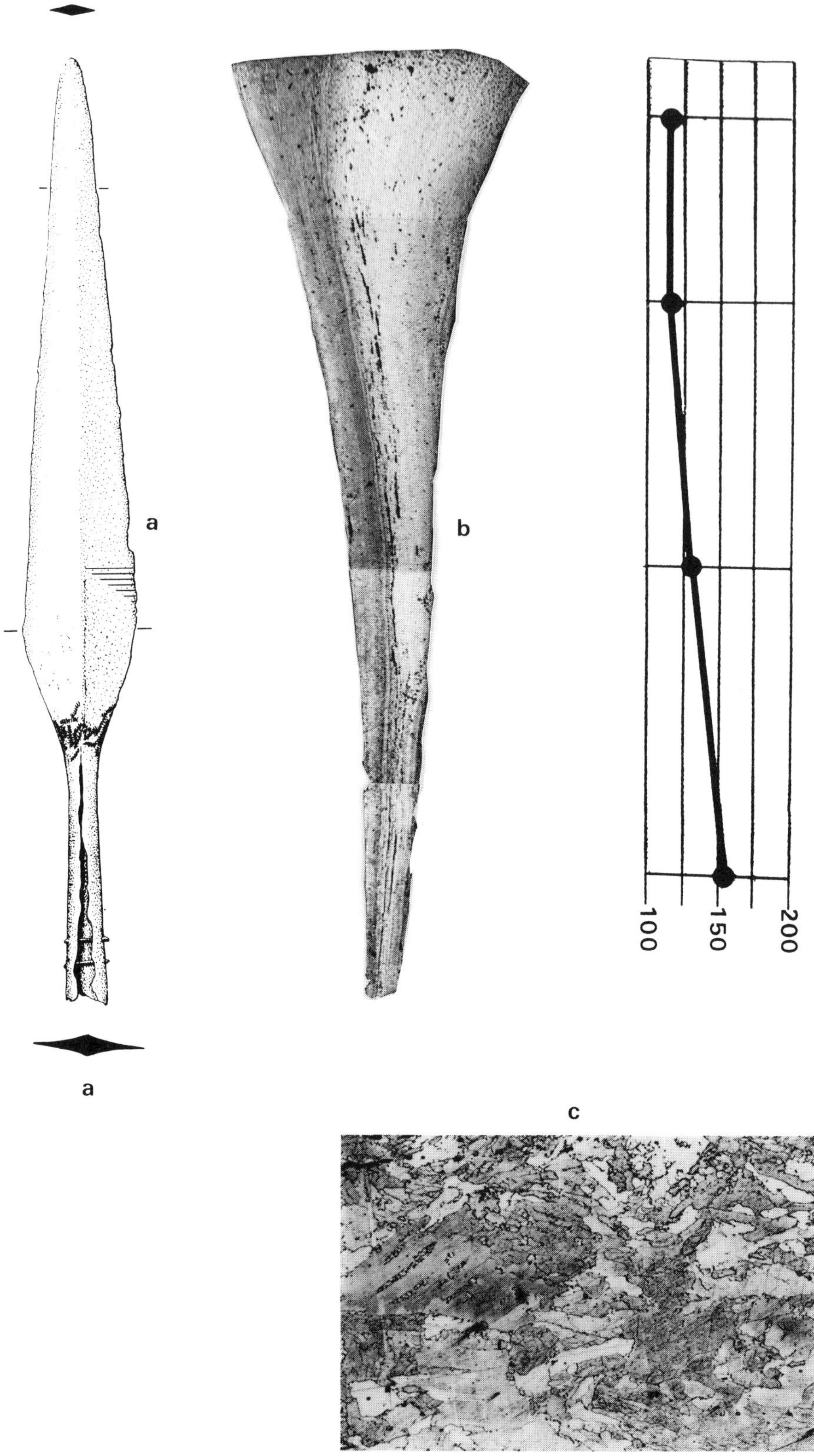

Plate 4.2.10 Examination 15: (a) BM 1880.8–2.117 spearhead from Lisnacrogher, Co. Antrim x⅓, (b) macrostructure, (c) distorted ferrite near cutting edge x100, (d) H_v profile. Etched 4% Nital.

likely that this piece originally was of heterogeneous carbon content, possibly quite low in entrapped slag.

(14) UM B3 spearhead, Lisnacrogher, Co. Antrim (*CIIAA* 282: Plate 4.2.9)

A section was taken from cutting edge to midrib (Plate 4.2.7a).

Non-metallic inclusions: small two- and three-phase inclusions are scattered throughout the section.
Macrostructure: Etching revealed no major segregations (Plate 4.2.9b).
Microstructure: The structure consists almost entirely of regular polyhedral ferrite grains of uniform size (GS 4–6), although towards the midrib, interstitial pearlite appears, associated with a refinement in ferrite grain size (GS 6–8: Plate 4.2.9c). There is an area

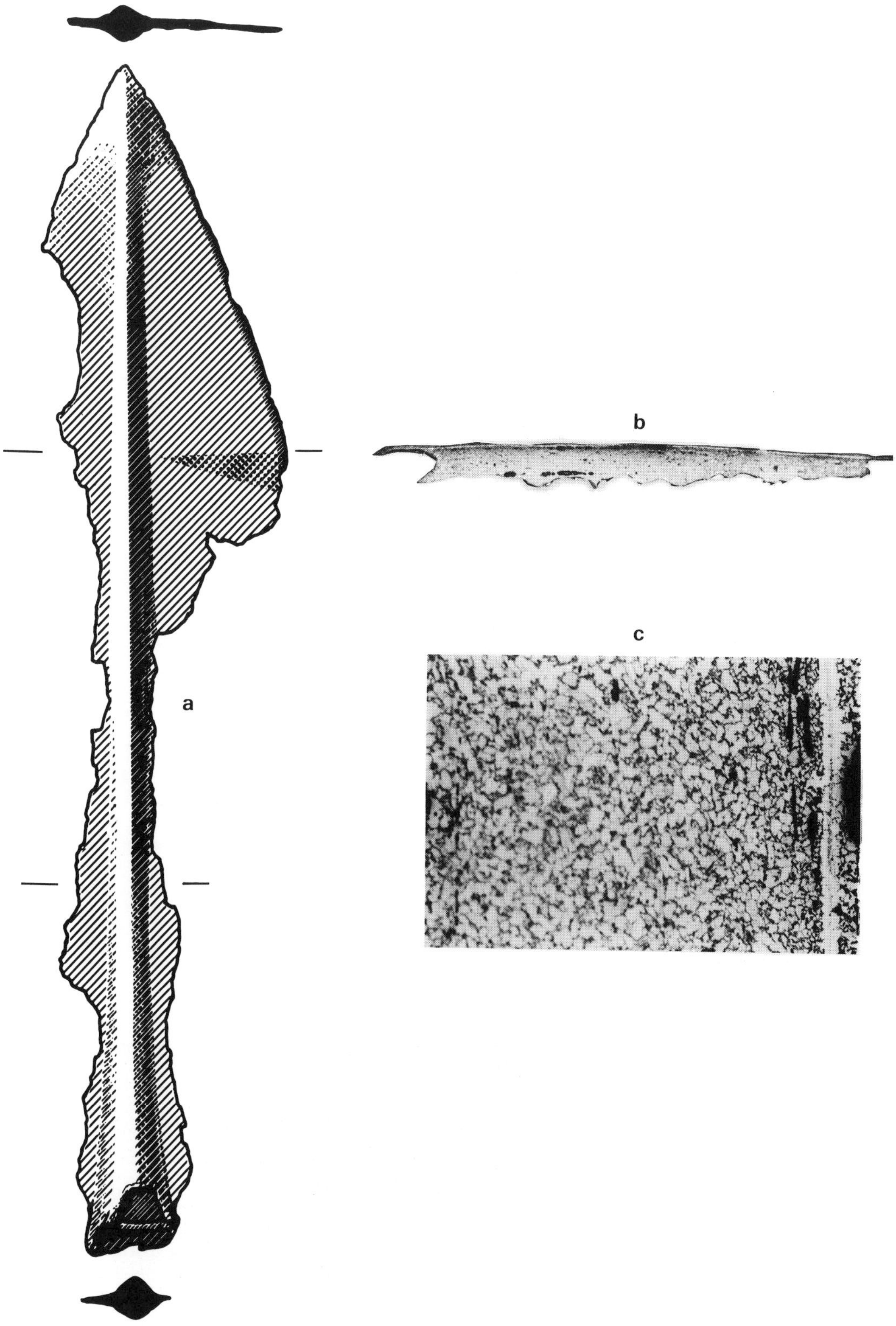

Plate 4.2.11 Examination 16: (a) NMI 1881–357:W1:WK40 spearhead 'River Shannon' x½, (b) macrostructure, (c) ferrite and interstitial pearlite, with weld seam x50. Etched 4% Nital.

of large ferrite grains (GS 1–2) near to the cutting edge (Plate 4.2.9d).
Hardness: mH_v ferrite 100–116: H_v 119–168.
Interpretation: Although finely shaped, the blade was made from stock that was virtually carbon-free, and allowed to cool normally. It would not have been a particularly effective weapon.

(15) BM 1880.8–2.117 spearhead, Lisnacrogher, Co. Antrim (*CIIAA* 281: Plate 4.2.10)

A section was cut from above the shoulders of the blade through to the midrib (Plate 4.2.10a).
Non-metallic inclusions: One- and two-phase inclusions, generally small and including one possible MnS

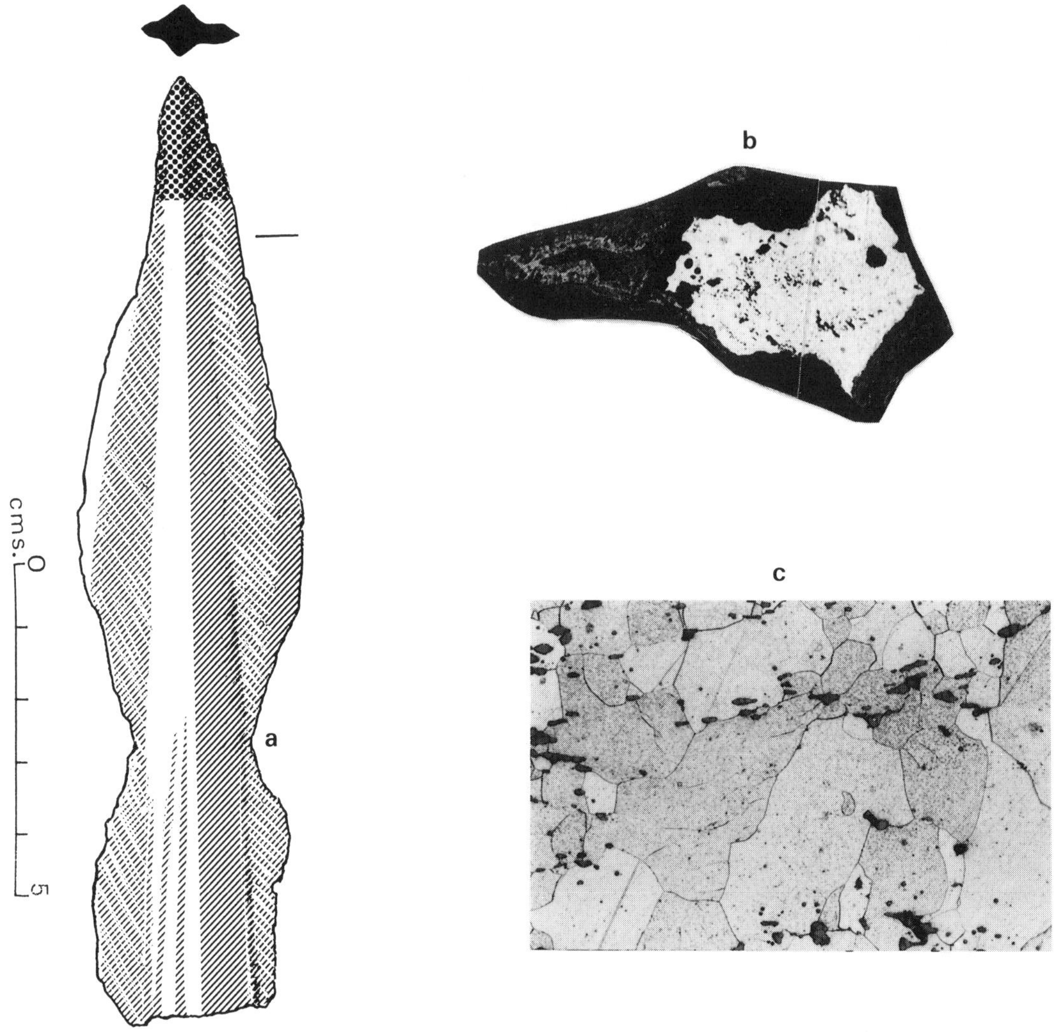

Plate 4.2.12 Examination 17: (a) NMI E79:2305 spearhead from Dún Ailinne, Co. Kildare, (b) macrostructure, (c) ferrite showing texture possibly resulting from age hardening x100. Etched 4% Nital.

inclusion, run in chains parallel to the long axis of the blade.
Macrostructure: Nital revealed some traces of C banding, while Oberhoffer revealed some banding attributable to P and/or As (Plate 4.2.10b).
Microstructure: There is some banding of ferrite and pearlite, with the areas of higher carbon content tending to Widmanstätten structure and a smaller grain size (ferritic areas GS 5; higher-C GS 8). Towards the cutting-edge, grains are elongated and somewhat distorted, indicating some degree of cold-working (Plate 4.2.10c).
Hardness: mH_v ferrite 102–132: H_v 112–153.
Interpretation: The spearhead was forged from one piece of metal, fairly rich in manganese, with some phosphorous, possibly indicating a bog-ore. It shows evidence of having been worked at high temperatures, followed by a fairly rapid cooling (but not sufficient to harden sufficiently). The elongation and distortion of ferrite near the cutting edge indicates some cold working as a final process, possibly in an attempt to harden the metal.

(16) NMI 1881–357:W1:WK40 spearhead, 'River Shannon' (Plate 4.2.11)

The blade has been seriously damaged by corrosion and, while a core of metal survives almost to the line of the cutting edge, all of the original surface has been lost. The corrosion capsule has, however, been preserved. A section was cut towards the midrib to take both surviving metal and contiguous corrosion (Plate 4.2.11a).
Non-metallic inclusions: One- and two-phase slag inclusions occur in relatively small numbers.
Macrostructure: Etching with both Nital and Oberhoffer indicated no significant segregations (Plate 4.2.11b).
Microstructure: The structure is primarily ferritic with fine pearlite occurring interstitially. There is little variation (GS 7–8) in ferritic grain-size. A single weld seam, probably the result of fold-welding, runs for the entire length of the section (Plate 4.2.11c: cf. Plate 4.2.11b).
Hardness: mH_v ferrite 225.

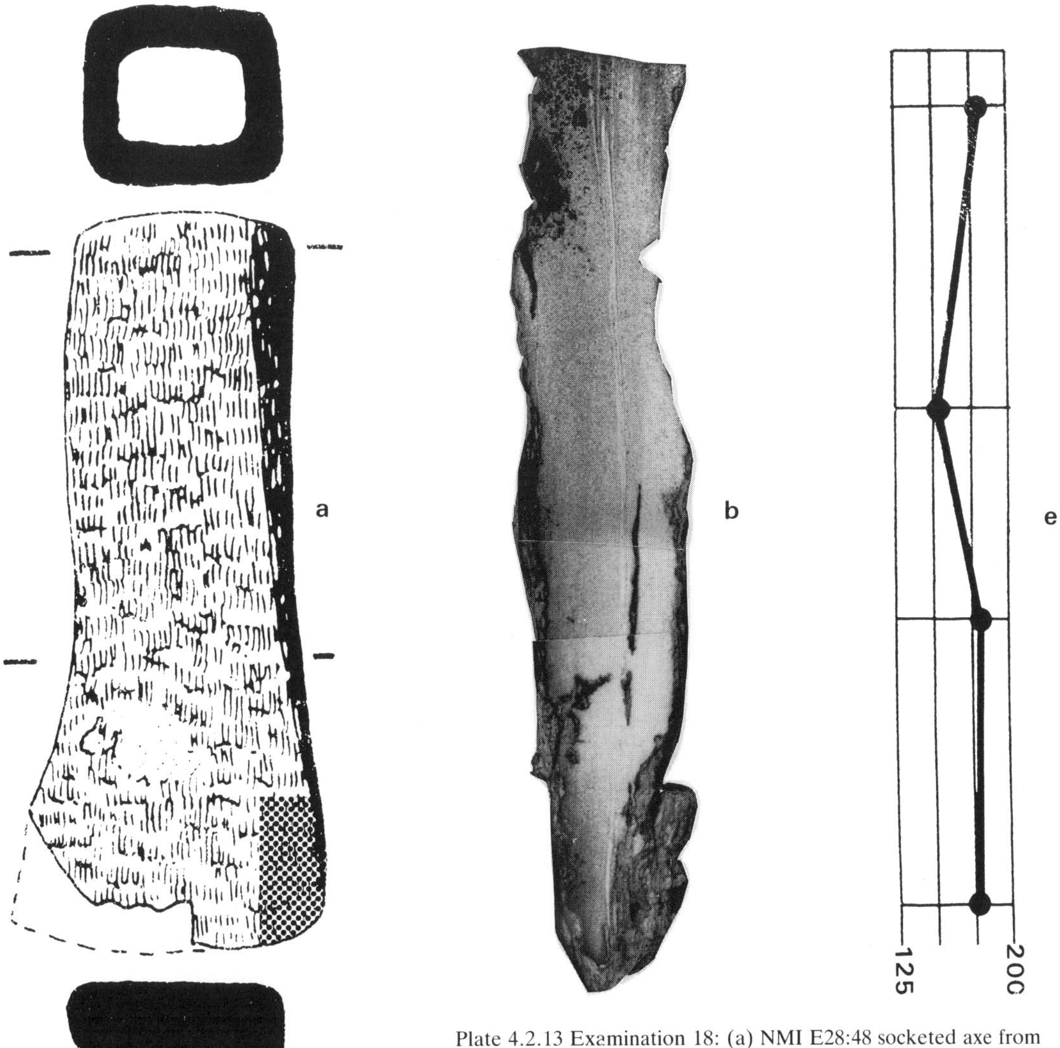

Plate 4.2.13 Examination 18: (a) NMI E28:48 socketed axe from Feerwore, Co. Galway x1/1, (b) macrostructure, (c) fine ferrite and interstitial pearlite, with one weld seam x100, (d) area of higher carbon content near top of section x100, (e) H_v profile. Etched 4% Nital.

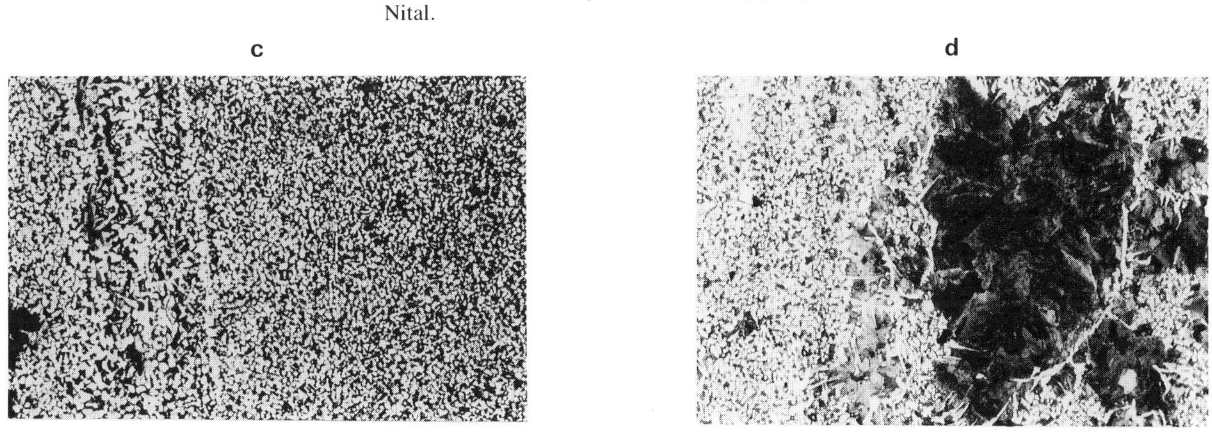

Interpretation: The blade was forged up from very low-C stock which had been subjected to of fold-welding. There was no attempt at heat-treatment. Despite its present condition, it is clear that this was once a most elegant and finely shaped spearhead, and an effective weapon.

(17) NMI E79:2305 spearhead, Dún Ailinne, Co. Kildare (Plate 4.2.12)

A section was taken across the blade and midrib 2 cm back from the point[10]. Only a core of metal survived with the rest of the shape represented by corrosion products (Plate 4.2.12a).

Non-metallic inclusions: Numerous, large two- and three-phase inclusions occur throughout the section.

Macrostructure: Etching with Nital and Oberhoffer revealed no significant segregations (Plate 4.2.12b).

Microstructure: The structure consists entirely of large polyhedral ferrite grains of regular size (GS 2–3) with some grain boundary cementite, but no pearlite in evidence. A significant number of the grains show the texturing associated with age precipitation after quenching (Plate 4.2.12c).

Hardness: mH_v 104–136.

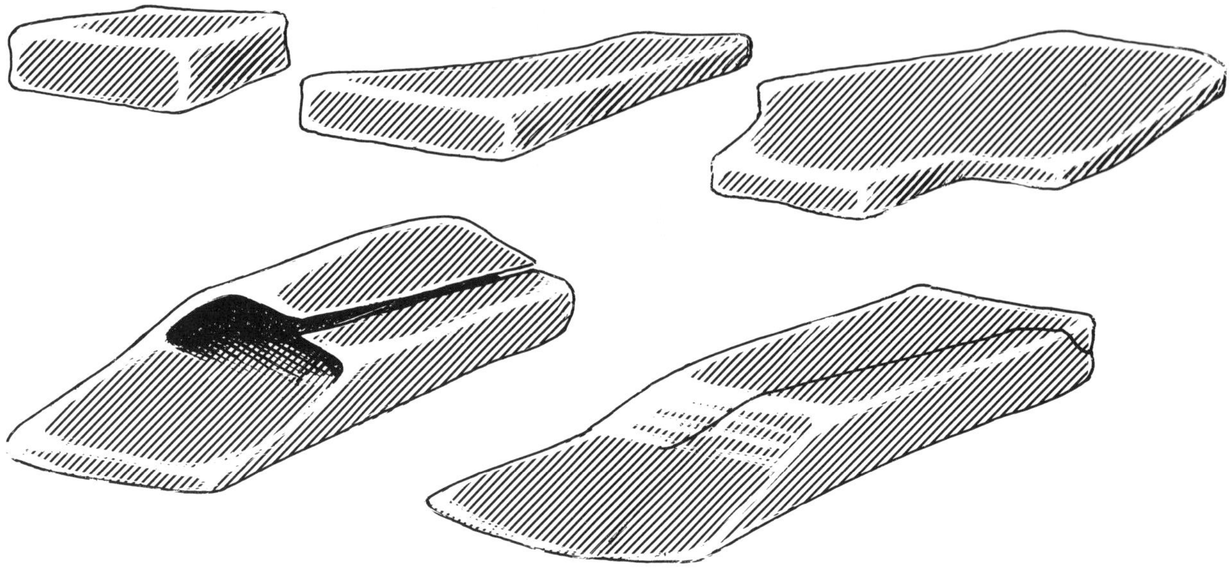

Fig. 4.2.1 Schematic diagram (not to scale) of the forging out of the Feerwore socketed axe.

Interpretation: This finely-wrought piece has been forged up from stock which in this section was uniformly free of carbon. The size of the slag inclusions and their quantity shows lack of success in expulsion of slag during consolidation of the bloom. The etch-pitting could be indicative of age hardening, although it must be noted that the microhardness of grains is low. It is possible that this piece was quenched in a wholly unsuccessful attempt to harden it. Precipitation could have been induced by an attempt to temper, or has happened over the two millennia or so of its interment. Despite what must have been originally a very attractive appearance, this would not have been a particularly good quality weapon.

(18) NMI E28:48 socketed axe, Rath of Feerwore, Co. Galway (*CIIAA* 578: Plate 4.2.13)

Although badly damaged by corrosion, much of the metal survives. A section was taken from the cutting edge parallel to the face of the blade (Plate 4.2.13a).
Non-metallic inclusions: These are principally one-phase dark-grey glassy inclusions, and are not particularly numerous.
Macrostructure: Etching with Nital showed a distinct concentration of carbon at the back of the section and towards the cutting edge, while etching with Oberhoffer clarified details of at least three weld-seams (Plate 4.2.13b).
Microstructure: The most obvious feature is the structure of the cutting edge, which shows very fine ferrite with fine (GS <8) pearlite (Plate 4.2.13c). Ferritic grain size is small throughout the section (GS 6–<8). Further into the body of the blade, the carbon content decreases, the structure showing pearlite with some Widmanstätten ferrite. Towards the body end of the section there are patches of very coarse pearlite with Widmanstätten ferrite (Plate 4.2.13d) which shows carbide spheroidisation. Three weld seams, one of which runs up to the cutting edge, are clearly visible (cf. Plate 4.2.13b).
Hardness: mH_v ferrite 198, pearlite 288–313: H_v 160–186.
Interpretation: Fig. 4.2.1 is a diagrammatic reconstruction (not to scale) of how the axe was made. The steep carbon gradient back from the cutting edge, indicates secondary carburisation of the cutting-edge region. However, once this had been achieved, no further improvements in the way of heat-treatments were applied. Despite this, the piece would have made a serviceable tool.

(19) NMI 1935:406 shaft-hole axe, Kilbeg Co. Westmeath (*CIIAA* 584a: Plate 4.2.14)

The effects both of corrosion and of overzealous cleaning in the past suggest a multi-component structure for this and the following three specimens (examinations 20–22). A section was cut back from the cutting edge parallel with the long axis of the blade (Plate 4.2.14a).
Non-metallic inclusions: Examination of the polished section taken from the cutting edge shows a small number of one- and two-phase inclusions.
Macrostructure: Etching with Nital revealed an even distribution of carbon (Plate 4.2.14b), while etching with Oberhoffer indicated only minor segregations.
Microstructure: The etched section shows an almost purely ferritic structure (GS 6–8), tending to Widmanstätten in places, with fine pearlite occurring interstitially (Plate 4.2.14c), and occasionally in larger colonies. Analysis showed the contents both of As and P to be unusually high (Table 4.2.2), but effects of this are not evident in the section.
Hardness: mH_v ferrite 233–268, pearlite 278–332: H_v 123–143.
Interpretation: This axe (and the next three) has been forged from two pieces of metal, one bent round to form the shaft-hole with its two ends welded to the faces of the second which formed the blade. The stock consisted of low carbon iron with elevated contents of As and P.

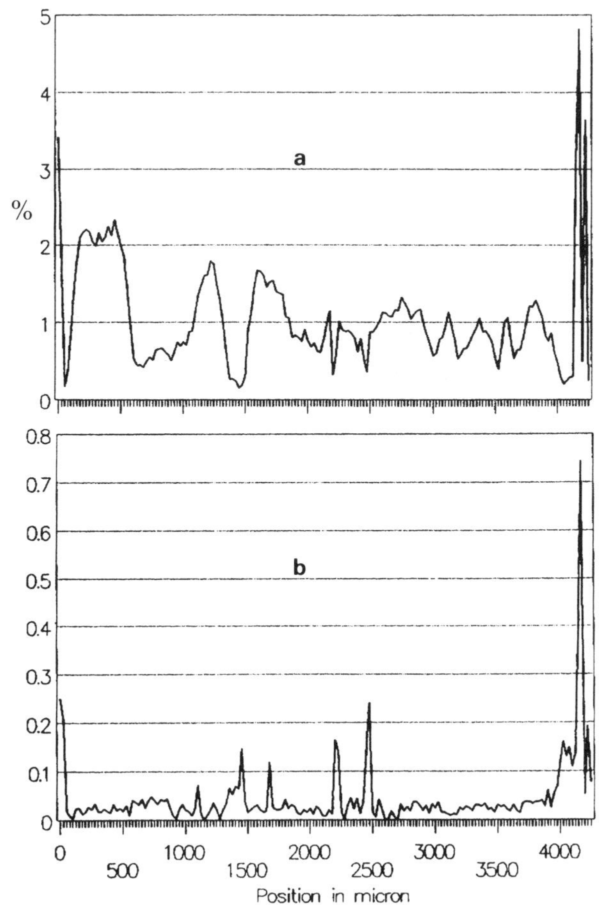

Fig. 4.2.2 Examination 20: microprobe scan for (a) As and (b) P.

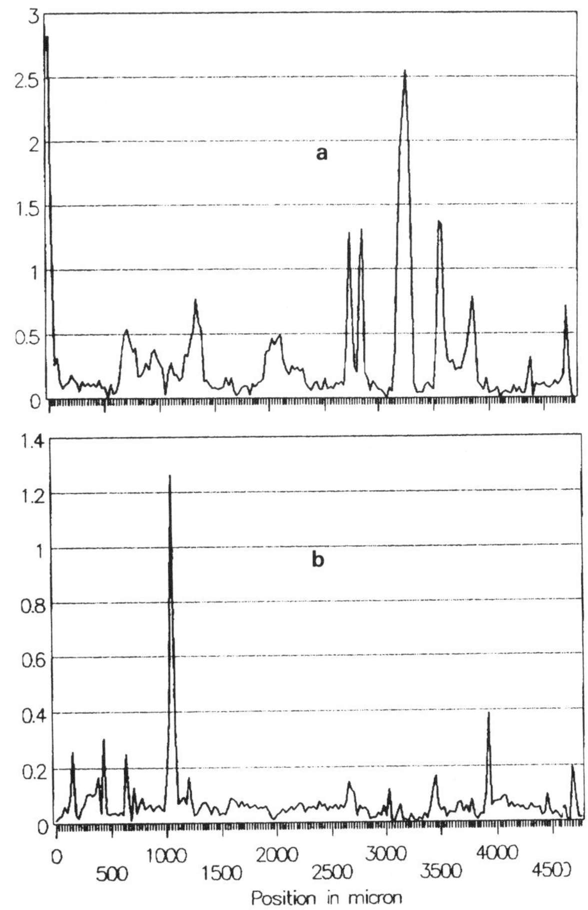

Fig. 4.2.3 Examination 21: microprobe scan for (a) As and (b) P.

(20) NMI 1930:407 shaft-hole axe, Kilbeg Co. Westmeath (*CIIAA* 584b: Plate 4.2.15)

A section was cut back from the cutting edge parallel to the long axis of the blade (Plate 4.2.15a).
Non-metallic inclusions: A number of small single-phase grey inclusions run in chains parallel to the long-axis of the section.
Macrostructure: Etching with Nital showed light and dark bands (Plate 4.2.15b), while Oberhoffer highlighted the highly laminar structure of the blade.
Microstructure: There is virtually no pearlite in any part of the section. The structure consists of alternate laminae comprisied of dark- (GS 6–7) and light-etching (GS 2–3) polyhedral ferrite grains in bands delimited by the slag lines (Plate 4.2.15c). The darker-etching ferrite is accompanied by interstitial pearlite. As with the previous axe, analytical contents of both As and P are unusually high. Etching with Oberhoffer indicates distinct segregations, with the higher concentrations corresponding to the larger-grained ferrite, where, also, acicular precipitates may be observed (Plate 4.2.15d). Microprobe scan across the section shows concentrations of As and P at the margins, concentrated by differential oxidation during forging (Fig. 4.2.2). However, in the interior, P and As concentrations tend to be separated (cf. Figs. 4.2.3–4.2.4 and Fig. 5.3.1).
Hardness: mH_v ferrite 176–182: H_v 129–182.
Interpretation: The construction technique was as for the previous specimen. The original stock could well have been a single piece of heterogeneous metal with areas low in carbon, P and As, and others which were carbon-free, with segregations of P and As. Alternatively, the starting point could have been two (or more) pieces of metal with these characteristics. In either case, fold-welding has produced the structure as we have it now, where at least 11 laminae can be observed. The section shows well the effects of differential surface enrichment through oxidation during fold-welding (cf. Chilton and Evans 1955).

(21) NMI 1935:408 shaft-hole axe, Kilbeg, Co. Westmeath (*CIIAA* 584c: Plate 4.2.16)

A section was cut back from the cutting edge parallel to the long axis of the blade (Plate 4.2.16a).
Non-metallic inclusions: Lines of two-phase inclusions run parallel to the long axis of the section.
Macrostructure: Nital etching revealed a highly banded structure (Plate 4.2.16b), which is accentuated by Oberhoffer.
Microstructure: The structure is almost entirely ferritic with interstitial fine pearlite, the ferrite tending to a Widmanstätten structure in places (Plate 4.2.16c). The laminae consist of ferrite of slightly different grain sizes (GS 4–6). As with the previous specimen, microprobe scan across the section shows evidence of surface enrichment of P and As, and of the different distribution of these elements (Fig. 4.2.3).
Hardness: mH_v ferrite 228–244: H_v 110–168.
Interpretation: The construction of this axe was the same as for the previous two specimens, while the original stock would have been as for NMI 1935–407. Fold-welding produced at least 8 laminae.

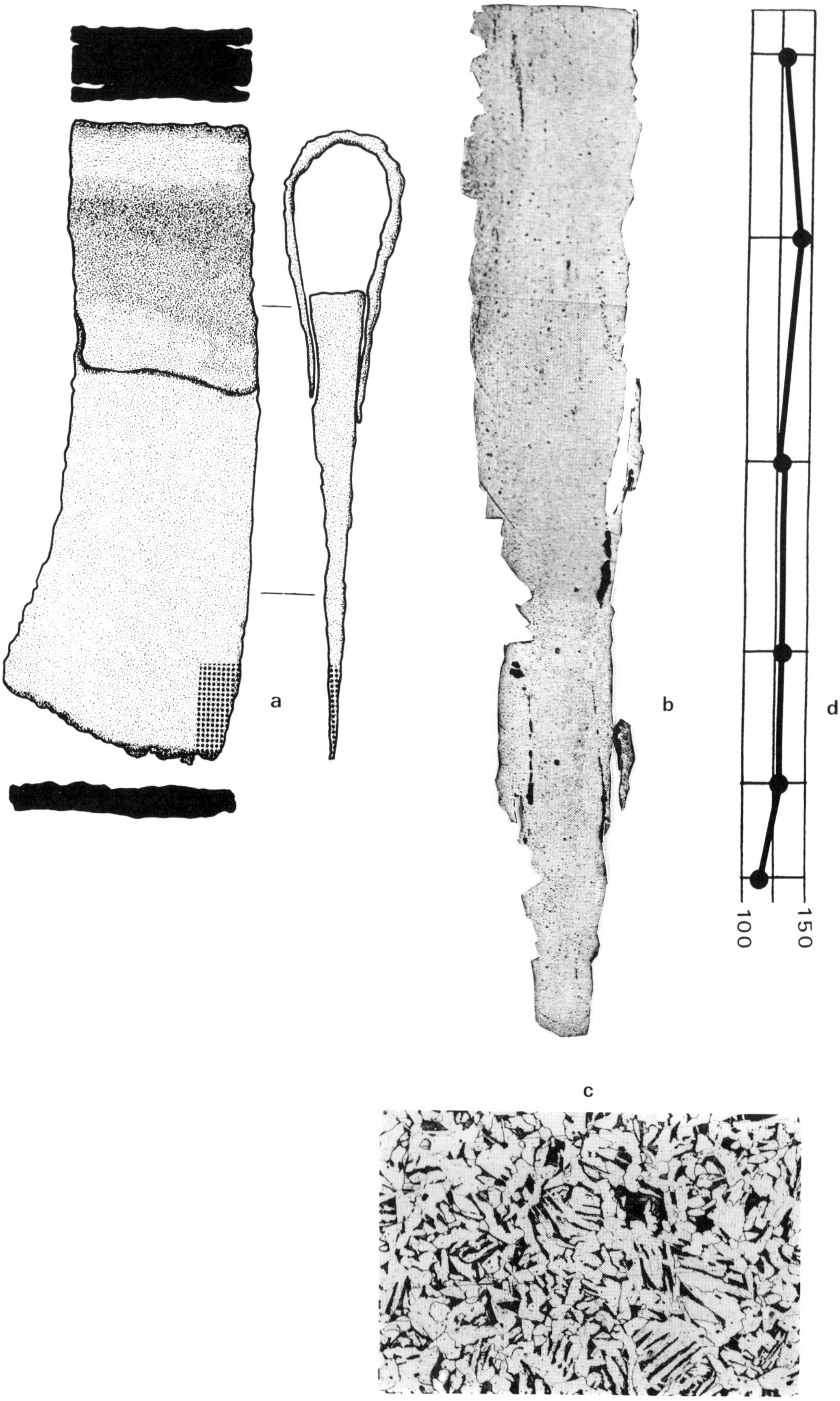

Plate 4.2.14 Examination 19: (a) NMI 1935:406 shaft-hole axe from Kilbeg, Co. Westmeath x½, (b) macrostructure, (c) Widmanstätten ferrite and pearlite x100, (d) H_v profile. Etched 4% Nital.

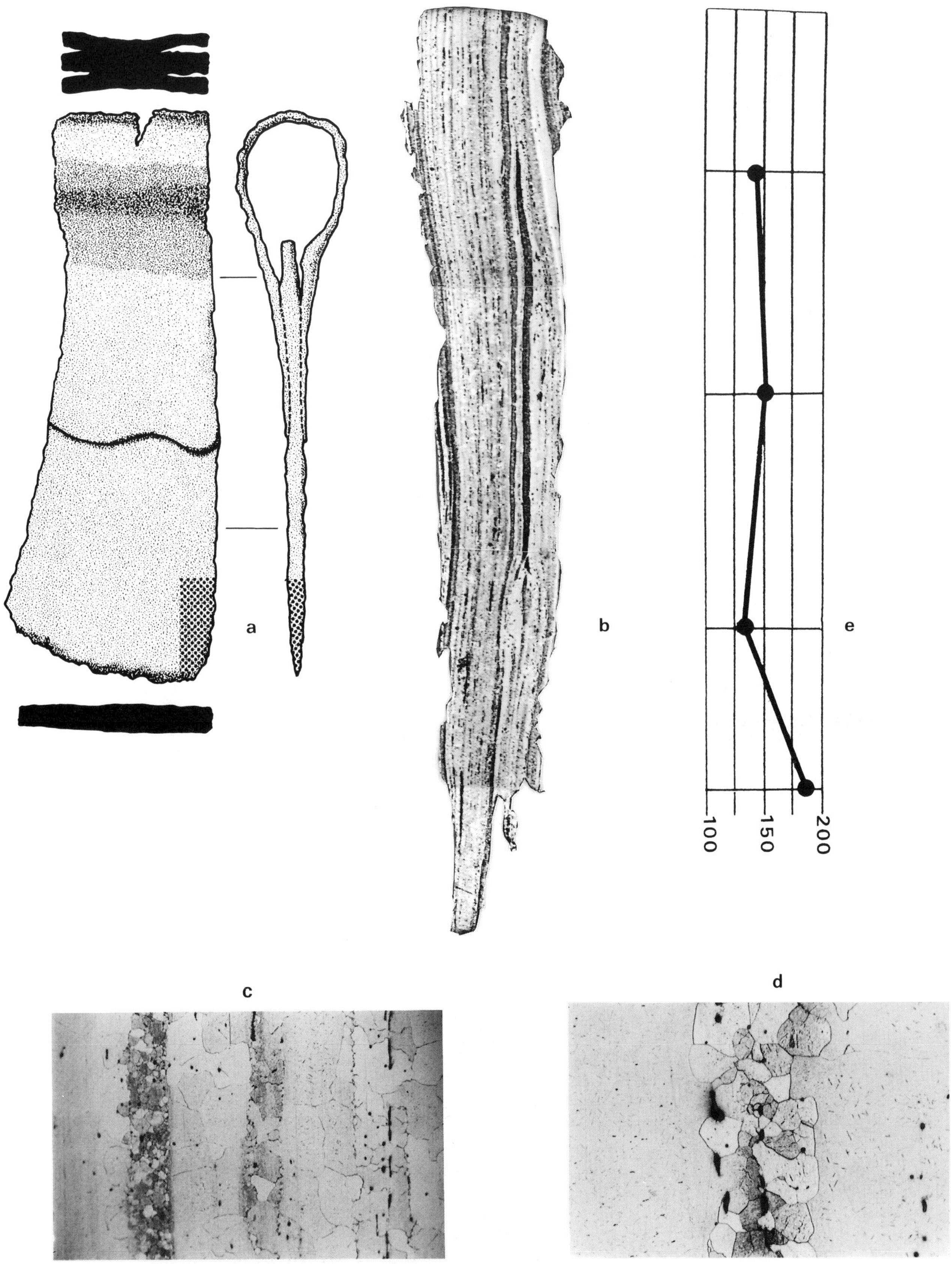

Plate 4.2.15 Examination 20: (a) NMI 1935:407 shaft-hole axe from Kilbeg, Co. Westmeath x½, (b) macrostructure, (c) banding of ferrite x100, (d) acicular precipitates (γ'-Fe_4N?) in larger ferrite grains x500, (e) H_v profile. Etched 4% Nital.

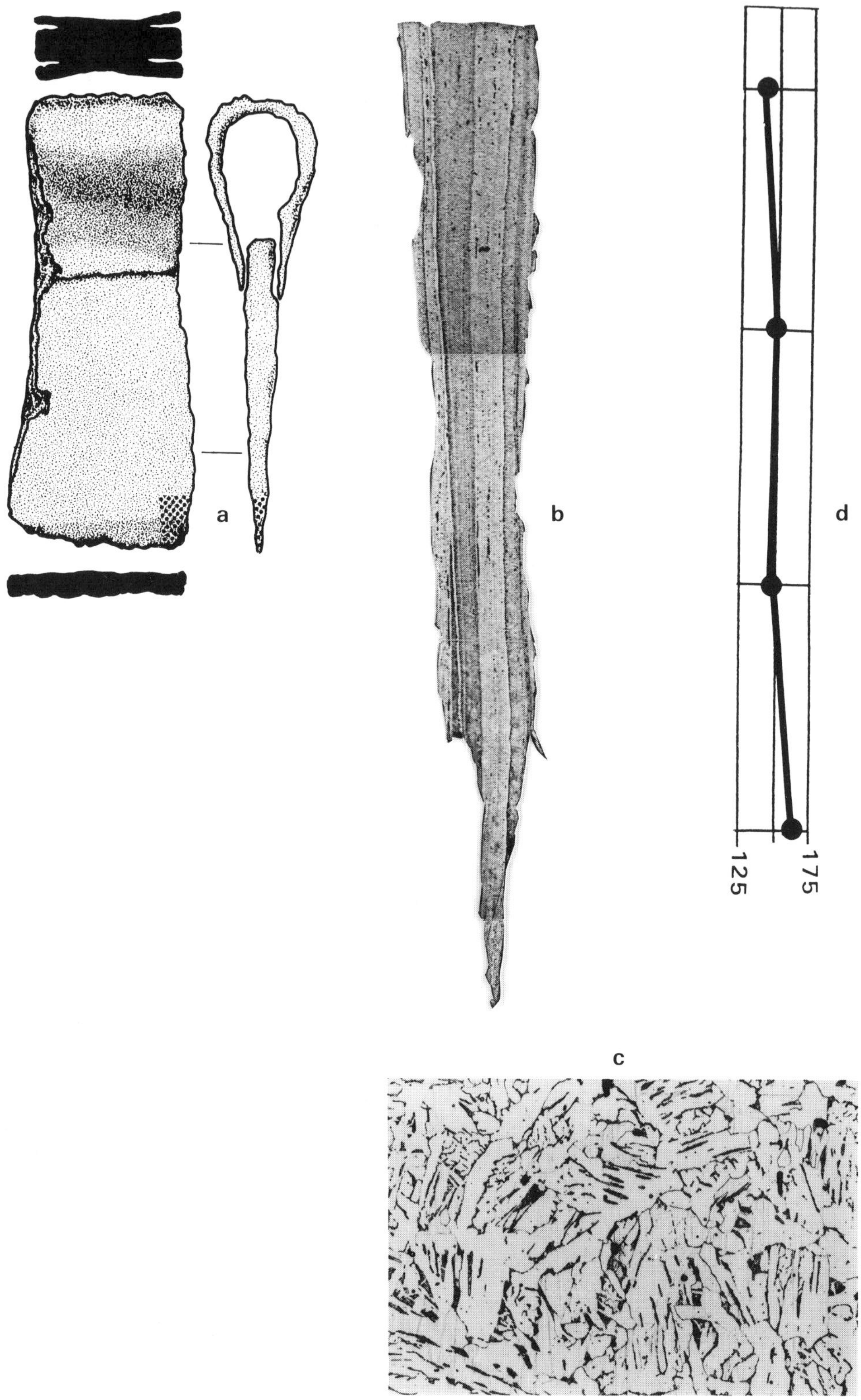

Plate 4.2.16 Examination 21: (a) NMI 1935:408 shaft-hole axe from Kilbeg, Co. Westmeath x½, (b) macrostructure, (c) Widmanstätten ferrite and pearlite x100, (d) H_v profile. Etched 4% Nital.

(22) NMI 1957:61 shaft-hole axe, Castlereban North, Co. Kildare (*CIIAA* 582: Plate 4.2.17)

A section was cut back from the cutting edge parallel to the long axis of the blade (Plate 4.2.17a).
Non-metallic inclusions: Lines of small one- and two-phase inclusions run in chains parallel to the long axis of the specimen.
Macrostructure: Etching with Nital showed banding (Plate 4.2.17b), which was accentuated by etching with Oberhoffer.
Microstructure: The structure is primarily ferritic (GS 4–6) with some interstitial pearlite (Plate 4.2.17c) which shows some spheroidisation of carbide. Some of the ferrite grains show α-veining (Plate 4.2.17d). The bands all have the same structure. Microprobe scan shows the effects of surface enrichment of P and As well (Fig. 4.2.4).

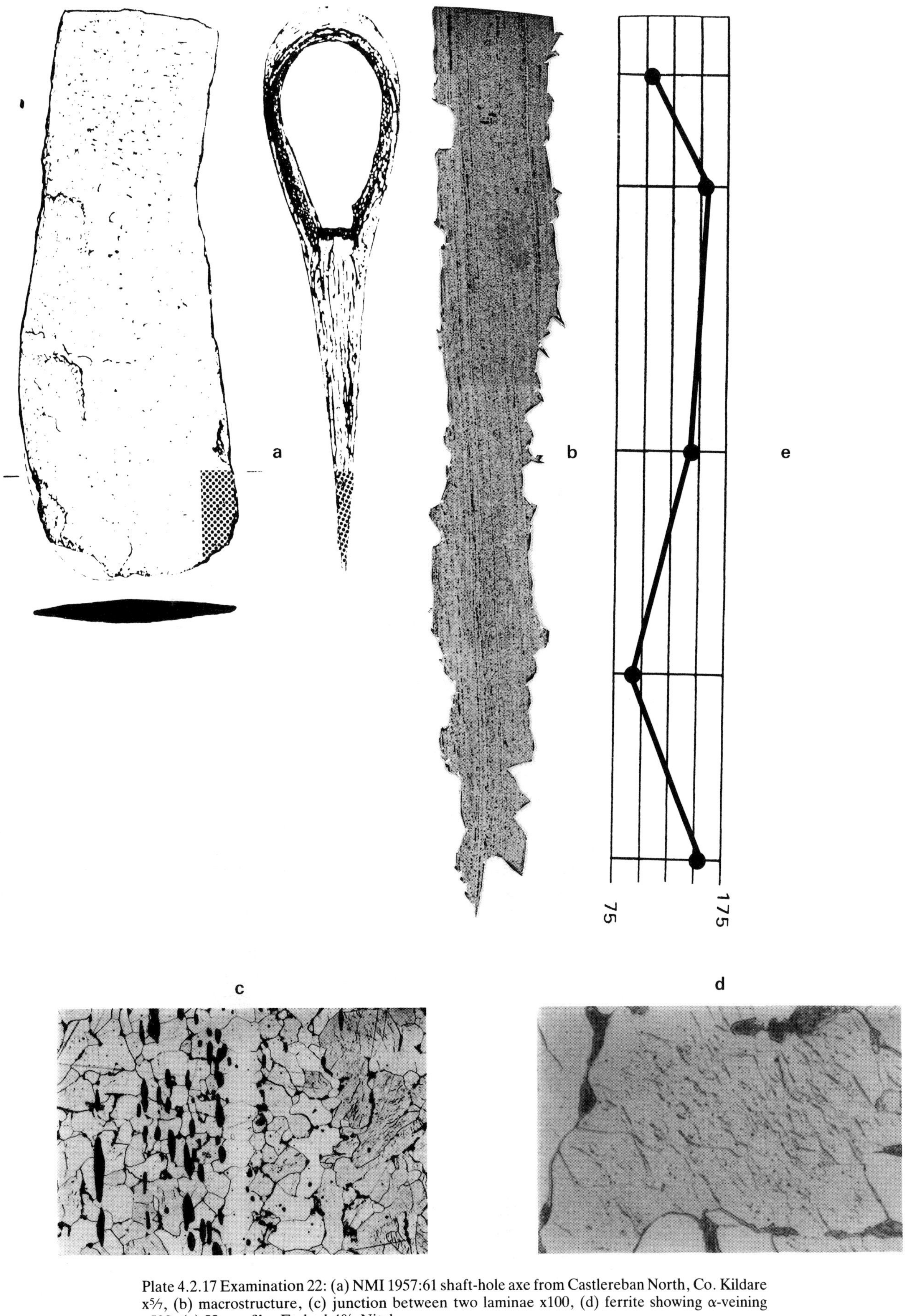

Plate 4.2.17 Examination 22: (a) NMI 1957:61 shaft-hole axe from Castlereban North, Co. Kildare x5/7, (b) macrostructure, (c) junction between two laminae x100, (d) ferrite showing α-veining x500, (e) H_v profile. Etched 4% Nital.

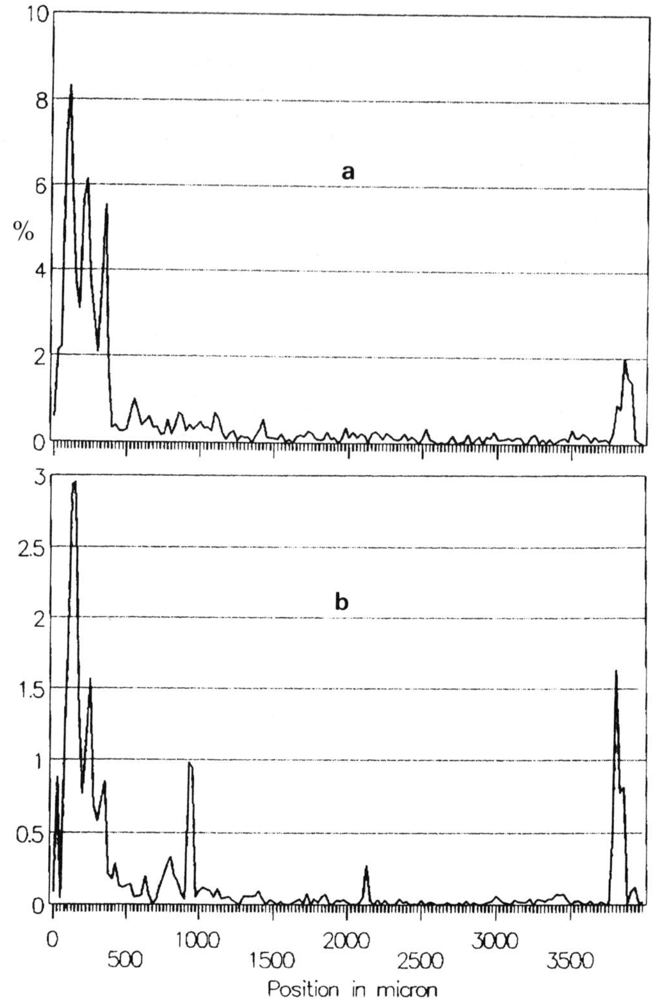

Fig. 4.2.4 Examination 22: microprobe scan for (a) As and (b) P.

Hardness: mH_v ferrite 230: H_v 88–163.
Interpretation: As with the previous three specimens, the axe was forged in two pieces, the stock for the blade being as for NMI 1935–407 and 408. Fold-welding produced at least 15 laminae.

(23) NMI E28:63 chisel, Rath of Feerwore, Co. Galway (*CIIAA* 587: Plate 4.2.18)

Although damaged by corrosion, this tool retains most of the original metal. A section was taken from the cutting edge running parallel to the body (Plate 4.2.18a).
Non-metallic inclusions: The polished section shows an abundance of entrapped slag, consisting entirely single-phase grey, glassy inclusions running in lines parallel to the long axis of the specimen.
Macrostructure: Etching showed no significant segregations (Plate 4.2.18b).
Microstructure: The structure is entirely ferritic (Plate 4.2.18c) with some grains showing grain boundary cementite. There is only minor variation in grain size (GS 5–6). No trace of distortion was observed at the cutting edge, although this has been severely eroded back by corrosion.
Hardness: mH_v ferrite 193: H_v 178–205.
Interpretation: The tool was made from carbon-free stock from which only part of the entrapped slag had been expelled. Although quite hard, it would not have been a particularly serviceable tool.

(24) NMI CC23:8/9 shears, Carbury Hill, Co. Kildare (*CIIAA* 590: Plate 4.2.19)

Two sections were cut, one from the left-hand blade (*section 1*), the other (*section 2*) from the spring (Plate 4.2.19a).
Section 1: *Non-metallic inclusions*: The section shows numerous one- and two-phase inclusions, some large.
Macrostructure: Nital and Oberhoffer show no significant segregations, although several weld seams were highlighted (Plate 4.2.19b).
Microstructure: The structure is principally ferrite with interstitial pearlite (GS 6–<8), with several areas of Widmanstätten ferrite. Although corrosion has removed much of the original surface, there is evidence of a carbon gradient that could be the result of secondary carburisation on one side of the section (Plate 4.2.19c). The weld seams visible in macrosection appear to be bloom welds.
Hardness: mH_v ferrite 162–214, pearlite 282: H_v 133–197.
Section 2: Non-metallic inclusions: There is a small number of two-phase inclusions scattered throughout the section.
Macrostructure: Carbon distribution is variable, but

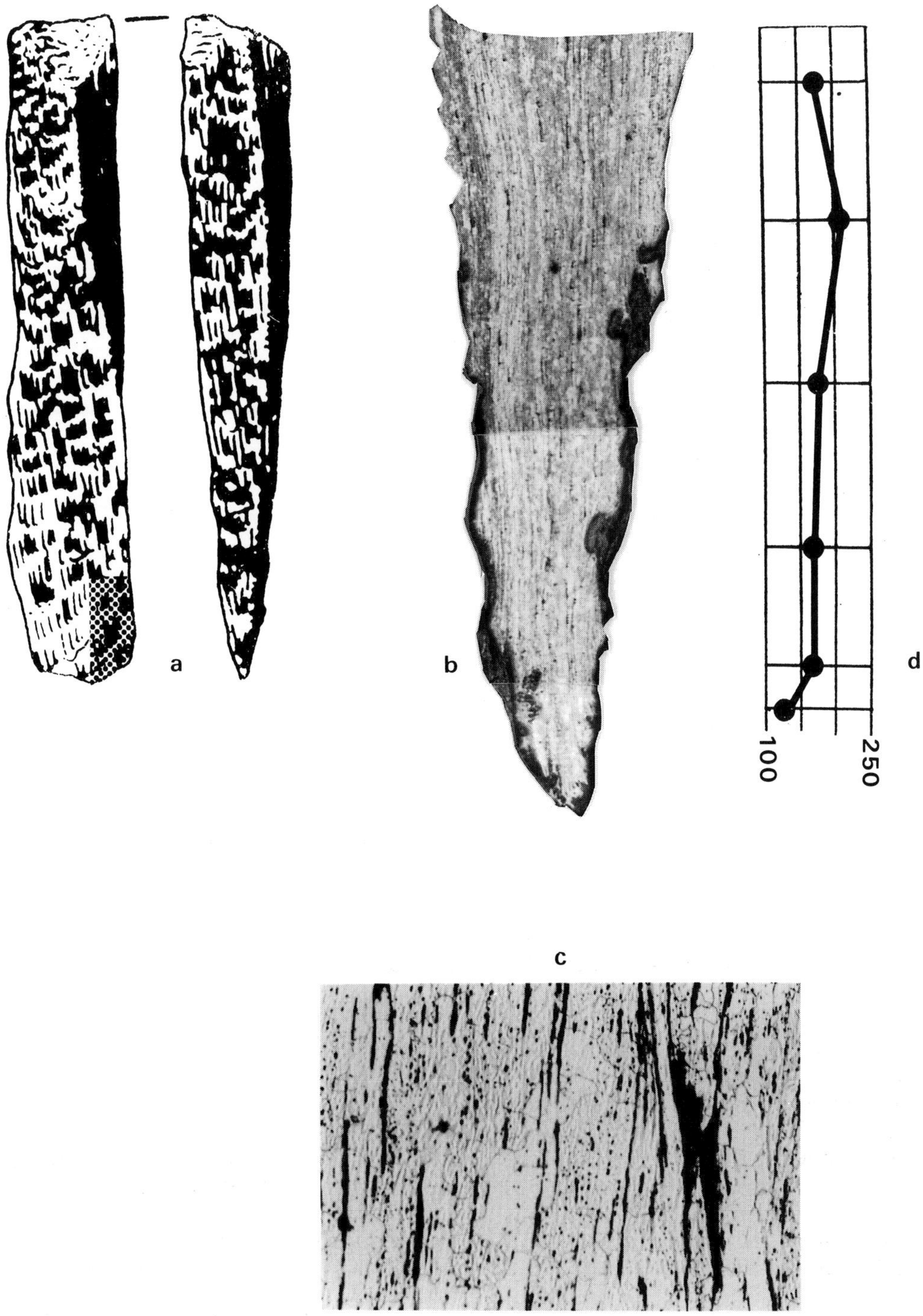

Plate 4.2.18 Examination 23: (a) NMI E28:63 chisel from Feerwore, Co. Galway x1/1, (b) macrostructure, (c) ferrite with massive slag inclusions x100, (d) H_v profile. Etched 4% Nital.

Oberhoffer reveals no other significant segregations (Plate 4.2.19d).
Microstructure: Carbon content is variable, with ferrite (GS 4–5) and interstitial pearlite, and a pronounced Widmanstätten structure in places (Plate 4.2.19e). No weld seams are visible either in macro- or micro-section. Pearlite is coarse, showing a tendency to spheroidisation.
Hardness: mH_v ferrite 175–186, pearlite 251.
Interpretation: The shears were forged from heterogeneous stock, retaining bloom welds in places. There are traces of secondary carburisation in the blade of the shears, but not in the spring. Neither area was subjected to heat-treatment.

(25) BM 80.8–2.119 Sickle, Lisnacrogher, Co. Antrim (*CIIAA* 585: Plate 4.2.20).

A section was cut from the cutting edge towards the back of the blade (Plate 4.2.20a).
Non metallic inclusions: Numerous large two-phase inclusions run parallel to the long axis of the specimen, and corrosion has penetrated along some of the slag lines.
Macrostructure: Etching indicates no significant segregations (Plate 4.2.20b).
Microstructure: The blade is composed almost entirely of large ferrite grains (GS 2–3). Towards one side is a patch of fine grained ferrite (GS 6–7) with

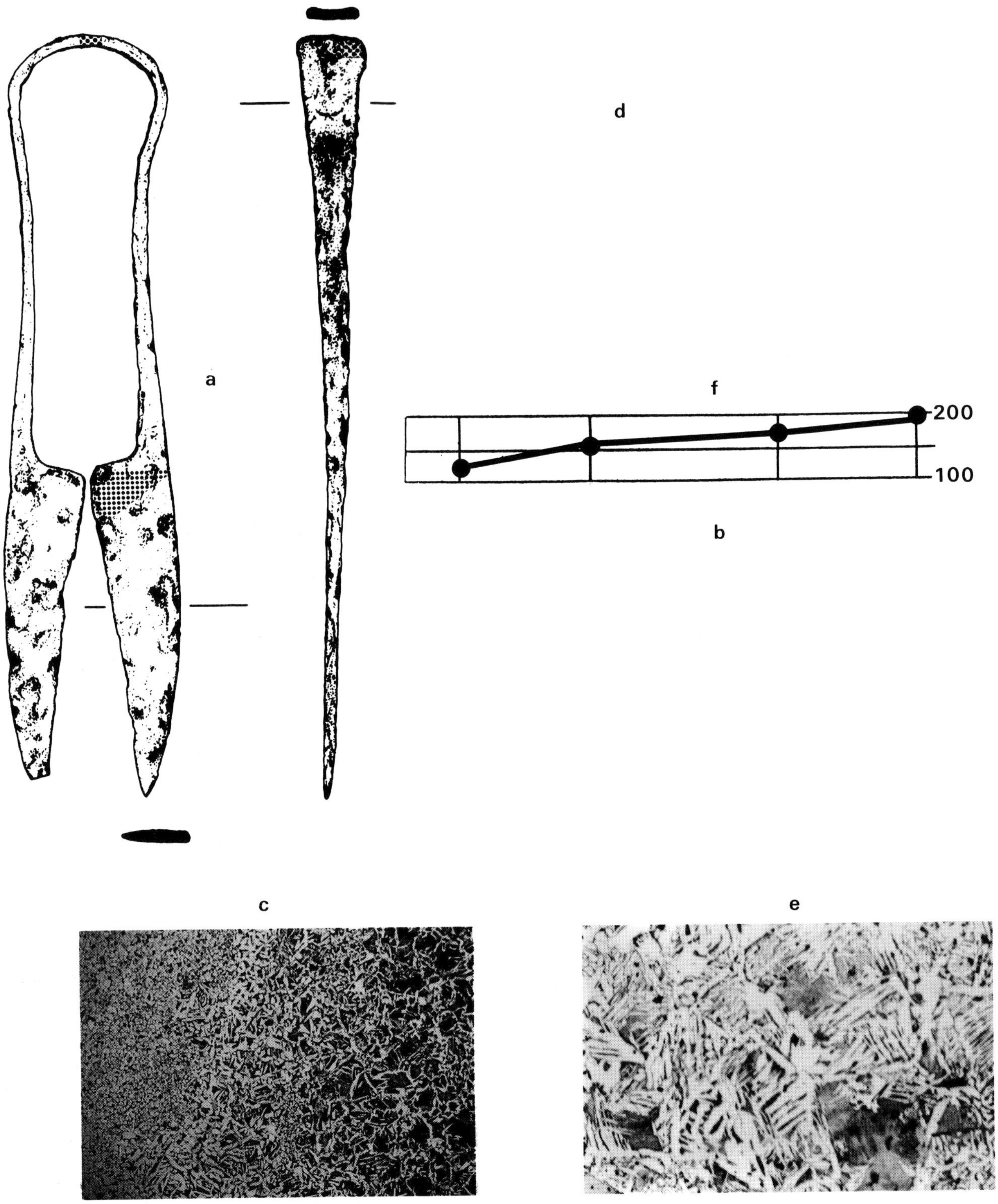

Plate 4.2.19 Examination 24: (a) NMI CC23:8/9 shears from Carbury, Co. Kildare x$^{7}/_{10}$, (b) macrostructure of section 1, (c) carbon gradient near cutting edge x50, (d) macrostructure of section 2, (e) Widmanstätten ferrite with pearlite x100, (e) H_v profile. (b)-(d) etched 4% Nital, (e) etched Picral.

some interstitial pearlite (Plate 4.2.20c). Examination of the corrosion envelope shows fragments of metal throughout, all ferritic, indicating that the small area of higher carbon content is primary.
Hardness: mH_v ferrite 133: H_v 82–92.
Interpretation: The sickle was forged from stock that was essentially carbon-free, but with small patches of higher carbon. The smith was not particularly sucessful in removing entrapped slag and this would not have been an especially serviceable tool.

(26) NMI E79:407 tanged, double-edged knife, Dún Ailinne, Co. Kildare (Plate 4.2.21)

A section was cut across the broken end of the knife (Plate 4.2.21a).
Non-metallic inclusions: Fairly numerous small two-phase inclusions running parallel to the long axis of the section are accompanied by some larger two-phase inclusions.
Macrostructure: Etching with Nital revealed a highly

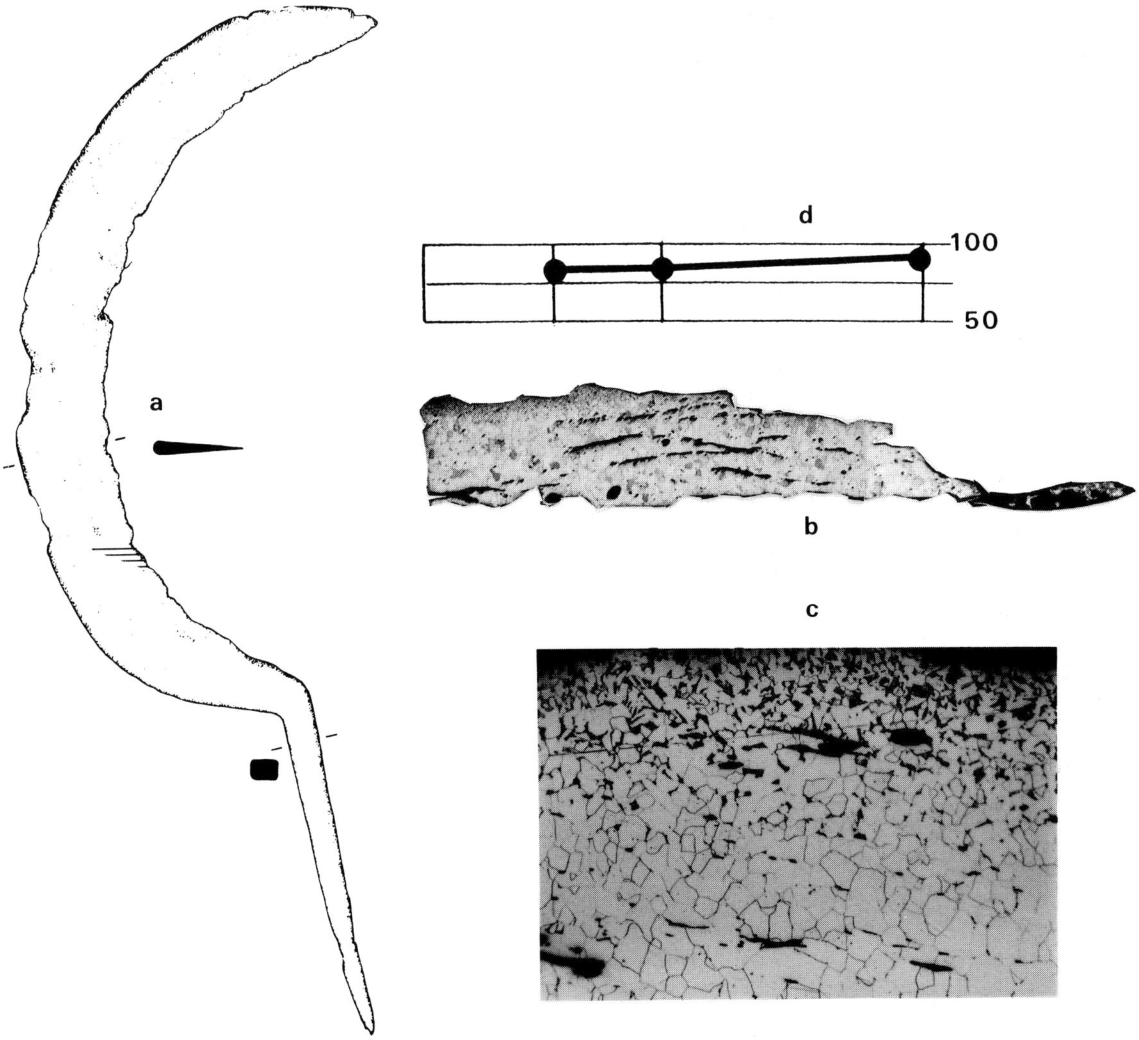

Plate 4.2.20 Examination 25: (a) BM 80.8–2.119 sickle from Lisnacrogher, Co. Antrim x½, (b) macrostructure, (c) slight carbon gradient x100, (d) H_v profile. Etched 4% Nital.

laminar appearance with a distinct concentration of carbon towards the centre of the section. Oberhoffer emphasised the laminar appearance (Plate 4.2.21b).
Microstructure: The structure varies from regular polyhedral ferrite (GS 4–5) with interstitial coarse pearlite, to very fine ferrite (GS <8) with coarse pearlite which shows spheroidisation of the carbide lamellae. The pronounced white lines (Plate 4.2.21c) are not accompanied by the chains of slag inclusions which are normally associated with weld seams. There are areas with closely packed multiple white lines which are associated with slag, and these are likely to be bloom welds. In addition to the central concentration of carbon, there is a distinct gradient in places towards the outside of the section.
Hardness: mH_v ferrite 157–234, pearlite 212–241: H_v 201–226.
Interpretation: This knife was forged from heterogeneous stock, which retained weld seams from the consolidation of the bloom. The banding of high and low carbon areas may be attributed to the presence of P and As in the bloom. The final treatment which the knife received was to be held above A_1 long enough for grain refinement to occur, but there was no attempt at heat-treatment.

(27) NMI E79:1926 round-sectioned rod, Dún Ailinne, Co. Kildare (Plate 4.2.22)

Three sections were cut, one longitudinal section from approximately the middle of the rod (*section 1*), one cross-sections at either end (*section 2*), and one longitudinal section (*section 3*) back from one end (Plate 4.2.22a). Sections 1 shows major penetration by corrosion.
Section 1: *Non-metallic inclusions*: .There are numerous, one- and two-phase inclusions throughout the section. Some of the inclusions are large.
Macrostructure: Etching with Nital revealed no segregation of carbon, but showed banding of ferrite (Plate 4.2.22b), an effect accentuated by Oberhoffer. Major penetrations of corrosion can be observed.
Microstructure: The structure comprises areas of large polyhedral ferrite grains with some grain boundary cementite (GS 2–3), and bands of small (GS 7–<8) polyhedral ferrite grains with minor amounts of coarse interstitial pearlite (Plate 4.2.22c). There are acicular precipitates in some of the larger grains (Plate 4.2.22d).
Hardness: mH_v large ferrite grains 180–187, small ferrite grains 192.

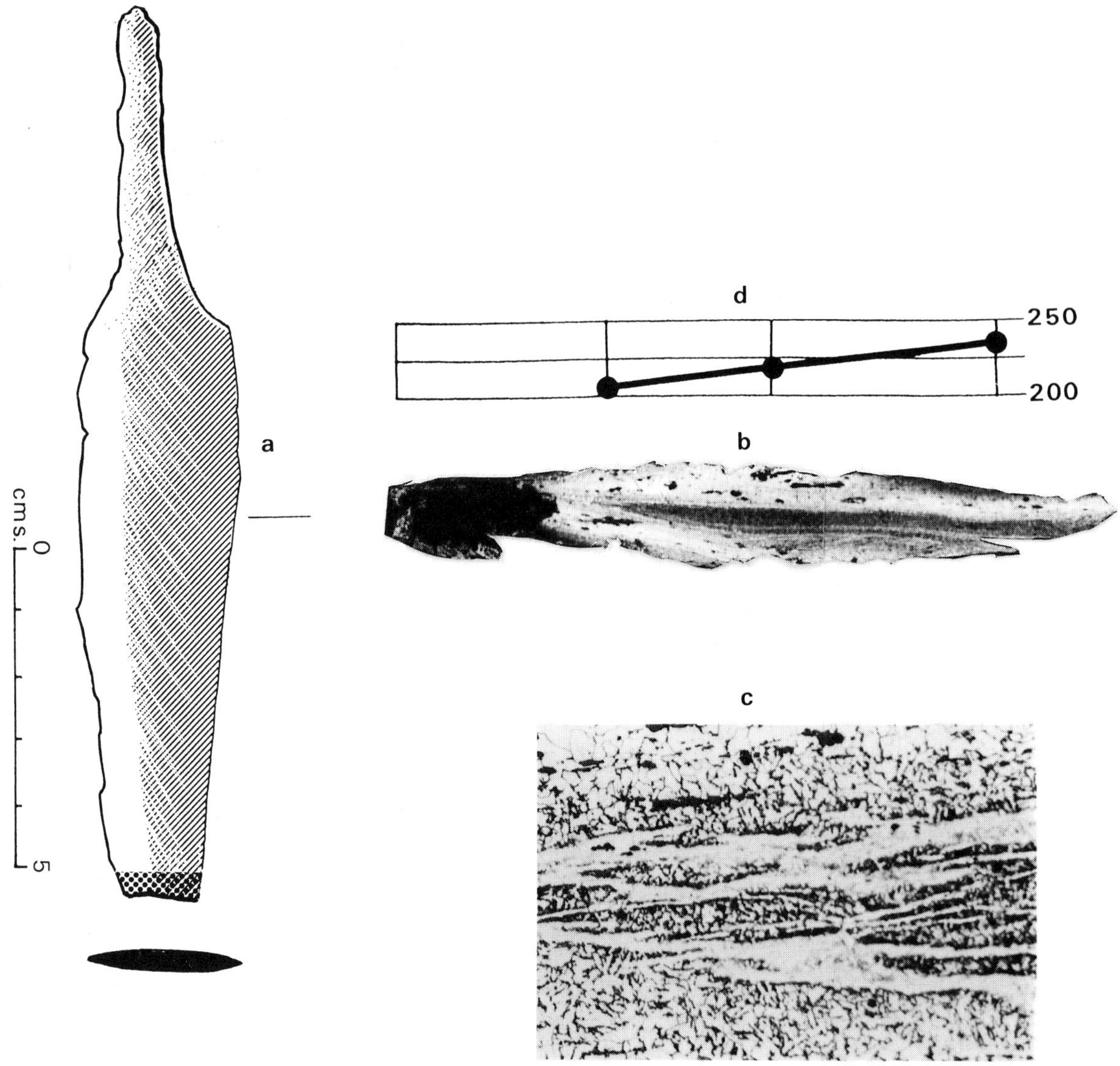

Plate 4.2.21 Examination 26: (a) NMI E79:407 tanged, double-edged knife from Dún Ailinne, Co. Kildare, (b) macrostructure (corrosion dark), (c) fine ferrite and interstitial pearlite, with banding x50, (d) H_v profile. Etched 4% Nital.

Section 2: *Non-metallic inclusions*: As for section 1. *Macrostructure*: Etching with Nital again showed banding of ferrite (Plate 4.2.22e), accentuated by Oberhoffer.
Microstructure: As with section 1, there is banding of larger (GS 4–6) and smaller (GS 7–8) ferrite grains. There is a small amount of interstitial coarse pearlite associated with the bands of small ferrite grains.
Hardness: mH_v large ferrite grains 151–206, small grains with interstitial pearlite 140–175.
Section 3: *Non-metallic inclusions*: As for section 1. *Macrostructure*: Etching with Nital showed the banding seen in sections 1 and 2 (Plate 4.2.22f). and this was even more pronounced on etching with Oberhoffer.
Microstructure: The structure consists of bands of larger polyhedral ferrite grains with some grain boundary cementite (GS 3–5), and smaller polyhedral grains (GS 7–<8) associated with minor amounts of coarse interstitial pearlite.
Hardness: mH_v large ferrite grains 148–208, small ferrite grains with pearlite 174–195.
Interpretation: The bar has been forged from stock of low carbon content but with significant segregations of P/As. Like carbon banding, banding of ferritic grain size results from heterogeneity of the bloom, but based on the distribution of P/As rather than carbon. The function of the artefact is unclear, although it is unlikely to have been particularly effective as a metalworking tool.

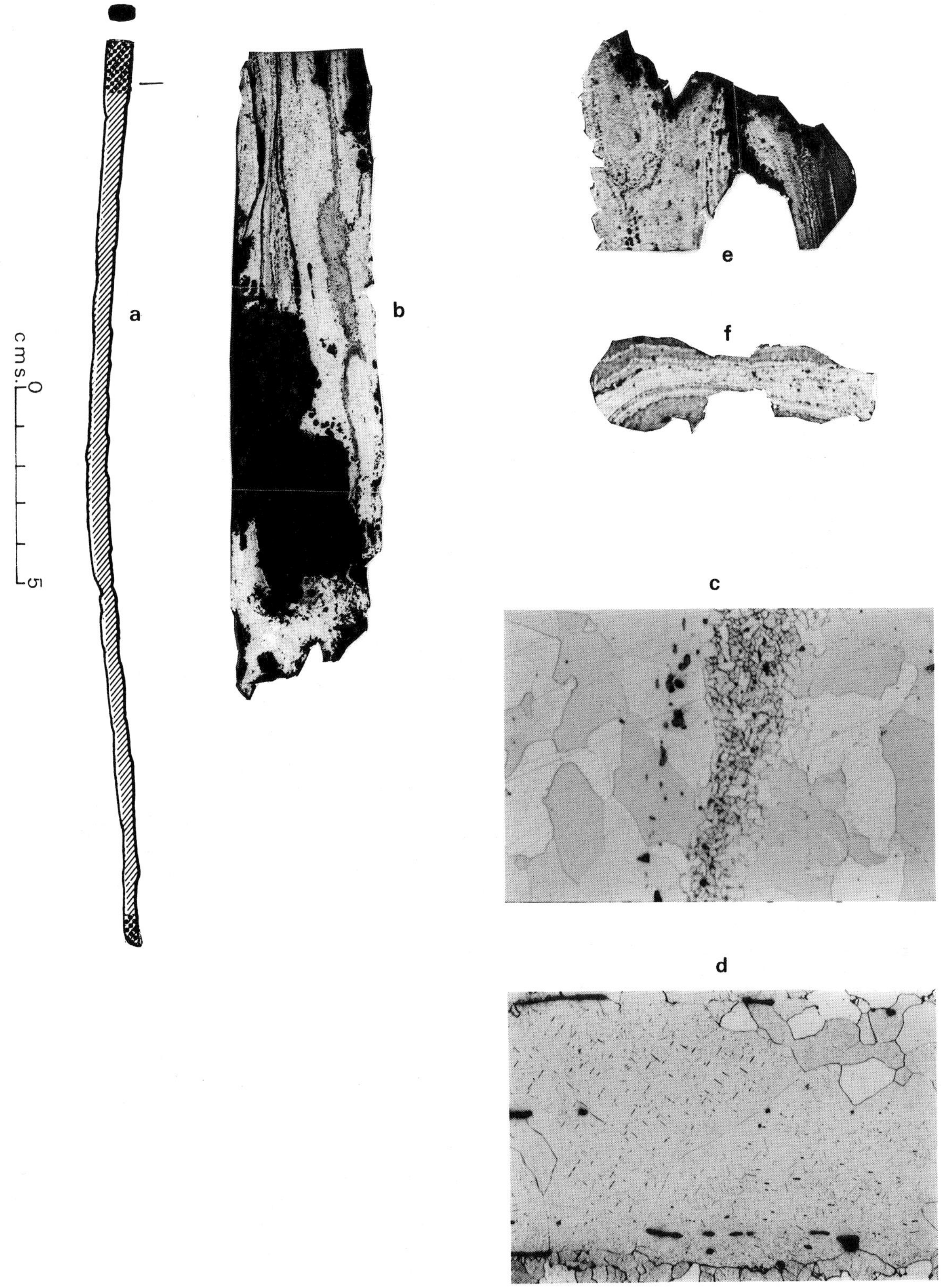

Plate 4.2.22 Examination 27: (a) NMI E79:1926 round-sectioned rod from Dún Ailinne, Co. Kildare, (b) macrostructure of section 1, (c) banding of ferrite in section 1 x100, (d) acicular precipitates (γ'-Fe_4N?) in enlarged ferrite grains x500, (c) macrostructure of section 2, (f) macrostructure of section 3. Etched 4% Nital.

Table 4.2.1

Summary of Metallographic results

Number	Artefact	Sample site	Wid	F/W	B/W	SEG	C/B	F/B	A	CAR	H/T	C/W	Carbon content%	Hv 1000gm	Structural components	Grain size	mHv 50gm
6 Plate 4.2.1	Sword Lisnacrogher Co. Antrim UM B2	*Section 1* cutting edge	–	–	–	+	–	–	–	–	–	–	0.00 – 0.01		ferrite pearlite	3 – 5 6 – 8 I	269
		Section 2 midrib area	–	–	?	+	–	–	+	–	–	–	0.00 – 0.01		ferrite pearlite cementite	>1 – 7 I GB	233
7 Plate 4.2.2	Sword Lisnacrogher Co. Antrim UM B5	blade cross- section	+	+	?	–	+	–	–	–	–	–	0.10 – 0.65	182 – 234	ferrite pearlite	7 – 8 I 4 – 5	315 448
8 Plate 4.2.3	Sword Lisnacrogher Co. Antrim BM 1880.8–2.116	cutting edge	–	–	–	–	–	–	–	–	–	–	0.00 – 0.01	129 – 140	ferrite pearlite cementite	6 – 7 I GB	100– 183
9 Plate 4.2.4	Sword Kildrinagh Co. Laois NMI WK10:W135 F468	cutting edge	+	+	+	–	+	–	–	+	?	–	0.00 – 0.70	115 – 162	ferrite pearlite	7 –<8 I 6 – 7	183 263– 369
10 Plate 4.2.5	Sword Kildrinagh Co. Laois NMI WK24:W137 P470	blade cross- section	–	–	–	–	–	–	–	–	–	–	0.00 – 0.01	99 – 113	ferrite pearlite	4 – 6 8 I	135– 178
11 Plate 4.2.6	Sword Banagher Co. Offaly NMI WK18:A130	cutting edge	+	–	+	–	?	–	–	–	–	–	0.00 – 0.50		ferrite pearlite	3 – 4 7 –<8 I 5 – 6	255 404
12 Plate 4.2.7	Sword(?) Toome Bar Co. Antrim UM 640–30	cutting edge	+	+	+	–	+	–	–	–	–	–	0.00 – 0.60	125 – 180	ferrite pearlite	1 – 2 7 –<8 I 5 – 7	194– 242 255– 295
13 Plate 4.2.8	Sword(?) Altanagh Co. Tyrone DoENI	X- section	?	–	–	–	–	–	–	–	–	–			remanent ferrite & carbide fragments		
14 Plate 4.2.9	Spearhead Lisnacrogher Co. Antrim UM B3	cutting edge	–	–	+	–	–	–	–	–	–	–	0.00 – 0.01	119 – 168	ferrite pearlite	1 – 2 4 – 6 6 – 8 I	100– 116
15 Plate 4.2.10	Spearhead Lisnacrogher Co. Antrim BM 1880.8–2.117	cutting edge	+	+	+	–	–	+	–	–	–	+	0.00 – 0.01	112 – 153	ferrite pearlite	5 – 8 I	102– 132
16 Plate 4.2.11	Spearhead River Shannon at Athlone Co. Westmeath NMI 1881–357: W1:WK40	cutting edge	–	?	–	–	–	–	–	–	–	–	0.00 – 0.05		ferrite pearlite	7 – 8 I	225
17 Plate 4.2.12	Spearhead NMI E79: 2305 Dún Ailinne Co. Kildare	cutting edge	–	–	–	–	–	–	–	–	?	–	0.00		ferrite cementite	2 – 3 GB	104– 136

I = interstitial GB = grain boundary
Wid = Widmanstätten structure F/W = fold-welding B/W = bloomery weld F/B = ferrite banding SEG = P/As seggregation C/B = carbon banding
A = acicular precipitates H/T = heat treatments C/W = cold working CAR = secondary carburisation
+ = present ? = uncertain occurrence – = absent or not detected

BM = British Museum DoENI = DoE Historic Monuments and Buildings Branch (NI) NMI = National Museum of Ireland (Dublin)
UM = Ulster Museum (Belfast)

Table 4.2.1 ctd.

Summary of Metallographic results

Number	Artefact	Sample site	Wid	F/W	B/W	SEG	C/B	F/B	A	CAR	H/T	C/W	Carbon content%	Hv 1000gm	Structural components	Grain size	mHv 50gm
18 Plate 4.2.13	Socketed axe Feerwore Co. Galway NMI E28:48	cutting edge	+	?	–	–	–	–	–	+	–	–	0.10 – 0.60	160 – 186	ferrite pearlite	6 –<8 4 – I	198 288– 313
19 Plate 4.2.14	Shaft-hole axe Kilbeg Co. Westmeath NMI 1935:406	cutting edge	+	–	–	–	–	–	–	–	–	–	0.00 – 0.01	123 – 143	ferrite pearlite	6 – 8 8 – I	233– 268 278– 332
20 Plate 4.2.15	Shaft-hole axe Kilbeg Co. Westmeath NMI 1935:407	cutting edge	–	+	–	+	–	–	+	–	–	–	0.00 – 0.01	129 – 182	ferrite pearlite	2 – 3 6 – 7 I	176– 182
21 Plate 4.2.16	Shaft-hole axe Kilbeg Co. Westmeath NMI 1935:408	cutting edge	+	+	–	+	–	–	–	–	–	–	0.00 – 0.01	110 – 168	ferrite pearlite	4 – 5 7 – 8 I	228– 244
22 Plate 4.2.17	Shaft-hole axe Castlereban Co. Kildare NMI 1957:61	cutting edge	–	+	–	+	–	–	–	–	–	–	0.00 – 0.01	88 – 163	ferrite pearlite	4 – 6 I	230
23 Plate 4.2.18	Chisel Feerwore Co. Galway NMI E28:63	cutting edge	–	–	–	–	–	–	–	–	–	–	<0.005 –	178 – 205	ferrite cementite	5 – 6 GB	193
24 Plate 4.2.19	Shears Carbury Hill Co. Kildare NMI CC23:8/9	*Section 1* blade	+	–	+	–	–	–	–	+	–	–	0.00 – 0.45	133 – 197	ferrite pearlite	6 –<8 8 – I	162– 214 282
		Section 2 spring	+	–	–	–	–	–	–	–	–	–	0.00 – 0.50		ferrite pearlite	4 – 5 4 – I	175– 186 251
25 Plate 4.2.20	Sickle Lisnacrogher Co. Antrim BM 1880.8–2.119	cutting edge	–	–	–	–	–	–	–	–	–	–	0.00 – 0.01	82 – 92	ferrite pearlite	2 – 3 6 – 7 I	133
26 Plate 4.2.21	Knife Dún Ailinne Co. Kildare NMI E79:407	X-section	–	+	?	+	–	–	–	–	–	–	0.00 – 0.01	201 – 226	ferrite pearlite	4–<8 8 – I	157– 234 212– 241
27 Plate 4.2.22	Rod Dún Aulinne Co. Kildare NMI E79:1926	X-sections	–	+	+	+	–	?	+	–	–	–	0.00 – 0.01		ferrite pearlite cementite	2 – 3 7 –<8 I GB	140– 208

I = interstitial GB = grain boundary
Wid = Widmanstätten structure F/W = fold-welding B/W = bloomery welds F/B = ferrite banding SEG = P/As segregation C/B = carbon banding
A = acicular precipitates H/T = heat treatments C/W = cold working CAR = secondary carburisation
+ = present ? = uncertain occurrence – = absent or not detected

BM = British Museum DoENI = DoE Historic Monuments and Buildings Branch (NI) NMI = National Museum of Ireland (Dublin)
UM = Ulster Museum (Belfast)

Table 4.2.2

Chemical analyses

Reg. no.	Mg	Al	Si	P	S	Ca	Ti	V	Cr	Mn	Co	Ni	Cu	Zn	As	Zr	W	Au	Pb
6 UM B2 /1	0.008	0.032	0.068	0.038	tr	0.027	tr	nd	nd	0.102	0.064	tr	tr	nd	0.014	nd	nd	nd	tr
/2	tr	tr	0.011	0.139	tr	0.078	nd	nd	tr	0.072	0.148	0.125	tr	nd	0.032	nd	nd	nd	nd
7 UM B5	nd	0.007	0.007	0.063	nd	0.010	nd	0.034	tr	0.027	0.136	0.022	0.004	tr	tr	nd	nd	tr	tr
8 BM 1880.8–1.116	0.005	0.007	0.027	0.175	tr	0.046	nd	nd	tr	0.096	0.023	tr	tr	nd	0.070	nd	tr	tr	tr
9 NMI WK10:W135: F468	tr	0.030	0.071	0.057	nd	0.033	nd	nd	0.060	0.026	0.011	0.018	0.007	nd	0.030	nd	tr	nd	nd
10 NMI WK24:W137 P470	0.003	0.029	0.102	nd	nd	0.001	nd	nd	0.089	0.022	0.009	0.024	0.010	tr	tr	nd	nd	nd	nd
11 NMI WK18:A130	tr	0.039	0.088	0.083	tr	0.012	nd	nd	0.076	0.053	0.011	0.030	0.012	nd	0.030	nd	nd	nd	nd
12 UM 640–30	tr	tr	0.039	0.036	tr	0.010	nd	tr	tr	0.011	0.007	0.004	0.003	nd	0.014	nd	tr	tr	tr
13 Altanagh	no analyses made on this wholly corroded blade																		
14 UM B3	0.001	0.020	0.065	0.079	tr	0.044	0.007	tr	tr	0.007	0.098	0.009	0.004	0.004	0.050	nd	nd	tr	tr
15 BM 1880.8–2.117	nd	0.014	0.042	0.089	tr	0.019	tr	tr	nd	0.011	0.008	0.040	tr	tr	0.055	nd	nd	tr	tr
16 NMI 1881–357:	0.001	0.008	0.089	0.072	0.014	0.045	tr	nd	tr	0.059	0.021	0.030	0.005	0.006	0.037	tr	tr	tr	0.028
*17 NMI E79:2305	—	—	—	0.093	—	—	nd	nd	0.007	0.061	0.011	0.010	nd	nd	0.068	—	nd	—	—
18 NMI E28:48	0.001	0.007	0.074	0.115	0.026	0.028	0.002	nd	tr	0.028	0.066	0.036	0.005	0.005	0.151	tr	tr	nd	0.029
19 NMI 1935:406	0.021	nd	0.013	0.656	tr	0.009	nd	nd	0.054	0.030	0.025	0.017	0.010	nd	0.726	nd	nd	nd	nd
20 NMI 1935:407	0.019	nd	0.034	0.632	nd	0.008	nd	nd	0.053	0.025	tr	0.017	0.006	nd	0.786	nd	nd	nd	nd
21 NMI 1935:408	0.044	nd	0.029	0.291	tr	0.027	nd	nd	0.040	0.065	0.006	0.040	0.065	nd	0.256	nd	nd	nd	nd
22 NMI 1957:61	0.001	0.004	0.059	0.159	0.013	0.025	0.001	nd	tr	0.007	0.036	0.021	0.006	0.004	0.243	tr	nd	tr	0.030
23 NMI E28:63	0.003	0.015	0.187	0.289	0.032	0.024	0.004	nd	tr	0.047	0.007	0.016	0.011	0.004	nd	nd	nd	tr	0.029
24 NMI CC23:8/9 /1	0.007	0.013	0.102	nd	0.021	0.056	nd	nd	tr	0.008	tr	0.028	0.006	0.022	0.372	tr	nd	tr	0.025
25 BM 1880.8–2.119	nd	0.003	0.039	0.056	nd	0.017	nd	nd	tr	0.018	0.052	nd	nd	nd	0.016	nd	nd	0.007	nd
*26 NMI E79:407	—	—	—	0.101	—	—	nd	nd	nd	0.023	nd	0.036	0.022	nd	0.099	—	nd	—	—
*27 NMI E79:1926	—	—	—	0.132	—	—	nd	nd	nd	0.030	nd	nd	nd	—	—	—	0.083	—	—

*= trace (P, Pb <0.02%, As <.01%, Au <0.005, others <0.001) nd = not detected * = analysis by XRF — element not sought*
BM = British Museum DoENI = DoE Historic Monuments and Buildings Branch (NI) NMI = National Museum of Ireland (Dublin)
UM = Ulster Museum (Belfast)

* * *

4.3 – Summary and Conclusions

The metallographic examinations of artefacts of EIA B date from Ireland show what is perhaps a disappointingly restricted repertoire of techniques for the manipulation of the properties of iron and steel. Thus there are only three definite examples of secondary carburisation and none of heat treatment, among the artefacts examined. It is clear that smelters were producing both iron and primarily-carburised iron, but it is difficult to make any judgement on the overall level of output of iron during EIA B on the basis of survivals. However, the axes of the Kilbeg, Co. Westmeath hoard (examinations 19–21) and the very similar Castlereban North, Co. Kildare axe (examination 22) show the use of small pieces of iron welded together. The axes of the Kilbeg hoard weigh in total 2.42 kg, the largest axe (NMI 1935–406) weighing 1.15 kg, the smallest (NMI 1935–408) weighing a mere 0.39 kg. This might indicate smelts producing only small blooms during the period.

Such small-scale production should not be construed, however, as meaning a shortage of iron, an untenable explanation advanced by Warner (1974, 46) to explain the small dimensions of swords of the period. Although iron is far less common than the other metals, the fact that it was used to produce fine and prestigious pieces, as well as mundane pieces suggests strongly that what was lacking were numbers of smiths, and production centres with the capacity for the sort of bulk output of contemporary continental industrial sites. It is worth noting in this context the five tons of iron nails buried by Roman legionnaries at their fort at Inchtuthill, Perth (Angus et al. 1962). The excavators argued convincingly that this was to deny such a mass of iron to local tribes, used to producing small quantities at a time and for whom this would have been an industrial windfall. They do not, however, suggest that local tribes could not produce iron, but rather that they could not in any way match Roman mass production output since they lacked the industrial organisation necessary. This also would appear to be the situation in Ireland for most of EIA B. But we must not deny the existence of non-ferrous metalworking foci, since as both Warner (1983) and B. Raftery (1984, 324) point out, the artistic tradition running through EIA B and expressed chiefly in non-ferrous metalwork is characterised by what Raftery describes as '. . . technical and artistic homogeneity . . . within the confines of rigid and mutually accepted stylistic parameters.' Indeed, the tightly-knit group of artefacts from north-east Ulster, dating to the earlier part of EIA B, and to which one could happily give the title 'Lisnacrogher tradition', shows that workshops housing highly skilled craftsmen, albeit limited in scale of production, were in operation, at this time. But stylistic homogeneity is not really in evidence when we come to look at ironwork.

We may see this as further evidence of a break with the Dowris tradition in which major concentrations of metalworking expertise undoubtedly existed. If we look to barbarian Europe, then we may see a critical difference between the production of iron in Ireland and its production there. Unlike the Irish, or the groups around Inchtuthill, the smiths who worked on or around *oppida*, had become meshed into industrial organisations which served markets demanding the capacity for mass production. This is most strikingly illustrated by those sites where vast quantities of iron were consumed in the construction of iron-nailed timber defences (e.g. Collis 1976, 19). With the growth of urbanisation in barbarian Europe, the production of iron was, in turn, integrated into a much wider industrial and economic infrastructure. This simply did not happen in Ireland, despite the fact that the potential for similar development had existed in the Dowris period. That it did not occur may be attributed to the restructuring of society, and of the industries serving it, which was taking place during the latter half of the first millennium BC and the earlier centuries AD in the aftermath of the Dowris collapse.

The rarity of secondary carburisation to improve the quality of bloomery iron, and the absence of heat-treatment, is disappointing. The metallographic evidence demonstrates that the swords of this period were not the super weapons described in the heroic literature of the Later Iron Age. Indeed, the mechanical properties which can be projected for the swords examined in this section invite comparison with those of the Gauls, described by Polybius and Plutarch (Maryon 1960; Lang 1987, 64), which became bent and blunted in battle. In fact, Irish smiths in both EIA B and LIA produced swords which could well have proved a serious embarrassment to the hero wishing to deal spectacular death with single blows. Thus, we may note that in the story of the battle of *Magh Tuired* (Chapter 7.4), the smith-god *Gobniu* promised to forge weapons that were infallibly lethal, but was forced to effect instant repair on those which were damaged during each day's fighting!

One technique which does show up in nine (possibly eleven – Table 4.2.1) of the 22 examinations of EIA B artefacts discussed here is fold-welding, which produced laminar structures in cross section. This is particularly prominent in two of the three axes of the Kilbeg, Co. Westmeath hoard, and the very similar axe from Castlereban North, Co. Kildare: these three pieces also show significant levels of P and As. Fold-welding can be construed as an attempt to even out the distribution of carbon in stock with uneven primary carburisation, and in some of the specimens, such as Lisnacrogher sword UM-B5, it has succeeded. In the case of the sword from Kildrinagh, Co. Laois (examination no. 9), the process had not obliterated bloom welds. This contrasts with the evidence from EIA A, where fold-welding has not been detected, while secondary carburisation and heat treatment were. Fold-welding can be seen in a significant number of artefacts of the British and continental Iron Ages and was obviously common practice not just in Ireland. The fact that the technique has not been observed in EIA A artefacts may well be due to the small numbers of samples available for examination. Most likely, it was an early introduction along with the more specialised techniques observed in artefacts of EIA A.

The small size of the La Tène style Irish swords relative to their British and Continental counterparts has already been noted. An explanation for this trend has been offered recently by Rynne (1984, 10) in discussing a newly-discovered sword from Dangan Td., Co. Galway. He suggests that this was the direct result of an inability to fabricate longer weapons of adequate mechanical properties to be serviceable. Unfortunately, this is as faulty in its logic as Warner's iron shortage hypothesis: the realistic limit to the length of an iron sword would have been the strength of the user. When we look to comparative evidence for the technology of La Tène swords in other regions (e.g. McGrath 1968; Tylecote and Gilmour 1986; Lang 1987; Pleiner forthcoming), the Irish weapons do not seem wholly out of place metallurgically, although falling into the lower end of the quality range. Various workers have found contemporary swords of much greater length than the Irish examples (>65 cm) with inhomogeneous ferrite/pearlite structures. There is, however, a significant proportion of foreign blades showing evidence of secondary carburisation of cutting edges, of the welding on of carburised components into the cutting edge region and of heat treatment[11]. In this context, we may anticipate the discussion in Chapter 7 of literary descriptions of ironwork and ironworking and note the continuous emphasis placed on hilt and scabbard furniture as opposed to the actual sword blade itself. It is quite possible that sword size was dictated by fashion, with the decorated scabbard and hilt, as display pieces, being more important than the blade, and with the serious killing being done at this time with the spear.

There is a very clear distinction to be made here between the lack of manipulation of the properties of iron as seen from these examinations, and the *shaping* and *jointing* of iron. Some of the swords and spearheads of EIA B are masterpieces of the smith's craft. Surviving ironwork of the period displays a competence in metal forming that is a legacy of the long tradition of hammering metal stretching back at least to the Later Bronze Age. We can see this in swords such as those from Lisnacrogher, Edenderry and Ballinderry, with their elegant triple-ribbed blades, and from the sword from Dangan Lower, Co. Galway sword with its close-spaced central ribbing, more than somewhat reminiscent of continental Hallstatt C swords. Enhancing blades in this manner shows a sensitivity and artistry matching that displayed by the engraved scabbard plates by which many such weapons would have been enclosed. Leaving aside their effectiveness as killing tools, such swords and their scabbards must have been strikingly attractive and sources of great pride to those who carried them.

The expertise of smiths of EIA B in shaping iron is shown further by the spearhead from the River Shannon (Plate 4.2.11), sadly much damaged by corrosion. Its shape may be determined, and again, the high quality of workmanship is evident in the symmetry of even the surviving portions of the blade and socket, and in the even taper of the midrib. Similar fine workmanship can be found in the spearhead from Corrofin, Co. Clare. The prominent midrib and leaf-shaped blades of the 'River Shannon' and Corrofin spearheads are matched in elegance by those from Mullaghmore, Co. Roscommon and Moyfin, Co. Meath (cf. Fig. 4.1.1). The Moyfin spearhead again has an elegant tapering midrib, but this time a much narrower blade (again, much damaged by corrosion). In each of these spearheads, the socket has been carefully forged round and closed by skilful welding that has left the thickness of the metal even throughout the cross section.

Even the more mundane ironwork of the period shows a good command of the skills of forging. The socketed axe from the Rath of Feerwore, Co. Galway (Plate 4.2.13) shows that the smith had mastered the knack of forging a right-angle, something which requires no little expertise in upsetting the metal to prevent the metal from thinning out at the bend (cf.

Fig. 4.2.1). The simple shapes of the axes from Kilbeg and Castlereban belie the skill which has gone into the attachment of socket to blade. The severe chemical cleaning to which they have been subjected has highlighted the sites of the welds, but it would seem that originally they would have blended nicely into the surfaces of the blade faces. The lines of the Carbury shears (Plate 4.2.19) and the Lisnacrogher sickle (Plate 4.2.20) again show pleasing symmetry which belies the indifferent quality of the stock from which they were forged.

We may yet conclude from the admittedly scant survivals of the period, that during EIA B iron became firmly established as the major industrial metal, even though its production was at a low level in comparison with contemporary barbarian Europe. Because of the lack of firmly dated pieces, it is not possible really to give any realistic quantitative assessment of the rate of increase in production from the start to the finish of EIA B. We know that at the start of the period, iron was comparatively scarce, while by LIA A, it was commonplace. We have seen that in EIA A, iron was used, largely experimentally, in what was in essence a Later Bronze Age industrial setting. The stimulus for this must have been through contact with foreign iron-using peoples. It is clear also that such contacts must have been at first hand, since such complex technology cannot have been learned simply through second-hand accounts. Contacts in prehistory, seen from finds of foreign-produced artefacts, are the results of movements of one or more persons. While obviously it can be argued that foreign artefacts in Ireland and Irish artefacts abroad result from the peregrinations of outsiders, it is more likely that both Irish and non-Irish alike plied the sea-lanes to and from Ireland, bringing novel artefacts and novel ideas. Such contacts are clearly attested for the Later Bronze Age in Ireland, a time when, elsewhere in the barbarian world, metalworkers were coming to grips with the new metallurgy, in turn as a result of contacts with the east Mediterranean (Chapter 2.1). It is by this route that the first knowledge of ironworking and artefact design, including the techniques of fold-welding, secondary carburisation and heat-treatment, would have reached Ireland. But were there further sources of stimulus to spur on the development of a more technologically advanced industry?

B. Raftery (1984, 333f) concludes that there is strong evidence for widespread contacts between Ireland, Britain and western Europe in the period up to the earlier centuries AD. For the rest of EIA B, Warner (1976, 1983) has provided equally clear evidence of continuing contact with the late- and post-Roman world. It does seem that in the later centuries BC, Irish smiths were at least as expert in the shaping of iron as their counterparts working in southern Britain and Atlantic Europe. From the metallurgical studies of Piaskowski and Pleiner in particular, we can see that contemporary central European smiths were regularly producing good quality artefacts whose manufacture often involved secondary carburisation and heat treatments. We do not have the same body of evidence for the Atlantic seaboard, and this is unfortunate, since it is to this region that we might look as a potential source of knowledge available to Irish smiths.

For Britain, the largest bodies of metallographic examinations are contained in the publications by Ehrenreich (1985) and by Tylecote and Gilmour (1986). In examinations of some 500 pieces from a number of Iron Age sites in Wessex (centering on Danebury, Hampshire), Ehrenreich was only able to identify one example of heat-treatment[12]. The work by McGrath (1968), by Lang (1987) and by Tylecote and Gilmour (1986) on British Iron Age swords does not indicate outstanding superiority over Irish smiths, although one could make the purely subjective judgement that overall the tendency is towards higher quality. While we might look to the Romans to have brought a higher standard of technology to Britain, Tylecote and Gilmour (1986, 93ff) clearly demonstrate quite the reverse, with average standards being fairly appalling! By the late Pre-Roman Iron Age, the shaft-furnace had reached Britain (Tylecote 1986, 132f) and was used there throughout the Roman period. In the light of present evidence at least, this technology certainly did not travel across the Irish sea. Thus, even with frequent contact between Ireland and pre-Roman, then Roman-dominated Britain, there was no immediately obvious source there of superior forging expertise for Irish smiths to draw on to advance their skills and the potential advantages of the shaft-furnace in increasing production went unheeded.

In terms of the organisation of production, there is simply no evidence in EIA B for the development of industrial concentrations which could output large quantities of iron (or of other materials) and of artefacts. Without such foci for external contact and internal cross-fertilisation, stimuli towards innovation and experimentation would have been restricted. The Lisnacrogher finds could well be the products of a smith or smiths working under royal patronage, although this workshop had a low output. We might also see the fine blades from Edenderry, Co. Offaly as indicative of another, similarly small-scale operation producing status weapons for local royalty. In each case, the workers here could have been open to learn through foreign contact. If these interpretations are correct, then Lisnacrogher, and its unidentified counterpart in Offaly, perhaps constitute the first evidence for the growth of patronage support for master craftsmen which is prominent in the Later Iron Age.

Irish polity certainly appears to have been in a state of flux in the earlier centuries AD (Warner 1988, 55), with no one grouping emerging as other than a political power by the later 1st millennium AD. There is an important distinction to be made here between the political power and influence which might have been exercised by any one of the many Irish kings of later prehistory and early Irish history, and the economic power which comes from centralised industrial and economic structures. There is much evidence for the former, none at all for the latter. Kingship in historic times rested on wealth, often expressed ostentatiously in the size of the royal residence, and kings received food and services from clients (e.g. Kelly, F. 1988, 26ff) who appear to have operated and produced independently from one another. There is no evidence that agriculture and industry had any central foci mirroring the central focussing of power in kingship.

If we take the somewhat dangerous course of

extrapolating back into late prehistory, then we might also see a similar expression of wealth and power through such projects as the linear earthworks of the Dorsey in Co. Armagh (Lynn 1982a) and the Black Pig's Dyke in Cos. Monaghan and Armagh (Walshe 1987), and the massive wooden trackway at Corlea, Co. Longford (Raftery, B. 1987). The earthworks were under construction in the last two centuries BC, while timber from Corlea has been dated by dendrochronology to 148 BC. These works required co-ordination and direction, and, for EIA B, stand out as examples of mass effort. However, imposing as these might be, their construction would have been comparatively simple to organise as 'one-offs', since the primary requirement would have been for labour, not for large amounts of manufactured materials (as was the case with the *murus gallicus*). The development of large-scale production in agriculture, and in metals and other materials, organised to service on a regular basis markets wider than the territories of individual rulers would have been quite a different undertaking. Although the evidence is scant, it is probable that EIA B was the period in which the pattern of dispersed, small-scale production that characterises LIA A became fixed in place.

Given the apparent lack of consistently high-quality industrial production in those regions with which the EIA B Irish would have been in regular contact, and the liklihood that production inside Ireland was dispersed and small-scale, it seems obvious that the opportunities for Irish ironworkers in EIA B to learn and to develop more advanced smithing techniques were strictly limited. Although the limited range of techniques displayed might be disappointing, the perceived drop in quality between EIA A and EIA B is not unexpected in the broader context. We have noted already (Chapter 2.1) that in the general European record, there is an initial horizon of high-quality material which is followed by a lowering of standards. This can be explained as resulting from the dilution of expertise, initially concentrated amongst workers in non-ferrous metals, as iron became more widely available, with smiths being less firmly grounded in metalworking than their pionneering predecessors. In barbarian Europe, this resulted from the growth in scale of production involving large numbers of workers. In Ireland, there would have been the same dilution of expertise, but with the additional complication of the dispersal of production, a situation in which development of skills through the interaction of craftsmen would have been difficult. Contact and exchange would have been localised and limited, and this explains, for example, the failure of Irish ironworkers to acquire and exploit the superior production capacity of the shaft furnace. The techniques of shaping metal, evident both in copper alloy and iron, show that it was not lack of imagination or ability that prevented the improvement of technological ironworking skills, but rather the inability to respond effectively to any available outside stimuli, rare as these seem to have been.

It is noticeable that one of the principal concentrations of EIA B material is in north-east Ulster with its major deposits of lateritic iron ores (Chapter 6.2). In similar circumstances in barbarian Europe, possession of such resources would probably have led to the establishment of a highly profitable monopoly. It was the monopoly on another precious commodity, salt, which formed the basis of the wealth of the inhabitants of the Hallstatt region. We might see the concentration of EIA B material in this region as evidence of the recognition of its metalliferous wealth. But the potential for the development of the large and centripetal socio-economic and political groupings required to exploit such assets had already disappeared by the end of Dowris C.

* * *

Notes to Chapter 4

1 B. Raftery (1984, 314), usefully discusses the doubts over the nature of the original find-spot.

2 Early writers such as Coffey (1910) and Armstrong (1923) identified a number of spearheads as belonging to the EIA B assemblage on the basis of typological parallels with La Tène Europe. Barry Raftery is much more austere in his criteria (*CIIAA* I, 84), but while appreciating his caution, I will include a number of artefacts into this period which do not appear in *CIIAA*.

3 Cf. the form of files from Manching, Bavaria (Jacobi 1974, Taf. 5) with the surviving drawing of the Carbury piece (*CIIAA* II, Fig. 153).

4 It must be remembered, however, that such marks have often arisen in modern times when finders have occasionally taken a file to an object 'to see what it is made of'!

5 The use of specially shaped punches to produce special marks on ironwork is not unknown in barbarian Europe. Thus, one such punch was used to produce the image of a crescent with a winged skull on the tang of a sword from Augsberg (Lang 1987, 71).

6 B. Raftery (1984) draws attention to early Irish *gaisced* 'weapons, arms, valour, feats of arms, skill at arms etc.' (*DIL* Fasc. G:33.36f: *gaibid gaisced* 'assumes arms'), and *gaiscedach* 'one bearing arms, a warrior, a champion' (*DIL* G:33.77f), which derive from *gae* 'spear, javelin' (*DIL* G:10.26ff), a term of venerable antiquity. As Mallory (1982, 108f) notes, swords which we identify as part of the EIA B assemblage would not be realistic candidates for the weapons which decapitated or split opponents from crown of head to navel, with a single stroke. It is obviously only a subjective judgement, but it seems to the writer that violent death by spear thrust is at least as common as death by the sword in those tales in early Irish literature which K. Jackson (1964) cited as providing a 'window' on the Irish Iron Age. The spear also seems to have been as prominent as the sword in the Welsh classic *Y Gododdin* – Jarman 1988). Even allowing for the greater susceptibility of iron to corrode to completion in comparison with copper

alloys, then the survivals of spearheads and butts outnumber those of swords by more than three to one in EIA B.

7 In his discussion of the finds from Lagore crannog, Co. Westmeath, Hencken (1950, 88ff) raised the question of the influence of Roman weaponry on Irish Iron Age swords. He suggested that a group of Irish weapons could be identified which are closely similar in shape to the Roman *gladius* or short stabbing sword, and the longer, more slender *spatha*. In view of the evidence for intercourse across the Irish Sea in the 2nd to 5th centuries AD (e.g. Warner 1976), including extensive Irish raiding in Britain, this is a quite reasonable hypothesis. It is marred only by the lack of good dating evidence in the form of members of the group from firmly dated archaeological contexts. In fact, the only dating of any kind at present is for specimens among the old finds at Lagore crannog, Co. Westmeath, which is unlikely to predate the earlier 7th century AD. For this reason, while recognising that these swords may well have their origins in EIA B, they will be discussed along with other swords of LIA A in the next chapter.

8 Both Warner (1974, 45) and Caulfield (1981, 208) follow Alcock's mistaken attribution of an axe from Deganwy to this type (Alcock 1972, 110). In fact, this axe is of shaft-hole type, not socketed (Scott 1974b, 48).

9 I am indebted to Ms. J. Lang, British Museum Research Laboratory for taking the samples from this weapon, and from the Lisnacrogher spearhead (BM 80.8–2.117) and sickle (BM 80.8–2.119), and for collaborating with me in the metallographic examinations.

10 I am indebted to Ms. K. Ryan and Ms. K. Moreau, MASCA for providing me with data for examinations 17, 26 and 27, and to Professor B. Wailes, University of Pennsylvania, the excavator of Dún Ailinne, for making samples available to me, and for permission to include the results here.

11 A major survey of metallographic studies on Iron Age swords has been undertaken by Radomír Pleiner, and the work is at present in press. I am indebted to him for the opportunity to read this work in manuscript and to make reference to it here in advance of its publication.

12 Ehrenreich's methodology might be open to criticism on the grounds that the millimetric sizes of the samples which he took were too small to detect more than those features which occupied a sufficiently large proportion of volume as to make it most likely that they would be represented in the sample. Set against this is the fact that he sited his samples at working edges and surfaces. The 'ideal' sample, an entire longitudinal half-section of an artefact, is something which is only available in the most exceptional of circumstances, and thus it is not impossible that Ehrenreich's table of technological interpretations is an incomplete representation of the range of features which might have been found had larger samples been taken.

Note Added in Press

Just as this book was going to press, an *iron* doorknob spearbutt from Co. Fermanagh was brought to the Ulster Museum (Warner pers. comm.).

CHAPTER 5

Consolidation and Development – Later Iron Age A

5.1 – Ironworking and infrastructure

The boundary dates used to define Irish Later Iron Age A are ca AD 500 at the lower end and ca AD 900 at the upper. As we shall see below, these do not constitute an especially effective demarcation, since some artefact types defy convenient allocation to chronological zones. Nevertheless, during this period, the archaeological record 'comes alive', with occupation sites in abundance, and assemblages, some rich indeed, of artefacts in a variety of media. Although in EIA A and B we see iron-use becoming established, it is in LIA A that we can demonstrate for the first time the industrial predominance of iron. Most sites produce evidence of the smelting and smithing of iron, and the first identifiable blacksmithing tools appear in excavation assemblages, albeit in very small numbers (cf. Chapter 1.5). Significant assemblages of industrial debris, and the not-infrequent finds of artefacts such as ploughshares in whose making sizeable amounts of iron were used, demonstrate that the capacity for iron production had grown. The numbers of artefacts attributable to the period, and the range of types which is greatly increased from EIA B, show that this rise in capacity was market-driven. It seems clear also that there was segmentation of the market for iron and iron products, and thus that production would have been shaped to meet the requirements of the different segments.

LIA A saw the emergence of literacy and the production of some of the finest works of art on vellum, in stone and in non-ferrous metals produced anywhere, and at any time, in the Old World. It was the time of the heroes (mythical, pseudo-historical and historical), of saints, scholars, and of jurists, and it was a period of paradoxes. There is evidence of extensive contact with the outside world set beside that for insularity. Wheel-thrown pottery was imported into largely aceramic areas. There was adherence to fragmentary socio-political and economic structures in the face of strong urbanising trends elsewhere.

Byrne (1973) opened his survey of early Irish polity by noting that '. . . many Irishmen can claim descent from kings . . .', while Binchy (1970) characterised early Irish society as '. . . tribal, rural, hierarchical and familial . . .'. Both views highlight usefully the major feature of the archaeological record of the period which is dispersal. Underpinning the patchwork of political groupings of LIA Ireland was an agricultural economy whose basic production unit was the small, self-supporting farmstead (Duignan 1944; Proudfoot 1961, 1970). We have already observed that Ireland breaks early with the urbanising trend evident elsewhere in contemporary barbarian Europe, with the population having distributed itself over the small scattered units which dominate the archaeological landscape from the second half of the 1st millennium AD onwards. There is a heirarchy of size and a wide range of form, and, from the small proportion of such sites excavated, a mere 0.5% or so, a hierarchy of wealth, although size and wealth correlate only imprecisely. In addition to secular sites, we have, with the introduction of Christianity, the appearance of monastic settlements. At first no more than a few cells clustering round tiny churches, some developed into major ecclesiastical centres. What is missing throughout the period, however, is any manifestation of the concept of the 'town' as a socio-economic and political unit. This obviously was quite foreign to the Irish, and they really did not come to terms with it until well into LIA B.

The scale of ironworking observed on the great majority of LIA A sites shows that it must have been minor in comparison with the effort invested in other activities, such as food production. In the majority of cases, it is usually more representative of work to meet the immediate needs of the occupants, rather than as the result of production to achieve a marketable surplus. The debris of smelting/smithing from a ringfort of the later 1st millennium AD frequently amount to no more than a few tens of kilos in weight (Appendix 1). A number of sites stand out in terms of the volume of metalworking debris which they have produced, and in this contrast we may detect two different levels of production and working. There are sites which produce no ironworking debris, but where iron is in use. On other sites, there was ironworking, but little or no consumption, while on others, ironworking and iron use went on together. It is thus clear that there were different forms of relationship between producers and consumers, and hierarchies of each.

One serious bar to interpretation of the evidence is the difficulty of distinguishing between smelting slags, those resulting from bloomsmithing, and those from forging (Chapter 6.4). To date, no Irish site assemblage of 'slags' has been studied in sufficient detail to allow meaningful distinctions to be made. Also, it is exceedingly difficult now to obtain precise data on quantities involved. In the absence of furnace remains, it is tempting to equate the bulk of ironworking remains with smithing. However, as studies elsewhere show (Chapter 6.4), the by-products of smelting and working iron form a continuum. Until such times as we have detailed studies which can produce realistic figures for the proportion of smelting, bloomsmithing and forging operations, the precise industrial status of many sites will remain unclear. One must thus enter the *caveat* that the interpretations of activity offered here are tentative only, and likely to be revised in the light of further research.

There is archaeological evidence (cf. Chapter 6.5) to correspond with that of contemporary writings (cf. Chapter 7.4) for the existence of secular ironworking centres. Two sites in particular, Lisleagh Co. Cork (Monk 1988) and Ballyvollen, Co. Antrim (Williams 1985a), have produced evidence of ironworking on what, for Ireland, can only be described as a large scale, with debris measured in hundreds of kilos. Other sites at Ballyvourney (O'Kelly 1952), Garannes, (Ó Ríordáin 1942), and Garryduff (O'Kelly 1964), all in Co. Cork, Knowth, Co. Louth (Eogan 1972), the later ringfort occupation at Clogher, Co. Tyrone (Warner 1972), and Lagore, Co. Westmeath (Hencken 1950), have yielded significant (though unquantified) concentrations of the debris of ironworking. At Ballyvourney, Garryduff, Knowth and Lisleagh, bowl furnace remains have been uncovered, while Clogher and Garryduff each also produced what may well have been smithing hearths. The scales of production on each of these sites might well have exceeded the level of production required purely for the use of the inhabitants, and it may well be that there was deliberate production for outside consumption.

While at Ballyvollen and Lisleagh, we see ironworking on a large scale, neither produced evidence of significant iron-use on site, and a reasonable assumption is that whatever the products, the bulk went for consumption elsewhere. In the absence of detailed study of the slag assemblages, activity perhaps is interpreted best as a combination of smelting and bloomsmithing to provide iron for working up into artefacts at other sites, and some smithing to produce finished artefacts for use elsewhere. Thus, these sites would represent the servicing of a market segment requiring iron stock for working elsewhere, possibly also tools and household ironwork. We must consider also the iron tribute supposedly paid to their overlords by the obscure *Na Cuirc* (Chapter 7.2). If we can see the *Cuirc* as specialist producers who smelted iron and consolidated blooms which were then distributed for processing elsewhere, then we have at least a model for the sort of group who might have worked at Ballyvollen and Lisleagh. In fact, we might see this reference as reflecting the provision of iron amongst the other goods and materials supplied to royalty by clients.

Ballyvourney is a problematic site since, although it has been interpreted as a smelting site, the form of the slag remains indicates that a significant proportion of the activity was concerned with bloomsmithing and/or smithing. If Ballyvollen and Lisleagh are to be interpreted as being involved in the main with smelting, then we may view Ballyvourney as a major forge of the type referred to in early law. This again would be a specialist site, probably producing some of its own iron, but also drawing upon production from other centres, to provide the raw materials for the making of artefacts.

Of what we may describe as the major industrial sites, Clogher, Garannes, Knowth and Lagore all have been identified with centres of political power. The extent of the evidence for non-ferrous and other industrial activities at Clogher, Garannes and Lagore in particular, indicates the presence of a range of craft professionals working under the patronage of the local dynasty. From the literature (Chapter 7.4), it is quite clear that the skilled blacksmith in LIA A was regarded, along with fellow craft specialists, as a person of substance and status. His forge was one of the focal points of the community. This derived from the quality of the products which they supposedly supplied the upper strata of society. The production of weaponry and of prestige objects such as the Lagore collar may be seen as the servicing of a segment at the upper end of the market. While undemonstrable at present, it is tempting to see the bulk of ironwork found at these sites as having been made there for the residents. Even with the difficulties in differentiating between smelting and smithing, it is tempting to see the iron for this production having been supplied from sites like Lisleagh and Ballyvollen.

Another mode of production is represented by sites producing small quantities of siliceous waste, or small blooms, raw or consolidated (Chapter 6.4). Here, while there was some smelting, the bulk of activity would have been the forging of new artefacts, probably also the repair of damaged ones, for use on site. In the 8th century tract *Críth Gablach* (*Críth Gabl.* §15.191: Chapter 7.2), the list of household possessions appropriate to one grade of farmer, the *mruigfher*, includes '. . . a sack of charcoal for irons . . .'. This is likely to refer fuel for the forge hearth, although it could refer also to fuel for smelting. Whatever the interpretation, sites like these, which produce the bulk of evidence for early ironworking (see Appendix 1), constitute the early equivalent of the 'do-it-yourself' segment of the market, possibly 'buying in' ore, or in most cases stock, and supplementing this with the use of scrap metal in the form of broken or badly worn artefacts. But this was not specialist activity in the sense in which we have been discussing it for the major production sites, but rather production integrated into domestic economy as part of the suite of activities required on a small, self-sustaining farmstead.

From both literary and archaeological evidence, the monastic settlements which grew out of the acceptance of Christianity were, from earliest times, establishments where secular skills were practised alongside spiritual virtues (O'Neill, H. 1967: cf. Chapter 7.4). Even where the call of God led the faithful to build in most inhospitable and inaccessible places, some craft work still went on, despite lack of natural resources. On the anchoritic foundation of Sceilg Mhiceal, lying some seven miles off the coast of Co. Kerry, 'slag' has been found (Chapter 6.5), although, unfortunately, we do not know whether or not the slag was from smelting or smithing. With the complete absence of iron ore and timber on what can only be described as a pinnacle of rock jutting out of the wild Atlantic ocean, either ore or iron was transported, along with charcoal, across this often treacherous stretch of water, there to be smelted or smithed. Similar finds of vitreous debris have been made at other island monastic foundations (Chapter 6.5), equally devoid of resources. If, however, we question the validity of interpretation of this activity on the grounds of transport problems, we need only remind ourselves of the effort involved in transporting to Church Island, Co. Kerry, all of the stone required to build the oratory (O'Kelly 1959, 118). Iron smelting and probably smithing was carried out on small monastic sites at Reask, Co. Kerry (Fanning 1981), and at Kilpatrick, Co. Westmeath (Swann 1973), where bowl furnaces have been recovered. There was also ironworking at Nendrum, Co. Down, alongside other craftwork (Lawlor 1925), from the earliest stages of occupation at Movilla Abbey, Co. Down (Ivens 1984a, 73), and from Cathedral Hill, Armagh (Brown and Harper 1984, 159). The production in religious establishments of fine works of art in non-ferrous metal, stone, and on vellum further shows the range of skills that were to be found at the disposal of the church. The story of Molua's dagger in Adomnan's *Life of Columba* (Chapter 7.3) describes the presence of *monaci scientes experti* at Iona in the saint's lifetime. As with their secular equivalents, the early monasteries integrated ironworking into their internal economies, possibly buying in ore and metal, or receiving raw materials as gifts, and recycling scrap.

The one area of the production chain which we have not considered is that of the mining and transport of ore, and the production of charcoal. We can do no more than speculate in the absence of identifiable early mines and the present impossibility of linking in Ireland ores and slags. There are scattered references to mining and miners in the literature, including a scrappy passage in the laws, and mention in the *Dindsenchas* of the activities of the *Síl ṁBuinne* in the region of Bantry, Co. Cork (Chapter 7.2), but these are vague. While ores may have been available and accessible in the vicinity of some sites (like Ballyvourney), it is not unlikely that there were groups who mined and processed ore for smelting on production sites. Equally, in many areas wood for charcoal would have been available close to hand, and we might think in terms of smelters also preparing their own fuel. It is infuriating to read that the *mruigfher* possessed charcoal, but not how he came by it! Miners and smelters may often have been one in the same, but in the light of our present knowledge, it is impossible to be more precise.

Leaving aside the difficult question of ore and fuel resources, we can see the market for iron broadly segmented into stock, everyday implements, prestige objects and weaponry. Production to meet these requirements took place at five levels, two involving the making of stock, and three for the forging of artefacts. The upper tier of iron smelting is represented by specialist sites, for which Ballyvollen and Lisleagh, can be offered as candidates. The iron which they made and processed into consolidated blooms or ingots of some form, would have gone to major forges of the type suggested to have existed at Clogher, Garannes, Knowth and Lagore, and also to meet the needs of the occasional smithing carried out on many smaller sites. On these smaller sites too, some smelting was carried out, again to meet needs, and this represents the lower tier of production.

The upper two tiers of smithing are represented on the one hand by a forge like Ballyvourney, and on the other by the activities of the smiths operating at Clogher, Garannes, Knowth and Lagore. While the distinction may be artificial, it seems useful to differentiate between those who might have supplied a wide local market, and those who would have restricted their work to servicing the top consumers. The status of Garryduff is unclear since, while it obviously had smiths and smelters working there, it is not identifiable at present with any grouping. The temptation is to class it as another of the sites on which smiths worked under patronage. The counterpart of the master smith working under patronage was the master smith in a monastic community, making high quality artefacts for use there. The lowest tier of smithing is represented by the occasional production and repair of artefacts on the smaller sites, secular and religious.

Magnusson (1986; 1987, 282ff) has provided a useful model of economic and industrial activity based on studies of post-Mediaeval communities in southern Sweden. With some modification, this has relevance for the lowest tier of production in LIA Ireland which, quantitatively, makes up the bulk of ironworking. In essence, it shows that most tasks could have been carried out as seasonally appropriate, with those over and above basic requirements being fitted into the periodic gaps between essential activities. Projecting this into early Ireland, we can see that the making of iron and iron artefacts (and associated activities) can be best interpreted as an occasional activity on many sites, fitting in to a wider pattern of economic activity. We have seen that the course of development from the end of the Later Bronze Age was away from the concentration of industrial production into the hands of comparatively small groups of professionals. Without the rise of urban centres, craft specialists could be responsible for a small proportion of output only. Magnusson's model of a cycle based on husbandry and harvest, around which other activities like hunting, crafts and building revolve, provides a useful basis for considering the craftwork on the majority of LIA A sites.

We may see most of the industrial activity on the monastic metalworkers as the equivalent of that on the lower-tier secular sites. The making or repair of a knife or an axe for use in the work of the community would be a part of its self-sufficient internal economy. Similarly, the fabrication of a piece of metalwork such as an iron bell for religious purposes, would have been an equally important part of the spiritual economy. If the industrial activity on sites such as Reask and Kilpatrick is seen as contributing to the overall needs of the members of the community, then at this level at least, we can see that the rural model for the integration of industrial activity into the economy can equally well be applied to that of the early religious communities. We must be conscious, however, of the presence of master smiths working in some monastic communities. These persons, prominent in the literature (Chapter 7.4) but impossible to identify in the archaeological record, would correspond to the secular master smiths working to nobility, thus being the monastic equivalents of upper-tier production.

The upper tiers of production are difficult to separate in practice, since distinction between a major forge serving a district as a commercial undertaking, and the workshop of a smith working under the patronage of a ruler must be made on subjective grounds. Thus, the metallurgically-productive sites of Clogher, Garannes, Knowth, and Lagore are identified on the basis of historical evidence as the seats of royal dynasties. No such evidence has been forthcoming, on the other hand, for the equally productive sites at Ballyvollen, Ballyvourney, Garryduff and Lisleagh. Furthermore, the heirarchy of size and wealth represented by these sites does not fit well with a simplistic ordering on the grounds of ironworking alone. Without the documentary links for Clogher with the Uí Crimthainn, for Knowth and Lagore with branches of the southern Uí Néill, and for Garannes with the Uí Echach, it would be easy to identify all four as production centres only. Herein lies the difficulty, since while patronage is an obvious interpretation for 'royal sites', the craftsmen there could also have been projecting their products onto a wider market. Similarly, Ballyvollen, Ballyvourney, Garryduff and Lisleagh could well be the unrecorded centres of local groups whose resident metallurgists gained a good enough living through patronage without needing to sell further afield than the seats of power of their patrons. It is thus necessary to guard against simplistic projection of single modes of craft-professionalism onto the evidence. While these broad divisions undoubtedly existed, it is likely that activity on any one site ascribed to the upper tiers contained elements both of patronage and the servicing of a wider clientage.

* * *

5.2 – LIA A ironwork

The range of other iron artefact types used during this period is far greater than we have been able to see for the Earlier Iron Age. In addition to weaponry, we also find blacksmiths' tools, horse-harness, personal ornaments, handles and bindings for wooden vessels, sheet iron vessels, agricultural implements including plough parts, knives and other general purpose tools all being made of iron. The laws include lists of ironwork appropriate to a well-appointed household, and recent excavations at Deerpark Farm (Lynn pers. comm.) have retrieved the entire assemblage attributed to one grade of farmer, the *mruigfher*. Although much later by comparison with other regions, we have now reached the stage at which iron had completely displaced copper-alloy as the primary industrial metal, with the latter now only being employed for decorative and non-utilitarian purposes. We might see some of the use of iron for personal ornament, such as the iron penannular brooches from Lagore and the iron bifid pins from Garryduff (cf. Fig. 1.5.3) not as the 'poor man's' bronze or silver, but rather perhaps as evidence of the belief in the prophylactic powers of the metal (cf. Chapter 7.2). LIA A was the time in which the scribes were recording for posterity the tales of heroic deeds on and off the field of battle, and ascribing to their central characters weapons whose physical properties would be the envy of modern metallurgists! As with

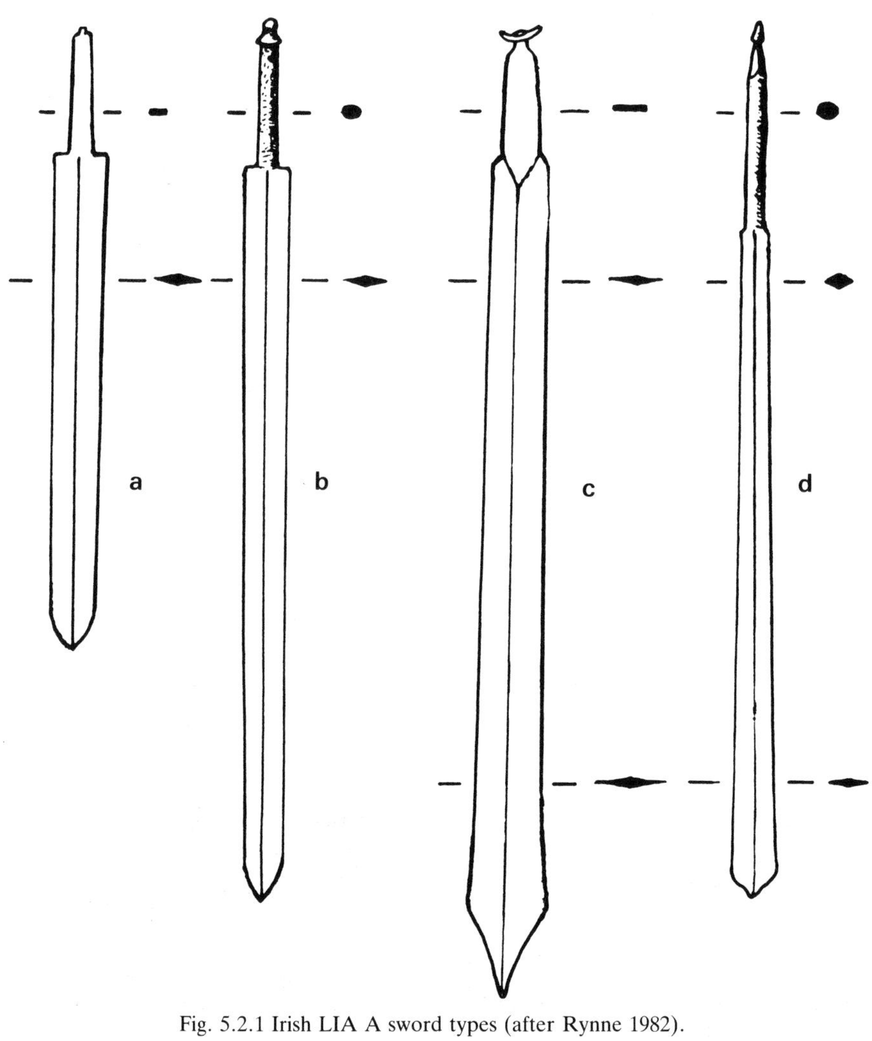

Fig. 5.2.1 Irish LIA A sword types (after Rynne 1982).

earlier periods, greatest emphasis shall be placed on the metallography of weaponry and of edged tools to see just how far the evidence of texts and the microscope correspond. As with the earlier periods, the seriation of iron artefact types in LIA A is fraught with uncertainty, although now we have better associations in which to base at least some typology and dating.

Rynne (1982, 1983), following on from Hencken (1950, 88ff), identifies a group of LIA A swords showing influence from Romanised Britain in the form of the *gladius* and *spatha*. These weapons are characterised by a parallel sided blade coming to an angular point, square shoulders, and a flat, rectangular- or round-sectioned tang (Fig. 5.2.1a and b). He sees these weapons as representing a progression from the La Tène style blades under the influence of Roman types. The *gladius*-derived swords, whose foreign ancestors were characterised by a short, broad stabbing blade, are represented by specimens such as those from Teeshan, Co. Antrim (Plate 5.3.2a) and Ballinlig, Co. Westmeath (Plate 5.3.3a). The *spatha*-derived swords have longer, more slender parallel-sided blades, and are represented by the swords from Lagore (Hencken 1950, 90, Fig. 25B) and Craigywarren crannog, Co. Antrim (Plate 5.3.1a). While both types probably saw service before the 5th century, the Lagore old finds (Hencken 1950, 90, Fig. 25C and D), if indeed from the crannog deposits themselves, are unlikely to predate the mid-7th century AD. Indications are thus that blade forms changed but slowly, since fighting styles (centering on the spear?) probably remained static, and that as a consequence, several different types were in use at the same time.

A purely insular type has a blade of simple lozenge-shaped section which widens gradually towards a blunt point, and a tang which may be round-sectioned or flat (Fig. 5.2.1c and d). The likely *floruit* for this type is the 6th/7th centuries, although the possibility that the Altanagh and Toome Bar objects (Chapter 4.1) are early (4th/5th centuries AD) representatives of the class cannot be dismissed. Rynne (1982, 1983) divides the type into two variants, the 'expanded ended' and 'crannog' swords, on the basis principally of tang and overall size. The 'expanded ended' form has a flat, rectangular-sectioned tang, while the 'crannog' form has a thick, round-sectioned tang, often roughened to provide better attachment of the grip. As Rynne notes (1982, 96), this latter is also a feature of many of the *spatha*-derived weapons, indicating a close developmental relationship. An example of the 'expanded ended form' recently recovered from the Lung river in Co. Roscommon (Rynne 1985) has two surviving copper alloy hilt plates bearing interlace ornament which can scarcely be earlier than the 7th century. The weapon from Ballinderry crannog no. 2, Co. Offaly (Plate 5.3.9a) was dated by the excavator to the start of the 7th century AD (Hencken 1950, 91), while that from Lagore crannog, Co. Westmeath is no earlier than the mid-7th century.

Along with the blades showing Roman influence, the 'expanded ended' and 'crannog' swords possess the most massive of the early Irish blades and were obviously intended as slashing weapons, with the centre of gravity lying inside the lower half of the

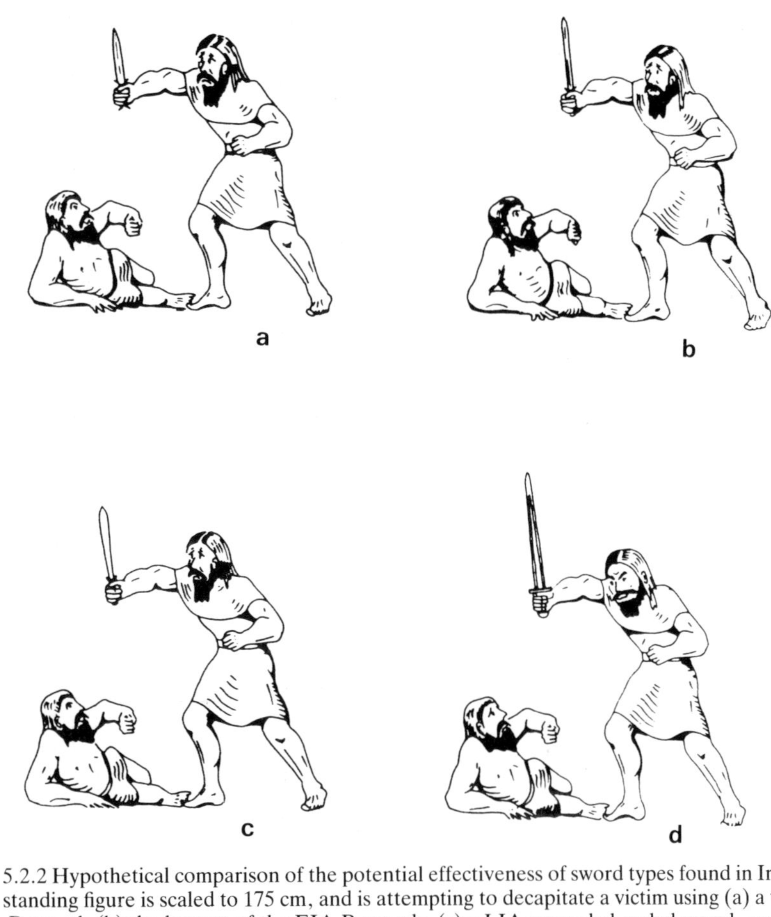

Fig. 5.2.2 Hypothetical comparison of the potential effectiveness of sword types found in Ireland. The standing figure is scaled to 175 cm, and is attempting to decapitate a victim using (a) a typical EIA B sword, (b) the longest of the EIA B swords, (c) a LIA expanded-ended sword, and (d) a Viking sword. (after Mallory 1982).

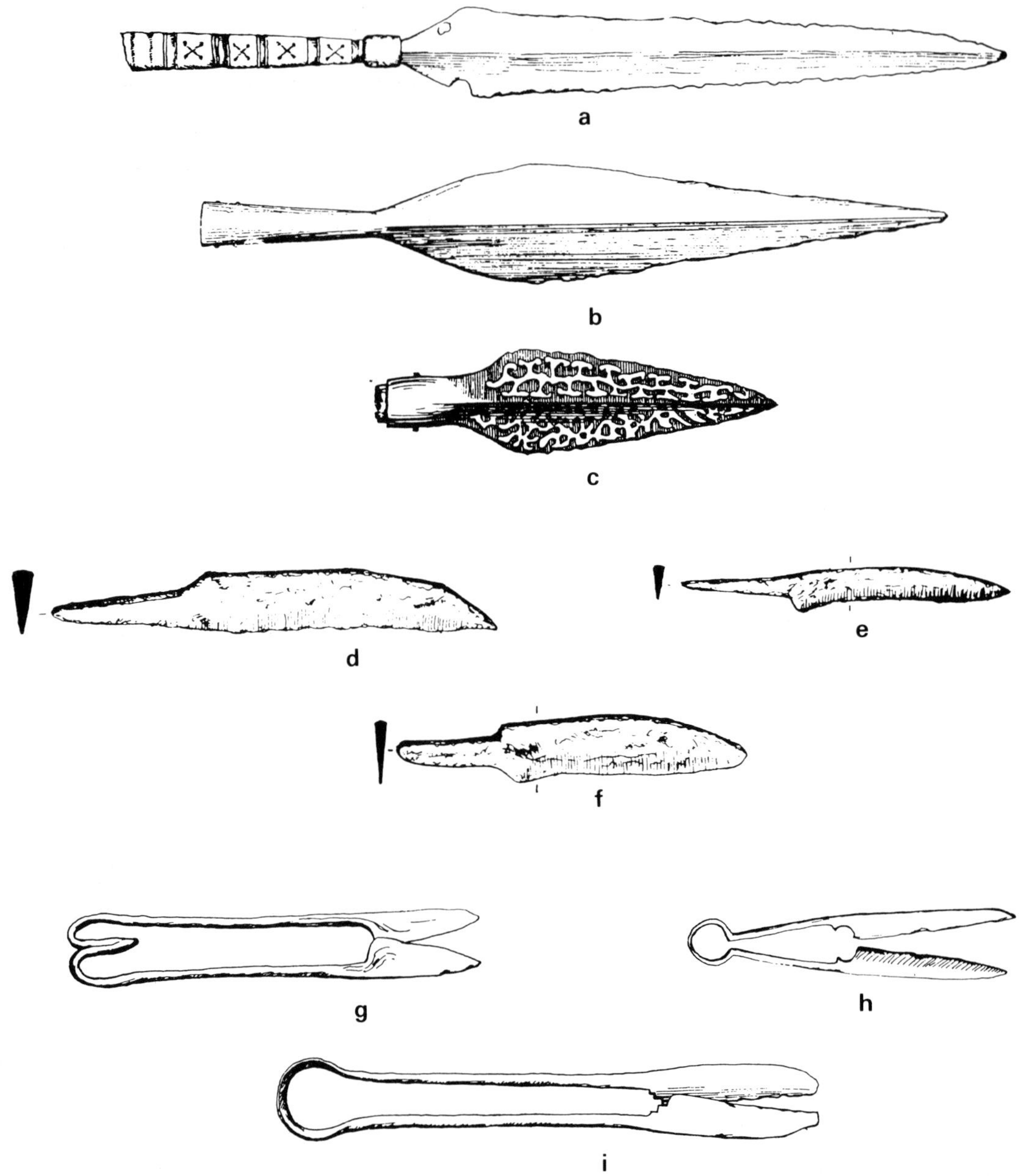

Fig. 5.2.3 Some iron artefact types from LIA A Ireland: (a)-(c) spearheads from Lagore, Co. Westmeath (all x$\frac{1}{3}$: after Hencken 1950), (d)-(f) tanged single-edged knives from Carraig Aille II, Co. Limerick (all x$\frac{1}{2}$: after Ó Ríordáin 1950), (g)-(i) shears from Lagore, Co. Westmeath (all x$\frac{1}{3}$: after Hencken 1950).

blade. These are the only native produced weapons which would have had a chance of producing anything approaching the sort of devastating damage which so delighted early Irish storytellers. Mallory (1982, 108) might seem to be overcautious in his comparison between the effectiveness of Irish Later Iron Age swords and those of Viking intruders (Fig. 5.2.2), although practical testing perhaps is not yet a feasible proposition! Yet swords such as those from Clanrolla, Co. Armagh (Plate 5.3.7a) and Ballinderry, in theory at least, could have cut through a significant thickness of tissue and bone if wielded with sufficient strength. Here the quality of the cutting edge obviously would have been a crucial factor. The comparatively blunt points of the longer Irish LIA swords would have been quite ineffective for stabbing. Although there is no liklihood of a developmental link between the two, in effect the expanded-end swords recall the fighting techniques for which their Later Bronze Age predecessors were designed.

Rynne offers a further class of 'grooved' or 'fullered' swords, based really on only one example provenanced to the excavations at Lagore, Co. Westmeath (Hencken 1950, 91, Fig. 26: Plate 5.3.11a), which he compares with two others from among the extensive collection of 'old finds' made from and around the site in the 19th and earlier 20th century. In fact, the 'fuller' on one of the swords which Rynne ascribes to this group in fact is a ridge left by the welding of a cutting edge to the central region (examination 31: Plate 5.3.4a). Nevertheless, the provenanced example does exist, even if it is not a particularly inspiring weapon, now weighing a mere 154 gm. Rather than see it as a representative of a further class (*pace* Rynne 1982, 96), we might see it as a child's toy weapon, or somesuch, rather than as a weapon with which a grown man seriously might have considered entering a fight. Another rather strange fullered blade was recovered from Killyliss, Co. Tyrone in a context dated to the end of LIA A (Ivens

1984b, 25f). The hilt has been lost through corrosion, while the point tapers down to what almost might be described as a 'carps tongue' (Plate 5.3.12a). Like the Lagore example, it is not really a weapon to be taken seriously. Bearing no resemblance to Rynne's fullered type, it further illustrates the variability in forms of the overall assemblage.

B. Raftery (1984, 62) notes the use of Old Irish *gaisced*, deriving from *gae* 'spear' for a warrior (cf. Chapter 4, note 6), which indicates the importance of this weapon. But if seriation of swords of LIA A is difficult, then that of the spearheads effectively is well-nigh impossible at the moment without a complete *corpus*, and this is a field for future research. The assemblage from Lagore, Co. Westmeath (Hencken 1950, 94 ff) shows well the scope and variety of form and potential influences (Fig. 5.2.3a–c) seen in Ireland from the 7th century onwards. A very general point of note is that the practice of leaving the edges of the socket unwelded, seen on Anglo-Saxon and Viking spearheads, is uncommon on those from Ireland not immediately identifiable as intrusive.

One distinctive type which calls for comment displays lunate depressions, bounded by thickened cutting edges, on either side of a fine midrib. At least three examples are known, one supposedly from a crannog at either Lankill or Drumadarragh, Co. Fermanagh (Jope 1955; Scott 1977b, 303f: Plate 5.3.18a), another from the River Shannon at Drumshanbo, Co. (Plate 5.3.19a), and a third from Monalty Lough, Co. Monaghan (NMI 1965:180). A blade with pronounced ribbing of not-dissimilar concept, though elongated, came from Ballinderry crannog no. 2, Co. Offaly (Hencken 1942, 47, Fig. 19 no. 529). Jope (1955, 16) noted parallels with post-Roman Germanic weapons, although Swanton (1974, 20) suggested that the technique displayed in fact was borrowed from Iron Age smiths[1]. It is worth also drawing attention to the spearheads, from Hillsborough, Co. Down (Plate 5.3.16a) and Lagore (Plate 5.3.17a), which have octagonally-facetted sockets. This feature is known on Scandinavian weapons and may indicate stylistic influence.

Perhaps the commonest iron artefact of the period is the tanged, single-edged iron knife which had been in use since at least EIA B. The form and variety of this simple tool (cf. Fig. 1.2.1) makes typological seriation by conventional archaeological criteria impossible. The range of shapes certainly has been influenced by sharpening, with blades made of softer metal suffering more metal loss than those with hard edges. The tang may be in line with the cutting edge, or with the back of the blade, or positioned between the two (Fig. 5.2.3d–f), seemingly without any consistent purpose. Another type which shows a similar wide range in form is the shears whose blades are all of roughly the same shape, but whose springs show a number of variants (Fig. 5.2.3g–i).

Axes of the period are all of the shaft-hole variety, with a wide range of shapes, from heavy, thick-butted forms through to those with comparatively slender blades: one example of a double-edged form was found at Clogher, Co. Tyrone (Plate 5.3.25a). Hencken suggested that an axe found at Lagore (1950, 106, Fig. 40A) was related to late Roman types in Britain and on the continent. Similar examples have come from Cahercommaun, Co. Clare and Lough Faughan, Co. Down (Fig. 1.5.2d), and in view of the potential influence of Roman types on sword blades, this is not an unreasonable suggestion. As suggested in Chapter 1.5, these might be smith's tools, with the blade acting as a sett for cutting, and the thick, flat butt doubling as a striking platform and as a coarse hammer.

Artefacts made from sheet iron are found in significant numbers during LIA A, perhaps the best known examples being the small ecclesiastical bells, some of which have been flashed with copper alloy, such as that from Oldcourt, Co. Cork (Murphy 1961). Their construction was simple, consisting of forging out a plate of metal and folding it over, with the seams being secured by rivets (Fig. 5.2.4). Vessels made of sheet iron are known from Lagore (Hencken 1950, 113, Fig. 46) and Deerpark Farm, Co. Antrim (Lynn pers. comm.). Iron attachments, including some attractively twisted handles (e.g. Hencken 1950, 114, Fig. 47) for both metal and wooden vessels were being made also.

Two pieces which illustrate well the work of master smiths, as well as the range of shaping skills attained, and the influences open to Irish smiths in LIA A, are the decorated hinged iron collar and chain from the earliest period of occupation at Lagore crannog, Co. Westmeath, and the remains of a copy from among the old finds at the site (Hencken 1950; Scott 1978b). There seems to be no reason to alter the excavator's dating to the mid-7th century AD of the phase from which the complete collar and chain were recovered (cf. Warner 1986). The most significant feature of the complete collar, a multi-piece construction (cf. Fig. 1.5.4) of pleasing design concept, is the herringbone pattern decoration with which one half is is embel-

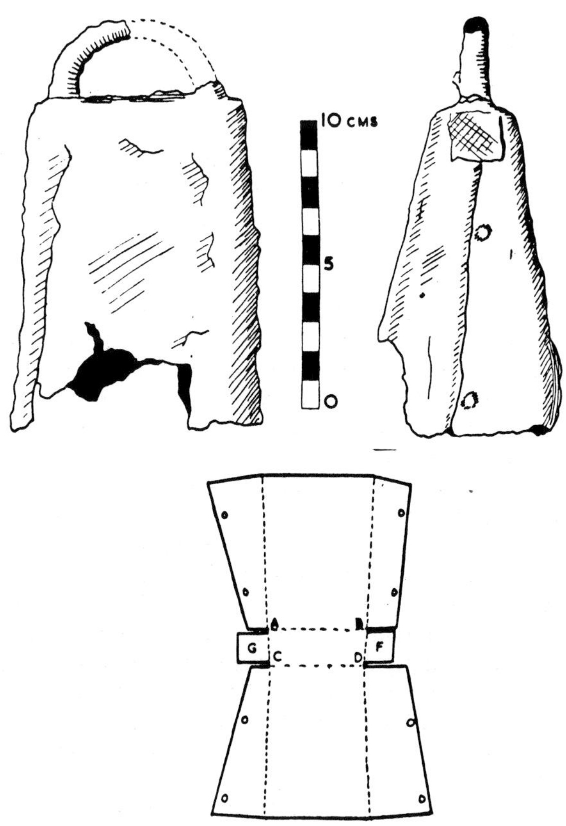

Fig. 5.2.4 Sheet-iron bell from Oldcourt, Co. Cork, and its construction (after Murphy 1961).

Plate 5.2.1 The complete decorated iron collar from Lagore, Co. Westmeath x½ (*Courtesy of the National Museum of Ireland*).

lished (Plate 5.2.1). The chain attachment is of interest, first because of the links which are in the form of a folded-over 'figure-of-eight', second because they taper in size down from the collar to the end. Chains constructed in this way (Fig. 1.5.4k) can only be lengthened easily from the end, and the type is comparatively rare in Ireland.

The form of the chain is best parallelled in Roman and post-Roman Britain (Scott 1978b, 223), while the collar itself is without precise comparanda. The twisted rod decoration is highly suggestive of the herringbone patterning of Anglo-Saxon and Scandinavian swords, although we might also look to decoration in non-ferrous metals, such as the trichinopoly panels on the underside of the Ardagh chalice (cf. Rynne 1987, 85f): it must be noted, however, that opposed twists occur also on the chains in the Broighter, Co. Derry hoard and these may be paralleled in the Roman world of the 1st–3rd centuries AD (Warner 1982, 29). A significant combination of the chain type and twisted rods is provided by the cauldron suspension gear from Sutton Hoo (Bruce-

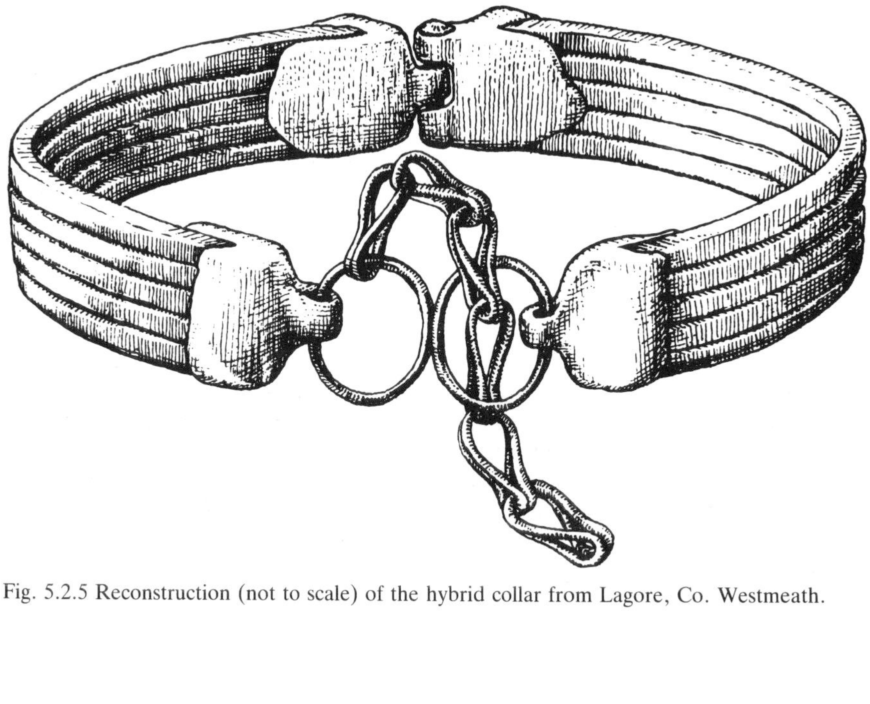

Fig. 5.2.5 Reconstruction (not to scale) of the hybrid collar from Lagore, Co. Westmeath.

Fig. 5.2.6 Plain slave-collar from Lagore, Co. Westmeath showing the closure mechanism (x¾).

Mitford 1973, 39 and 91) which is now dated to the middle of the 7th century AD. There is a strong Irish connection at Sutton Hoo through the hanging bowl escutcheons (Warner 1987), themselves indicative of wider contacts and influences. The assemblage of old finds from Lagore includes several large single-edged knives reminiscent of the Anglo-Saxon scramasax (Hencken 1950, 90, Fig. 25E–I).

The Lagore collar, as we have it now, has been restored in antiquity by combining the herringbone-decorated half with another half in which the twists in the rods all run the same way. The joint made at the back hinge is crude in comparison with the quality of workmanship shown by the rest of the piece. The arguments for the constructional sequence are detailed elsewhere (Scott 1978b), and it will suffice to note here that it seems likely that the original collar, represented by the herringbone-decorated panel, was made under outside influence. A second collar, represented by the other half and cannibalised to repair the original, is likely to have been the work of a local smith, who may also have produced the hybrid collar represented now only by the remains of its chain attachment (Fig. 5.2.5). This collar can be reconstructed from the surviving fragment as showing similar concept. A plain, one-piece collar, again from among the old finds at this site, has a similar closure mechanism, and the same type of chain attached (Fig. 5.2.6). The hybrid collar was probably made by a smith influenced by the form of the original, but unaware of herringbone pattern-welding and the need for a backing plate for strength. It shows the influence of imported ideas worked up by a native smith passed on to another worker who, while skillful, did not fully understand them.

The decorated collar has been described as a 'slave collar', although a more reasonable suggestion for such a piece is that it was used for hostages. But the most attractive proposition is that the collar was for a dog. A passage in *Táin Bó Cúalnge* which relates how the infant Cú Chulainn got his name, tells how a fierce dog belonging to Culann Cerd, and of which he was inordinately proud, came to be killed. In this passage, the dog is described as being *chained* up because of its temperament (*Táin* 856–871). At Lydney Park, Glos. was found a small bronze figurine of a dog, identified as an Irish wolfhound whelp, and which is depicted as wearing a large collar (Wheeler 1932, 40, 88–89 and Plate XXV). Both of the two-piece collars show significant investment of time and skill, indicative of the work of smiths in the top tier of production.

* * *

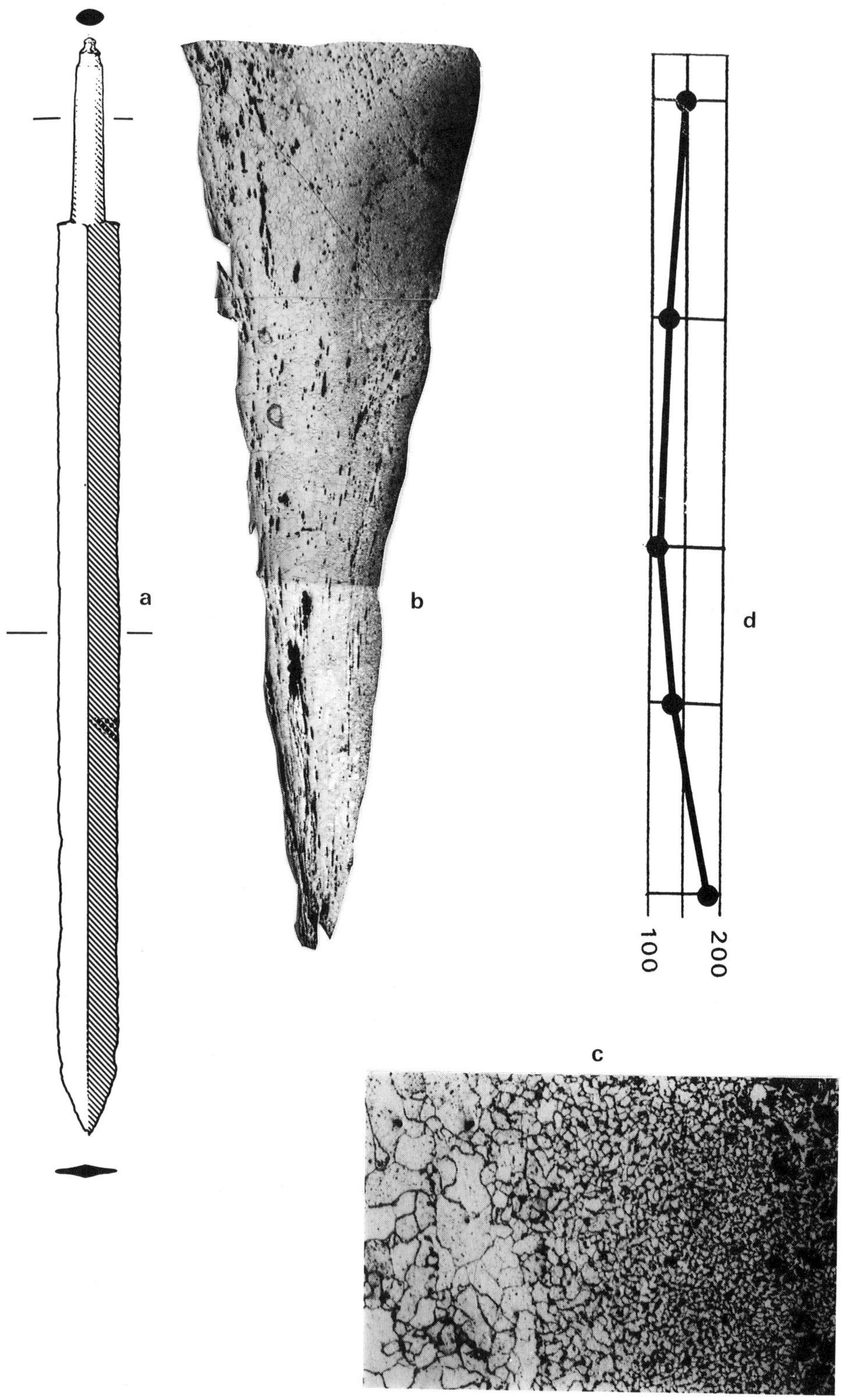

Plate 5.3.1 Examination 28: (a) NMI 1902:58 sword from Craigywarren crannog, Co. Antrim x¼, (b) macrostructure, (c) gradient from low- to high-C x100, (d) H_v profile. Etched 4% Nital.

5.3 – Metallographic examinations of LIA A iron artefacts

(28) NMI 1902:58 sword, Craigywarren crannog, Co. Antrim (Plate 5.3.1)

A section was taken from cutting-edge to midrib (Plate 5.3.1a).

Non-metallic inclusions: Numerous inclusions, principally one-phase but with some two-phase.
Macrostructure: Etching with Nital revealed an uneven distribution of carbon (Plate 5.3.1b). Oberhoffer indicated minor segregations.
Microstructure: There is a gradient from large polyhedral ferrite grains (GS 1–3) on one side of the section

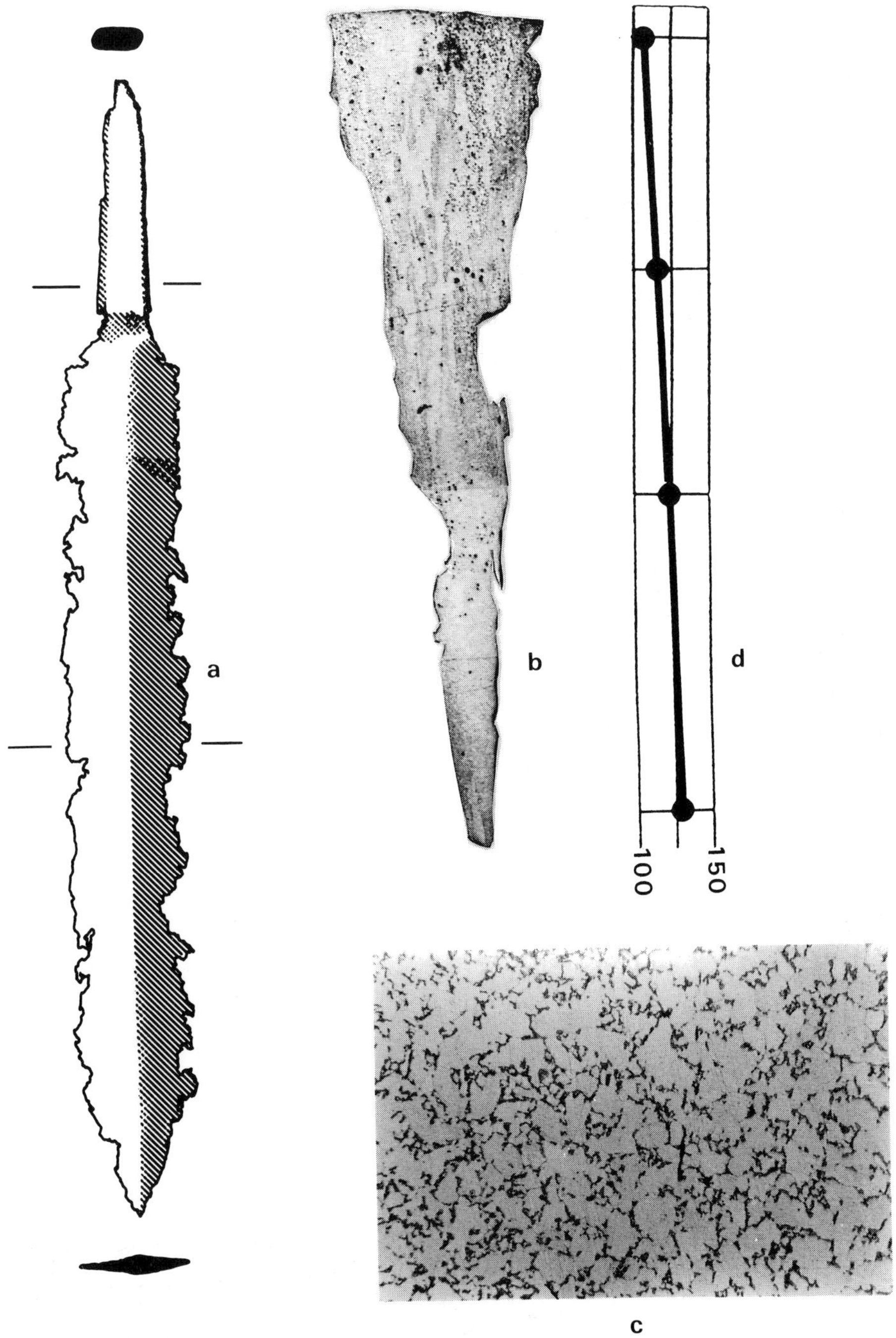

Plate 5.3.2 Examination 29: (a) UM AL1:1988 sword from Teeshan crannog, Co. Antrim x½, (b) macrostructure, (c) ferrite and interstitial pearlite x100, (d) H_v profile. Etched 4% Nital.

to fine ferrite (GS 6–7) with interstitial coarse pearlite. Some of the ferrite tends to a Widmanstätten structure (Plate 5.3.1c), while pearlite shows signs of spheroidisation.
Hardness: mH_v ferrite 152–232, pearlite 321: H_v 119–186.
Interpretation: This finely shaped piece was forged from inhomogeneous stock, and was worked for some time above A_3, before being allowed to cool slowly.

(29) UM AL1:1988 sword, Teeshan, Co. Antrim (Plate 5.3.2)

A section was taken from cutting-edge to midrib (Plate 5.3.2a).
Non-metallic inclusions: Numerous two- and three-phase inclusions.
Macrostructure: Etching revealed no significant segregations (Plate 5.3.2b).
Microstructure: The structure consists entirely of small polyhedral ferrite grains (GS 6–7), with coarse interstitial pearlite (Plate 5.3.2c).
Hardness: mH_v ferrite 146–199: H_v 106–131.
Interpretation: Although nicely shaped, the sword has been forged from stock low in carbon, and in which much bloomery slag remains.

(30) NMI WK5:W7:F355 sword, Ballinlig, Co. Westmeath (Plate 5.3.3)

A section was taken from cutting-edge to midrib (Plate 5.3.3a).

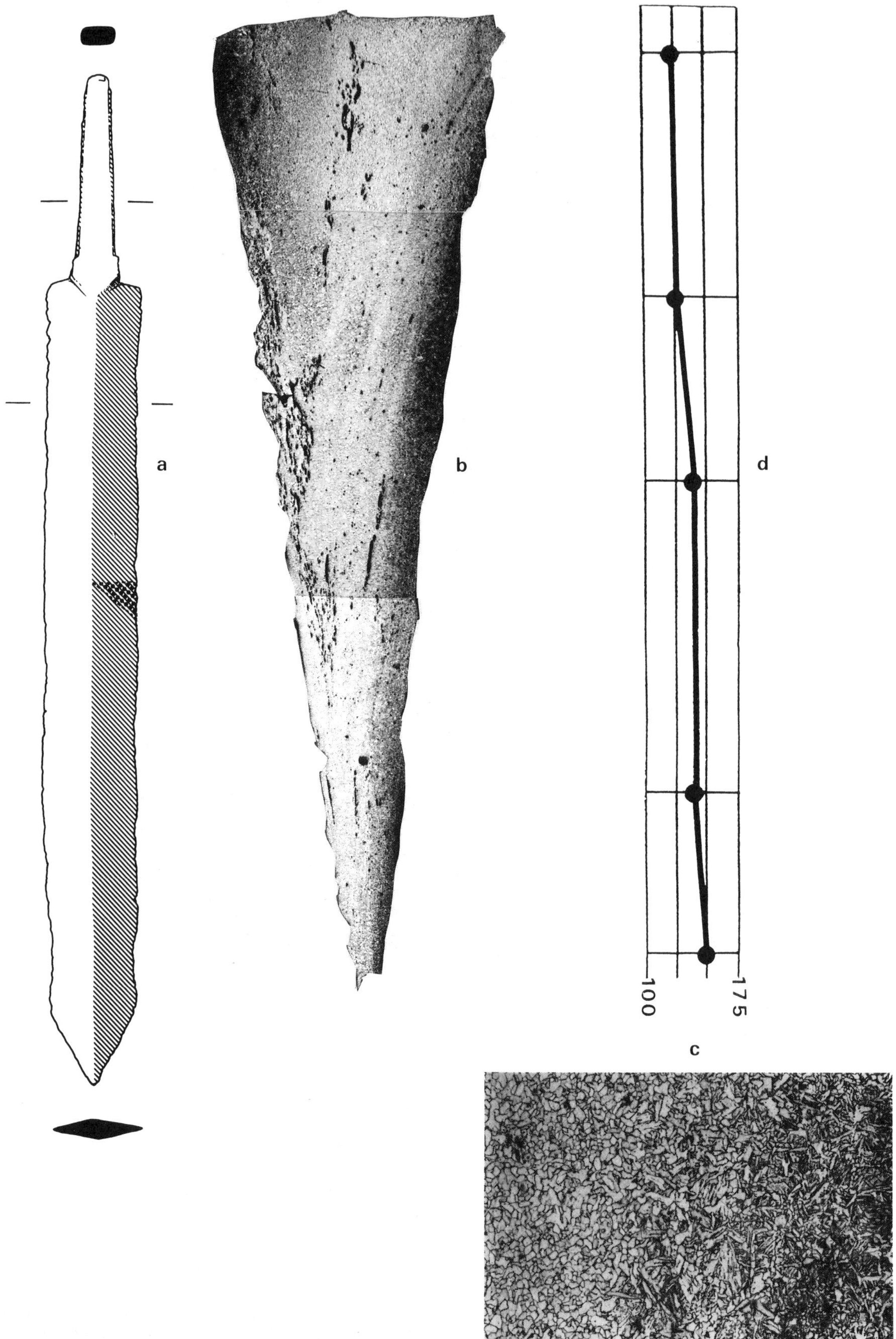

Plate 5.3.3 Examination 30: (a) NMI WK5:W7:F355 sword from Ballinlig, Co. Westmeath x⅓, (b) macrostructure, (c) carbon gradient x50, (d) H_v profile. Etched 4% Nital.

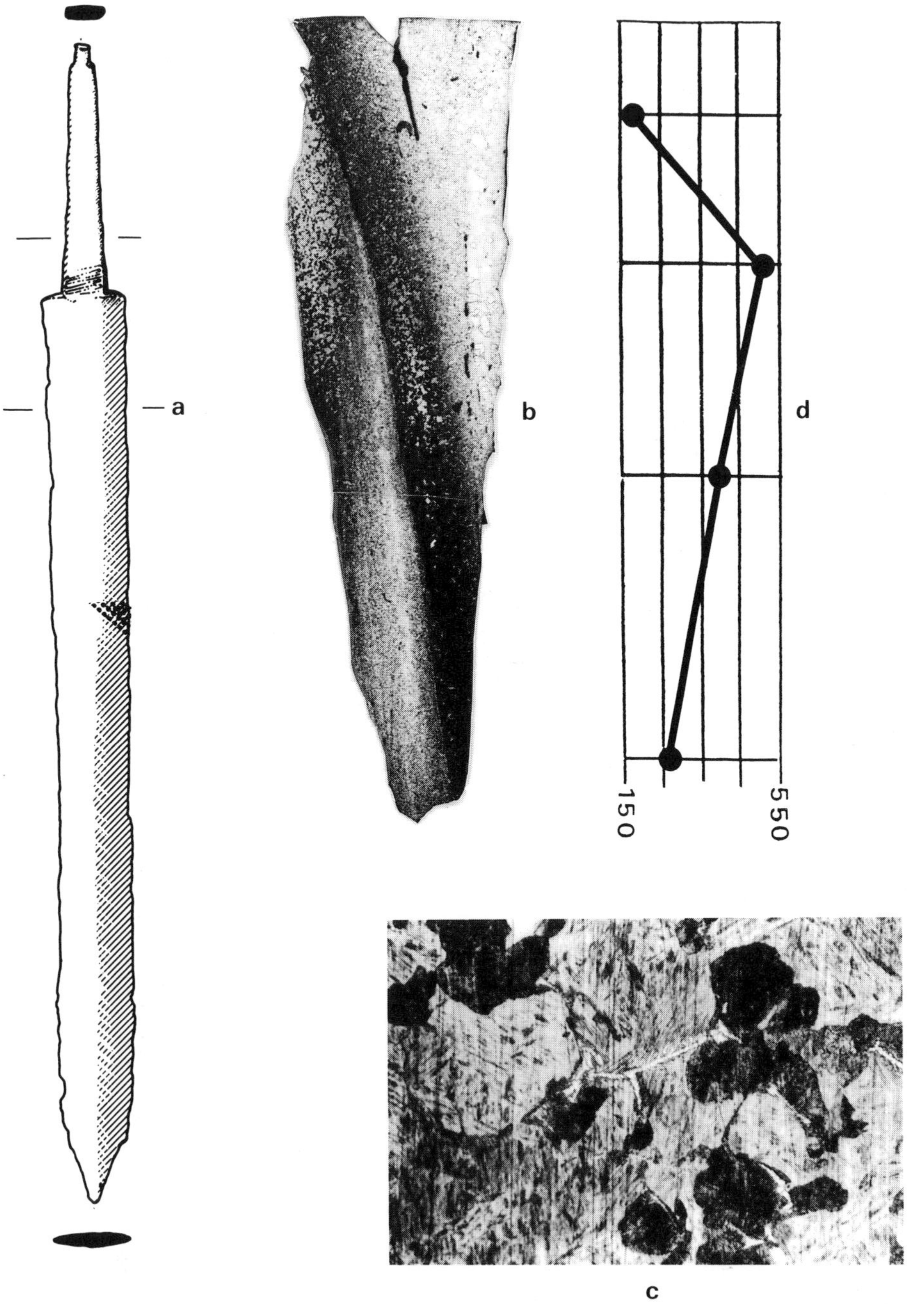

Plate 5.3.4 Examination 31: (a) NMI WK3:P855 sword from Lagore, Co. Westmeath x1/3, (b) macrostructure, (c) martensite, troostite and remanent ferrite x500, (d) H_v profile. (b) etched 4% Nital, (c) etched Bisulphite.

Non-metallic inclusions: A small number of one- and two-phase inclusions occur in clusters.
Macrostructure: Etching with Nital showed some concentration of C towards both blade faces, while Oberhoffer indicated no other significant segregations (Plate 5.3.3b).
Microstructure: The structure in the centre of the section consists of polyhedral ferrite grains (GS 4–5), with some grain boundary cementite. Towards both faces of the blade there is a refinement in ferritic grain size (GS 7–8), followed by a distinct carbon gradient to give Widmanstätten ferrite with pearlite (Plate 5.3.3c) which is coarse, showing advanced spheroidisation of carbide.
Hardness: mH_v ferrite 135–227, pearlite 250: H_v 120–159.
Interpretation: This well-shaped blade was forged up from stock which was virtually carbon-free and low in included slag. It was then carburised, and the final heating and cooling regime caused some refinement of grain size.

(31) NMI WK3:P855 sword, Lagore crannog, Co. Westmeath, sword (Plate 5.3.4)

A section was taken from cutting-edge to midrib (Plate 5.3.4a).
Non-metallic inclusions: A few small single-phase grey inclusions are scattered throughout the section surface. At one point near the midrib there are a number of large inclusions, also single-phase and grey.
Macrostructure: Nital showed carbon distribution to be highly uneven, and revealed one major weld seam, but Oberhoffer indicated no other major segregations (Plate 5.3.4b).

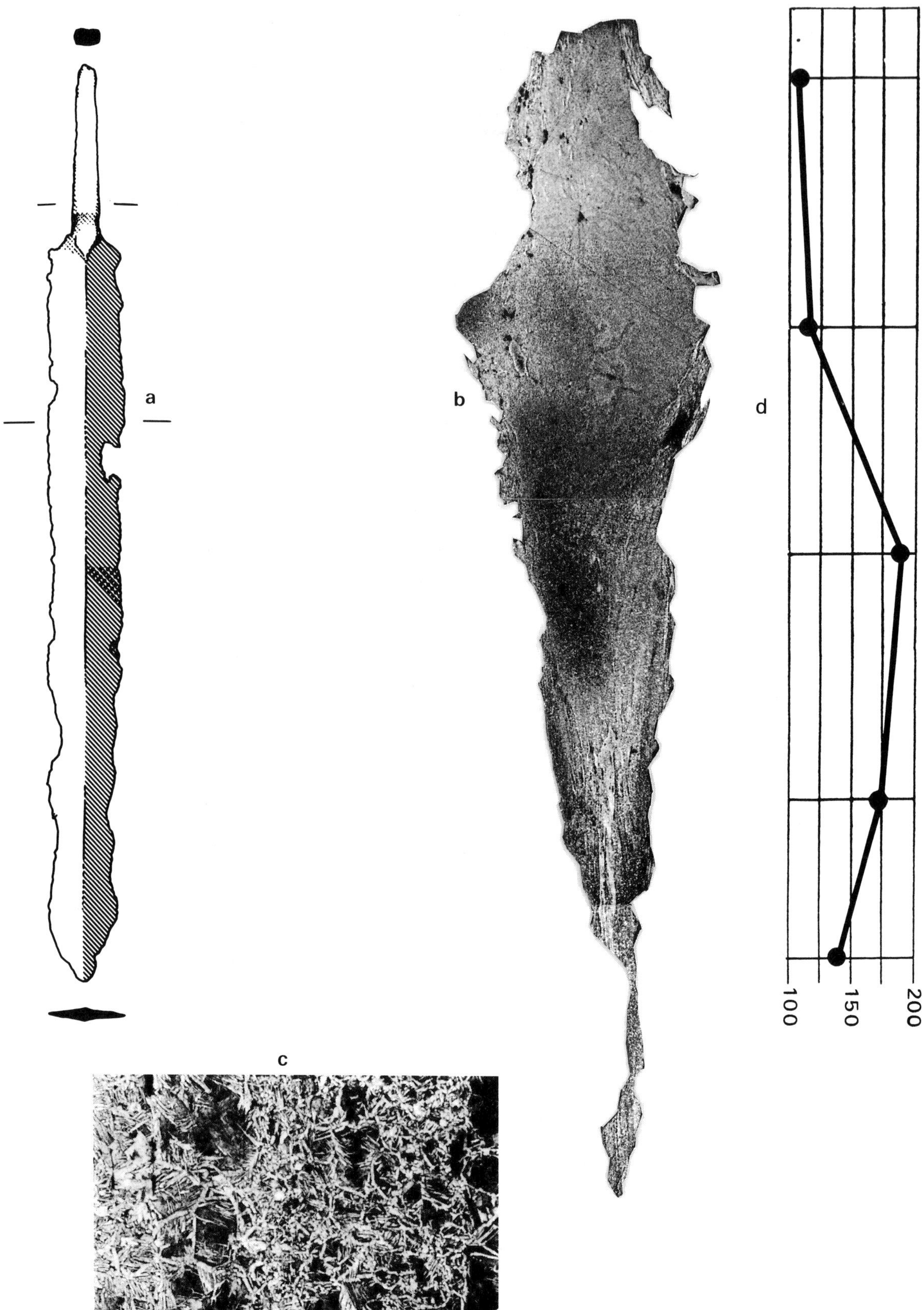

Plate 5.3.5 Examination 32: (a) UM 184–1952 sword from 'Northern Ireland' (no locality) x$\frac{1}{3}$, (b) macrostructure, (c) Widmanstätten ferrite and pearlite in the centre of the section x100, (d) H_v profile. Etched 4% Nital.

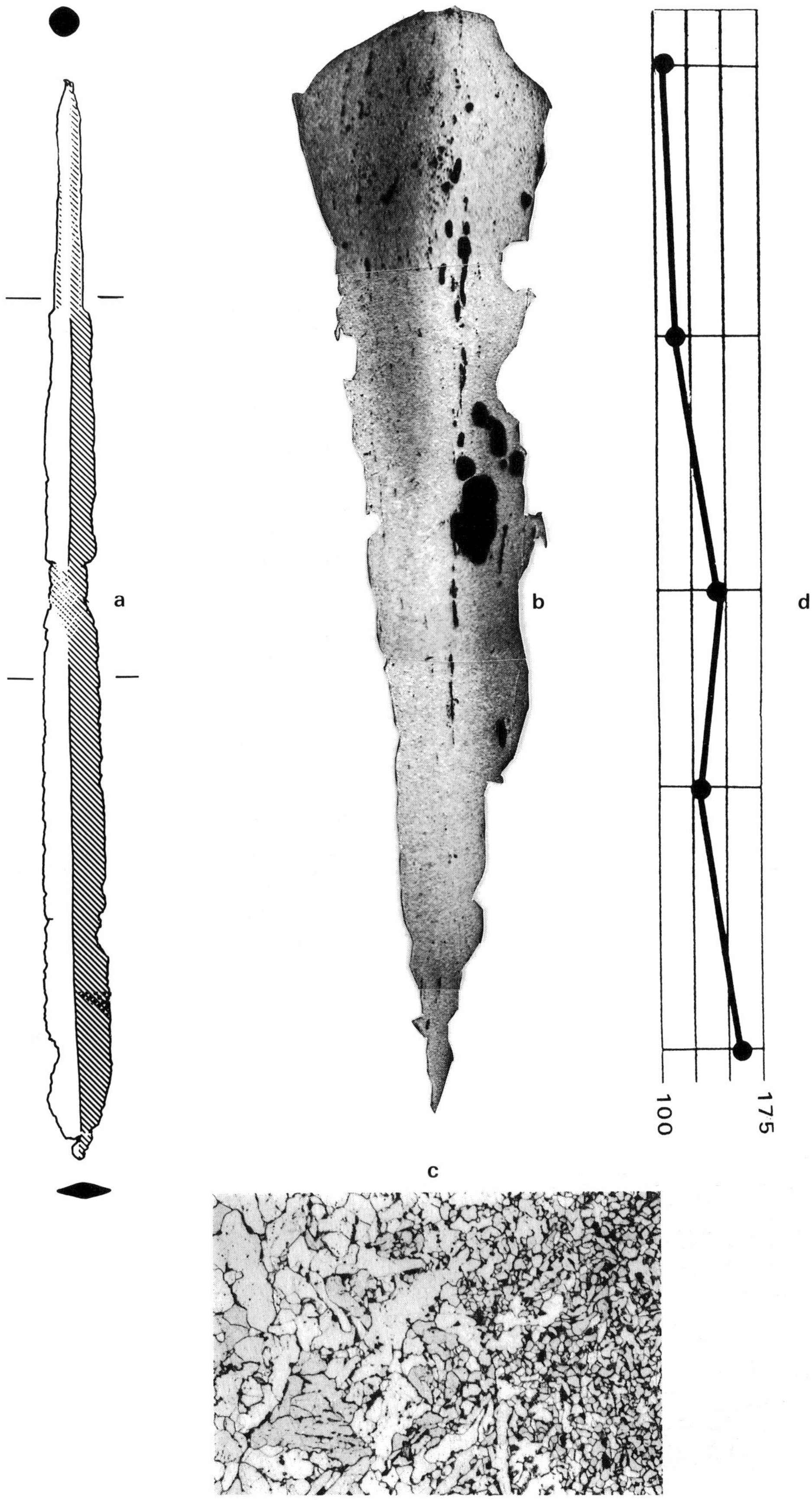

Plate 5.3.6 Examination 33: (a) NMI 1905–30 sword from Toome Bridge, Co. Antrim x¼, (b) macrostructure, (c) ferrite and interstitial pearlite x100, (d) H_v profile. Etched 4% Nital.

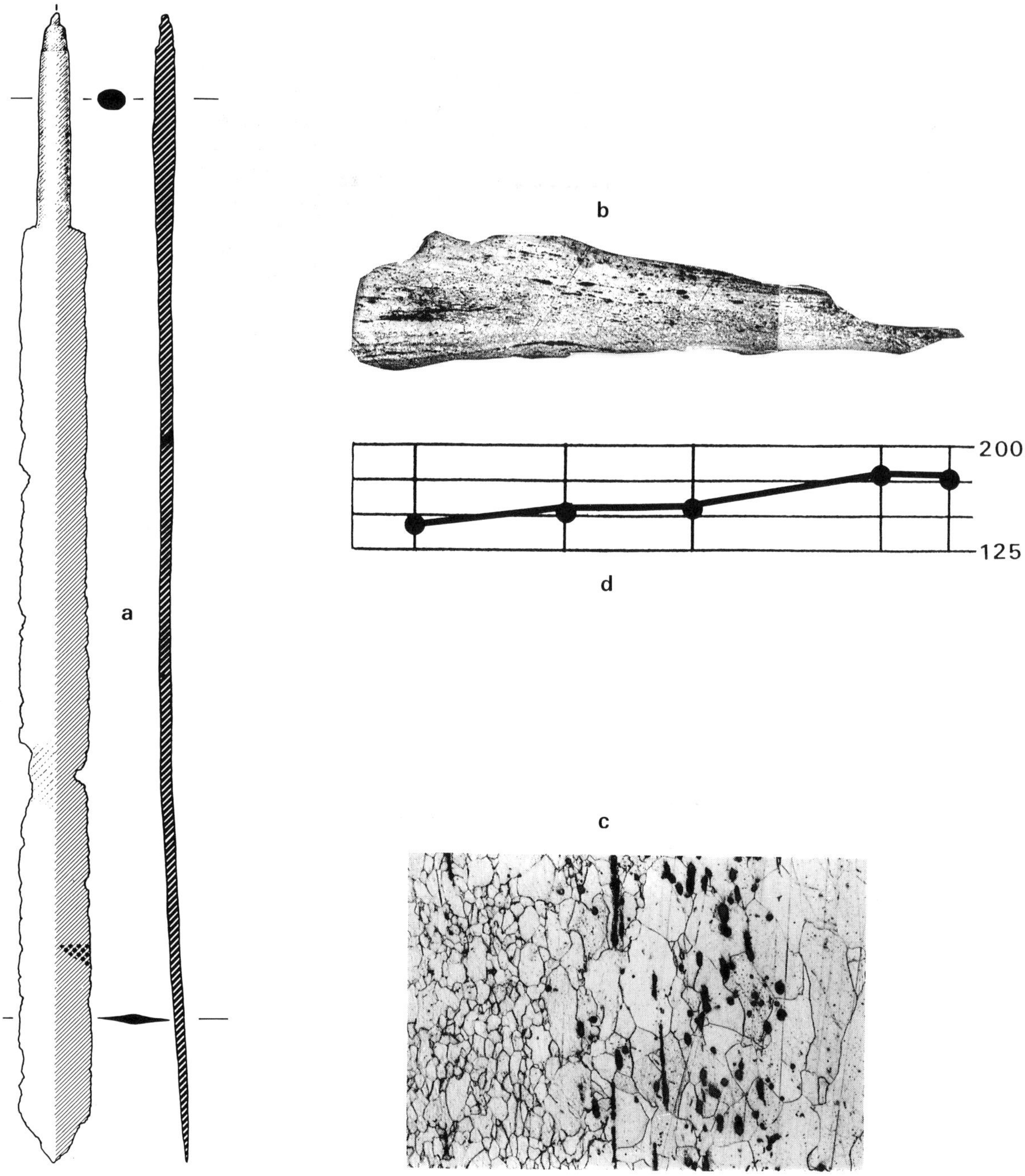

Plate 5.3.7 Examination 34: (a) ARM 1–1947 sword from Clanrolla, Co. Armagh x¼, (b) macrostructure, (c) ferrite with slag inclusions in the centre of the section x100, (d) H_v profile. Etched 4% Nital.

Microstructure: The structure ranges from polyhedral ferrite near the midrib (GS 5–6), through martensite and troostite, to martensite with a remanent ferrite net near to the cutting edge (Plate 5.3.4c) but not at it, and the structure there is fine ferrite with interstitial pearlite. One major weld seam runs obliquely from cutting edge to midrib, and there has been some diffusion of carbon across it.
Hardness: mH_v ferrite 288, pearlite 312–427, troostite 616, martensite 1305: H_v 175–515.
Interpretation: The main body of the blade appears to have been forged out of stock with uneven primary carburisation. An area near the edge was fullered down, and a piece of steel was welded on to form the cutting edge. The welding operation was prolonged sufficiently for some diffusion across the seam. After cooling, the blade was heated to above A_3, but not held there for sufficient time for the transformation to go to completion, before being quenched.

(32) UM 184–1952 sword, 'Northern Ireland' (Plate 5.3.5)

A section was taken from cutting-edge to just past the midrib (Plate 5.3.5a).
Non-metallic inclusions: There are numerous inclusions, all one-phase.
Macrostructure: The etched structure showed uneven distribution of carbon, with some banding near to the cutting edge (Plate 5.3.5b). Etching with Oberhoffer indicated no other significant segregations.
Microstructure: Towards the centre of the section, at the midrib, there is a well developed Widmanstätten structure (Plate 5.3.5c). This grades into ferrite (GS

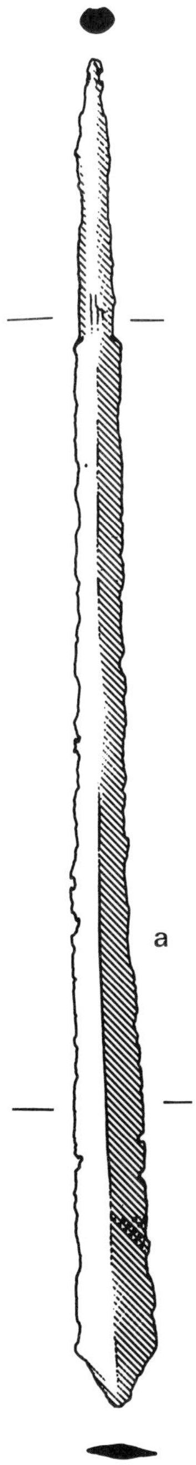

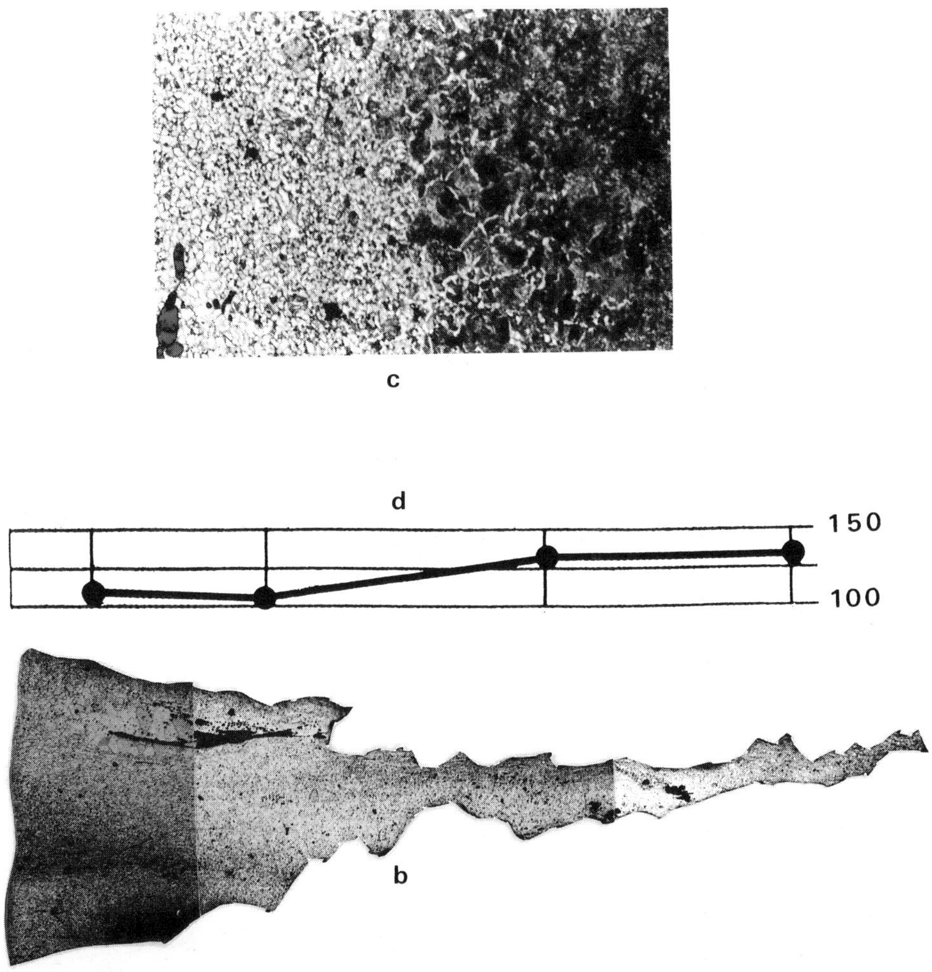

Plate 5.3.8 Examination 35: (a) NMI WK24:W6:F502 sword from the River Corrib at Galway, Co. Galway x¼, (b) macrostructure, (c) gradient from low- to high-C x100, (d) H_v profile. Etched 4% Nital.

6–8) with interstitial pearlite with banding banding between ferrite (GS 5–6) and ferrite with interstitial pearlite (GS 6–8). Traces of bloom welds may be observed.
Hardness: mH_v ferrite 182–304, pearlite 276–328: H_v 108–189.
Interpretation: The blade was made from inhomogeneously carburised stock which underwent some decarburisation of the surface during forging.

(33) NMI 1905–30 sword, Toome Bridge, Co. Antrim (Plate 5.3.6)

A section was taken from cutting-edge to midrib (Plate 5.3.6a).
Non-metallic inclusions: There are only a few small two- and three-phase inclusions.
Macrostructure: Etching revealed no significant segregations (Plate 5.3.6b).
Microstructure: The structure consists primarily of ferrite (GS 4–5 and 7–8) with interstitial coarse pearlite (Plate 5.3.6c).
Hardness: mH_v ferrite 143–283: H_v 110–158.
Interpretation: The sword was forged up from reasonably homogeneous low carbon stock and allowed to cool normally.

(34) ARM 1–1947 sword, Clanrolla, Co. Armagh (Plate 5.3.7)

A V-shaped section was taken from cutting edge to midrib (Plate 5.3.7a).
Non-metallic inclusions: Numerous one- and two-phase slag inclusions, some large, orientated along the long axis of the specimen.
Macrostructure: Nital showed some banding (Plate 5.3.7b), accentuated by Oberhoffer.
Microstructure: The structure consists almost entirely of ferrite, with bands of fine ferrite (GS 6–7) with coarse interstitial pearlite, beside a band of large (GS 1–3) ferrite grains (Plate 5.3.7c) with some grain

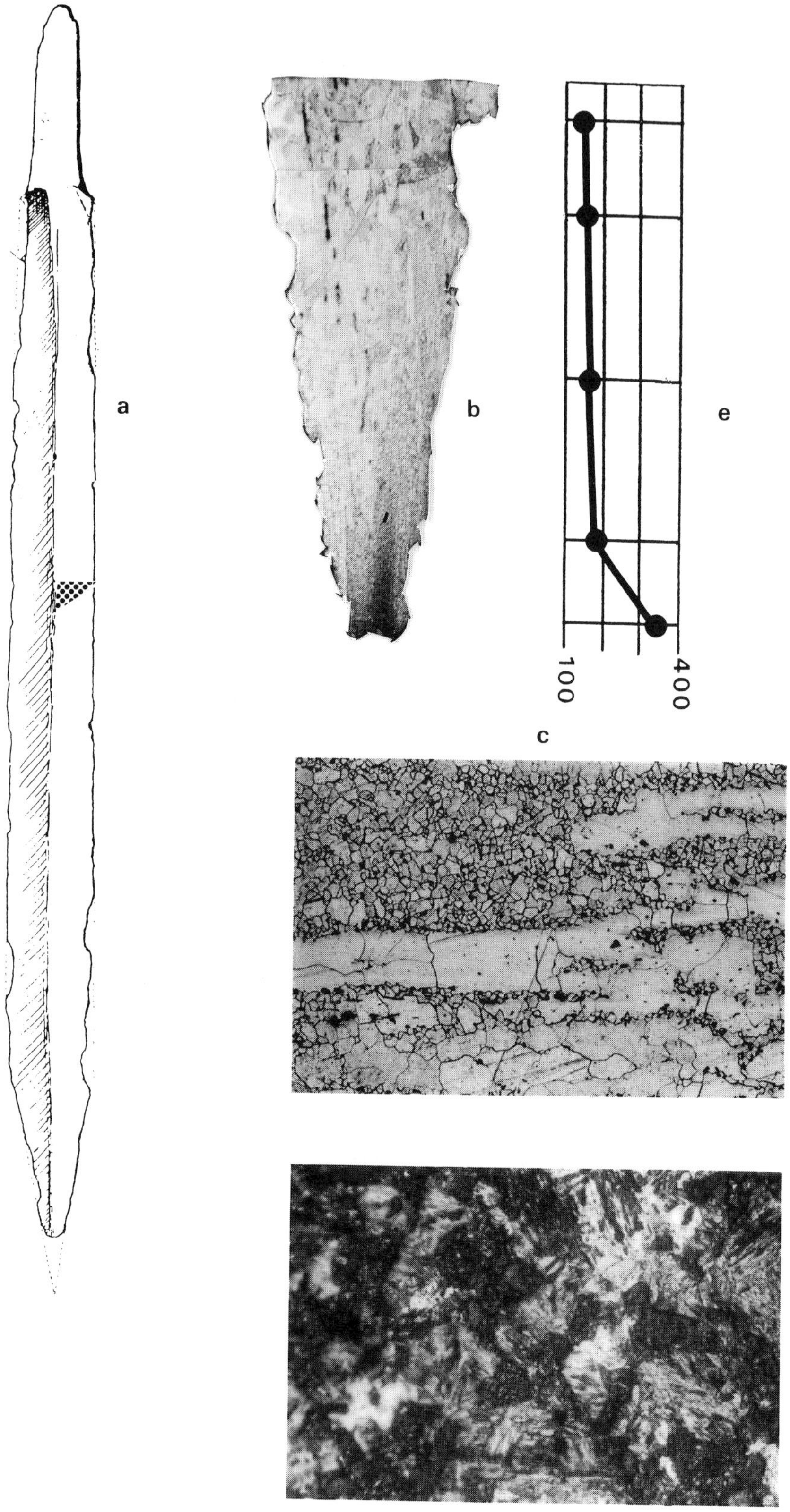

Plate 5.3.9 Examination 36: (a) NMI E6:45A sword from Ballinderry crannog no. 2, Co. Offaly x⅓, (b) macrostructure, (c) variation in ferritic grain size x100, (d) martensite and troostite at the cutting edge x500, (e) H_v profile. (b) and (c) etched 4% Nital, (d) etched Bisulphite.

boundary cementite. Grain size diminishes towards the cutting-edge and there is some distortion of the ferrite grains there (Plate 5.3.7d).
Hardness: mH_v ferrite 222–259: H_v 140–182.
Interpretation: This nicely shaped piece shows that it was forged from stock virtually free from carbon, and with an uneven distribution of P which has led to some banding. Despite its looks, it would not have been a particularly good weapon.

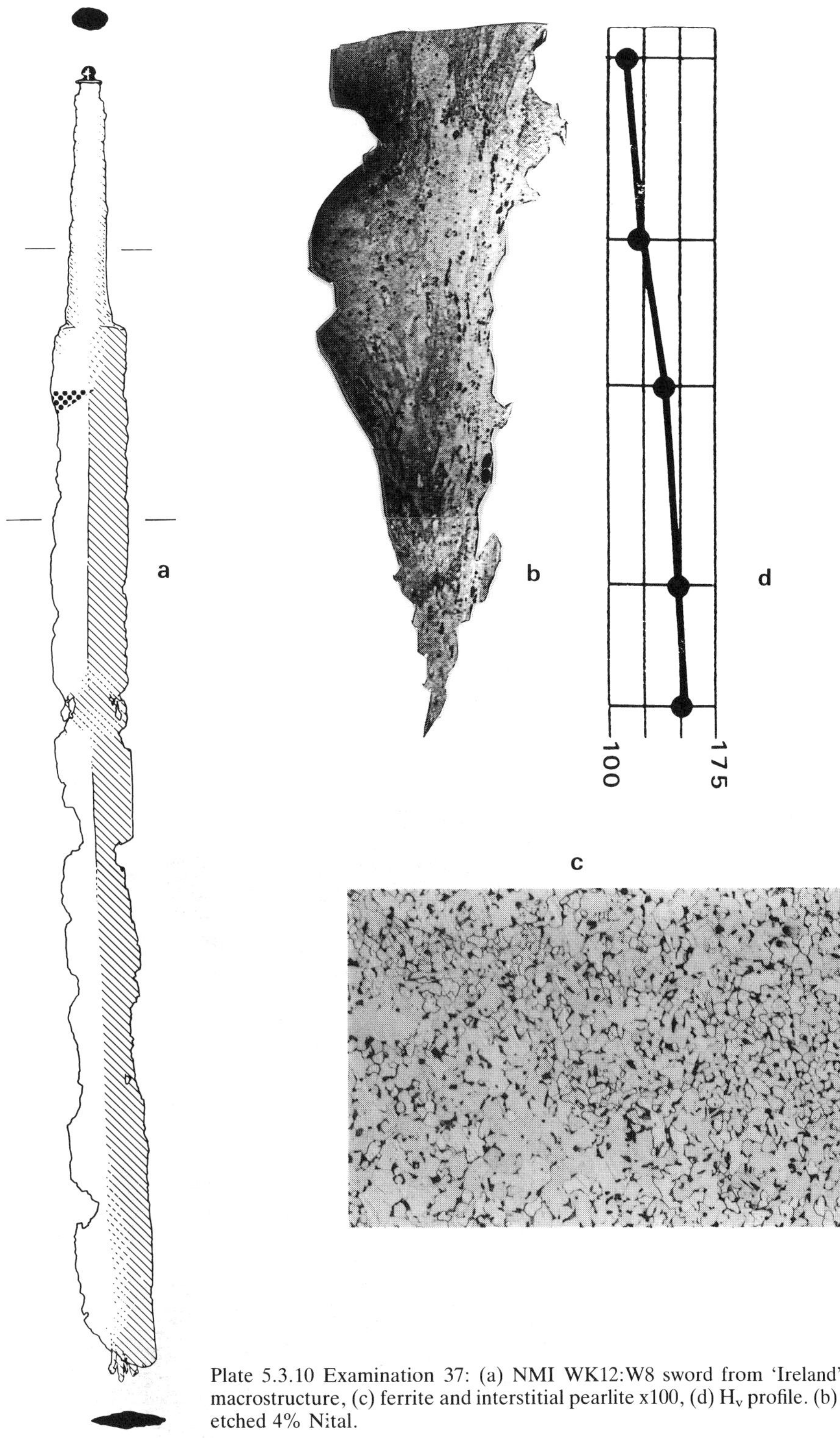

Plate 5.3.10 Examination 37: (a) NMI WK12:W8 sword from 'Ireland' (no locality) x⅓, (b) macrostructure, (c) ferrite and interstitial pearlite x100, (d) H_v profile. (b) etched Oberhoffer, (c) etched 4% Nital.

(35) NMI WK24:W6:F502 sword, River Corrib at Galway, Co. Galway (Plate 5.3.8)

A section was cut from cutting-edge to midrib, near to the point of maximum width (Plate 5.3.8a).
Non-metallic inclusions: Numerous two- and three-phase inclusions, some large.
Macrostructure: Etching with Nital indicated a concentration of C round the margins of the specimen, suggesting carburisation (Plate 5.3.8b). Oberhoffer indicated no other segregations.
Microstructure: The structure towards the centre of the specimen is ferrite (GS 5–6), changing towards both faces of the blade through ferrite and pearlite (GS 7–8) to near eutectoid, with a ferrite net in places (Plate 5.3.8c). The structure at the cutting-edge is Widmanstätten ferrite with pearlite.
Hardness: mH_v ferrite 145–182, pearlite 252–300: H_v 110–141.
Interpretation: The blade was forged from carbon-free stock retaining much slag, and was then carburised, somewhat unevenly. There is no trace of the application of any heat-treatment.

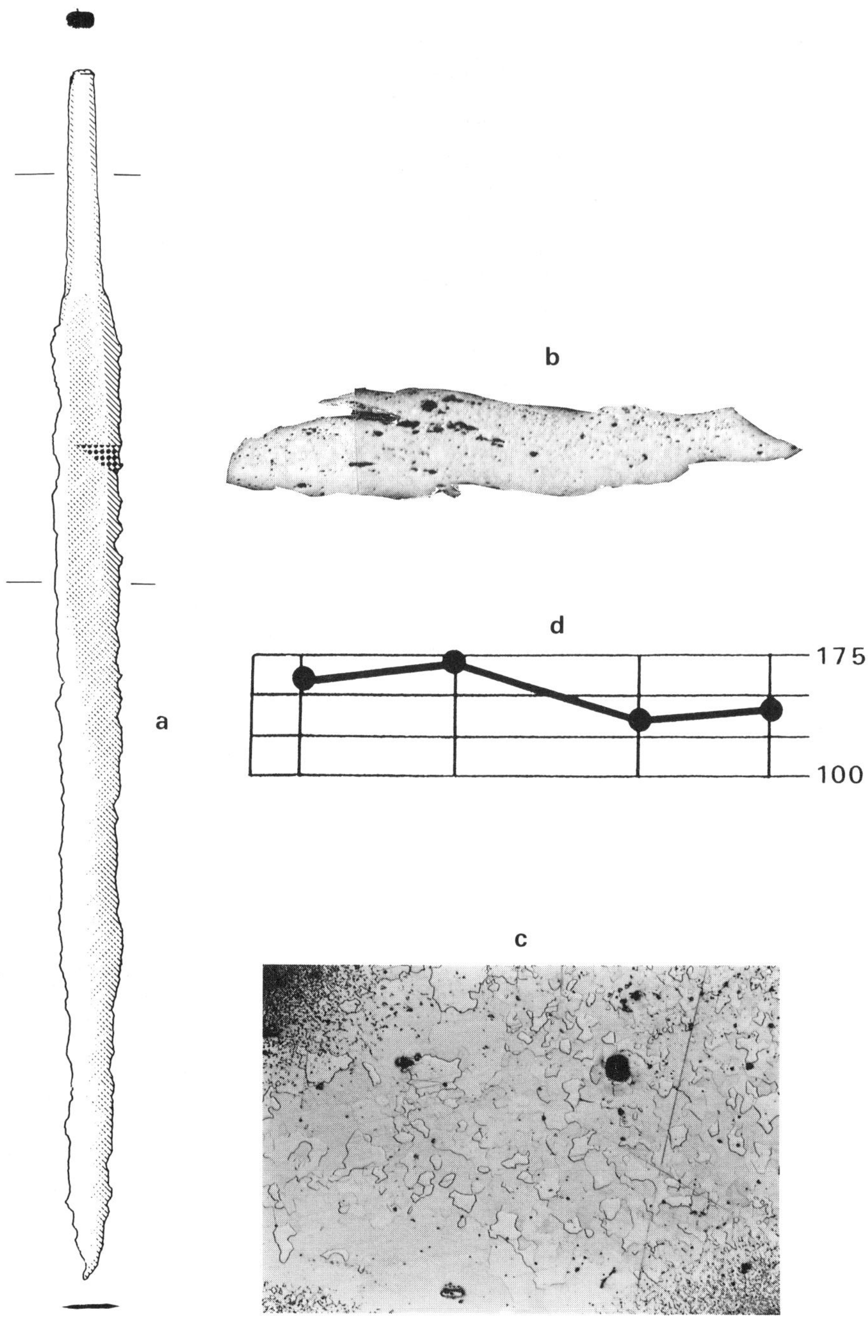

Plate 5.3.11 Examination 38: (a) NMI E14:1059 sword from Lagore crannog, Co. Westmeath x$\frac{1}{3}$, (b) macrostructure, (c) structure near upper face of section x100, (d) H_v profile. Etched 4% Nital.

(36) NMI E6:45A sword, Ballinderry crannog no. 2, Co Offaly (Plate 5.3.9)

A section was taken from cutting-edge to midrib (Plate 5.3.9a).
Non-metallic inclusions: There are numerous large two- and three-phase inclusions.
Macrostructure: Etching with Nital indicated wide variation in carbon content and grain size distribution (Plate 5.3.9b). Oberhoffer indicated minor segregations.
Microstructure: Towards the midrib area the structure is primarily ferritic. Areas of large ferrite grains growth (GS 1–3) lie beside areas where the ferrite grains are much smaller (GS 6–7) and are accompanied by interstitial fine pearlite (Plate 5.3.9c). The carbon content rises markedly towards the cutting-edge, indicating deliberate secondary carburisation. The structure at the cutting edge is martensite and troostite with remanent ferrite (Plate 5.3.9d).
Hardness: mH_v ferrite 225–264, troostite 585, martensite 722: H_v 148–368.
Interpretation: The blade has been forged up from low-C stock, which was then carburised. The bloom originally contained minor segregations which have formed the ferritic streaks observed in the section. The blade was heated to above A_3 prior to quenching, but was not kept there sufficiently long for the α-γ transformation to go to completion, as may be seen from the remanent ferrite with the martensite resulting from the quenching operation.

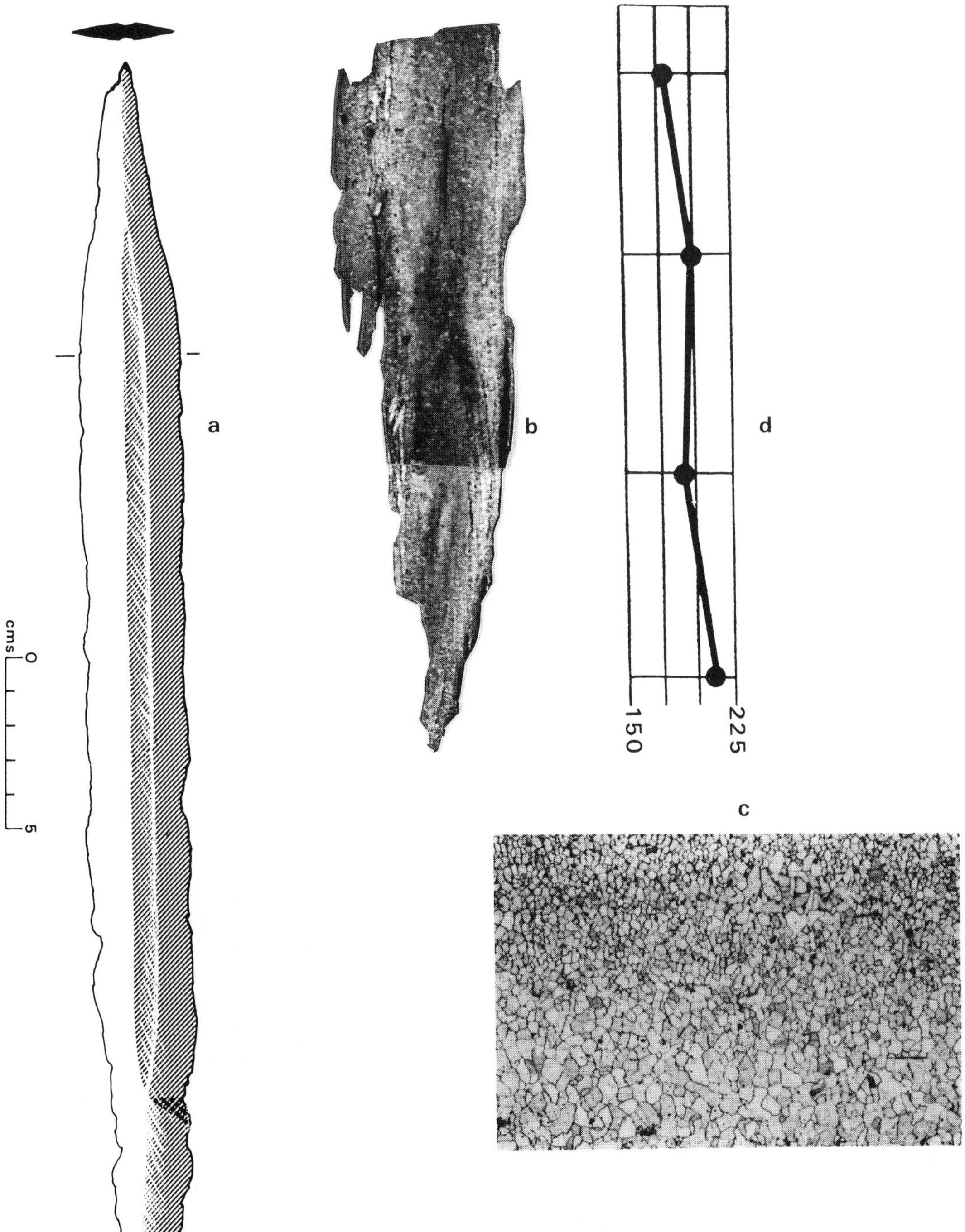

Plate 5.3.12 Examination 39: (a) DoENI sword from Killyliss, Co. Tyrone, (b) macrostructure, (c) ferrite with interstitial pearlite x100, (d) H_v profile. (b) etched Oberhoffer, (c) etched 4% Nital.

(37) NMI WK12:W8 sword, 'Ireland' no locality (Plate 5.3.10)

A section was taken from the cutting edge to the midrib (Plate 5.3.10a).
Non-metallic inclusions: Groups of small two-phase inclusions run parallel to the long axis of the specimen.
Macrostructure: Etching with Nital indicated an uneven distribution of carbon and of grain size (Plate 5.2.10b), while Oberhoffer accentuated these differences.
Microstructure: The dominant feature is the variation in ferritic grain size, with areas of ferrite (GS 4–6) with interstitial coarse pearlite beside areas of fine ferrite (GS 7–<8), also with coarse interstitial pearlite (Plate 5.2.10c). There are also some scattered areas of pearlite with Widmanstätten ferrite.
Hardness: ferrite 136–208, pearlite 262–385: H_v 110–153.
Interpretation: The sword has been forged up from heterogeneous stock and air cooled. While nicely shaped, it would not have been a particularly effective weapon.

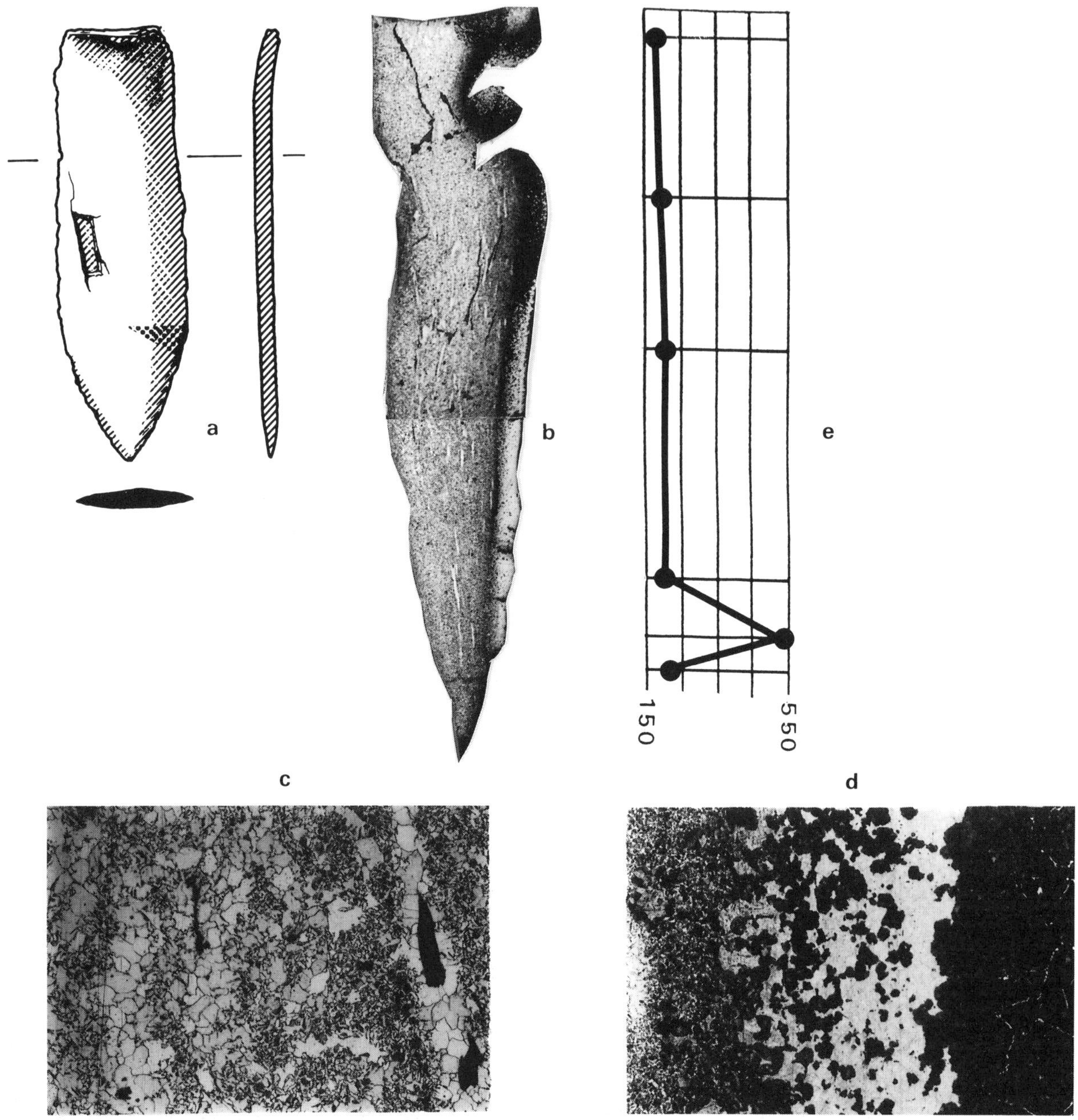

Plate 5.3.13 Examination 40: (a) NMI E21:230 broken tip of sword from Lough Gara, Co. Sligo x½, (b) macrostructure, (c) structure in the centre of the section x100, (d) martensite and troostite grading into tempered martensite with the remains of a cementite net x100, (e) H_v profile. Etched 4% Nital.

(38) NMI E14:1059 sword, Lagore, Co. Westmeath (Plate 5.3.11)

A section was taken from cutting edge to midrib (Plate 5.3.11a).
Non-metallic inclusions: Numerous two- and three-phases inclusions, some large, occur throughout the section.
Macrostructure: Etching revealed no significant segregations (Plate 5.3.11b).
Microstructure: The structure is almost entirely ferritic (GS 4–5), except for a small patch of higher carbon content towards one face of the blade (Plate 5.3.11c).
Hardness: mH_v ferrite 121–190: H_v 133–172.
Interpretation: The sword was forged from stock of almost free of carbon and containing numerous large inclusions, and cooled normally. It could not have been a weapon to take seriously!

(39) DoENI Sword, Killyliss, Co. Tyrone (Plate 5.3.12)

A V-shaped section was taken from cutting-edge to midrib (Plate 5.3.12a).
Non-metallic inclusions: Strings of small single-phase inclusions run parallel to the long axis of the section.
Macrostructure: Nital and Oberhoffer etching showed no significant segregations (Plate 5.3.12b.
Microstructure: The structure is entirely fine ferrite (GS 7–<8) with interstitial pearlite (Plate 5.3.12c). There are traces of bloom welds.
Hardness: mH_v ferrite 266–271, pearlite 360: H_v 175–214.
Interpretation: The sword has been forged up from stock of low carbon content which is fairly evenly distributed. It was held above A_1 long enough for grain refinement to occur.

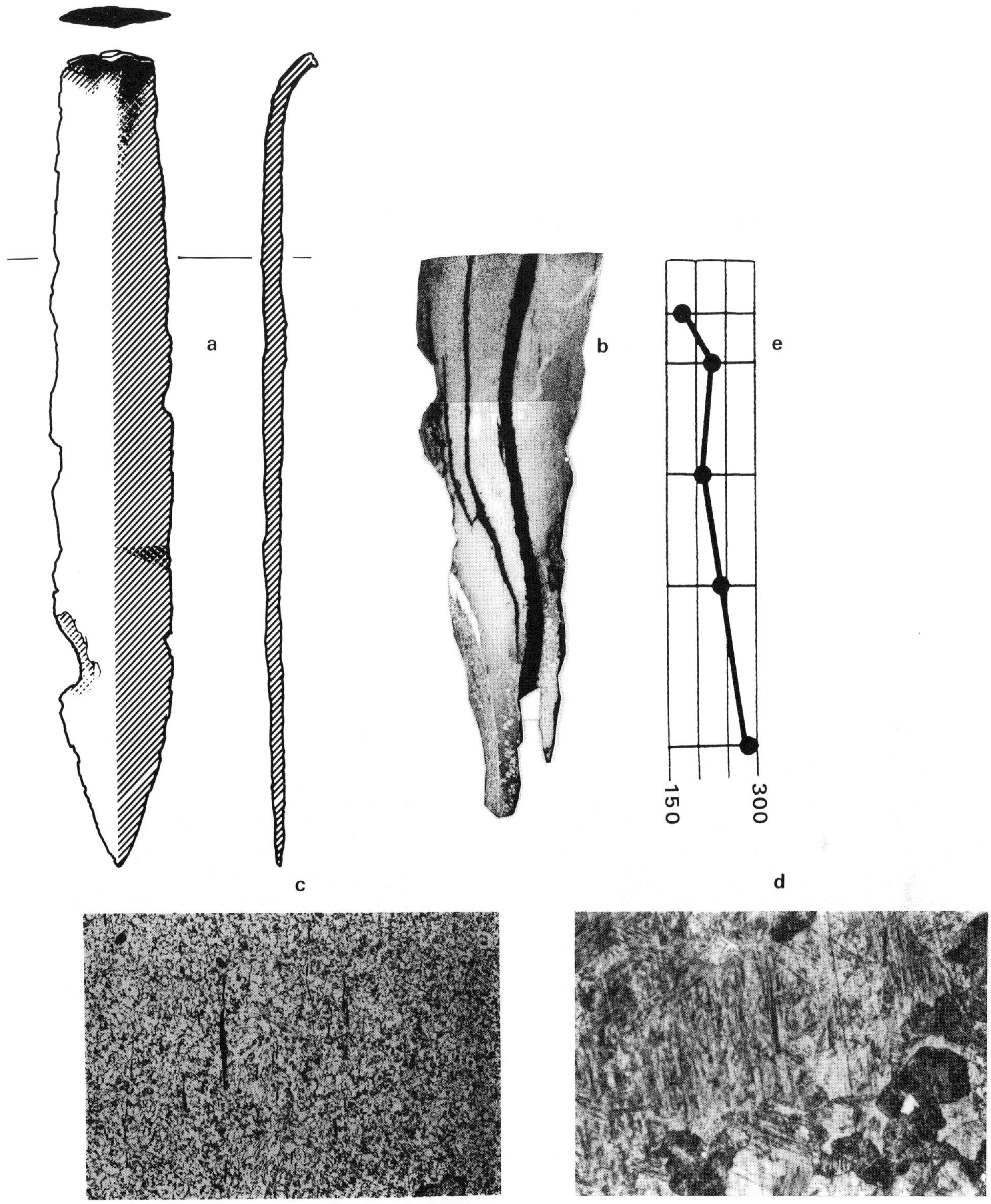

Plate 5.3.14 Examination 41: (a) NMI E148:1472 broken end of sword from Randalstown, Co. Westmeath x½, (b) macrostructure, (c) fine ferrite with interstitial pearlite at the centre of the section x100, (d) martensite and troostite at the cutting edge x500, (e) H_v profile. (b)-(d) etched 4% Nital, (e) etched Bisulphite.

(40) NMI E21:230 sword tip, Lough Gara, Co. Sligo (Plate 5.3.13)

This weapon is represented only by the broken tip of the blade. A section was taken from cutting-edge to midrib, just below the break (Plate 5.3.13a).
Non-metallic inclusions: Numerous large, two-phase inclusions are distributed with no obvious orientation throughout the section.
Macrostructure: Nital indicated an uneven distribution of carbon (Plate 5.3.13b), while Oberhoffer suggested other segregations.
Microstructure: The structure is exceedingly heterogeneous, ranging from very fine ferrite (GS <8) with interstitial pearlite to martensite and troostite with remanent ferrite. Towards the centre of the section there are patches of larger ferrite grains (GS 4–6: Plate 5.3.13c). There is a steep carbon gradient particularly towards one of the faces of the blade (cf. Plate 5.3.13b), where the structure grades from mar-

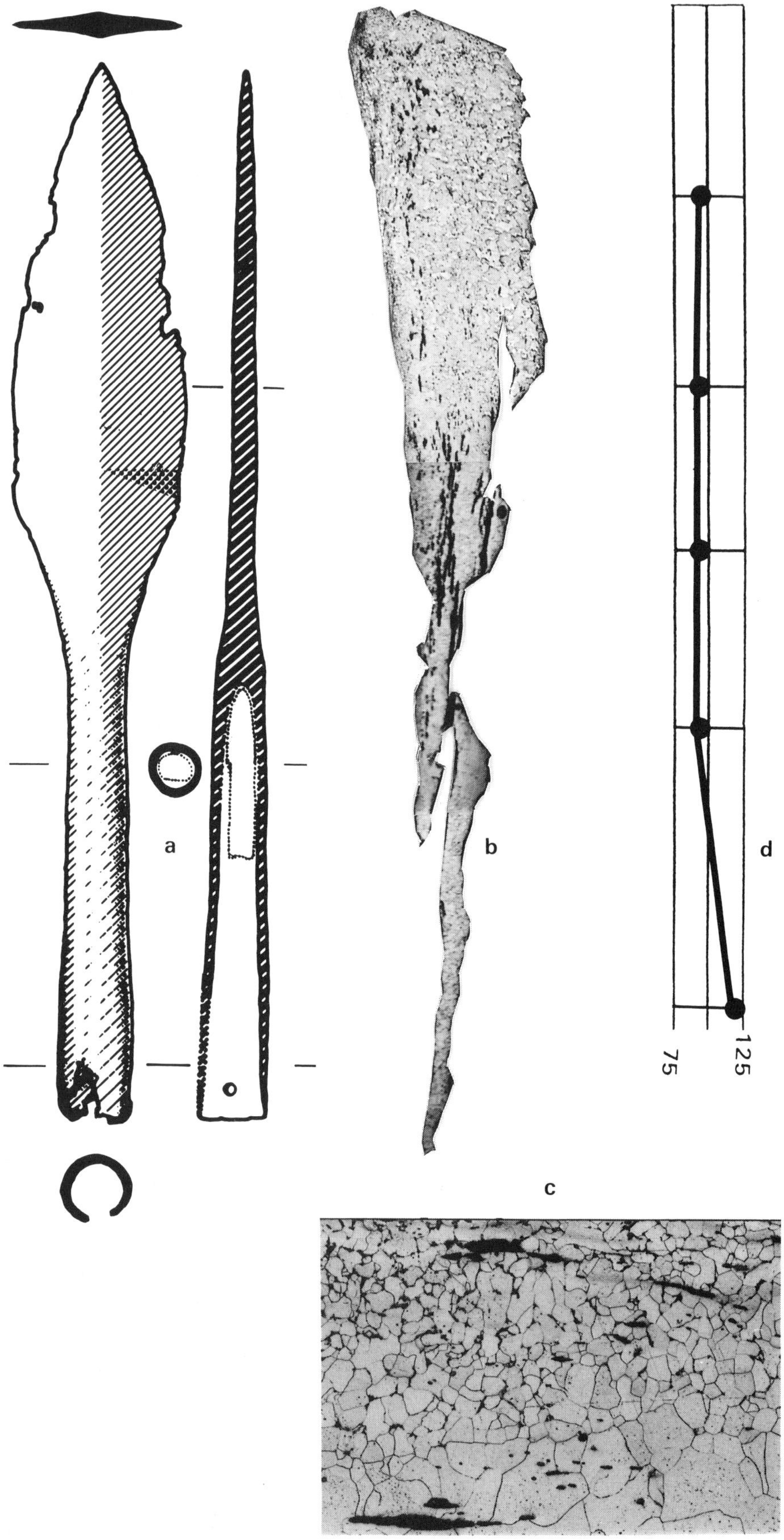

Plate 5.3.15 Examination 42: (a) ARM 2–1947 spearhead from Clanrolla, Co. Armagh x4/7, (b) macrostructure, (c) ferrite and interstitial pearlite with slag inclusions x100, (d) H_v profile. Etched 4% Nital.

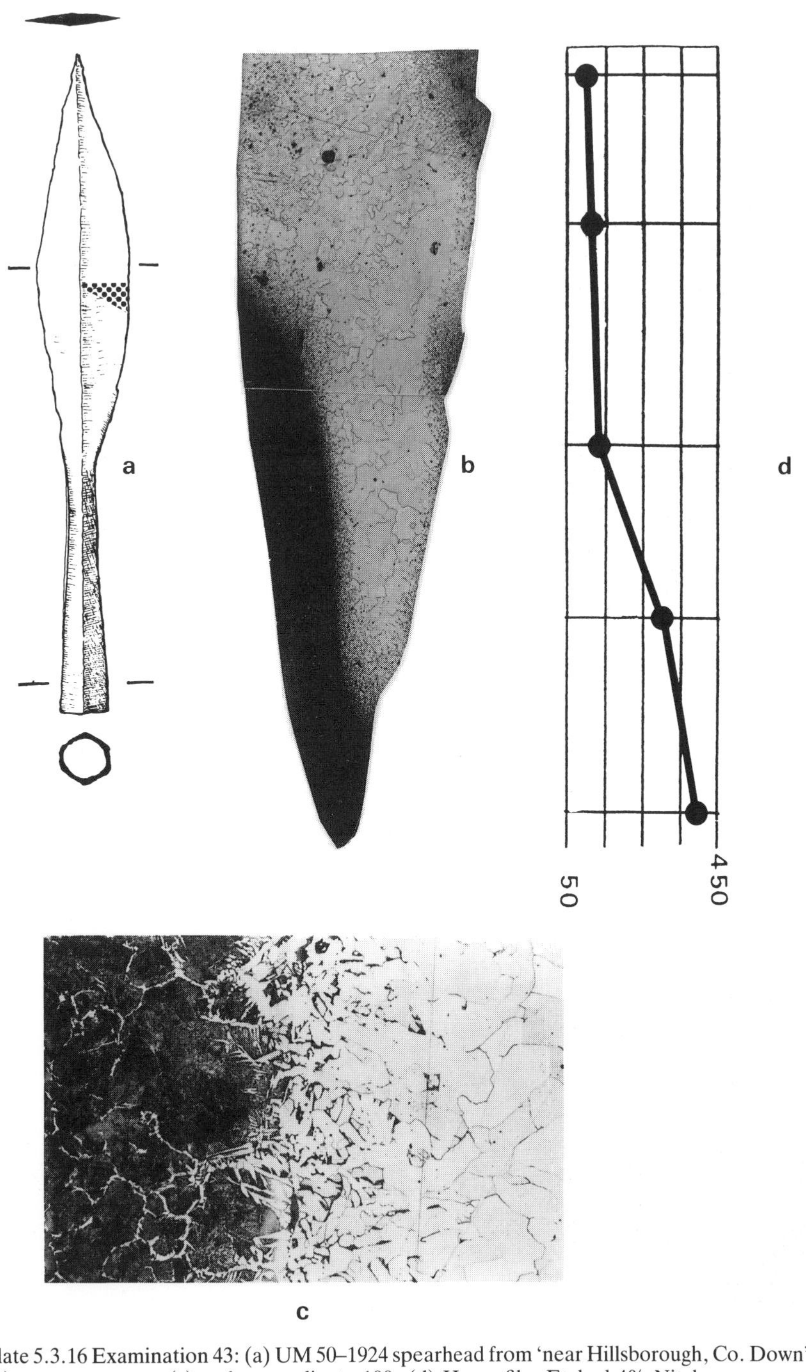

Plate 5.3.16 Examination 43: (a) UM 50–1924 spearhead from 'near Hillsborough, Co. Down' x⅓, (b) macrostructure, (c) carbon gradient x100, (d) H_v profile. Etched 4% Nital.

tensite with some remanent cementite to martensite at the section margins (Plate 5.3.13d). There are a number of quench cracks in these regions. Towards the cutting edge the structure is ferrite with interstitial pearlite.

Hardness: mH_v ferrite 190–260, pearlite 338, troostite 476, martensite 1374: H_v 186–551.

Interpretation: The blade was forged from stock of low but uneven carbon distribution is which there were some local segregations of P. It was carburised and the outermost regions reached a hypereutectoid composition with formation of a cementite network. It was reheated to above A_3 but for insufficient time for the complete formation of austenite, and then quenched, leading to cracking. This was followed by a short temper at low temperature, again for insufficient time for tempering of all of the martensite. Despite the cracking, this could have been quite an effective weapon.

(41) NMI E148:1472 sword, Randalstown, Co. Westmeath (Plate 5.3.14)

This sword[2] is represented only by the end of the blade and point which is broken off from the rest of the blade. A section from cutting edge to midrib was taken near the point (Plate 5.3.14a).

Non-metallic inclusions: The unetched section showed deep penetration of corrosion along lines of large and single-phase inclusions.

Macrostructure: Nital etching revealed an inhomogeneous structure, with two main welds (Plate 5.3.14b), while Oberhoffer indicated no other segregations.

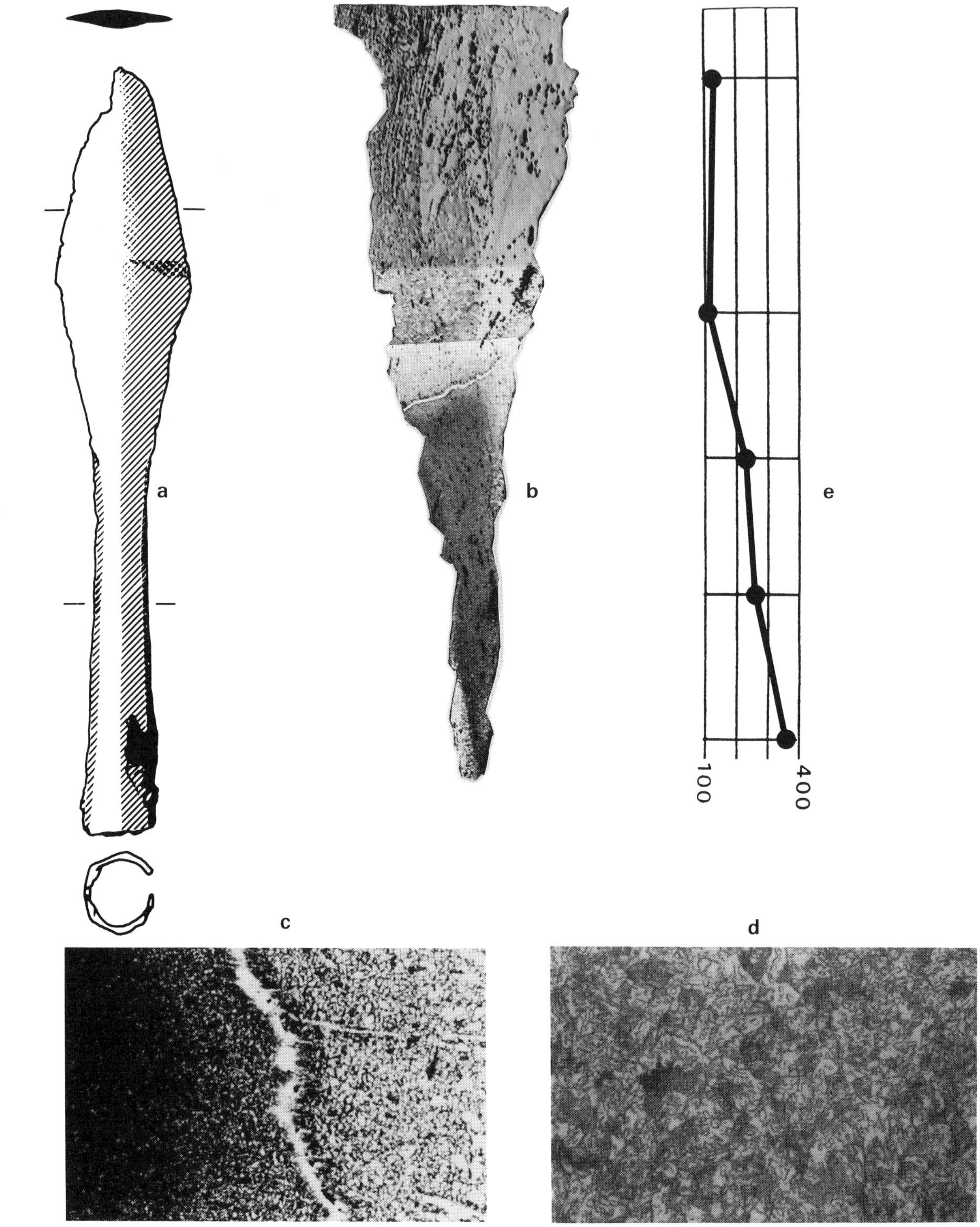

Plate 5.3.17 Examination 44: (a) NMI WK5:W50:C29 spearhead from Lagore crannog, Co. Westmeath x½, (b) macrostructure, (c) area of weld seam between high-C edge and low-C back x50, (d) spheroidised carbides in a ferrite matrix x500, (e) H_v profile. (b)-(c) etched 4% Nital, (d) etched Picral.

Microstructure: At least three separate components were joined across two seams which, in this section, have been completely obliterated by corrosion. All of the components have a structure near the midrib of very fine ferrite (GS 7–<8) with interstitial fine pearlite (Plate 5.3.14c). Moving towards the cutting edge, the outer surfaces show a carbon gradient, and the carbon content rises steeply at the cutting edge. The high carbon areas show a structure of martensite, troostite and remanent ferrite (Plate 5.3.14d).

Hardness: mH_v ferrite 266, pearlite 352, troostite 463, martensite 903: H_v 186–276.

Interpretation: The sword was made by fold-welding reasonably homogeneous low carbon stock. The welds were poorly executed, with the seams containing much scale along which corrosion has penetrated deeply. The blade was carburised, and then heated above A_3 but not held there for long enough to allow the transformation to go to completion. Finally, it was quenched. It would have been a reasonably effective weapon.

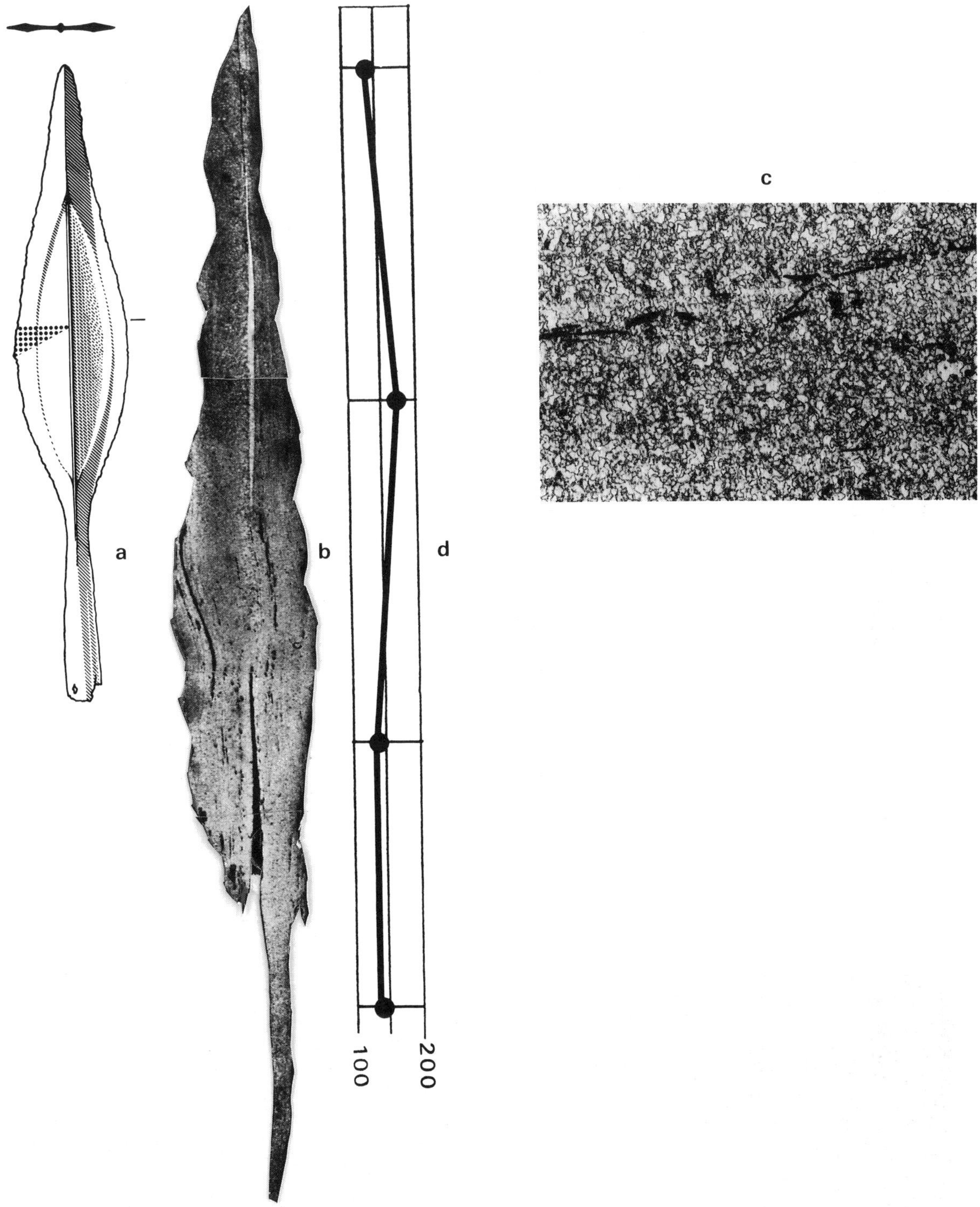

Plate 5.3.18 Examination 45: (a) UM 7–24 spearhead from 'a crannog in Co. Fermanagh' x⅓, (b) macrostructure, (c) ferrite with interstitial pearlite x100, (d) H_v profile. Etched 4% Nital.

(42) ARM 2–1947 spearhead, Clanrolla, Co. Armagh (Plate 5.3.15)

A V-shaped section was taken from cutting edge to midrib (Plate 5.3.15a).
Non-metallic inclusions: Inclusions are large, numerous, and two- and three phase.
Macrostructure: Nital and Oberhoffer indicated a largely homogeneous structure (Plate 5.3.15b).
Microstructure: Towards the midrib area, the structure is ferrite (GS 3–5) with some grain boundary cementite in places. On one side of the section there is a small area of ferrite (GS 5–6) with interstitial fine pearlite (Plate 5.3.15c). Damage by corrosion has separated the cutting edge region from the midrib, and there is an indication that this might have been welded on originally. The structure here, however, is polyhedral ferrite (GS 2–3). Acicular precipitates may be seen with the larger ferrite grains.
Hardness: mH_v ferrite 138: H_v 93–123.
Interpretation: The spearhead was forged from iron that was largely carbon-free. Because of the ingress of corrosion it is difficult to be sure whether or not there was fold-welding or the addition of a piece at the cutting edge. If there was welding, then the former seems the more likely.

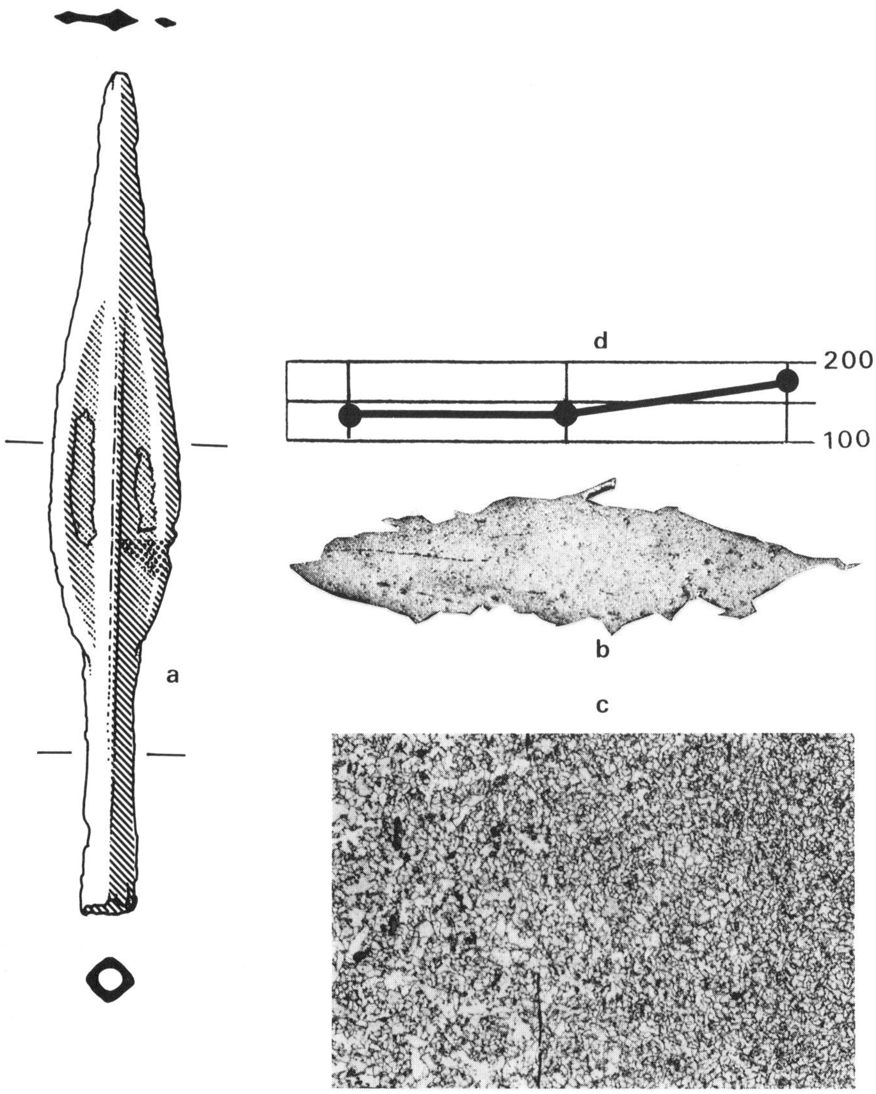

Plate 5.3.19 Examination 46: (a) NMI 1939–270 spearhead from the River Shannon at Drumshanbo, Co. Leitrim x½, (b) macrostructure, (c) fine ferrite with interstitial pearlite x100, (d) H_v profile. Etched 4% Nital.

(43) UM 50–1924 spearhead, 'near Hillsborough, Co. Down' (Plate 5.3.16)

A V-shaped section was taken from cutting edge to midrib (Plate 5.3.16a).
Non-metallic inclusions: There are only a few small single-phase inclusions.
Macrostructure: Etching with Nital revealed a concentration of carbon towards the cutting edge (Plate 5.3.16b). Oberhoffer indicated no other significant segregations.
Microstructure: In the carbon-free area which extends from the midrib down towards the cutting edge, the structure is purely ferritic with large grains (GS >1–2), some of which show acicular precipitates. Towards the edges of the upper part of the section there is a distinct carbon gradient from the outside, with smaller ferrite grains (GS 5–7) and interstitial fine pearlite. Approaching the cutting edge, there is a very steep carbon gradient, where the structure changes from ferrite, through fine grained ferrite with pearlite (Plate 5.3.16c), to heavily spheroidised carbide in a ferrite matrix, finally with tempered martensite at the cutting edge itself.
Hardness: mH_v ferrite 124–191, pearlite 413, spheroidised carbide 528, martensite 645: H_v 99–390.
Interpretation: The blade was forged up from stock that was virtually carbon-free, and with an uneven distribution of P. The cutting edges were carburised and then quenched and tempered. As well as being a very attractive weapon, this would have been very effective also.

(44) NMI WK5:W50:C29 spearhead, Lagore, Co. Westmeath (Plate 5.3.17)

A V-shaped section was taken from cutting edge to midrib (Plate 5.3.17a).
Non-metallic inclusions: Towards the midrib area, two-phase slag inclusions are numerous. Closer to cutting edge, inclusions are less common, and single phase.
Macrostructure: Etching with Nital indicated a high-C cutting-edge welded onto an iron core (Plate 5.3.17b). Oberhoffer accentuated differences in ferritic grain size in the iron core.

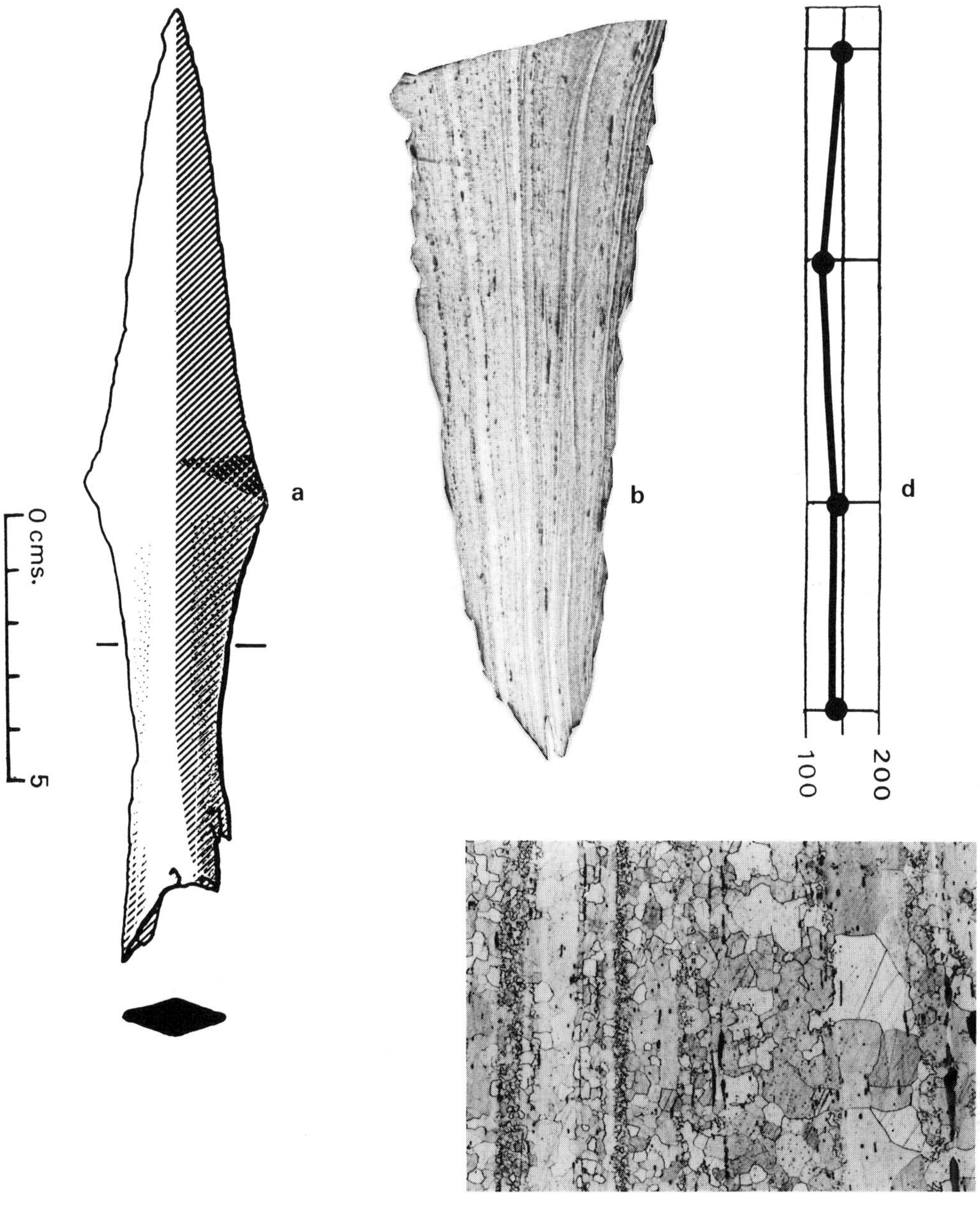

Plate 5.3.20 Examination 47: (a) UM CL:2422 spearhead from Clogher, Co. Tyrone, (b) macrostructure, (c) banding of ferrite x100, (d) H_v profile. Etched 4% Nital.

Microstructure: A weld seam is clearly visible between the carbon-free midrib area and the high-C cutting-edge. Ferrite grains are regular and polyhedral but segregated into areas of large (GS 1–3) and small grains (GS 4–5), decreasing further in size (GS 6–7) towards the weld seam, where there has been carbon diffusion from the high-C cutting-edge and fine pearlite appears (Plate 5.3.17c). Across the seam and towards the cutting-edge, the structure is a mixture of very coarse pearlite with ferrite and spheroidised cementite in a ferrite matrix (Plate 5.3.17d). At the cutting edge, there are some small patches of martensite.
Hardness: mH_v ferrite 120–160, pearlite 225, spheroidised carbide 4532, martensite 749: H_v 125–257.
Interpretation: The body of the spearhead was forged up from carbon-free stock containing much entrapped slag. Two high-C cutting-edges were welded on. The blade was quenched and then tempered at high enough temperature and for sufficient time for considerable spheroidisation of cementite (i.e. close to A_1). The small area of martensite is difficult to explain, unless the cutting edge was allowed briefly to reach a temperature above A_3 followed by quenching. This would have been a very effective weapon.

(45) UM 7-24 spearhead, 'Co. Fermanagh' (Plate 5.3.18)

A V-shaped section was taken from cutting edge to midrib (Plate 5.3.18a).
Non-metallic inclusions: Numerous one- and two-phase inclusions run parallel to the long axis of the specimen.
Macrostructure: Etching with Nital showed a homo-

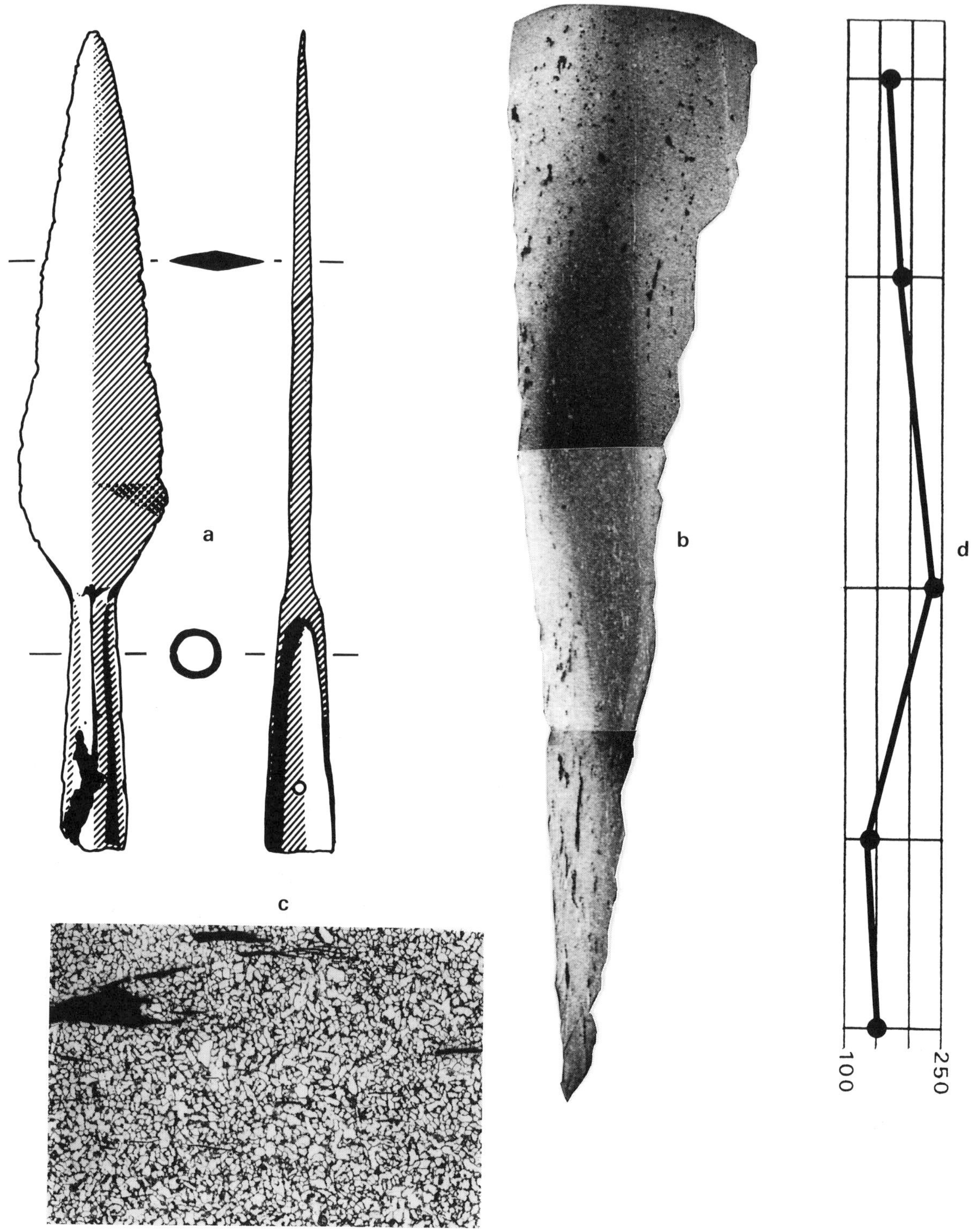

Plate 5.3.21 Examination 48: (a) NMI 1932:7211 spearhead from Ballinderry, Co. Offaly x½, (b) macrostructure, (c) ferrite and interstitial pearlite with slag near the cutting edge x100, (d) H_v profile. Etched 4% Nital.

geneous structure with traces of at least three weld-seams (Plate 5.3.18b), accentuated by Oberhoffer.
Microstructure: The structure consists primarily of fine-grained ferrite (GS 6–8) with interstitial fine pearlite (Plate 5.3.18c). The lines of at least three weld-seams are clearly visible, although the slag inclusions and structure are the same throughout the section.
Hardness: mH_v ferrite 180, pearlite 287–368: H_v 143–175.
Interpretation: This finely-shaped piece was forged from a piece of evenly carburised stock, which had been fold-welded, and held finally at heat for long enough for grain refinement to occur.

(46) NMI 1939:270 spearhead, River Shannon at Drumshanbo, Co. Leitrim (Plate 5.3.19)

A V-shaped section was taken from cutting edge to midrib (Plate 5.3.19a).
Non-metallic inclusions: There are only a few small one-phase inclusions.
Macrostructure: Both Nital and Oberhoffer indicate a homogeneous structure (Plate 5.3.19b).

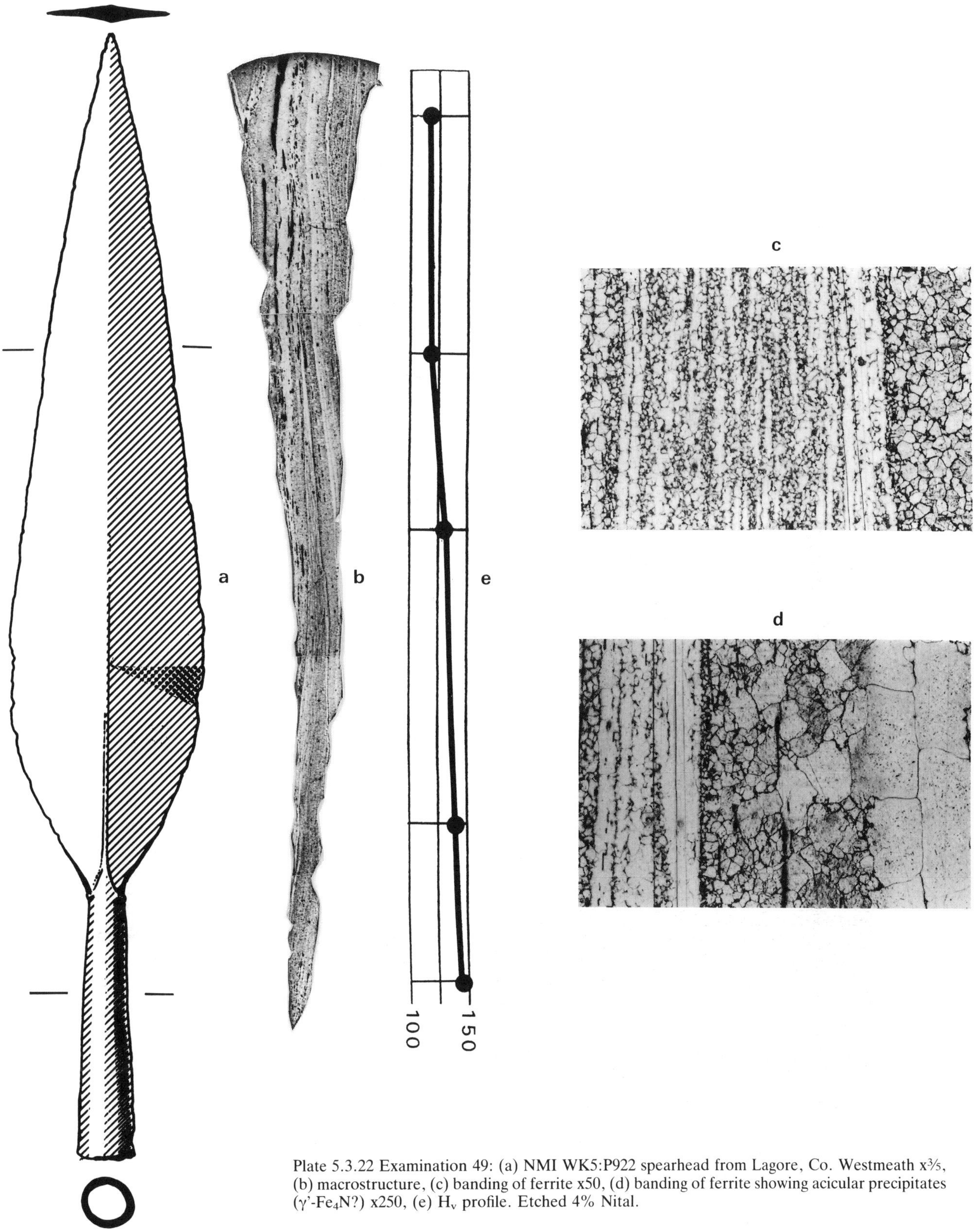

Plate 5.3.22 Examination 49: (a) NMI WK5:P922 spearhead from Lagore, Co. Westmeath x3/5, (b) macrostructure, (c) banding of ferrite x50, (d) banding of ferrite showing acicular precipitates (γ'-Fe_4N?) x250, (e) H_v profile. Etched 4% Nital.

Microstructure: The structure consists of very fine-grained ferrite structure (GS 8–<8) with interstitial coarse pearlite (Plate 5.3.19c). The distribution of carbon is uniform, but low.
Hardness: mH_v ferrite 291–306: H_v 140–180.
Interpretation: The blade was forged from a good low-C metal and held long enough at heat for grain refinement to occur, followed by air cooling. This would have been an attractive and quite effective weapon.

(47) UM CL:2422 spearhead, Clogher, Co. Tyrone (Plate 5.3.20)

A V-shaped section[3] was taken from cutting edge to midrib (Plate 5.3.20a).
Non-metallic inclusions: Numerous inclusions in chains run parallel to the long axis of the section. Inclusions vary in different parts from single-phase light- and dark grey to two-phase.
Macrostructure: Etching with Nital showed distinct

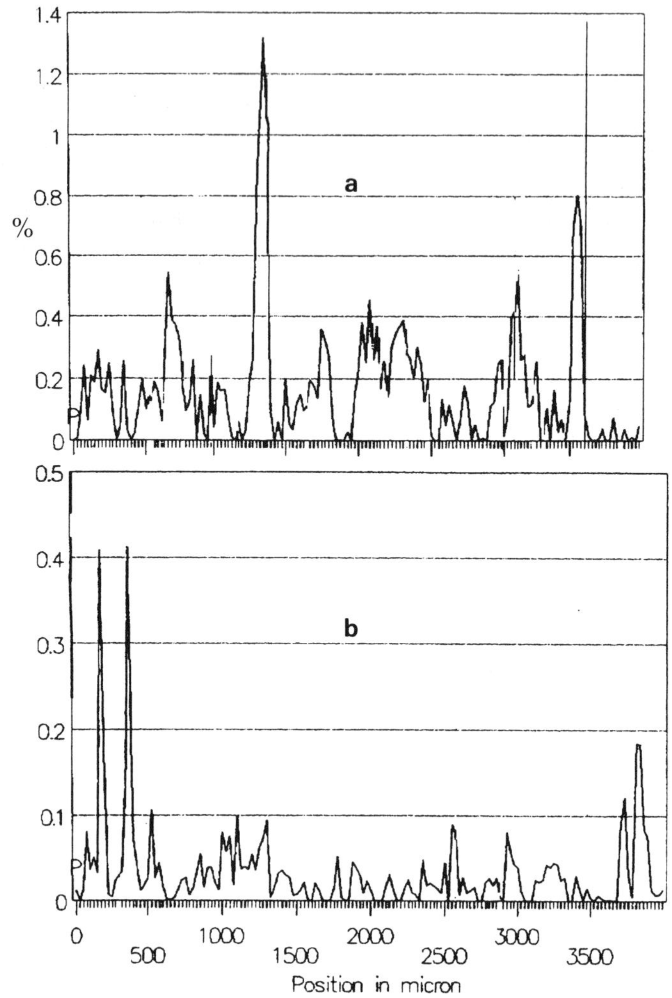

Fig. 5.3.1 Examination 49: microprobe scan for (a) As and (b) P.

banding, which was pronounced on etching with Oberhoffer (Plate 5.3.20b).
Microstructure: The structure consists of at least 20 fine laminae of ferrite (GS 3–5) alternating with ferrite (GS 7–<8) with interstitial pearlite (Plate 5.3.20c).
Hardness: mH_v ferrite 195–268: H_v 110–148.
Interpretation: The laminae observed represent the result of fold-welding together stock made from probably more than one piece of low-C metal. Slag inclusions vary so markedly in places as to indicate at least three different pieces of stock.

(48) NMI 1932:7211 spearhead, Ballinderry, Co. Offaly (Plate 5.3.21)

A section was taken from cutting edge to midrib (Plate 5.3.21a).
Non-metallic inclusions: There are a few single-phase inclusions of varing size scattered with no obvious orientation throughout the section.
Macrostructure: Nital indicated an uneven distribution of carbon, tending to concentrate towards the centre of the section (Plate 5.3.22b), while Oberhoffer indicated no other segregations.
Microstructure: Towards the outside of the section the structure is very fine grained ferrite (GS 7–<8) with coarse interstitial pearlite (Plate 5.3.21c). Towards the centre, the carbon increases to give massive and very coarse pearlite with considerable spheroidisation of carbide, and in places spheroidised carbide in a ferrite matrix.
Hardness: mH_v ferrite 239–257, pearlite 408–498: H_v 148–245.
Interpretation: The piece was made from heterogeneous stock, and probably underwent decarburisation during forging.

(49) NMI WK5:P922 spearhead, Lagore, Co. Westmeath (Plate 5.3.22)

A V-shaped section was taken from cutting edge to midrib (Plate 5.3.22a).
Non-metallic inclusions: Numerous one- and two-phase inclusions in chains run parallel to the long axis of the section.
Macrostructure: Etching with both Nital and Oberhoffer showed pronounced banding (Plate 5.3.22b).
Microstructure: The structure consists of bands of fine ferrite (GS 7–8) with interstitial fine pearlite (Plate 5.3.22c), alternating with bands of large ferrite grains (GS 2–3), some with acicular precipitates (Plate 5.3.22d). At least 9 major laminae are visible in the section. Microprobe scan shows surface enrichment of P and As, as well as the different distribution of these elements (Fig. 5.3.1).
Hardness: mH_v ferrite 241–262: H_v 119–148.
Interpretation: This weapon was forged by fold-welding from stock with segregations of P and As, and of low-C content.

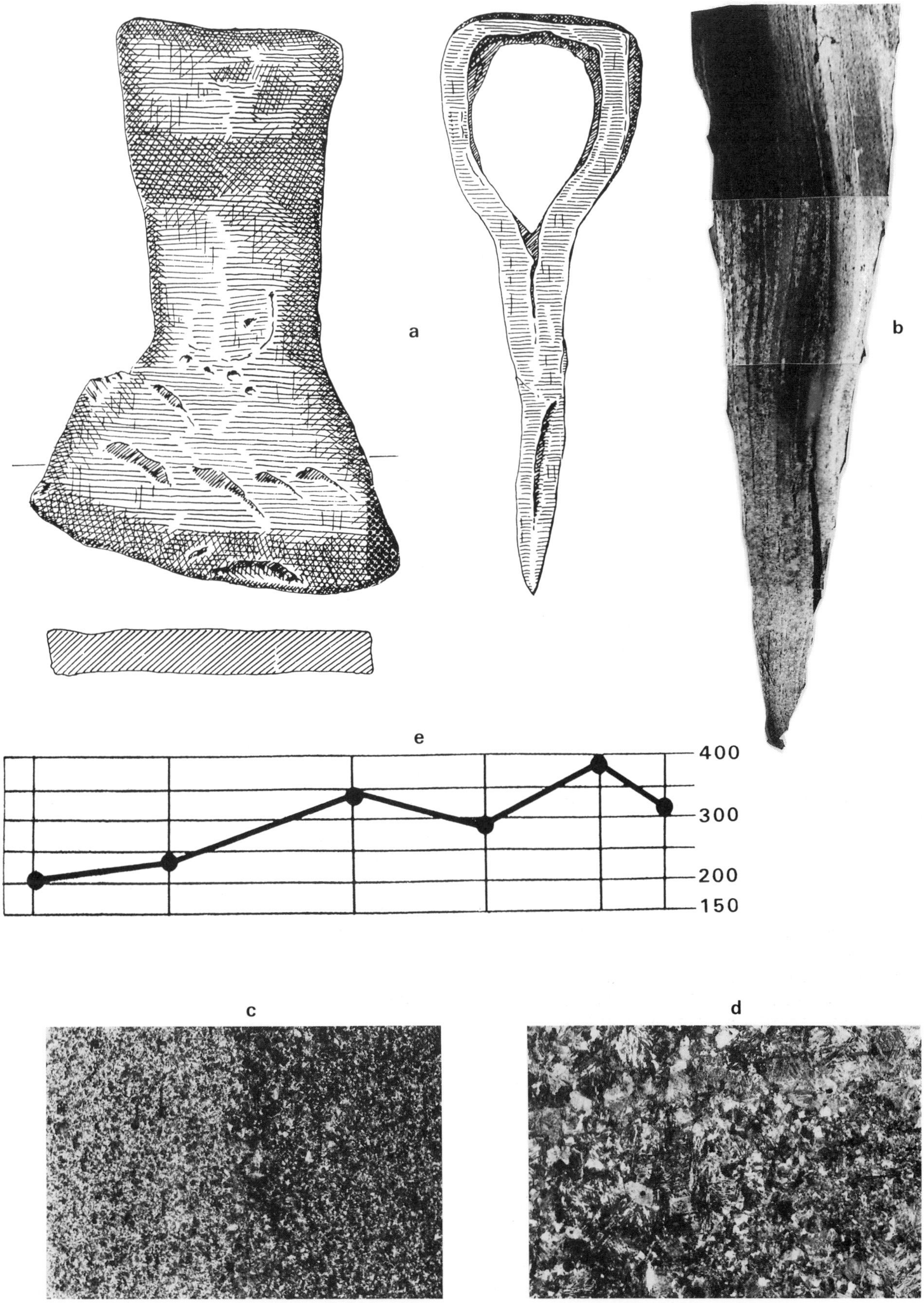

Plate 5.3.23 Examination 50: (a) DoENI DPF: 2859 shaft-hole axe from Deerpark Farm, Co. Antrim x⅔, (b) macrostructure, (c) banding between low- and high-C areas x50, (d) martensite, troostite and remanent ferrite near the cutting edge x100, (e) H_v profile. Etched 4% Nital.

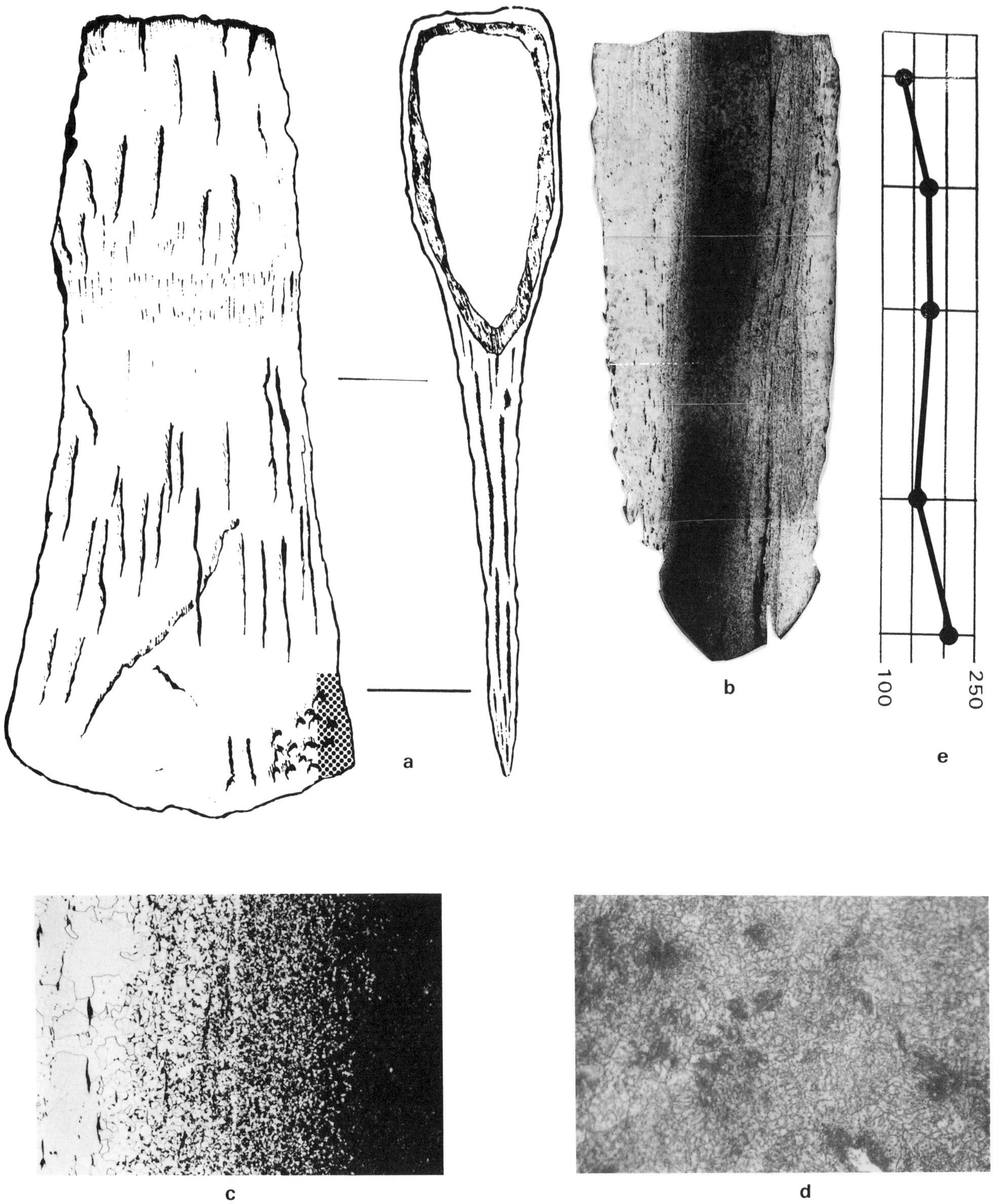

Plate 5.3.24 Examination 51: (a) UM 108–1907 shaft-hole axe from Ballynahinch, Co. Dowm x⅔, (b) macrostructure, (c) area of weld seam between low- and high-C components x50, (d) spheroidised carbide and ferrite x500, (e) H_v profile. (b)-(c) etched 4% Nital, (d) etched Picral.

(50) DoENI DPF:2859 shaft-hole axe, Deerpark Farm, Co. Antrim (Plate 5.3.23)

A section[4] was cut back from the cutting edge (Plate 5.3.23a).

Non-metallic inclusions: Inclusions are numerous, sometimes large, two- and three-phase, and obviously orientated along the long axis of the section.

Macrostructure: Etching with Nital revealed a highly uneven distribution of carbon (Plate 5.3.23b), although Oberhoffer indicated no other major segregations.

Microstructure: The most obvious features are the carbon banding throughout the section and a major welding seam running back from the cutting edge and along which corrosion has penetrated (cf. Plate 5.3.23b). At the top of the section, the structure ranges from ferrite (GS 5–6) with interstitial fine pearlite to very fine ferrite (GS <8) with fine pearlite. Moving down the section, there is pronounced

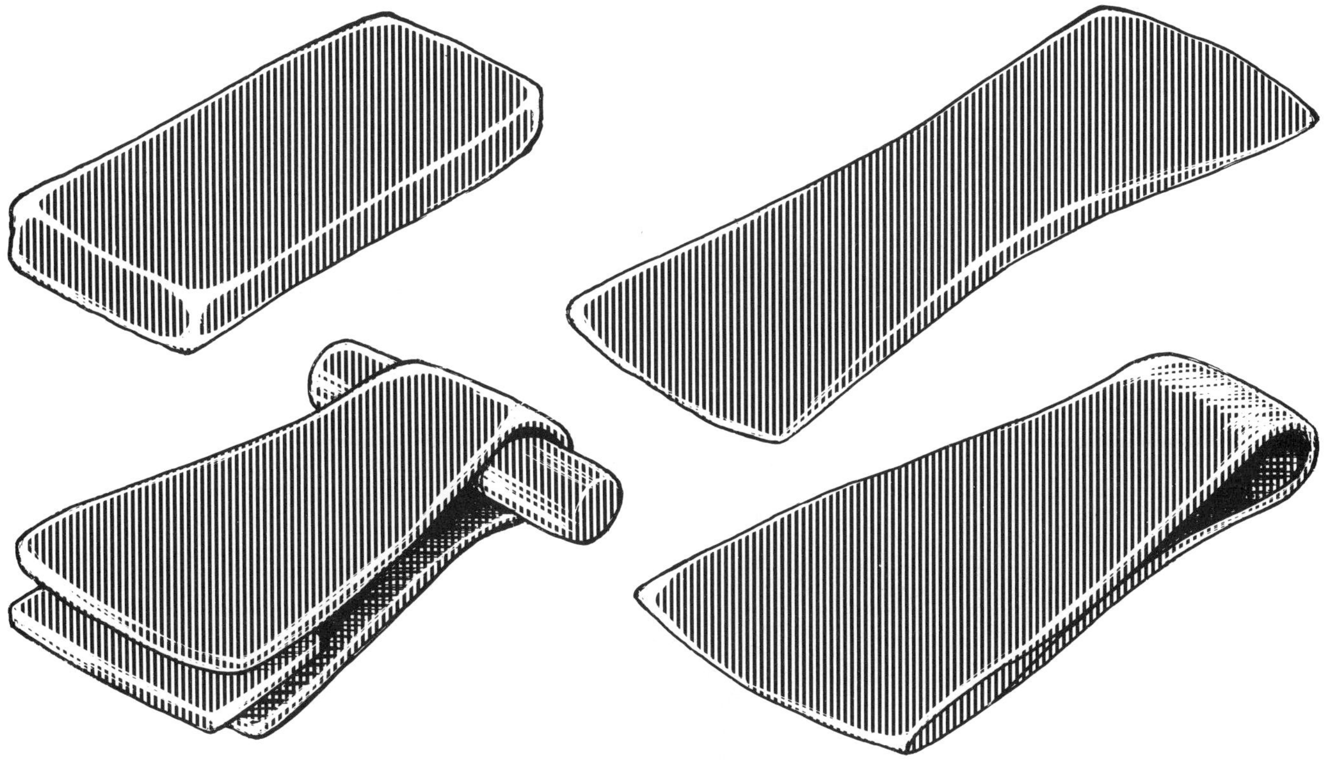

Fig. 5.3.2 Construction of the Ballynahinch axe (not to scale).

banding (Plate 5.2.23c) of fine ferrite (GS<8) with interstitial pearlite and pearlite (GS 7–<8) with a Widmanstätten ferrite net. Around the middle of the section martensite appears at the margins (Plate 5.3.23d), and the high carbon regions of the interior show martensite, troostite and remanent ferrite. At the cutting edge, the structure is martensite with the remanents of a Widmanstätten ferrite net. The main weld seam has been penetrated for about a third of its length by corrosion. Further up, there is significant diffusion across it. A number of other minor seams, also running parallel to the long axis of the section, are visible.
Hardness: mH_v ferrite 257, pearlite 380, troostite 534, martensite 717: H_v 189–390.
Interpretation: The axe was made from one sheet of stock of very uneven carbon content. It was folded over to form the socket, and the blade was forged out and folded back on itself, with the line of fold becoming the cutting edge (cf. the construction of the Rathtinaun axe, Fig. 3.3.2). The two halves of the blade were welded closed, and allowed to cool fairly rapidly. The blade was then reheated to above A_3, but was not held there long enough for the α-γ transformation to take place. It was subsequently quenched, but there is no evidence of secondary tempering. Some autotempering of martensite has taken place in the interior. This would have been a very serviceable tool, and still retains a sharp cutting edge.

(51) UM 108–1907 shaft-hole axe, Ballynahinch, Co. Down (Plate 5.3.24)

A section was taken running back from the cutting edge, parallel to the long axis of the blade (Plate 5.3.24a).
Non-metallic inclusions: These are not common, but where they occur, they are large and three-phase. Very small one-phase inclusions are associated with weld seams.
Macrostructure: Nital showed a composite construction (Plate 5.3.24b), while Oberhoffer indicated no significant segregations.
Microstructure: The section comprises two side-plates of large ferrite grains (GS 2–3), decreasing in size (GS 8–<8) towards the weld seams (cf. Plate 5.3.24b) across which there has been significant diffusion of carbon (Plate 5.3.24c). These delimit the central plate which has a high carbon structure comprising heavily spheroidised pearlite with ferrite (Plate 5.3.24c). At the cutting edge there is a region of higher carbon content.
Hardness: mH_v ferrite 188, pearlite 281–387 spheroidised carbide 456: H_v 148–210.
Interpretation: The axe was made (Fig. 5.3.2) from one piece of low carbon metal which was folded over round a bar to form the socket. A piece of high carbon metal was inserted between the two faces of the blade and the components were welded together. During welding, there was appreciable diffusion of carbon across both seams. The blade was held around A_1 for long enough to allow for the degree of carbide spheroidisation observed.

(52) UM CL:1847 shaft-hole axe, Clogher, Co. Tyrone (Plate 5.3.25)

A section was taken running back from the cutting-edge, parallel to the long axis of the blade (Plate 5.3.25a).
Non-metallic inclusions: These are few in number, and where they occur, they are small and one- and two-phase.

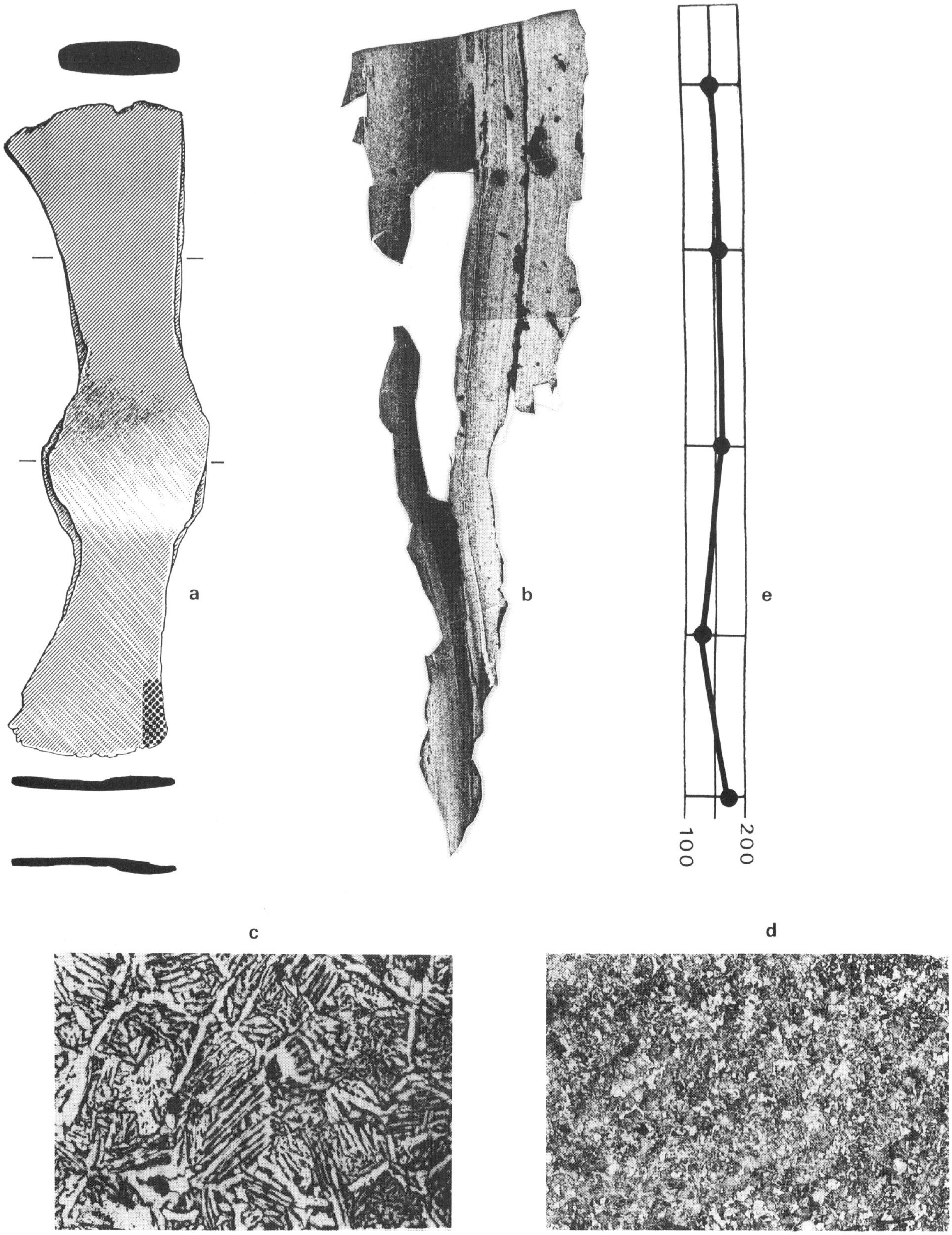

Plate 5.3.25 Examination 52: (a) UM CL:1847 shaft-hole axe from Clogher, Co. Tyrone, (b) macrostructure, (c) Widmanstätten ferrite x100, (d) fine ferrite and pearlite x100, H_v profile. Etched 4% Nital.

Macrostructure: Nital revealed a very uneven carbon distribution, while Oberhoffer did not indicate any other segregations. Traces of two major weld seams are visible, and there has been considerable penetration of corrosion along one of them (Plate 5.3.25b).
Microstructure: One of the components comprises ferrite (GS 4–5) with interstitial coarse pearlite. The structure of the other two components is ferrite and pearlite with banding. Towards the back of the section there is Widmanstätten ferrite with coarse pearlite (Plate 5.3.25c), while closer to the cutting edge the structure grades into fine ferrite (GS 8 – <8) and very coarse pearlite (Plate 5.3.25d).
Hardness: mH_v ferrite 130–220, pearlite 280–293: H_v 138–172.
Interpretation: The axe was forged from at least three

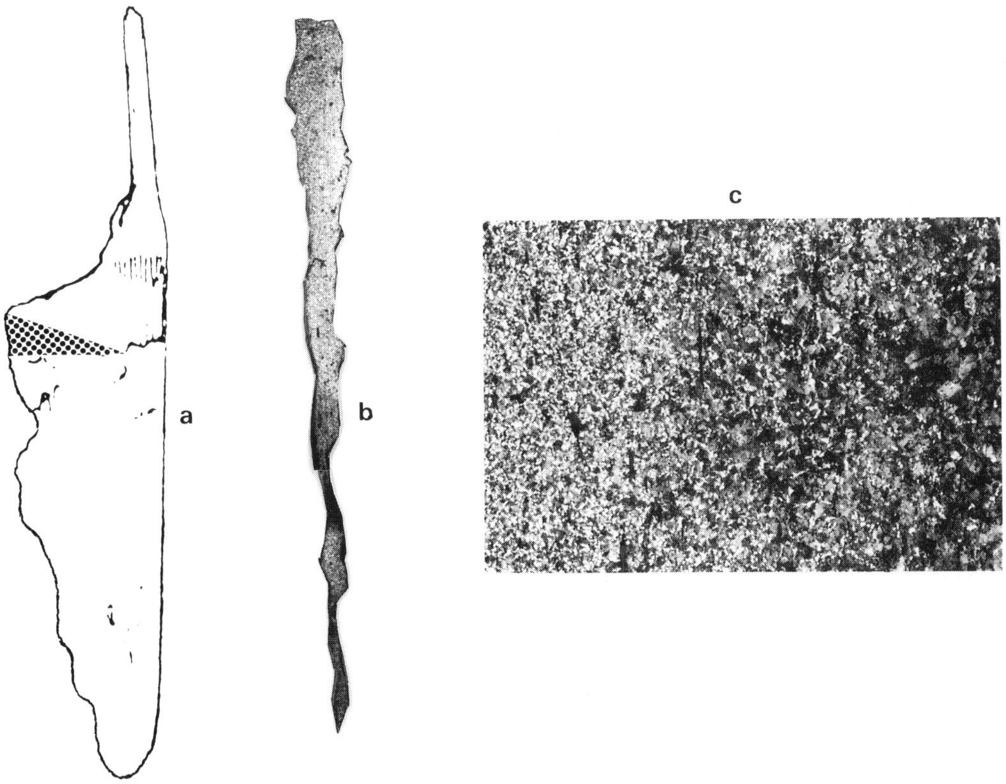

Plate 5.3.26 Examination 53: (a) NMI E6:468 blade of shears from Ballinderry crannog no. 2, Co. Offaly x1/1, (b) macrostructure, (c) ferrite and pearlite x100. Etched 4% Nital.

pieces of unevenly carburised stock. There was no attempt at heat-treatment. It would have been a reasonably effective tool.

(53) NMI E6:468 blade of shears, Ballinderry, Co. Offaly (Plate 5.3.26)

A V-shaped section was taken from the cutting edge towards the back of the blade (Plate 5.3.26a).
Non-metallic inclusions: There are virtually no inclusions, those that occur being very small and one-phase.
Macrostructure: Nital showed a somewhat uneven distribution of carbon (Plate 5.3.26b), but Oberhoffer indicated no other segregations.
Microstructure: The structure is very fine grained ferrite (GS 7–<8) with interstitial coarse pearlite (Plate 5.3.26c). There are several patches of higher carbon content which show heavy spheroidisation of the carbide lamellae of the pearlite.
Hardness: mH_v ferrite 131–203, pearlite 306–345.
Interpretation: The blade has been made from unevenly carburised stock and worked at heat for long enough to allow a significant degree of spheroidisation of carbides. There has also been appreciable grain size refinement. This would have been quite an effective tool.

(54) NMI W49 shears, Strokestown crannogs, Co. Westmeath (Plate 5.3.27)

A V-shaped section was taken from the cutting edge of one blade (*section 1*), and a second section (*section 2*) was cut from the spring (Plate 5.3.27a).
Section 1: *Non-metallic inclusions*: Large, one-phase inclusions run parallel to the long axis of the specimen.
Macrostructure: Nital indicated a high carbon edge welded to a low-C back (Plate 5.3.27b), while Oberhoffer showed no other significant segregations.
Microstructure: There is a distinct seam between what was an area of virtually carbon-free metal, and a high-C cutting edge, with appreciable diffusion of carbon across the seam (Plate 5.3.27c). The low carbon area comprises ferrite (GS 3–4), grading into ferrite with fine pearlite at the weld seam. The structure of the high carbon area at the weld seam shows depletion of carbon due to diffusion, with fine ferrite (GS 7–8) and fine interstitial pearlite grading into martensite and ferrite, then to tempered martensite for the remainder of the length of the section.
Hardness: mH_v ferrite 185–247, martensite 639–831.
Section 2: *Non-metallic inclusions*: Small, single-phase inclusions run in lines parallel to the long axis of the section.
Macrostructure: Nital showed distinct banding, accentuated by Oberhoffer (Plate 5.3.27d).
Microstructure: Apart from a small area showing a Widmanstätten structure, the section shows ferrite (GS 5–7) with interstitial pearlite. At least 8 laminae can be observed (Plate 5.3.27e).
Hardness: mH_v ferrite 252–257: H_v 189–229.
Interpretation: The shears were forged from a fairly homogeneous low-C stock containing appreciable amounts of entrapped slag: the stock was subjected to fold-welding. The blades were made by welding on two high-C steel pieces and were then quenched and tempered. This would have been a very effective cutting tool, although the spring might not have given such good service.

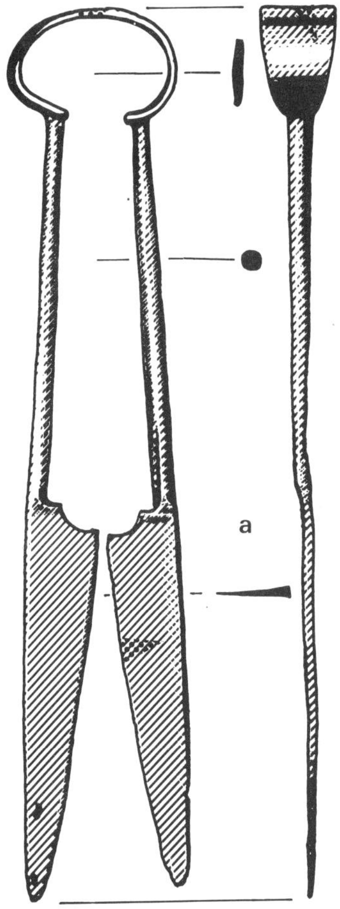

a

d

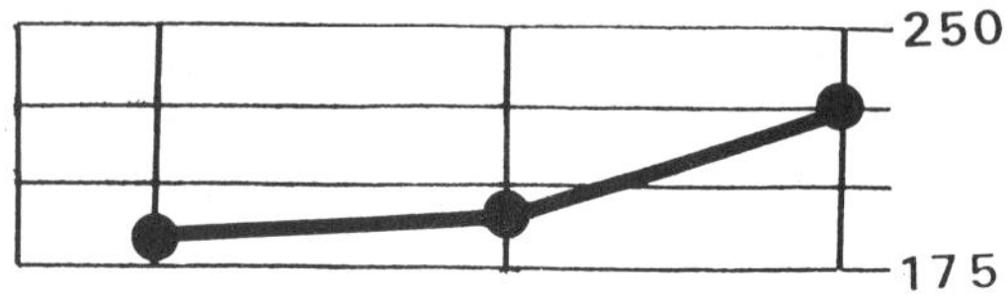

f

b

Plate 5.3.27 Examination 54: (a) NMI W49 shears from Strokestown, Co. Roscommon x½, (b) macrostructure of section 1, (c) area of weld seam between low-C back and high-C edge x50, (d) macrostructure of section 2, (e) banding of ferrite and interstitial pearlite with slag inclusions x100, (f) H_v profile of section 2. Etched 4% Nital.

c

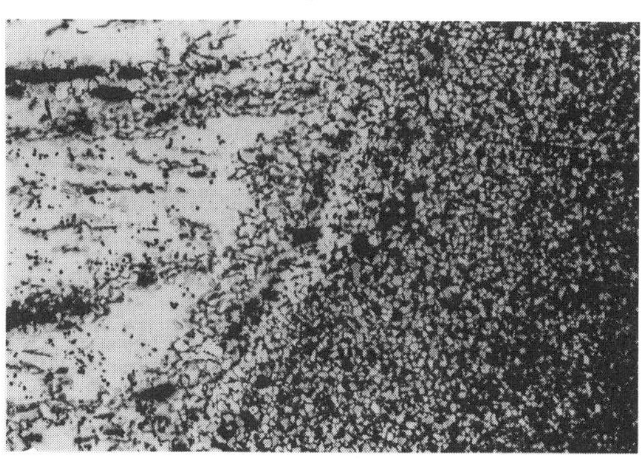

e

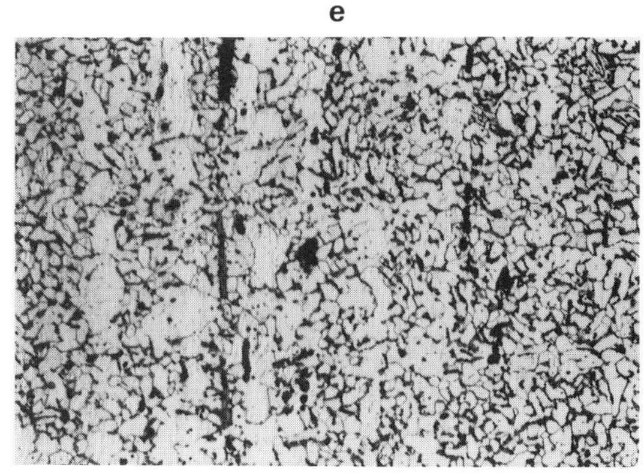

(55) Plough share, White Fort, Drumaroad, Co. Down (Plate 5.3.28)

A section was cut from the cutting edge (Plate 5.3.28a), above a longitudinal split which runs from the junction of the shaft and blade for just under half the length of the blade.

Non-metallic inclusions: On one side of the section inclusions are numerous, large and three-phase. Towards the other side is a prominent line of small, light-grey one-phase inclusions, while at the far side, inclusions are dark grey, one-amd two-phase.

Macrostructure: Etching with Nital shows an uneven distribution of carbon (Plate 5.3.28b), but Oberhoffer indicates no major segregations.

Microstructure: The most prominent feature of the section is a weld seam which runs parallel with the long axis (Plate 5.3.28c). To the left of the seam the structure is purely ferritic (GS 2–4) with large slag inclusions. On the other side of the seam (across which there has been significant diffusion of carbon – cf. Plate 5.3.28b), the structure shows Widmanstätten ferrite often enclosing large colonies of pearlite showing advanced spheroidisation of the carbide lamellae (Plate 5.3.28d). The carbon content falls off towards the cutting edge.

Hardness: mH_v ferrite 170, pearlite 227: H_v 112–162 (ferrite area), 159–172 (pearlite area).

Interpretation: The body of the share was forged from iron devoid of carbon and retaining much slag. To the blade was welded a plate of heterogeneously-carburised iron. The spheroidisation of cementite and the coarse Widmanstätten ferrite indicates prolonged forging. The weld between the two

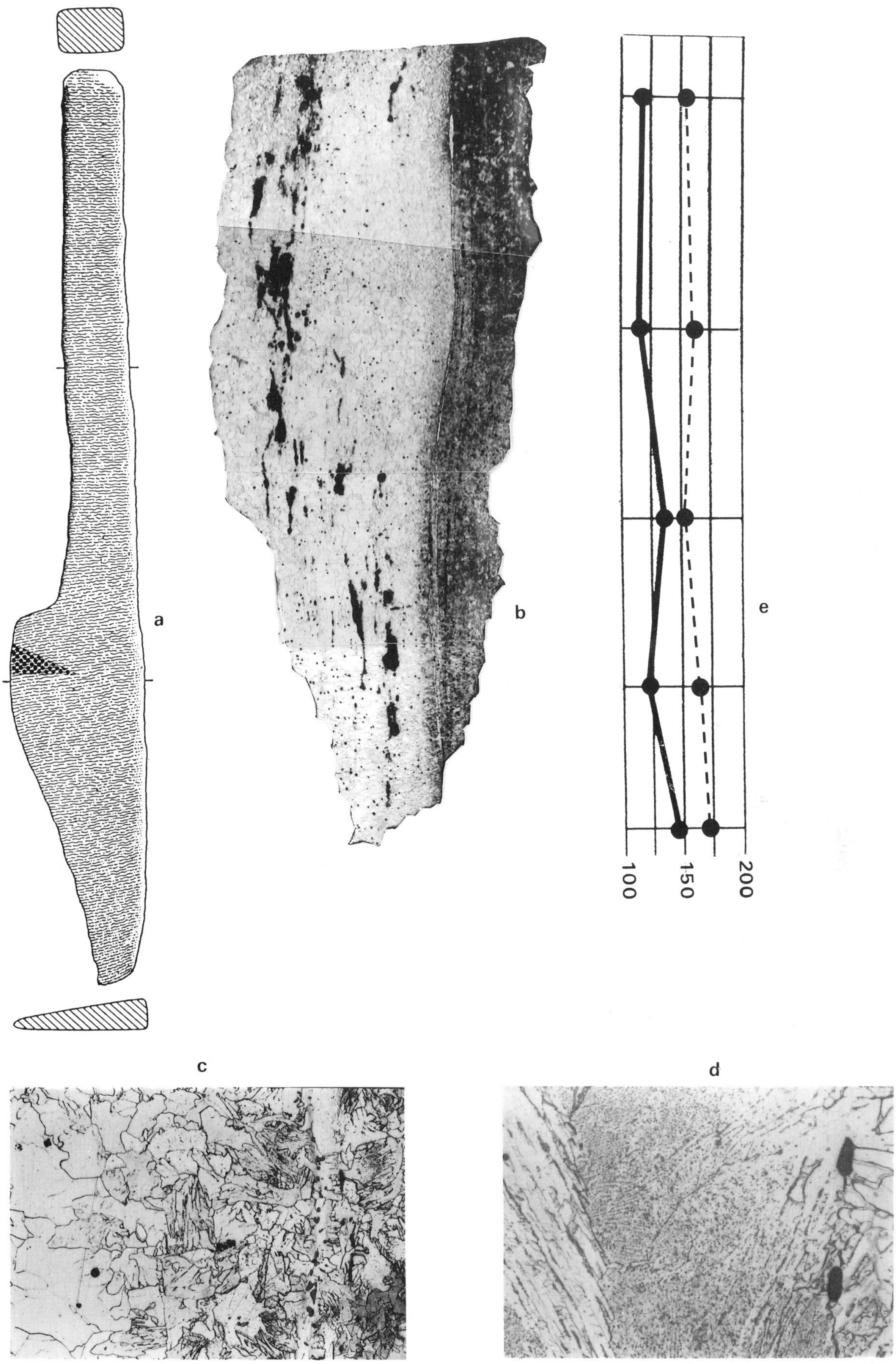

Plate 5.3.28 Examination 55: (a) DoENI plough share from White Fort, Drumaroad, Co. Down x⅓, (b) macrostructure, (c) weld sem between low- and high-C areas x100, (d) pearlite showing spheroidisation of carbides x500, (e) H_v profile - solid line ferritic area, dotted line pearlitic area. (b)-(c) etched 4% Nital, (d) etched Picral.

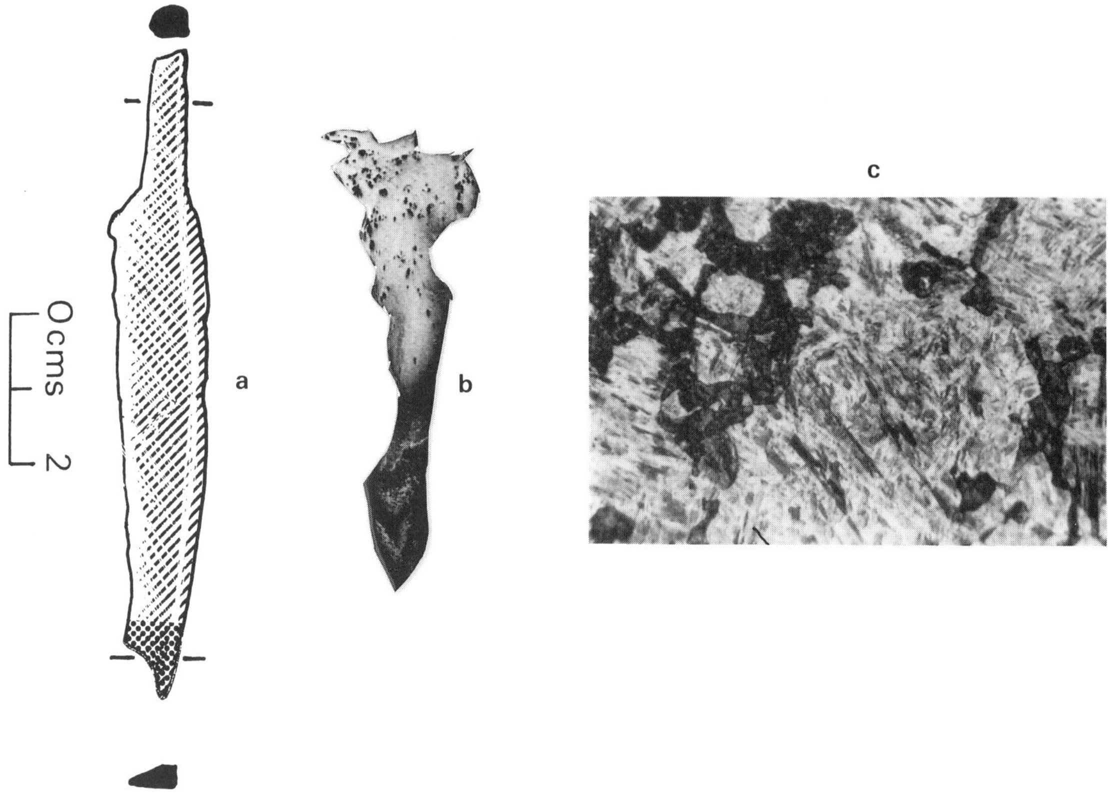

Plate 5.3.29 Examination 56: (a) DoENI 76/259a tanged single-edged knife from Gransha, Co, Down, (b) macrostructure, (c) martensite and troostite near the cutting edge x500. (b) etched 4% Nital, (c) etched Bisulphite.

components was imperfect for a significant part of its length, and this has been exaggerated by corrosion. It is possible that the decrease of carbon towards the cutting edge indicates decarburisation resulting from prolonged forging. This could not have been a particularly effective implement.

(56) DoENI 76/259a single-edged, tanged knife, Gransha, Co. Down (Plate 5.3.29)

A section[5] was cut across the blade near the tip (Plate 5.3.29a).

Non-metallic inclusions: Slag inclusions are not

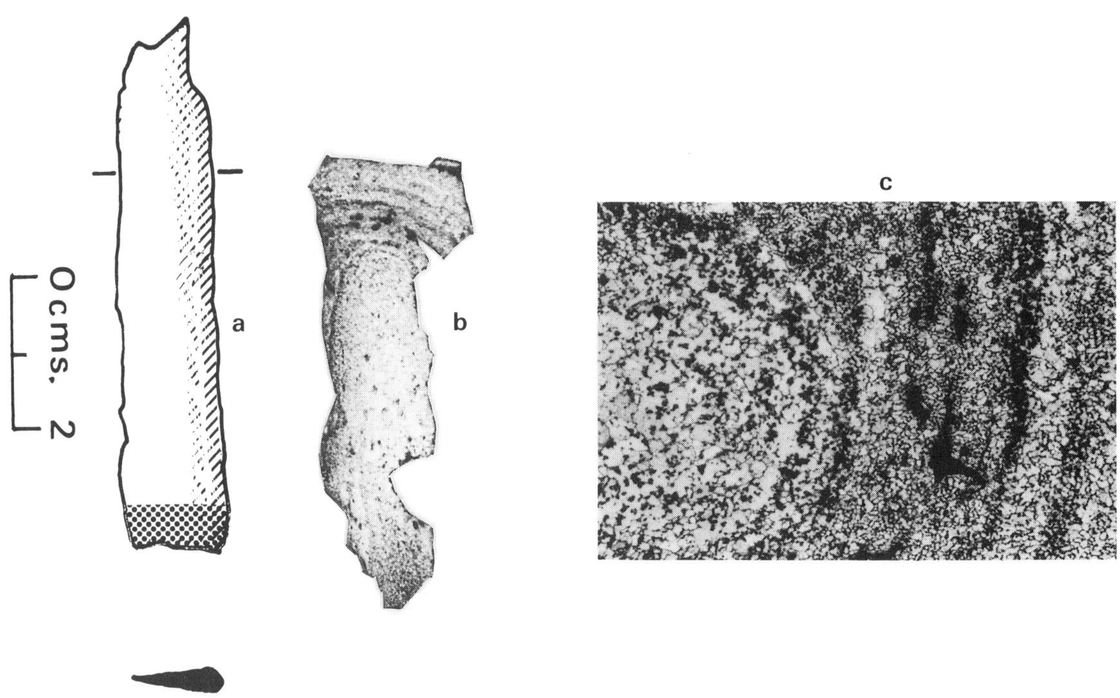

Plate 5.3.30 Examination 57: (a) DoENI 76/180 tanged single-edged knife from Deerfin, Co. Antrim, (b) macrostructure, (c) banding of ferrite at the back of the blade x50. Etched 4% Nital.

common, but where they occur they are large two- and three-phase towards the back of the section, changing abruptly to one-phase towards the cutting edge. The inclusions towards the back have a variety of orientations.
Macrostructure: Nital indicated a high carbon edge welded to a low-C back (Plate 5.3.29b). Oberhoffer revealed no significant segregations.
Microstructure: There is a weld seam between an area of heterogeneous carbon distribution and a high carbon cutting edge (cf. Plate 5.3.29b). The back of the blade shows banding of larger (GS 3–4) and smaller (GS 6–7) ferrite grains with interstitial fine pearlite, and areas of higher carbon content where the ferrite has a Widmanstätten form (Plate 5.3.29c). There has been diffusion across the weld seam with the high carbon cutting-edge where the structure is martensite, troostite and the remanents of a Widmanstätten ferrite net (Plate 5.3.29d). There is one major quench crack near the edge along which corrosion has penetrated.
Hardness: mH_v ferrite 166, pearlite 262, troostite 631, martensite 983.
Interpretation: The back of the knife was forged from heterogeneous stock to which a high carbon cutting-edge was welded. The cutting edge region was quenched after heating for a length of time insufficient to allow the α-γ transformation to go to completion. This would have been quite an effective cutting tool, despite the quench cracking.

(57) DoENI 76/180 single-edged, tanged knife, Deerfin, Co. Antrim (Plate 5.3.30)

A section was cut across the badly corrosion-damaged blade, near to the tip (Plate 5.3.30a).
Non-metallic inclusions: Towards the top of the section these are large and one-phase, aligned roughly at right-angles to the long axis of the section. Further down, numbers and size of inclusions decrease.
Macrostructure: Nital shows a banded structure curving up from the former cutting edge and running round roughly parallel to its back (Plate 5.3.30b). Oberhoffer indicates minor segregations.
Microstructure: The centre of the section is ferrite (GS 3–4). Surrounding this area is a series of bands of ferrite (GS 6–7) and ferrite (GS 6–7) with interstitial coarse pearlite (Plate 5.3.30c).
Hardness: mH_v ferrite 173–199.
Interpretation: The piece was forged from stock containing minor segregations of P, and with a subsequently uneven, but low, distribution of C. It appears to have been fold-welded after forging, although in the surviving metal core, traces of weld-seams are really impossible to detect. There are no coherent remanent structures in the corrosion capsule, and so the structure near to the original surface is impossible to reconstruct. It would not have been a particularly effective implement.

(58) DoENI 76/182 single-edged, tanged knife, Deerfin, Co. Antrim (Plate 5.3.31)

A section was cut across the blade near the tip (Plate 5.3.31a).
Non-metallic inclusions: There are few inclusions, and those which occur are single-phase.
Macrostructure: The knife has been badly damaged by corrosion, but in the surviving metal core, Nital indicated a fairly even distribution of carbon, while Oberhoffer showed no significant segregations (Plate 5.3.31b).
Microstructure: The structure consists of tempered martensite with some traces of remanent Widmanstätten ferrite (Plate 5.3.31c), grading to martensite at the cutting edge. At least one quench crack is visible near the cutting edge.
Hardness: mH_v martensite 727–866.
Interpretation: The knife was made from high carbon metal and quenched, resulting in a martensitic structure and at least one crack which would have constituted a flaw. Quenching seems to have been brief enough to have allowed some autotempering of martensite in the thicker part of the section. Despite the crack, this would have been a very serviceable tool.

(59) DoENI 76/203 single-edged, tanged knife, Deerfin, Co. Antrim (Plate 5.3.32).

A section was cut across the blade near the tip (Plate 5.3.32a).
Non-metallic inclusions: Large one- and two-phase inclusions are numerous throughout the surviving metal.
Macrostructure: Etching with Nital indicated mainly carbon-free iron (Plate 5.3.32b), while Oberhoffer indicated no significant segregations.
Microstructure: The structure is almost entirely ferritic (GS 2–4: Plate 5.3.22c) with some grain boundary cementite. There are small patches of smaller-grained ferrite (GS 5–6) with interstitial fine pearlite
Hardness: mH_v ferrite 155: H_v 110–129.
Interpretation: The knife was forged from a piece of iron virtually free of carbon, and in which much entrapped slag remained. This would not have been a very effective implement.

(60) DoENI 77/226 single-edged, tanged knife, Ballywee Rath A, Co. Antrim (Plate 5.3.33).

A section was taken across the blade near to the tip (Plate 5.3.33a). The knife has been badly damaged by corrosion, and only a core of metal survives.
Non-metallic inclusions: Inclusions are fairly numerous, large and one-phase.
Macrostructure: Etching with Nital showed an uneven distribution of carbon (Plate 5.3.33b), but Oberhoffer indicated no other significant segregations.
Microstructure: At the centre of the section, the structure is ferrite (GS 5–6) with interstitial pearlite. Towards one side, ferritic grain size decreases (GS 7–<8) while the carbon content increases, with some Widmanstätten ferrite and pearlite (Plate 5.3.33c). This grades into tempered martensite at the interface between surviving metal and corrosion. Examination of the corrosion products on the other side of the section shows some remanent carbides.
Hardness: mH_v ferrite 158, pearlite 218, martensite 910.

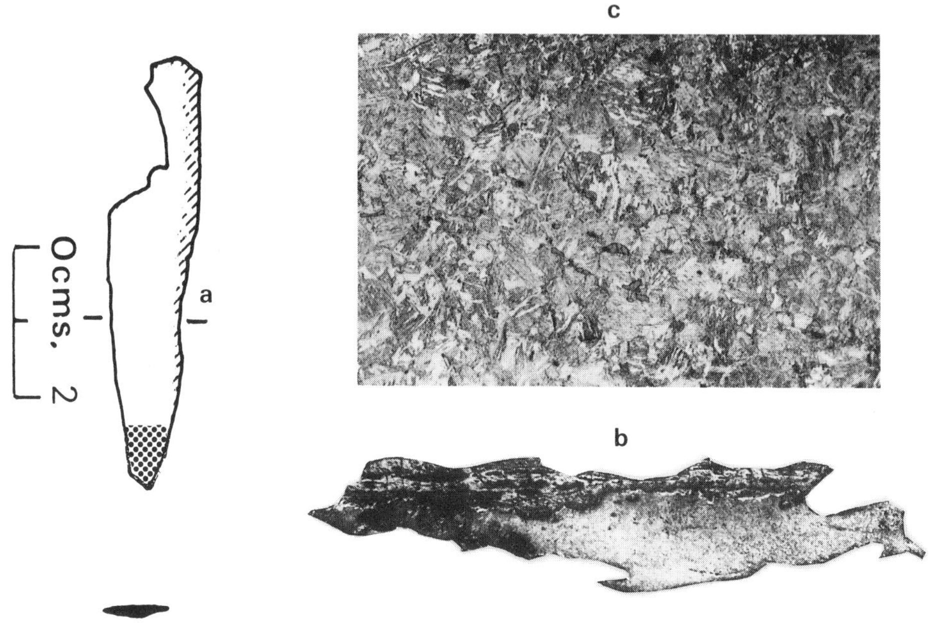

Plate 5.3.31 Examination 58: (a) DoENI 76/182 tanged single-edged knife from Deerfin, Co. Antrim, (b) macrostructure, (c) martensite x100. (b) etched 4% Nital, (c) etched Picral.

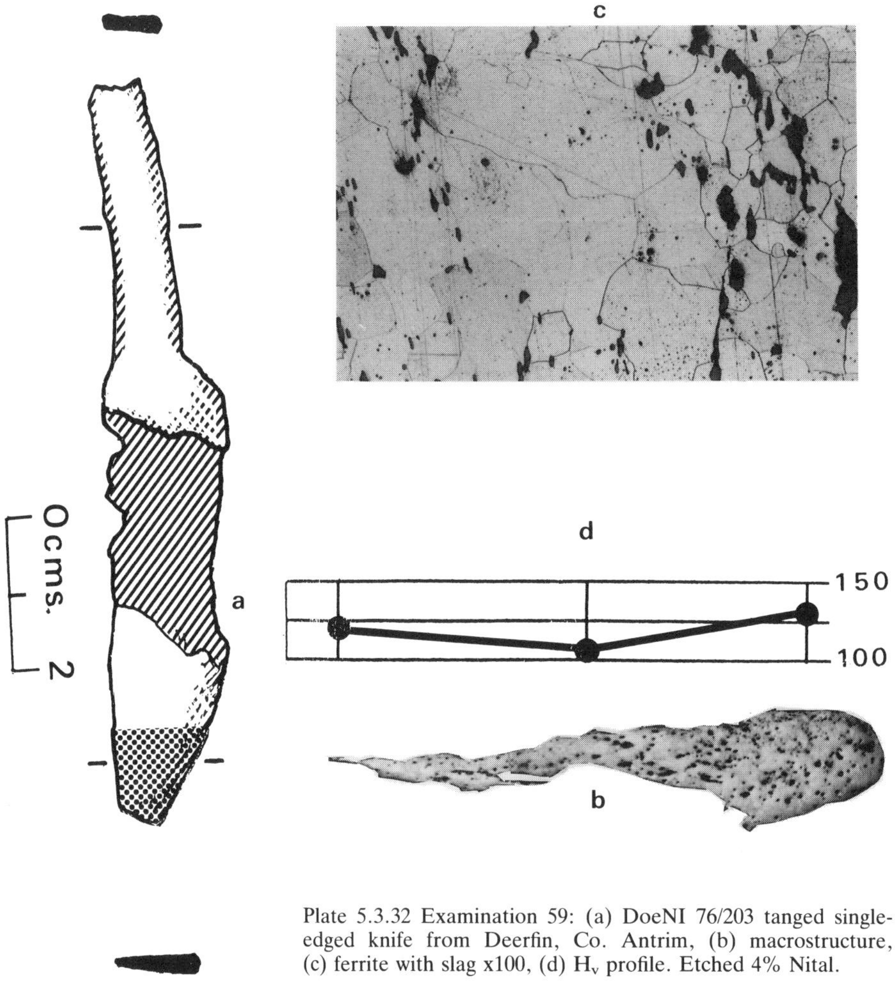

Plate 5.3.32 Examination 59: (a) DoeNI 76/203 tanged single-edged knife from Deerfin, Co. Antrim, (b) macrostructure, (c) ferrite with slag x100, (d) H_v profile. Etched 4% Nital.

Interpretation: The knife was forged from stock low in carbon. From the surviving structures it is likely that the carbon gradient observed results from secondary carburisation. It was then quenched. This would have been quite an effective tool.

(61) DoENI single-edged, tanged knife, Ballywee Rath A, House A, Co. Antrim (Plate 5.3.34).

A section was cut across the blade near the tip (Plate 5.3.34a).

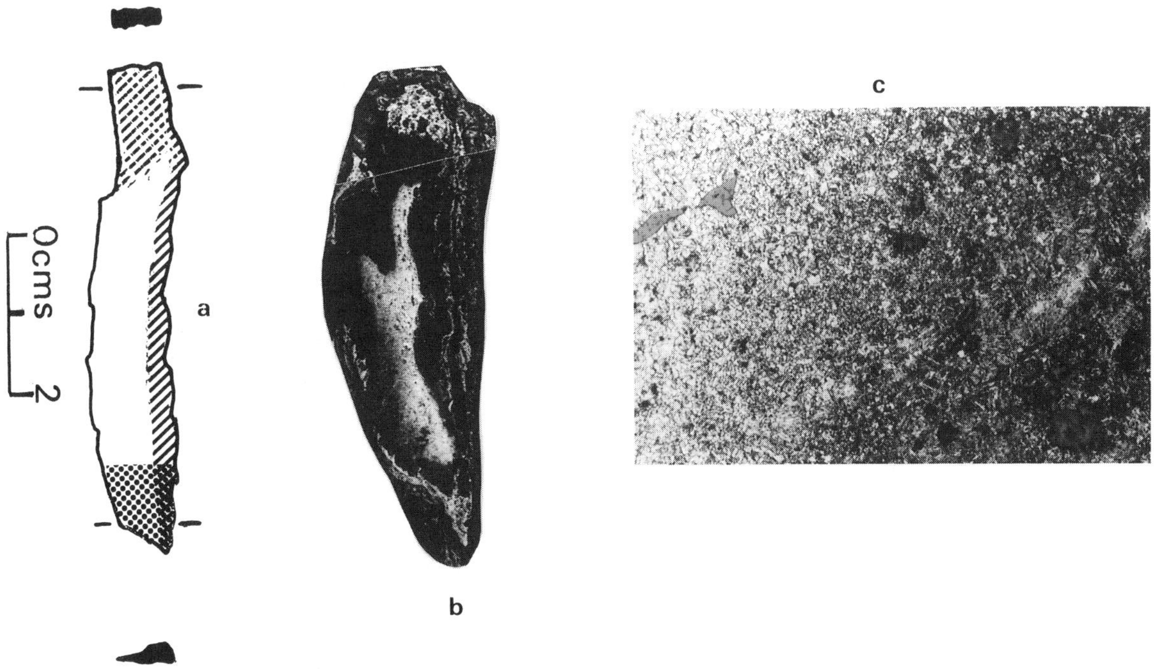

Plate 5.3.33 Examination 60: (a) DoENI 77/226 tanged single-edged knife from Ballywee Rath A, Co. Antrim, (b) macrostructure, (c) gradient from low- to high-C areas x100. Etched 4% Nital.

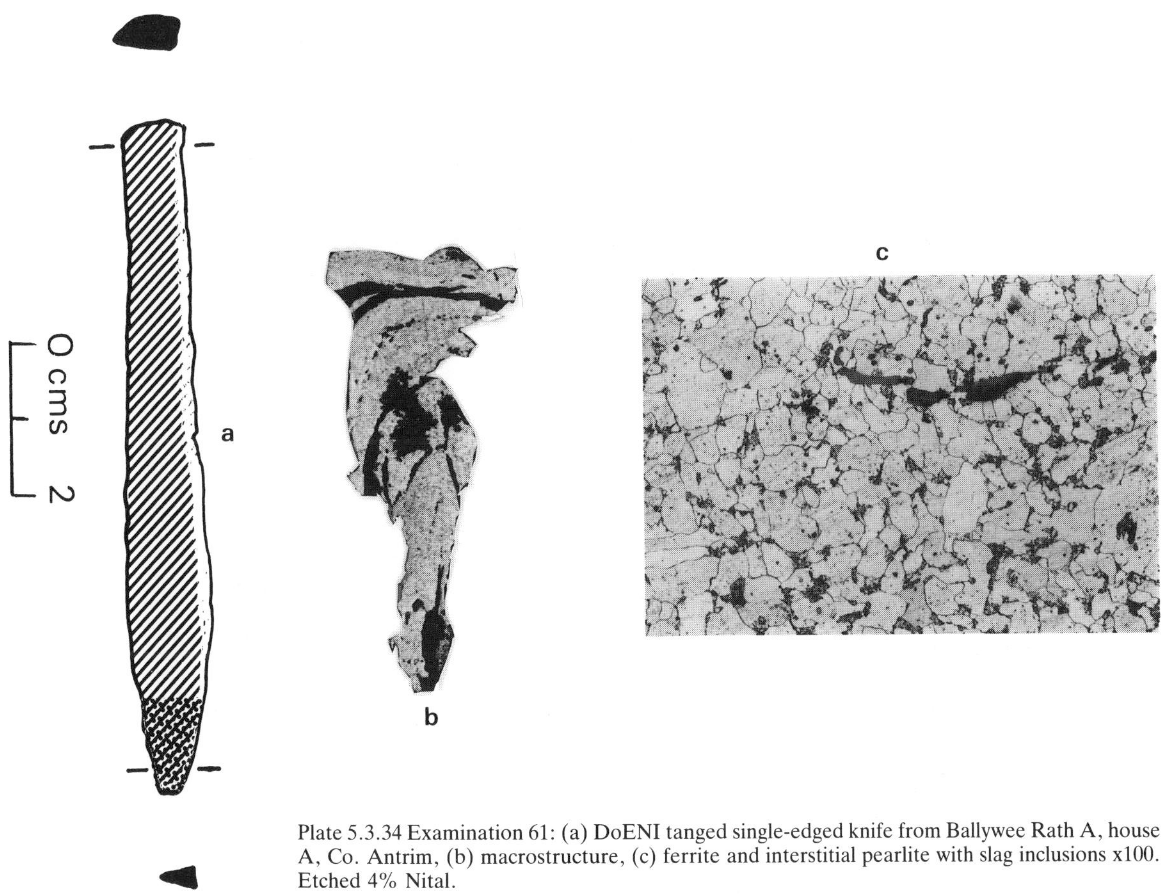

Plate 5.3.34 Examination 61: (a) DoENI tanged single-edged knife from Ballywee Rath A, house A, Co. Antrim, (b) macrostructure, (c) ferrite and interstitial pearlite with slag inclusions x100. Etched 4% Nital.

Non-metallic inclusions: In the body of the surviving metal core these are comparatively scarce, with the main inclusions showing up in badly executed weld seams.
Macrostructure: Nital indicated at least four pieces of low carbon metal welded together (Plate 5.3.34b), while Oberhoffer revealed no significant segregations. At least three weld seams are prominent in the section.
Microstructure: The structure throughout is fine ferrite (GS 5–6) with very coarse interstitial pearlite showing spheroidisation of carbide (Plate 5.3.34c). Four pieces of metal welded together can be observed, although corrosion may have destroyed evidence of others. The seams are all marked by massive inclusions, and accentuated by the ingress of corrosion. The joints between the components could never have been closed completely. Towards the cutting

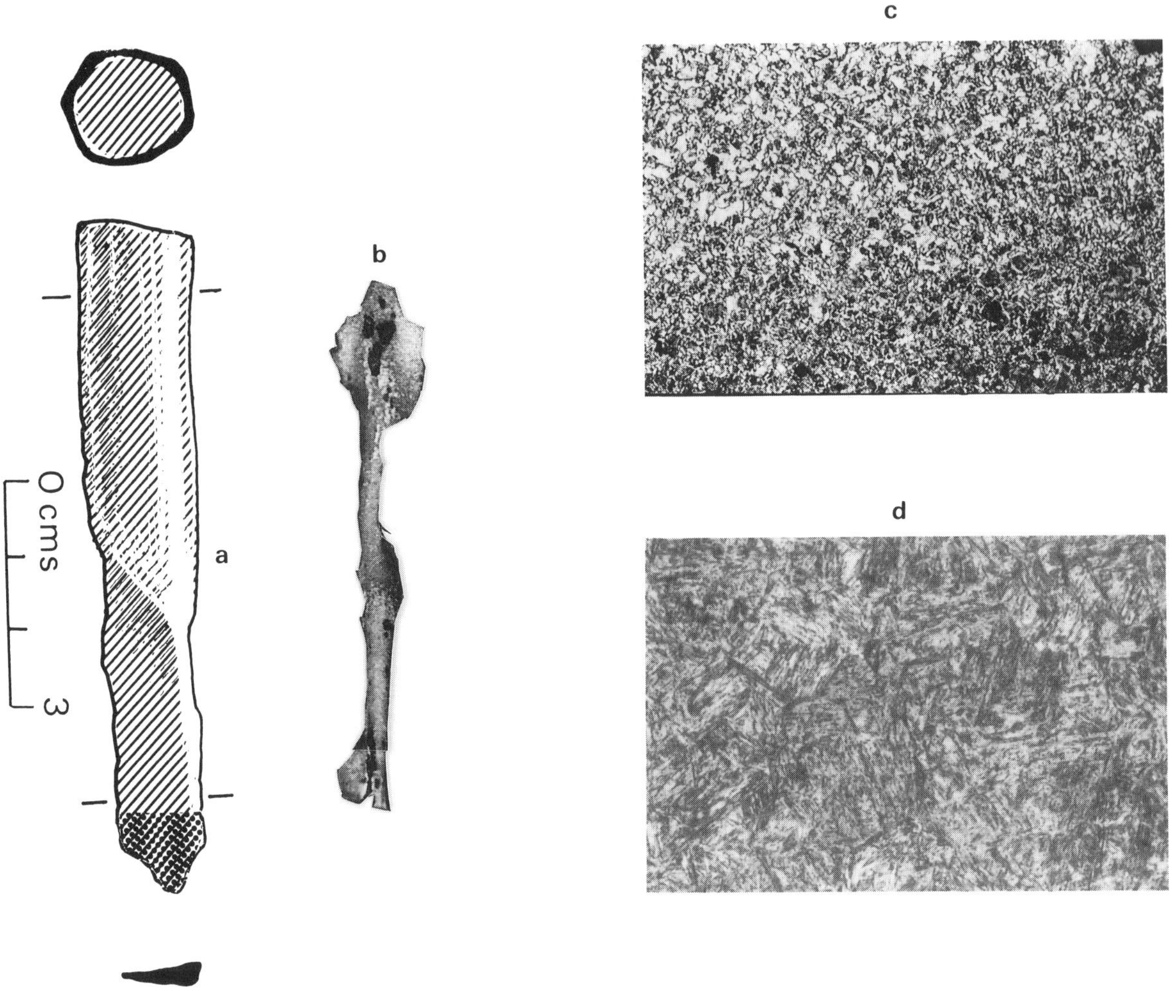

Plate 5.3.35 Examination 62: (a) DoENI socketed single-edged knife from Lough Faughan, Co. Down, (b) macrostructure, (c) ferrite and pearlite at back of blade x100, (d) martensite at cutting edge x500. (b) and (c) etched 4% Nital, (d) etched Bisulphite.

edge there is a patch of higher carbon metal with very heavily spheroidised carbides.
Hardness: mH_v ferrite 167.
Interpretation: The smith attempted to duplicate a multi-piece construction of the type employed often by Saxon and Scandinavian smiths. However, he used uniformly low-carbon material, and was not particularly competent at welding. While there might have been some minimal patterning on the blade surface, this would not have been a particularly attractive or effective implement.

(62) DoENI Single-edged, socketed knife, Lough Faughan, Co. Down (Plate 5.3.35).

A section was taken across the blade near the tip (Plate 5.3.36a).
Non-metallic inclusions: These are scarce, but where they occur they are large, single-phase.
Macrostructure: Nital revealed an uneven distribution of carbon (Plate 5.3.35b), but Oberhoffer indicated no other significant segregations. The knife has been severely affected by corrosion.
Microstructure: Towards the back of the blade, the structure is very fine ferrite (GS 8–$<$8) with interstitial fine pearlite (Plate 5.3.35c). Some ferrite tends to Widmanstätten form. Towards the centre of the section there is a distinct carbon gradient from outside to inside (cf. Plate 5.3.35b). Further down, the carbon content rises markedly throughout, and the structure is well developed martensite at the cutting edge (Plate 5.3.35d).
Hardness: mH_v ferrite 341, martensite 1260.
Interpretation: The blade was forged from a piece of stock originally with a heterogeneous carbon distribution. The cutting-edge was carburised, while the back of the blade probably was shielded with a clay coating. The piece was then heated to above A_3 which allowed all of the edge region to undergo the α-γ transition, which was also beginning towards the back. Quenching of the piece produced a martensitic structure at the cutting edge, and small areas of martensite in the high carbon region of the back.

Table 5.3.1

Summary of Metallographic results

Number	Artefact	Sample site	Wid	F/W	B/W	SEG	C/B	F/B	A	CAR	H/T	C/W	Carbon content%	Hv 1000gm	Structural components	Grain size	mHv 50gm
28 Plate 5.3.1	Sword Craigywarren crannog Co. Antrim NMI 1902:58	cutting edge	+	–	–	–	–	–	–	–	–	–	0.00 – 0.10	119 – 186	ferrite pearlite	1 – 3 6 – 7 I	152– 232
29 Plate 5.3.2	Sword Teeshan crannog Co. Antrim UM AL1:1988	cutting edge	–	–	–	–	–	–	–	–	–	–	0.10	106 – 131	ferrite pearlite	6 – 7 I	146– 199
30 Plate 5.3.3	Sword Ballinlig Co. Westmeath NMI WK5:W7: P855	cutting edge	+	–	–	–	–	–	–	+	–	–	0.00 – 0.40	120 – 159	ferrite pearlite	4 – 5 7 – 8 8 – I	135– 227 250
31 Plate 5.3.4	Sword Lagore Co. Westmeath NMI WK3:P855	cutting edge	–	–	–	–	–	–	–	–	+	–	0.00 – >0.35	175 – 515	ferrite pearlite troostite martensite	5 – 6 6 – I	288 312– 427 616 1305
32 Plate 5.3.5	Sword 'Northern Ireland' UM 184–1952	cutting edge	+	–	–	?	+	–	–	–	–	–	0.20 – 0.70	108 – 189	ferrite pearlite	5 – 6 6 – 8 4 – I	182– 304 276– 328
33 Plate 5.3.6	Sword Toome Bridge Co. Antrim NMI 1905–30	cutting edge	–	–	–	–	–	–	–	–	–	–	0.1	110 – 158	ferrite pearlite	4 – 5 7 – 8 I	143– 283
34 Plate 5.3.7	Sword Clanrolla Co. Armagh ARM 1–1947	cutting edge	–	–	–	–	–	+	–	–	–	–	0.00 – 0.05	140 – 182	ferrite pearlite cementite	1 – 3 6 – 7 I GB	222– 259
35 Plate 5.3.8	Sword River Corrib at Galway, Co. Galway NMI WK24:W6: F502	cutting edge	+	–	–	–	–	–	–	+	–	–	0.00 – 0.70	110 – 141	ferrite pearlite	5 – 6 7 – 8 6 – I	145– 182 252– 300
36 Plate 5.3.9	Sword Ballinderry crannog no. 2 Co. Offaly NMI E6:45A	cutting edge	–	–	–	–	–	–	–	+	+	–	0.10 – >0.20	148 – 368	ferrite pearlite troostite martensite	1 – 3 6 – 7 I	225– 264 585 722
37 Plate 5.3.10	Sword 'Ireland' no locality NMI WK8:W12	cutting edge	+	–	–	?	–	–	–	–	–	–	0.00 – 0.40	110 – 153	ferrite pearlite	4 – 6 7 –<8 6 – I	136– 208 262– 385
38 Plate 5.3.11	Sword Lagore Co. Westmeath NMI E14:1059	cutting edge	–	–	–	–	–	–	–	–	–	–	0.00 – 0.10	133 – 172	ferrite pearlite	4 – 5 I	129– 190
39 Plate 5.3.12	Sword Killyliss Co. Tyrone DoENI	cutting edge	–	–	?	–	–	–	–	–	–	–	0.15	175 – 214	ferrite pearlite	7 – <8 I	266– 271 360

I = interstitial GB = grain boundary E = eutectoid S = spheroidised carbide in a ferrite matrix N = network
Wid = Widmanstätten structure F/W = fold-welding B/W = bloomery welds F/B = ferrite banding SEG = P/As segregation C/B = carbon banding
A = acicular precipitates H/T = heat treatments C/W = cold working CAR = secondary carburisation
+ = present ? = uncertain occurrence – = absent or not detected
ARM = Armagh County Museum DoENI = DoE Historic Monuments and Buildings Branch (NI) NMI = National Museum of Ireland
UM = Ulster Museum

Table 5.3.1 ctd.

Summary of Metallographic results

Number	Artefact	Sample site	Wid	F/W	B/W	SEG	C/B	F/B	A	CAR	H/T	C/W	Carbon content%	Hv 1000gm	Structural components	Grain size	mHv 50gm
40 Plate 5.3.13	Sword tip Lough Gara Co. Sligo NMI E21:230	cutting edge	–	–	–	+	–	–	–	+	+	–	0.10 – >0.80	186 – 551	ferrite cementite pearlite troostite martensite	4 – 6 <8 N 8 – I	190– 260 338 476 1374
41 Plate 5.3.14	End of sword Randalstown Co. Westmeath NMI E148:1472	cutting edge	–	+	–	–	–	–	–	+	+	–	0.01 – >0.20	186 – 276	ferrite pearlite troostite martensite	7 –<8 8 – I	266 352 463 903
42 Plate 5.3.15	Spearhead Clanrolla Co. Armagh ARM 2–1947	cutting edge	–	?	–	–	–	–	+	–	–	–	0.00 – 0.01	93 – 123	ferrite pearlite cementite	2 – 5 5 – 6 I GB	138
43 Plate 5.3.16	Spearhead 'nr Hillsboro Co. Down' UM 50–1924	cutting edge	–	–	–	–	–	–	+	+	+	–	0.00 – >0.30	99 – 390	ferrite pearlite cementite martensite	>1 – 2 5 – 7 8 – I S	124– 191 413 528 645
44 Plate 5.3.17	Spearhead Lagore Co. Westmeath NMI WK5:W50: C:29	cutting edge	–	–	–	–	–	–	–	–	+	–	0.00 – >0.30	125 – 257	ferrite pearlite cementite martensite	1 – 3 4 – 7 8 – I S	120– 160 225 453 749
45 Plate 5.3.18	Spearhead Lankill or Drumadarragh crannogs Co. Tyrone UM 7–24	cutting edge	–	+	–	–	–	–	–	–	–	–	0.10	143 – 175	ferrite pearlite	6 – 8 8 – I	180 287– 368
46 Plate 5.3.19	Spearhead River Shannon at Drumshanbo Co. Leitrim NMI 1939:270	cutting edge	–	–	–	–	–	–	–	–	–	–	0.10	140 – 180	ferrite pearlite	8 –<8 I	291– 306
47 Plate 5.3.20	Spearhead Clogher Co. Tyrone UM CL:2422	cutting edge	–	+	–	–	–	+	–	–	–	–	0.01	110 – 148	ferrite pearlite	3 – 5 7 –<8 I	195– 268
48 Plate 5.3.21	Spearhead Ballinderry Co. Offaly NMI 1932:7211	cutting edge	–	–	–	–	–	–	–	–	–	–	0.01 – 0.70	148 – 245	ferrite pearlite cementite	7 –<8 4 – I S	239– 257 408– 498
49 Plate 5.3.22	Spearhead Lagore Co. Westmeath NMI WK5:P922	cutting edge	–	+	–	–	–	–	+	–	–	–	0.00 – 0.05	119 – 148	ferrite pearlite	2 – 3 7 – 8 I	241– 262
50 Plate 5.3.22	Axe Deerpark Farm Co. Antrim DoENI DPF:2859	cutting edge	+	+	?	–	+	–	–	–	+	–	0.00 – >0.50	189 – 390	ferrite pearlite troostite martensite	5 – 6 7 –<8 6 – I	257 380 534 717
51 Plate 5.3.23	Axe Ballynahinch Co. Down UM 108–1907	cutting edge	–	–	–	–	–	–	–	–	?	–	0.00 – >0.60	148 – 210	ferrite pearlite cementite	2 – 3 8 –<8 I S	188 281– 387 456

I = interstitial GB = grain boundary E = eutectoid S = spheroidised carbide in a ferrite matrix N = network
Wid = Widmanstätten structure F/W = fold-welding B/W = bloomery welds F/B = ferrite banding SEG = P/As segregation C/B = carbon banding
A = acicular precipitates H/T = heat treatments C/W = cold working CAR = secondary carburisation
+ = present ? = uncertain occurrence – = absent or not detected
ARM = Armagh County Museum DoENI = DoE Historic Monuments and Buildings Branch (NI) NMI = National Museum of Ireland
UM = Ulster Museum

Table 5.3.1 ctd.

Summary of Metallographic results

Number	Artefact	Sample site	Wid	F/W	B/W	SEG	C/B	F/B	A	CAR	H/T	C/W	Carbon content%	Hv 1000gm	Structural components	Grain size	mHv 50gm
			structural features														
52 Plate 5.3.25	Double-edged axe Clogher Co. Tyrone UM CL:1847	cutting edge	+	–	–	–	–	–	–	–	–	–	0.01 – 0.45	138 – 172	ferrite pearlite	4 – 5 8 –<8 4 – I	130– 220 280– 293
53 Plate 5.3.26	Shears blade Ballinderry crannog no. 2 Co. Offaly NMI E6:468	cutting edge	–	–	–	–	–	–	–	–	–	–	0.20 – 0.40		ferrite pearlite	7 –<8 7 – I	131– 203 306– 345
54 Plate 5.3.27	Shears Strokestown crannogs Co. Westmeath NMI W49	*Section 1* cutting edge	–	–	–	–	–	–	–	–	+	–	0.00 – >0.20		ferrite pearlite martensite	3 – 4 7 – 8 I	185– 247 831
		Section 2 spring	+	+	–	–	–	–	–	–	–	–	0.00 – 0.15	189 – 229	ferrite pearlite	5 – 7 I	252– 257
55 Plate 5.3.28	Plough share White Fort Drumaroad, Co. Down DoENI	cutting edge	+	–	–	–	–	–	–	–	–	–	0.00 – 0.50	112 – 172	ferrite pearlite	2 – 4 4 – I	170 227
56 Plate 5.3.29	Tanged knife Gransha Co. Down DoENI 76/259a	X-section	+	–	–	–	–	+	–	–	+	–	0.00 – >0.35		ferrite pearlite troostite martensite	3 – 4 6 – 7 5 – I	166 262 631 983
57 Plate 5.3.30	Tanged knife Deerfin Co. Antrim DoENI 76/180	X-section	–	?	–	?	+	–	–	–	–	–	0.00 – 0.05		ferrite pearlite	3 – 4 6 – 7 I	173– 199
58 Plate 5.3.31	Tanged knife Deerfin Co. Antrim DoENI 76/182	X-section	+	–	–	–	–	–	–	–	+	–	>0.20		ferrite martensite	N	 727– 866
59 Plate 5.3.32	Tanged knife Deerfin Co. Antrim DoENI 76/203	X-section	–	–	–	–	–	–	–	–	–	–	0.00 – 0.05	110 – 129	ferrite pearlite cementite	2 – 4 5 – 6 I GB	155
60 Plate 5.3.33	Tanged knife Ballywee Co. Antrim DoENI 77/226	X-section	+	–	–	–	–	–	–	+	+	–	0.01 – >0.20		ferrite pearlite martensite	5 – 6 7 –>8 8 – I	158 218 910
61 Plate 5.3.34	Tanged knife Ballywee Co. Antrim DoENI	X-section	–	–	–	–	–	–	–	–	–	–	0.00 – 0.05		ferrite pearlite	5 – 6 I	167
62 Plate 5.3.35	Socketed knife Lough Faughan Co. Down DoENI	X-section of blade	+	–	–	–	–	–	–	–	–	–	0.05 – >0.50		ferrite pearlite martensite	8 –<8 I	341 1260

I = interstitial GB = grain boundary E = eutectoid S = spheroidised carbide in a ferrite matrix N = network
Wid = Widmanstätten structure F/W = fold-welding B/W = bloomery welds F/B = ferrite banding SEG = P/As segregation C/B = carbon banding
A = acicular precipitates H/T = heat treatments C/W = cold working CAR = secondary carburisation
+ = present ? = uncertain occurrence – = absent or not detected
ARM = Armagh County Museum DoENI = DoE Historic Monuments and Buildings Branch (NI) NMI = National Museum of Ireland
UM = Ulster Museum

Table 5.3.2

Chemical analyses

Reg. no.	Mg	Al	Si	P	S	Ca	Ti	V	Cr	Mn	Co	Ni	Cu	Zn	As	Zr	W	Au	Pb
28 NMI 1902:58	tr	0.049	0.044	0.022	tr	0.058	nd	nd	0.062	0.014	nd	0.026	0.012	nd	tr	nd	nd	nd	nd
29 UM AL1:1988	0.006	0.028	0.112	tr	tr	0.031	tr	nd	0.063	0.010	tr	0.066	0.012	tr	tr	nd	nd	nd	nd
30 NMI WK5:W7:F355	0.022	0.033	0.073	0.025	0.006	0.055	nd	nd	0.059	0.039	0.077	0.019	0.007	nd	0.017	nd	nd	tr	tr
31 NMI WK3:P855	0.004	0.013	0.198	0.162	0.056	0.060	0.004	nd	tr	0.057	0.006	0.022	0.018	0.008	nd	0.005	nd	nd	0.032
32 UM 184–1952	tr	0.004	0.699	0.090	0.003	0.058	tr	tr	0.001	0.081	0.013	0.010	0.004	0.008	0.152	nd	0.003	tr	nd
33 NMI 1905–30	nd	0.002	0.013	0.032	0.001	nd	tr	nd	nd	0.004	0.053	0.050	nd	nd	0.013	tr	tr	tr	tr
34 ARM 1–1947	0.005	0.019	0.339	0.071	0.035	0.059	0.002	nd	tr	0.016	0.041	0.015	0.008	0.007	nd	0.006	nd	nd	0.033
35 NMIWK24:W6:F502	tr	0.018	0.110	nd	tr	0.111	nd	nd	0.065	0.016	tr	0.011	0.009	nd	tr	nd	nd	nd	nd
36 NMI E6:45A	0.003	0.002	0.019	0.081	0.003	0.146	nd	tr	tr	0.011	0.020	0.006	0.006	nd	0.072	nd	tr	tr	nd
37 NMI WK12:W8	0.003	0.015	0.092	0.048	0.018	0.017	0.001	nd	tr	0.015	nd	0.007	0.008	0.008	nd	0.006	nd	tr	0.034
38 E14:1059	tr	tr	0.055	0.045	tr	0.087	nd	nd	nd	0.019	0.044	0.057	tr	nd	tr	nd	nd	tr	tr
39 Killyliss	0.003	0.023	0.118	0.330	0.103	0.115	0.002	nd	tr	0.015	0.113	0.329	0.014	0.008	nd	0.004	nd	nd	0.031
40 NMI E21:230	0.004	0.002	0.028	0.128	0.026	0.028	nd	nd	tr	0.050	nd	0.012	0.019	0.018	nd	0.006	nd	nd	0.034
41 NMI E148:1472	0.009	0.025	0.257	nd	0.051	0.131	nd	nd	tr	0.022	tr	0.016	0.025	0.025	nd	0.006	nd	nd	0.042
42 ARM 2–1947	0.014	0.079	0.073	0.020	tr	0.062	0.003	nd	tr	0.065	0.031	0.064	0.004	nd	0.108	nd	nd	nd	nd
43 UM 50–1924	nd	0.028	0.037	0.080	tr	0.016	tr	0.001	tr	0.011	0.009	0.007	0.003	nd	0.034	nd	nd	tr	tr
44 WK5:W50:C29	0.002	0.011	0.071	0.054	0.066	0.079	0.001	tr	tr	0.004	tr	0.016	0.010	0.009	0.055	0.005	nd	nd	0.037
45 UM 7–24	0.007	0.039	0.031	0.067	0.005	0.137	0.001	0.015	0.007	0.059	0.028	0.084	0.075	nd	0.016	nd	nd	tr	0.050
46 NMI 1939:270	0.015	0.020	0.511	0.035	0.003	0.067	0.001	0.001	0.001	0.009	0.001	0.004	0.004	tr	tr	nd	tr	tr	0.020
47 UM CL:2422	0.003	0.008	0.048	0.054	0.004	0.015	0.006	0.001	0.015	0.018	0.034	0.014	0.006	nd	0.102	nd	0.002	tr	nd
48 NMI 1932:7211	tr	0.007	0.019	0.027	tr	0.020	tr	0.002	0.001	0.045	0.014	0.009	0.005	tr	0.044	nd	tr	tr	0.024
49 NMI WK5:P922	tr	tr	0.021	0.099	0.002	0.057	nd	tr	tr	0.030	0.016	0.047	0.006	nd	0.159	nd	nd	tr	tr
50 DoENI DPF:2859	0.007	0.019	0.035	0.038	tr	0.032	0.001	tr	tr	0.083	0.049	0.018	0.008	nd	tr	nd	0.001	tr	tr
51 UM 108–1907	nd	0.004	0.021	0.029	tr	0.014	nd	tr	nd	nd	0.002	0.004	0.005	nd	0.018	nd	0.004	0.074	tr
52 UM CL:1847	nd	0.002	0.007	0.023	nd	0.003	nd	nd	0.001	0.003	0.019	0.001	0.003	nd	0.225	nd	tr	tr	tr
53 NMI E6:468	0.003	0.036	0.189	0.234	0.256	0.227	nd	nd	tr	0.047	tr	0.010	0.270	0.020	nd	0.005	nd	nd	0.031
54 NMI W49/1	0.002	0.018	0.140	nd	0.032	0.042	0.012	nd	tr	0.042	nd	tr	0.006	0.012	nd	0.009	nd	nd	0.033
/2	0.002	0.010	0.062	0.130	nd	0.038	nd	nd	tr	0.002	tr	0.048	0.023	0.007	nd	0.006	tr	nd	0.031
55 DoENI Drumaroad	tr	0.025	0.038	0.075	0.003	0.017	tr	tr	0.026	0.048	0.020	0.019	0.007	tr	0.079	nd	tr	tr	tr
56 DoeNI 76/259a	0.004	0.013	0.082	0.111	0.003	0.024	tr	tr	tr	0.015	0.041	0.032	0.008	tr	0.046	nd	tr	0.005	tr
57 DoeNI 76/180	0.005	0.039	0.096	0.656	0.003	0.067	0.002	tr	nd	0.024	0.013	0.007	0.008	tr	0.084	nd	nd	nd	nd
58 DoENI 76/182	0.006	0.019	0.135	0.736	tr	0.064	tr	nd	nd	0.034	0.033	0.018	0.008	tr	tr	nd	nd	tr	nd
59 DoENI 76/203	0.021	0.033	0.085	0.377	tr	0.049	0.002	0.002	0.001	0.014	0.013	0.011	0.006	0.002	tr	nd	0.002	tr	tr
60 DoENI 77/226	tr	0.044	0.045	0.210	tr	0.055	0.003	0.003	0.002	0.022	tr	0.027	tr	tr	0.088	nd	tr	0.005	0.049
61 DoENI Ballywee	0.025	0.007	0.041	0.081	tr	0.011	nd	nd	tr	0.011	0.038	0.014	0.004	nd	0.020	nd	nd	nd	nd
62 DoENI Lough Faughan	nd	tr	0.244	0.439	tr	0.004	nd	nd	nd	nd	nd	0.006	0.008	nd	0.022	nd	nd	nd	nd

tr = trace (P, Pb <0.02%, As <.01%, Au <0.005, others <0.001) nd = not detected
ARM = Armagh County Museum DoENI = DoENI Historic Monuments and Buildings Branch (NI) NMI = National Museum of Ireland
UM = Ulster Museum

* * *

5.4 – Summary and Conclusions

We can observe a greater variety in the assemblage of ironwork in LIA A over that ascribed to EIA B. At first sight, the range of techniques employed is predominantly the same as in the preceding period. This is not particularly surprising after all, since the increase in quantities of iron and in the range of forms is due in no small part to the appearance of types which did not require more sophisticated techniques of production to make them serviceable. However, we do see for the first time the welding on of high carbon cutting edges to low carbon backs, and the regular use of heat treatments. Of the 35 edged weapons and tools described here, 2 have been carburised only, 5 have been both heat treated and carburised, and a further 7 have been heat treated only. Thus 40% of the artefacts examined show deliberate attempts to exploit the superior qualities of steel, as compared with the virtual absence of heat treatment, and the scarcity of secondary carburisation in EIA B. Set against this, there is still a very high proportion of artefacts which show the use of poor metal. But qualitatively at least we can see clear evidence for overall improvement in the handling of iron and steel.

While much of the metal being worked was still heterogeneous, there is a high proportion of material showing even primary carburisation. The evidence points to a tendency to produce metal of low carbon content, some of which was then subjected to secondary carburisation. This contrasts with the situation in EIA B where secondary carburisation was quite rare. This suggests a better understanding in LIA A of how to manipulate and to improve the qualities of bloomery iron. However, the population samples are still too small to allow more detailed evaluation of trends. Chemical analysis also indicates the possibility of some changes in ore exploitation, since from Table 5.3.2 it would seem that the use of high phosphorous and arsenic ores is somewhat less common in LIA A than in EIA B. Again, because the sample size is so small in relation to the size of the extant LIA A assemblage, the interpretation of the chemical analyses can only be treated as tentative.

In addition to the rise in technological standards, we can see that Irish smiths continued on from EIA B as highly competent *shapers* of iron. There is evidence that some of them were able to learn from foreign artefact designs. The Lagore collar shows a very pleasing adaptation of herringbone patterning presumably from Anglo-Saxon models. Similarly, the parallel-sided swords indicate outside influence, though not necessarily of the fighting style which they imply. The 'sandwich' construction of the Ballynahinch, Co. Down axe (examination 50: Fig. 5.3.2) would be unremarkable if found in the wider European context, but in Ireland indicates response to foreign stimulus. However, it is the borrowing of *shapes*, not of *techniques*, which is prominent. Nevertheless, given the lack of heat treatments in EIA B, and thus the period separating EIA A and LIA A, it is difficult to interpret the reappearance of heat treatments in LIA A as entirely the result of continuity from an earlier period. We must look also in the earlier part of the 1st millennium AD to a contribution from outside, probably from Britain, despite the fact that Tylecote and Gilmour (1986) demonstrate that much ironwork there, especially edged artefacts, was of fairly low quality in the earlier part of the 1st millennium AD.

Advances in technology are often seen in terms of the search for a better cutting edge. In this respect, some Irish smiths did make improvements by carburising and by welding on steel edges which were subsequently heat treated. Thus they were applying the basics of more complex constructions. It is interesting to note the emphasis on improving the cutting edge, and to contrast this with one example of the lack of similar upgrading of other working parts. While the shears from Strokestown (examination 53) had steel edges welded onto low iron backs, the spring of the shears showed no attempt at either carburisation or heat treatment. Tempered steel made a very good spring, and since we have evidence of knowledge of tempering (cf. Table 5.3.1), it is somewhat surprising to find tempering of a shears' blade, but not of the spring. What we do not appear to find is the combination of techniques that was employed by foreign smiths to produce pattern-welded blades. While in other regions such pieces belong to the upper end of the market, and as such may be atypical, they do demonstrate the abilities of contemporary smiths outside Ireland, and thus represent an application missing from the repertoire of smiths here. Along with the lack of shaft-furnace production, the prominent absence of native Irish pattern-welding requires an attempt at explanation.

In a closely-argued paper, Mallory (1982, 107f) proposed that the models for the super-swords of the early Irish tales, capable of one-blow decapitations and bifurcations, were most unlikely to have been Irish EIA B forms, and less likely to be native Irish LIA A than intrusive Scandinavian. Clearly the *gladius*-derived swords would have been most effective as stabbing rather than slashing weapons, and can be excluded from consideration on this count. The *spatha*-derived sword from Craigywarren crannog, which in theory would have been long enough and heavy enough to inflict serious damage as a slashing weapon, was made of virtually pure iron (examination 28). Of the three with expanded ends, the heaviest of the LIA A swords, the Clanrolla example is of soft iron (examination 34). However, the River Corrib specimen showed carburisation of the cutting edge (examination 35), while that from Ballinderry crannog 2 had been carburised and heat treated (examination 36), and both of these would have been effective weapons. Thus, two out of four of the most likely candidates for a native Irish model for the hero's sword of the story-tellers could possibly have approached their performance specifications. In addition, the Lough Gara (examination 40) and Randalstown (examination 41) specimens had both been carburised and heat treated. The spearhead from Clanrolla (examination 42) is most kindly described as less than adequate metallurgically. However, the carburised and heat treated specimen from Hillsborough (examination 43), and the heat treated specimen from Lagore, with its welded-on steel cutting edges (examination 44), as well as having been attractive pieces, would have been very effective also. Thus, while Mallory's thesis is persuasive, there are grounds for suggesting that by the 7th century AD at least, there were native-produced weapons which could have given rise to tales of prodigious feats of arms. The later intrusive Viking weaponry obviously would have reinforced the image, but the original inspiration may well have been Irish warriors wielding Irish weapons. Nevertheless, the actual blades themselves, while nicely proportioned, would not have been in themselves particularly eye-catching, and it is interesting to note the emphasis placed in the literature on sword furniture as opposed to blades.

The work of Tylecote and Gilmour (1986) demonstrates just how common was multi-piece patterned blade construction in Anglo-Saxon England. The various Annals record martial contact during LIA A, and yet we have yet to find an identifiably native-produced blade of the period with genuine pattern-welding. The knives from Deerfin (examination 57) and Ballywee (examination 61) might owe their construction to such foreign inspiration, but the execution and the materials used were both of low standard. Looking further afield, and bearing in mind the strong possibility of influence from the late- and post-Roman world, there is simply nothing amongst the Irish assemblage of either EIA B or LIA A to correspond to the sumptuous patterning shown, say, by the *scramasax* blades examined by Tylecote and Gilmour, the swords from Nydam and Illerup in Denmark (Thomsen 1989), and those produced by smiths of the Przeworsk culture in Poland (Biborski and Kaczanowski 1989: Kędzierski and Stępinski 1989).

It is inconceivable that the Irish had not encountered pattern-welded blades before the intrusion of the Vikings, and yet we have no evidence of smiths here responding to the challenge of copying them. There is one possible explanation and it lies in the degree of skill shown by Irish smiths in welding together iron and steel. To produce patterning using pieces of iron and steel requires quick, precise welding in which there is insufficient time for significant diffusion of carbon across the join. From the metallographic evidence we can see that Irish smiths, by and large, worked a weld for much longer than, say, their Germanic contemporaries, with the result that there was usually appreciable diffusion across weld seams. Given that in some of the best examples, pattern components may be only a few millimetres thick, it seems likely that prolonged forging above A_3 would lead to their effect being lost as carbon diffused from high- to low-C areas. Thus, while Irish smiths could weld together iron and steel to produce effective and functional tools and weapons, a lack of appreciation of the speed and precision needed may have prevented the copying of foreign pattern-welded artefacts. There may well be examples of 'failed' pattern-welding in iron and steel yet to be uncovered by metallographic examination.

It is evident that there were some foreign influences on LIA A ironworking, but that what were taken up were only the most important elements, heat treatments and edge-welding, and then seemingly by a comparatively small proportion of craftsmen only. In contrast, all over contemporary Europe and in the southern part of Britain at least, there were blacksmiths deploying 'state-of-the-art' technology, not just on a one-off basis, but often as parts of wider mass-production organisations. There is a clear stratum of excellence across much of Europe showing that master-craftsmen usually learned quickly of innovations and were not slow in applying them in their own work. At this point, we move beyond the application of technique to produce an effective artefact, into combinations of technique with sophisticated design of shape and pattern. While mundane and low quality ironwork continued to be produced in abundance, there were now those whose knowledge and mastery of technique allowed them to use iron artefacts as vehicles for their artistry. By the earlier 1st millennium AD, some European smiths had elevated that artistry in iron to the levels previous achieved by workers in non-ferrous metals.

It is noticeable also that in tandem with the rise in quality of smithing amongst the top tier of foreign workers, comes improvement in production capacity, represented by the variants of the shaft furnace. In addition to offering the potential for greater control over the composition of the bloom, it offered also a greater output than the bowl furnace. The shaft furnace is the dominant structure for iron smelting across Europe and Scandinavia from the later 1st millennium BC through to the earlier 2nd millennium AD. Since it was commonplace in Britain from at least early Roman times onwards, it most unlikely that the Irish would not have encountered it during their contacts across the Irish sea. But they did not adopt it.

This brings us back once again to the questions of communication and dissemination of knowledge raised in Chapters 2 to 4, and to the differences between the role of ironworking in Ireland and elsewhere. There is good evidence that craftsmen here working in other fields were interacting with their foreign counterparts, and one might cite as just one example the influence of Irish non-ferrous metalworking, with its enamelling and millifiori glass, on contemporary work in Britain, returning applications of technique learned from outside the country. But this cross-fertilisation was really in terms of *styles* rather than basic technology. Irish workers whose EIA B predecessors had become familiar with the techniques of producing different colours of glass would have had little difficulty in figuring out how foreign composite millifiori rods were made, without the need for tuition. In contrast, an Irish smith could have worked out the basic constructional scheme of a pattern-welded sword, but would not been able to reproduce it because he was not fully aware of the underlying technology.

One of the principal channels for the dissemination and interchange of knowledge in LIA was through monastic foundations. I have argued elsewhere (Scott 1978b, 225ff) that one route for the transmission of style embodied in the Lagore collar was through connections with the Irish monastic settlement at *Cnobheresburgh*, on lands granted to St. Fursey around AD 630 by the Christian Wuffinga king Sigeberht whose relatives were buried at Sutton Hoo[6]. Irish monks travelled extensively both in Britain and on the continent in LIA A, founding religious establishments in a number of countries. And since crafts were an integral part of monastic economies, they cannot have failed to have seen at first hand the range of techniques being employed in foreign establishments. Their links with their native island were never severed, and there was a flow of knowledge back home. Clerics were also mobile locally, crossing political boundaries usually with impunity, and thus the transmission of ideas should have presented no problems. The situation is less clear for lay-smiths, although master-smiths seem in law to have been afforded the same freedom of passage as other members of the learned classes (Kelly, F. 1988, 4).

F. Kelly suggests that there is a uniformity in the laws due to the free passage afforded to lawyers (1988, 2), and O'Rahilly wrote of common dialect in the Old and Middle Irish periods that '. . . there can be no doubt that such local differences as then existed were trivial in comparison with those of later times.' (O'Rahilly 1932, 248). Overall in LIA there is a significant degree of uniformity of occupation sites, of technology, of decorative styles in the various media, and of industry that bespeaks what, in archaeological terms, is a broad cultural unity. This really is not surprising for an island the size of Ireland, even if politically it was a patchwork of small groupings whose codified boundaries placed barriers to major movements of people, though not of ideas easily communicable visually and/or orally. Complex ideas – religion, law, literacy – flourished and spread, seemingly without regard for territorial demarcation. Why then was advanced iron technology unable to do the same? A complete answer to this question probably will always elude us because of the complexity of the evidence with which we are dealing, and explanation will have to involve a number of factors. Two main areas may be considered briefly here, the nature of transmission of ideas, and the reasons for adoption/rejection of technical innovation.

As far as we can see, the basic technologies of non-ferrous metalworking (smelting, alloying and casting) were the same in LIA as in at least EIA, if not the Later Bronze Age. Thus, what we see in the development of fine metalwork in LIA is not fundamental technological and innovation in most instances. Rather, what is prominent is the development of applications by skilled and sensitive craftsmen, drawing on long-established technology. A new design of artefact, a new combination of decorative elements, sometimes tends to give rise to ideas of a new technology, when in fact it is an innovative application of an old technology. Even if the movement of people was as restricted as might be suggested by early literature, the movement of artefacts is unlikely to have been so limited. The uniformity of artistic styles throughout the country in LIA testifies to the mobility of small pieces which could be copied by competent craftsmen using existing technology[7]. Novel motifs could be disseminated by small numbers of mobile individuals, in the same way, for example, that legal and religious ideas could be. We might term these 'portable concepts' to underscore that they could be transferred from one part of the island to another with ease. In the case of art styles, ideas were completely portable, not requiring a knowledgeable carrier. The receivers of 'portable concepts' would have required only visual inspection and/or oral transmission, readily comprehensible in terms of past experience.

Novel technological ideas, on the other hand, require a much greater degree of direct contact between transmitters and receivers. We might describe these as 'transferable concepts' to emphasise the interaction needed. We have already referred to the absence of the shaft-furnace and of pattern-welding from Ireland. We can also note the lack of native wheel-thrown pottery in LIA A, even though substantial numbers of wheel-thrown vessels were being imported from outside. Even though it is impossible that at least some Irish were aware of them, none of these novelties could have been taken up by Irish craftsmen without explanation and practice under tuition. It is one thing to describe the wheel-throwing of pottery to someone unfamiliar with it, and quite another for that person to use the wheel to make pots. A period of apprenticeship would have been required before the receiver could have put new knowledge to effective use. Thus, while there would have been people familiar with the handling of clay, with smelting of metals, and the forging and welding of iron and steel, wheel-throwing, using the shaft-furnace and producing intricate patterns in iron and steel required additional information whose application required practice and training. Implantation of these new techniques into Ireland thus would have involved fairly free movement of craftsmen both across the Irish sea and within the island. There were forces resisting such movements of people, hence of transferable ideas, and these would have required pressure to overcome them. The most obvious pressure that could have been applied was market-driven demand. It is not unreasonable to suppose that the absence of the techniques we have been discussing is because there was not sufficient demand to exert that pressure.

It is too easy to be overcritical of Irish production standards in the light of the sophistication of contemporary foreign smiths. We must always ask if an artefact would have been effective in the form in which it was produced, and not demand higher technological standards which might only have led to a slight improvement in performance as opposed to a major improvement in appearance under the microscope. In LIA A, we are seeing for Ireland a far greater range of human activities represented in the archaeological record than for any previous period. One may feel, intuitively at least, that in earlier periods the inhabitants of Ireland had an equally diverse material (and spiritual) life. The evidence, however, is qualitatively much more informative for LIA A than for earlier prehistory, allowing us to reconstruct with clarity the cycles of activity which dominated the activities of kings and commoners alike. The sum total of excavation evidence to date, coupled with that of the written record, shows beyond dispute that the overwhelming majority of the population of Ireland in LIA A was heavily involved not just with individual technologies, but with a wide range of domestic production ranging from crop growing to house construction.

Ironworking in LIA A is perceptible as but one activity among many contributing to patterns of daily life. Having taken over the role of copper-alloy as the primary utility metal, it ceased to be an expensive novelty, moving gradually from being the 'prima-donna' of metals to the role of workhorse. Copper-alloy on the other hand moved from being the main material for tools and weapons, becoming a mainly artistic medium. It is difficult to draw comparisons between the two processes, since our knowledge of copper-alloy production during the Later Bronze Age and EIA A and B is woefully inadequate. But as a purely subjective view, the excavation evidence indicates that by LIA A, copper-alloy production was restricted to more specialist sites, while iron-working took place on the great majority of domestic sites.

When ironworking became simply one of the activities that made up the normal domestic cycle, there would have been little need or opportunity in most cases for trying to develop an acceptable technology further than was necessary to produce effective artefacts. From the analysis of excavation evidence, it would appear that the number of sites supporting specialist craftsmen, potential innovators and receivers of innovations, was small, hence that the numbers of potential innovators was also small. Existing smelting technology, however unsatisfactory it might appear now in the light of comparative evidence from abroad, seems to have fulfilled the demands of a dispersed population. While pattern-welded blades would surely have delighted warrior and poet alike, existing technology and the skills of those who made hilts and scabbards sufficed. Ironworking was only one facet of local, small-scale economies which were not really geared towards technological innovation, if for no other reason than the fact that the markets which producers served were so small. Had early Irish society and economy followed the same trends as in other regions, it is inconceivable that we would not see here the same overall range of skills having been learned and applied as in other regions.

One interesting development, seemingly of LIA A date, is the appearance of ironworking of some form at Neolithic tombs, observed within the structures at Largantea, Co. Derry (Herring 1938), Aghnaskea, Co. Lough (Evans, E.E. 1936, 1940) and Creevykeel, Co. Sligo (Hencken 1939), and in close association at Taylorsgrange, Co. Dublin (Keeley 1987) and Knowth, Co. Louth (Eogan 1972). As we shall see in Chapter 7.4, there is good early literary evidence for mystique surrounding the work of the blacksmith and his metal. It is interesting to speculate that if burial places of earlier periods were afforded supernatural association then the siting of smithies(?) in or around them could be taken as archaeological evidence supporting that of the texts.

It will be evident from the content of this chapter that there are many problems of interpretation still to be tackled. Perhaps because of the size and complexity of the surviving evidence, it is more difficult to isolate out iron and its working from other activities. Really until we have detailed study of site assemblages, coupled with study of non-ferrous metalworking technology, the picture will remain only a sketch.

* * *

Notes to Chapter 5

1 A further parallel, worth noting at least in passing, is with the form of the Halstatt D/La Tène I dagger blades from southern Britain, where lunate depressions on either side of a thin central midrib are also to be found (Jope 1961, esp. 325ff). While ascribing this type to the Later Iron Age, one must be prepared to anticipate the possibility of an example turning up in a context dated to EIA B or even to EIA A at some future time.

2 I am most grateful to the excavator, Mr. E.P. Kelly, Irish Antiquities Division, National Museum of Ireland for the opportunity to examine this blade and present the results in advance of his own publication of the site.

3 I am most grateful to the excavator, Mr. R.B. Warner of the Ulster Museum, for the opportunity to examine this piece, and the axe described in examination 51, and to present the results in advance of his own publication of the site.

4 I am most grateful to the excavator, Dr. C.J. Lynn, Historic Monuments Branch, DoENI, for the opportunity to examine this piece, and to present the results in advance of his own final publication of the site.

5 I am most grateful to Dr. C.J. Lynn for making this knife, and the others described in examinations 57–62, available for study, and to Mark Hall for collaboration in their metallographic examination.

6 However, pattern-welding is a style of working whose primary application was to weaponry – swords and spears, and thus probably would have been of little interest to monks. For this reason, it seems unlikely that the combination of techniques represented by the pattern-welded blade would have been carried to Ireland through monastic contacts. Indeed, one could conjecture that the Lagore collar, taking as it does elements of pattern-welding, symbolises the dissection out of decorative elements, acceptable to early Christians, from a martial and unacceptable complex. Even if the Lagore collar was for hostages and not for dogs, the early church condoned hostage-taking, as well as outright slavery, while condemning by and large the taking of life.

7 This is not to suggest in any way that there were no technological innovations in non-ferrous metalworking in LIA A Ireland, despite the paucity of our information. It is rather that in the state of present knowledge, they appear unimportant in comparison with stylistic innovations.

* * *

CHAPTER 6

Ores, Mining, Ironworking Residues and Structures

6.1 – From ore to artefact

The main stages of the process of obtaining iron artefacts from iron ores may be summarised as:

prospecting	locating suitable ores
mining	retrieving ore
ore processing	concentrating, roasting
smelting	reduction of ore to metal
bloomsmithing	consolidation of iron/slag sponge
forging	shaping of artefacts from the consolidated bloom

Some of these stages have secondary activities associated with them, including:

operation	**secondary activities**
mining	tool making, spoil removal, timberworking, making containers for carrying ore, transport
smelting	clay gathering and preparation, timber cutting and transport, charcoal burning, furnace building, making tuyères
forging	tool making, charcoal burning, disposal of products

With the exception of prospecting, the different stages, and many of the secondary activities associated with them, might be expected to leave some identifiable traces in the archaeological record.

operation	**archaeological survivals**
mining	mine workings, tools, containers
ore processing	ore roasting pits, charcoal, ore fines, gangue, roasted ore
smelting	furnace remains, tuyères, charcoal, unreduced ore, 'slags', raw bloom
bloomsmithing	hearth remains, tuyères, charcoal, anvil, hammers, 'slags', spheroidal gromps, hammer-scale consolidated blooms
forging	hearth remains, tuyères, charcoal, tools, 'slags', hammer-scale, semi-products

Thus, from the smelting process we might hope to find on a site the remains of the furnace including highly fired and vitrified clay, tuyères (the nozzles of bellows), fragments of unreduced ore, the siliceous waste from the process, 'slag', and occasionally, the raw bloom, the sponge of iron particles and slag. As with most things in archaeology, however, things are seldom as simple as this, and the identification of the remains of the various stages and their interpretation is fraught with difficulties.

Even within a small territory, various activities may have been carried out in different places (for example, mining and ore processing separate from charcoal burning, both in different localities to smelting and smithing), with the result that the assemblage from one site represents only a part of the overall process. Some finds will define activity, so that ore fines (a fine powder of ore particles produced during roasting) indicate ore processing, the raw bloom indicates smelting, gromps (small pieces of slag containing iron, often highly carburised: Nosek 1975) indicate bloomsmithing, and hammer scale (iron oxide formed when iron is heated in air), semi-products and tools indicate forging. Some finds, on the other hand, are ambiguous. The tuyère, for example, could have acted as a nozzle for the bellows supplying air both to the smelting furnace and to the smithing hearth. 'Slags' were produced during the smelting process by the reaction of iron oxides with the gangue (waste) minerals present, accumulating at the bottom of the furnace in lumps or blocks. But the consolidation of the sponge of iron particles and slag that is the primary product of a smelt expelled slag in a semi-molten state, and some of this again accumulated in the bottom of the hearth used for heating. If the base of the smelting furnace and smithing hearth were of the same shapes and dimensions, then it will be difficult to distinguish visually between smelting and bloomsmithing residues. In short, unless we are dealing with a site on which one single stage of the process (e.g. smelting) was carried out, the assemblage of industrial debris is likely to be confusing.

The Irish evidence is problematic for a variety of reasons, and in a survey such as this, it is not possible to offer detailed descriptions, explanations or interpretations. The following sections of the chapter are intended as a guide only to the evidence for ironworking on Irish sites, and to indicate the areas requiring further research.

* * *

6.2 – Irish iron ores – a general survey

In 1589, when English capitalists were turning to Ireland as a potential new source of exploitable resources, Richard Payne, assessed Irish iron ores and woods in his *Description of Irelande* thus:

> *There is verie riche and great plentie of iron stone . . . and wood sufficient to mayntayne divers Iron workes (with good husbandrie) for ever.*

By 1631, however, the most successful of the English ironmasters, Richard Boyle, first Earl of Cork, complained:

> *I think that God hath ordained my ironworks to be an endless trouble to me.*

And in 1672, William Petty, the last of the major English entrepreneurs who had hoped to make a fortune from exploiting iron production in Ireland, wrote plaintively:

> *Oh the ironworks, the ironworks in Kerry, invented in hell (which they resemble) have nipped me cruelly, and the misery is I must go on.*

By the middle of the 18th century, the last of the blast furnaces had ceased operation, and an enterprise begun in such optimism some century and a half previously petered out. The causes undoubtedly included the political upheavals of the mid 17th century (in which many English-owned concerns were destroyed), but centred principally on the savage deforestation (Mitchell 1976, 196f) which raised fuel prices enormously. The traveller and natural historian Gerard Boate graphically described the situation in his *Naturall Historie of Ireland* (1652, 67–68):

> *Through the aforesaid causes* [use of timber for barrel staves and ironworking] *Ireland hath been made so bare of woods in many parts that the inhabitants do not only want wood for firing . . . but even timber for building . . .*

Also, even though the island is well endowed with iron ores, ultimately they could not compete in quality with foreign resources (Scott 1985). And because there has been no significant industrial production of iron in Ireland since the early 18th century, information on ore resources, types and compositions is severely limited.

Surveys of economic resources were made in the 19th and early 20th centuries by workers such as Kane (1844), Holdsworth (1858), Kinahan (1878), Cole and Hallissy (1921), and Cole (1922). Kane, Kinahan and Cole all compiled lists of ore deposits, noting those reputed to have been worked in earlier times. The distribution of ores is shown in Fig. 6.2.1, taken from these three surveys, with additions for Northern Ireland taken from Groves (1952) and the *Minerals Deposits Map* of the *Geological Survey of Northern Ireland*. Although a number of writers make reference to Irish iron ore deposits in passing (e.g. Dubordieu 1812; Portlock 1845; Kilroe 1907; Fowler and Robbie 1961; Wilson and Robbie 1966), the largest body of information comes from Groves' wartime survey (published in 1952) of iron ore resources in Northern Ireland. This contains detailed compositional data on Ulster iron ores, in particular Co. Antrim laterites (and the associated bauxite which was extensively mined during World War II for aluminium to be used in the production of aircraft). Groves' study was supplemented by Eyles' work on the formation and composition of the Antrim laterites (Eyles 1952), by Wilson (1964) on the history of their working, and comparative compositional studies by Landgren (1948).

Iron ore resources potentially available to early ironworkers here have received scant attention. One finds frequent, usually vague, suggestions in archaeological literature that iron ores (usually bog ores) were available in the vicinity of a site, without firm supporting evidence. There are very few sites where iron ore has definitely been found in association with smelting debris. Thus, although very many sites of the period under review, have produced ironworking wastes, we know virtually nothing about the ores from which these derived. It is only possible here to

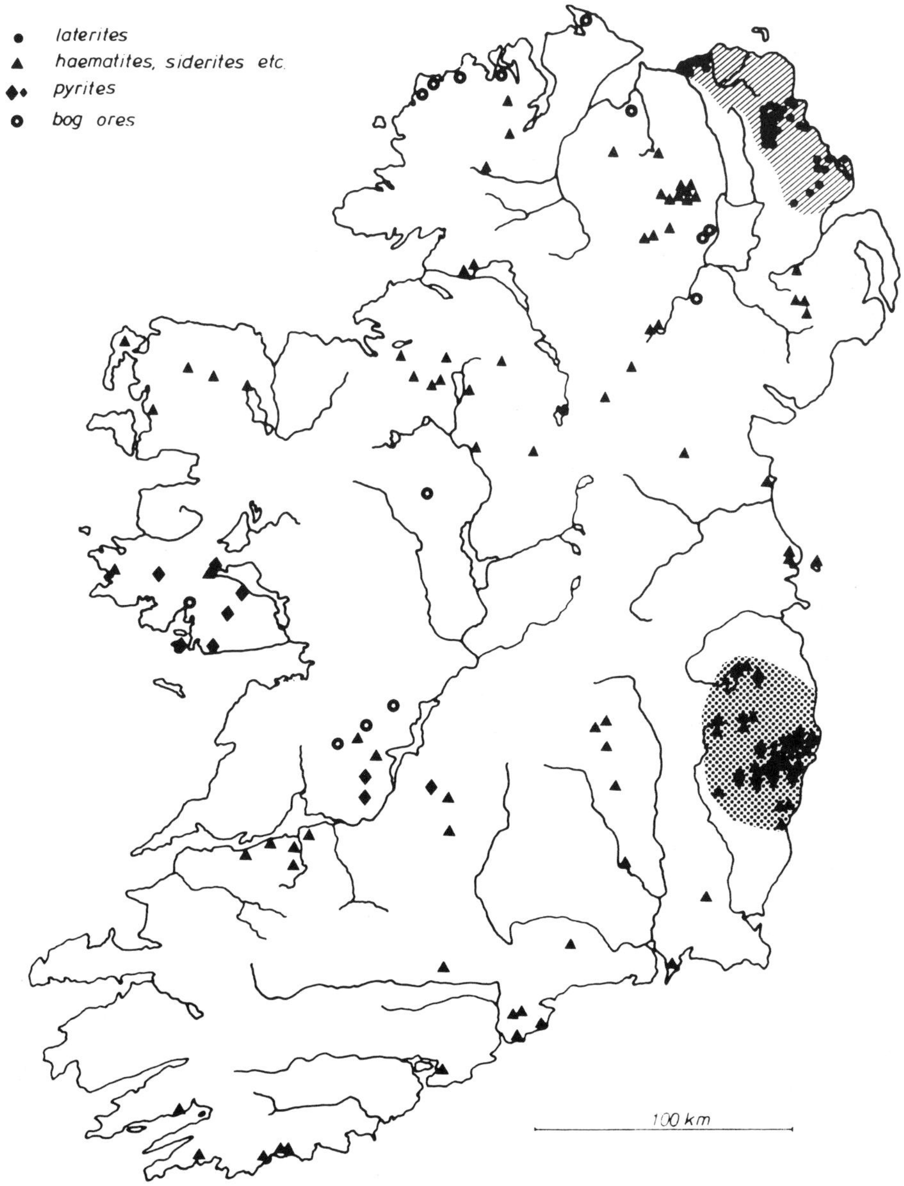

Fig. 6.2.1 Distribution of Irish iron ores: lightly-shaded area denotes major outcrops of the interbasaltic laterites, dark-shaded area denotes the metalliferous Co. Wicklow complex.

draw attention to this neglected area, and to indicate the main deposits from which the iron for our native iron artefacts could have been smelted.

(a) bog ores

These deposits are described elsewhere (e.g. Arrhenius 1967, 9–11) as hydrated iron(III)oxide, $Fe_2O_3.xH_2O$, often associated with high levels of MnO and P_2O_5. In addition, Arrhenius also notes significant contents of Ni, Co, V, Cr, Cu and Mo in Scandinavian bog ores. The mode of formation is suggested to involve bacteriological action, and the precipitation of iron oxides held in colloidal suspension in the anaerobic conditions in peat bogs and lake muds. Presumably the minor- and trace-element content will be significantly influenced by the local lithology.

Although it is conventional wisdom that in Ireland, bog iron ores were widely exploited in antiquity, they are by no means as ubiquitous as usually is suggested. It is certainly true that extensive deposits, particularly in Co. Donegal, were exploited for export to England in the last century to be used in gas purification and paint making (e.g. Kilroe 1907, 159–160; Cole 1922, 88–89). But while references to its occurrence are scattered through Geological Survey memoirs, there is no single survey of its modern occurrence. Payne, in 1589, noted:

> *There is verie rich and great plentie of iron stone, and one sort more than we have in England which they call Bogge Myne* [i.e. bog ore], *of the which a Smith will make iron at his forge presentlie.*

The same term *bogge myne* for bog iron ore was also used by Gerard Boate when he included it as one of the three main iron ores used by the contemporary English-owned iron industry that time (Scott 1985). He wrote (Boate 1652, 70):

> . . . *the first sort* [of ore] *is called bog-mine . . . found in low and boggie places, out of the which it is raised with very little charge . . . This oar is very rich of metal and that very good and tough . . .*

Theoretically, bog iron ore can occur in any peat

deposit where conditions for its formation are suitable. So far, only a few substantial occurrences have been identified in recent times, including around Dunfanaghy in Co. Donegal, in the Cookstown region of Co. Tyrone, and at the Bord na Móna works at Ballyshannon, Co. Donegal and Kinnegad, Co. Roscommon. A particular problem raised by any discussion of bog ores is their destruction through peat digging. Peat has long been used as a fuel in Ireland (its cutting was one of the many activities which the early Irish jurists sought to circumscribe). Although deposits can build up over quite short periods, it seems inevitable that continuous working, modern industrialised exploitation, and agricultural development has modified our bogs so that the distribution of deposits today may bear little or no relationship to that of antiquity. There is, however, no consistent recording of its occurrence during peat cutting operations by Bord Na Móna, nor does it appear to have been noted in the palaeocological studies of Irish peat bogs carried out since Jessen's major survey (Jessen 1949).

(b) siderite

Siderite consists of iron(III) carbonate, Fe_2CO_3, and has a maximum iron content of 48%. In Ireland it occurs principally as nodules intercalated in Carboniferous coal measures. The siderites which outcrop in the area around Lough Allen in Co. Leitrim, gave their name to the hill known as Sliabh an Iarann, which literally means 'mountain of iron'. Today, nodules of this ore may be observed as weathered-out boulders in stream beds. These were heavily worked by the Elizabethan entrepreneurs in the 17th century, and had a good reputation for quality. Boate (1652, 71) described it thus:

> *It . . . is digged out of the mountains in several parts of the kingdom . . . in lumps the bigness of a man's head . . . The iron coming from this oar is not brittle . . . but tough and in many places as good as Spanish iron.*

(c) bedded haematites and limonites

Iron is present either as iron(III) oxide, Fe_2O_3, haematite, the ore having a maximum iron content of 70%, or as hydrated iron(III) oxide (*goethite*), with a maximum iron content of 63%. They are found in altered Silurian strata at Deehommed, Co. Down, and in the Lower Carboniferous series at Lissan and Kildress near Cookstown in Co. Tyrone. The Lissan and Kildress ores were mined for Rennie's blast-furnace works near Cookstown.

TABLE 6.2.1

Analyses of Irish iron ores

	1	2	3	4	5	6	7	8
Fe_2O_3	72.67	51.44	95.44	68.45	72.50	83.33	20.11	76.36
FeO	.26	1.11	1.00	tr	1.48	2.25	—	8.05
SiO_2	18.20	28.09	1.00	14.62	6.80	1.55	20.10	6.33
Al_2O_3	4.22	2.78	.28	1.88	9.44	3.50	7.29	1.50
MgO	.72	1.69	tr	.53	.39	.55	—	.46
CaO	.19	4.83	tr	nd	.45	.40	—	.76
MnO	.05	.59	tr	1.07	.08	.22	28.00	1.13
TiO_2	.24	.26	—	.24	3.60	4.85	0.36	.12
P_2O_5	.05	.05	.06	.86	.09	.13	—	.34
Cr_2O_3	—	–	—	–	.04	—	–	–
CuO	—	–	—	–	.01	.01	—	–
T0TAL	95.88	90.84	97.75	87.65	94.88	96.92	75.86	95.05

1 – 2 = haematite, Deehommed, Co. Down
3 – 4 = haematite, Barrow Mine, nr. Cookstown, Co. Tyrone
5 = laterite, Min. of Production mines in Co. Antrim
6 = laterite, Glenravel, Co. Antrim
7 = bog ore, Ballyvourney, Co. Cork
8 = ore, Ballyhenry, Co. Antrim

(d) pyritic ores

Pyrites, FeS_2, maximum iron content 46%, occur principally in the metalliferous Co. Wicklow complex. Gossans (weathered and oxidised pyritic ores) would presumably have been available in the region, although the principal mining activity recorded for the area was in the 18th and early 19th centuries, when pyrites were mined for export for the manufacture of sulphuric acid.

(e) laterites

Here, iron is present as Fe_2O_3. The ores occur over a very wide area of Co. Antrim, extending into Co. Londonderry. They were formed by the tropical weathering of the Lower Tertiary basalt flows of north-east Co. Antrim, giving a series grading from iron ore, through ferruginous bauxite, to bauxite (an ore of aluminium). Analyses of these ores which were extensively worked in the 18th and 19th centuries for export, show elevated levels of Al and Ti, while the trace levels of V are said by Eyles (1952, 55) and Groves (1952, 230) to be high for British iron ores in general.

Although over 100 excavated sites have produced evidence of some sort for ironworking, there are only a few recorded finds of ore (see Appendix 1). Bog ores were identified at Ballyvourney, Co. Cork, Lough Faughan, Co. Down and Aghnaskea A, Co. Louth, while limonite was recorded from Garryduff I and Oldcourt, Co. Cork, and haematite was found at Ballyhenry, Co. Antrim. Ironstone nodules for smelting might also have come from Nendrum, Co. Down, Freestone Hill, Co. Kilkenny, and Kiltera, Co. Waterford. At Garryduff, there was evidence for ore roasting, while at Lagore, Co. Westmeath, Hencken (1950, 233) describes deposits in a series of hearths which included iron oxides, magnetite and charcoal, possibly fines resulting from roasting. It seems not unlikely that ores and fines have been uncovered on excavations, but not recognised.

* * *

6.3 – Mining

If our information on the compositions of Irish iron ores is decidedly sketchy, our knowledge of early mining is virtually non-existent. Those mining locations of which we do know are those used after the start of the Elizabethan period exploitation, but few sites other than the extensive Co. Antrim and Co. Wicklow workings (mainly 17th-19th centuries) have been located. Probably the most useful documentary sources for these are the surveys of Kane, Kinahan, Cole and Groves. Although the one surviving passage in the laws relating to mining is scrappy and difficult, this, along with other incidental references (Chapter 7.2), does at least provide glimpses of pre-16th century mining in a way that the existing archaeological evidence does not. Cole (1922) makes mention of a recent mining operation breaking into 'old men's workings', but no further useful information is given and the site of the mine is not recorded.

At the hillfort on Freestone Hill, Co. Kilkenny (Raftery, B. 1969, 41–44), Bersu located a number of pits in and around the hill which he interpreted as indicative of mining activity by the earliest inhabitants (EIA B). These have recently been reinterpreted as solution hollows produced by chemical weathering of the dolomitic limestone of which the hill is made up. It is interesting to note, however, that of the 59 stone 'finds' preserved from the second occupation period, 22 are identified by Jackson as sideritic or clay ironstone (Raftery, B. 1969, 59–61, 74–79, Figs. 26–28), all deriving from Carboniferous strata. Of these, 14 appear to show some deliberate working, while the remainder are natural nodules. Although it would appear that iron ore nodules were used for making various stone implements at this site, at least some may have gone into smelting operations to which finds of 'slag' might bear witness. A statement attributed to Bersu (Raftery, B, 1969, 44) that limonites and haematites frequently occurred in the occupation deposits requires further investigation.

Excavations conducted by Macalister at the LIA monastic site of Kiltera, Co. Waterford (Macalister 1938, 13), where finds of slag debris and possible ore fragments from the site showed ironworking activity, uncovered a conical-sectioned pit which he suggested might be a storage pit. Iron-ore bearing strata occur close to the surface in this area, and the structure (no plan given: but cf. the 'reconstructed' section offered in Proudfoot 1970, 40, Fig. 1) might possibly be interpreted as analogous to the shallow mine of the 'bell pit' type so common in the Sussex Weald (e.g. Straker 1931, 101f).

Early Irish law stipulates that one of the features which increased the value of land was an iron mine (see Chapter 7.2). something which would have been of particular interest to owners of tracts of bog. It is clear that the activities of the 17th, 18th and 19th century ironworkers will have obliterated the majority of 'old men's workings', while continuous and widespread digging of peat for fuel, coupled with the regeneration of peat bogs, and more recent reclamation of bogs for farming, will have confused earlier patterns of bog ore distribution. Nevertheless, in view of our understanding of copper ore mining during the Bronze Age (Jackson, J. 1978, 113, Table 2), one would hope that similar evidence of early iron mining activity will be forthcoming as a result of detailed field work.

* * *

6.4 – Smelting and smithing slags

By far the commonest indicator of ironworking in the archaeological record of Ireland is the siliceous debris, usually loosely classed as 'iron slag', which is included very frequently in the finds assemblages from excavated EIA B/LIA sites (see Appendix 1). Quantities vary from fragments weighing a few grams, to assemblages weighing many kilos. While this is normally taken as indicative of smelting, this is misleading, since are major problems in distinguishing between slags resulting from smelting activities, and those produced by subsequent operations. Thus, descriptions of plano-convex slag blocks as 'furnace bottoms', common in Irish archaeological literature, and implying that they derive solely from ore reduction, are not helpful. As will become evident, analysis and examination of slags is a very difficult field, and for this reason, the subject will be discussed only briefly here.

In the main, iron ores do not consist solely of iron compounds, but also include in their makeups compounds of oxygen, silica, aluminium and calcium in particular, as well as those of other elements. The production of iron from its ores by the bloomery process required that the metallic iron reduced from iron oxides be separated from the unwanted materials. Now, while minerals like quartz and alumina have very high melting points (over 1700°C), the presence of iron oxide acted as a flux, with the slags formed usually having melting points in the range ca 1100–1250°C, a temperature achieved comparatively easily in all of the different forms of early smelting furnaces. During their time at heat in the smelting furnace, slags also reacted with ash from the charcoal (introducing sodium and potassium which reduce melting points significantly), and with the clay lining (silicates), so that the final bulk composition of a bloomery smelting slag reflects not only that of the original ore, but also of these other materials. The slags which result are made up of mixtures (often very complex) of silicate minerals and iron oxides, sometimes also with droplets of metallic iron (Plate 6.4.1a,b).

In furnaces more advanced that the simple bowl type used in Ireland, provision was made for the removal of slag by allowing it to flow downwards (e.g. into a basal pit) and outwards – *tapping*. Tapped slags have a very characteristic texture with flow patterns on their surfaces (e.g. Tylecote 1986, 131, Fig. 75). It is the complete absence of such slags from Ireland, coupled with the structural evidence (sect. 6.5), that makes it certain that during the period under review, smelting was carried out in simple bowl furnaces with or without superstructures. For this reason, tapped slags will not be discussed further.

Nearly every slag assemblage shows a variety of morphology, and it might be thought that a simple classification by shape could be sufficient for basic distinctions to be made. For example, Sperl (1980, 14ff) distinguishes (amongst others) between the bloom (type A1: *Luppe*) and furnace bottom (type A2: *Ofensau*), slag formed during the smelting process (type B: *Ofenschlacke*), and smithing slags (type D; *Schlackenkuchen*) including the consolidated bloom (type D3: *Schmeideluppe*). There is a general European consensus that, as a rough guide, plano-convex slag blocks grade in size from large pieces (>30cm in diameter) produced in the smelting furnace, down to small cakes 10–15cm in diameter resulting from smithing. For example, Tylecote (1986, 133f) distinguishes between 'furnace bottoms'

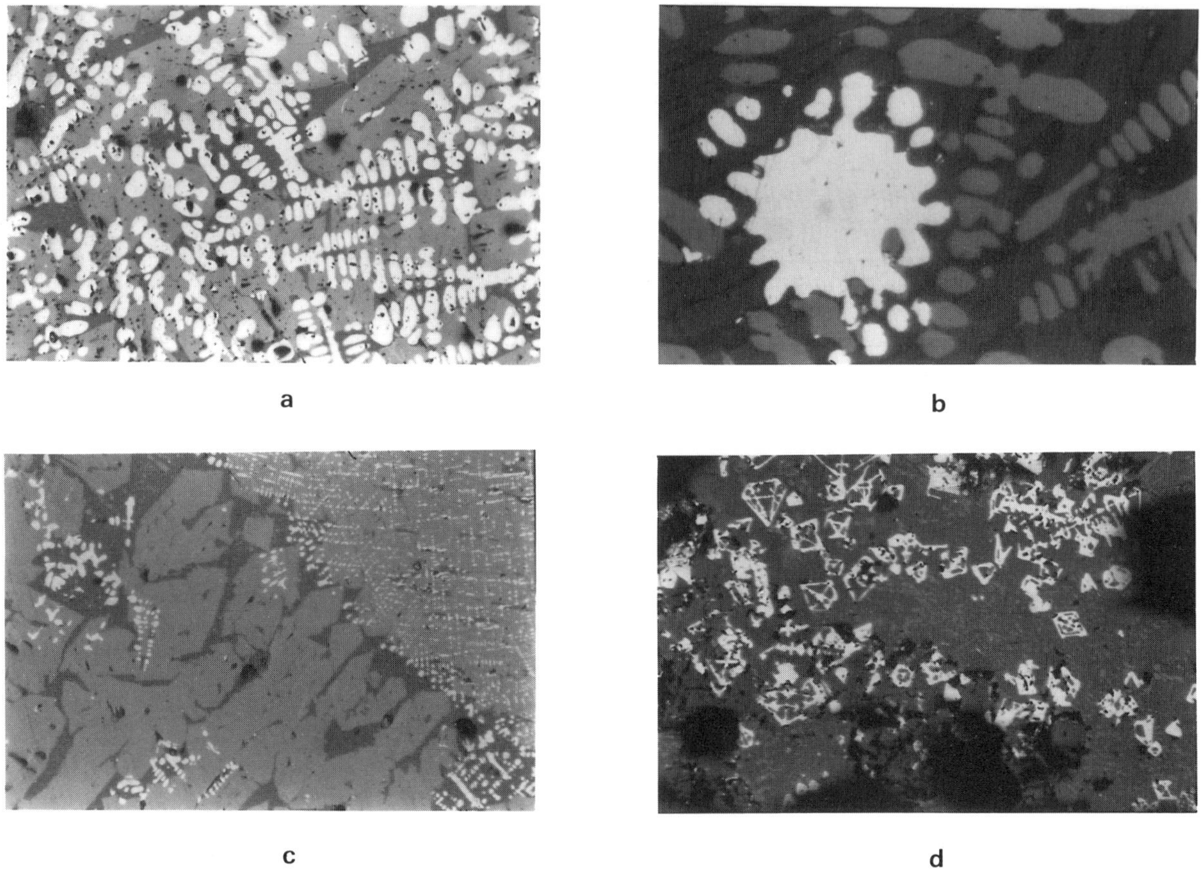

Plate 6.4.1 Structures found in slags from Irish sites: (a) slag from Lisleagh, Co. Cork showing wüstite dendrites (light) with fayalite crystals (grey) in a glassy matrix x100, (b) slag from Clogher, Co. Tyrone showing a droplet of metallic iron (ferrite) with wüstite dendrites (light grey) and fayalite (dark grey) x250, (c) slag from Altanagh, Co. Tyrone showing a possible cooling front x100, (d) slag from Lisleagh, Co. Cork showing magnetite (light) in a complex silicate matrix x100.

(FB) and 'smithing hearth bottoms' (SHB) on the basis of size, with FB's tending to the top of the size range and SHB's to the bottom. But because of the fact that Irish smelting was restricted to the bowl furnace, even this simple scheme presents problems. Thus, for the Irish archaeologist in the field, the situation is fraught with uncertainty when trying to identify origin, and hence anticipate structural evidence, from slag morphologies.

The form of the bowl furnace meant that the slag formed during the smelting process accumulated towards the bottom, below the bloom. If it was allowed to solidify before the furnace was opened, then it formed a plano-convex lump whose diameter and depth was determined by the furnace dimensions and the size of the charge of ore. While still viscous, the slag frequently became peppered with fragments of unburnt charcoal and sometimes also picked up a skin of clay on the underside. Because it was allowed to cool comparatively slowly, the formation of minerals was more likely to approach equilibrium conditions, and such slag lumps are usually reasonably homogeneous. While it can be suggested that smelting slags should tend to be much more homogeneous than those resulting from other processes, and to show a lower degree of oxidation, this is a broad (and somewhat crude) generalisation only.

Furnaces, however, do not seem always to have been allowed to cool before clearing out, and under these circumstances, once the raw bloom had been recovered, the remaining debris seems to have been raked out to one side and allowed to cool. Thus, while still often including charcoal and clay, they are randomly shaped, lacking the plano-convex form of slags cooled in the furnace. The more rapid cooling, will make such slags liable to have mineralogies showing deviation from equilibrium conditions.

The primary product of the smelting process, the raw bloom, was a spongy mass of reduced iron particles and slag (cf. Plate 1.4.2a). To obtain useable stock, the raw bloom had to be reheated to welding heat (around 1200°C), and hammered to weld the iron particles together and to squeeze out the slag. While much of the slag ended up as small particles scattered round whatever sort of anvil was used, some undoubtedly ended up at the base of the hearth used for reheating. Since this slag resulted directly from the smelting process, it is likely to be indistinguishable mineralogically and chemically from slag left in, or raked out of the furnace, as conditions at the base of the smithing hearth would be as reducing as in the smelting furnace. The forging of the bloom also threw off small particles of iron, encapsulated in slag, to cool around the anvil. First recognised on Polish sites, these are known as *gromps* (Polish *grąpie*: Nosek 1975).

Smithing slags resulted from the forging up of artefacts from stock at heat, during which oxidation of the iron produced *hammer-scale* which could fall into the smithing hearth, and be scattered around the anvil along with other debris. The deliberate use of a flux such as sand to clean surfaces preparatory to welding will also have produced a respectable slag (since, after all, silica and iron oxide are primary components of smelting slags) which, again, would form below the tuyère and could end up at the bottom of the hearth. At first sight, it might seem that smithing slags should show a significantly higher degree of oxidation than smelting slags. But this cannot be taken as other than a broad generalisation, since we may expect a proportion of these slags to have ended up under reducing conditions at the base of the hearth. Since it seems highly probable that in Ireland, bloomsmithing and smithing hearths in many cases were re-used bowl smelting furnaces, the resulting agglomerations of slag from them will be very similar in form and dimensions to that resulting from smelting. Equally, smithing slags might be detectable in some instances by the appearance of magnetite (Plate 6.4.1c), coupled with high silica levels, and mineralogical heterogeneity indicating disequilibrium conditions. Plate 6.4.1d shows what appears to be a zoning in mineral size due to a piece of slag having been subject to different cooling rates.

To summarise, while it is likely that FBs and SHBs can be distinguished on the basis of size at either end of the scale, there will be FBs as small as the average SHB and SHBs as large as the average FB. On Irish sites where perhaps no more than one or two smelting furnaces were in operation at any one time, the sizes of assemblages may be a poor guide as to the scope of ironworking.

While the detailed chemistry of slag formation during the bloomery smelting process and subsequent cooling is complex and difficult, studies of ores and slags (e.g. Morton and Wingrove 1969, 1971; Tylecote et al. 1971; Sperl 1980; Bachmann 1982; Kresten 1984; McDonnell 1986: Keesmann 1989) have clarified some at least of the factors governing their formation. It has been common for analyses of slags to be achieved by taking bulk samples covering all of the phases present. This provides useful information, and is perhaps the easiest way to obtain data. However, Kresten (1984) and McDonnell (1986) demonstrate the need to go beyond the analysis of slag fragments (*bulk analysis*), and to look in detail at the compositions of individual phases within slags as to determine more accurately their mineralogical histories. Only bulk analyses have been used in this study, and the conclusions drawn from the data can be taken as tentative only.

The chemical analysis of slags produces results in terms of the oxides of the various elements present, the principal constituents usually being those of iron, silicon, aluminium and calcium. Since any iron (III) oxide present is likely to result from the re-oxidation of iron(II) oxide, the composition of the slag as formed in the reducing zone of the furnace may be considered in terms of FeO only. Under furnace conditions, these oxides combine chemically to form minerals and, in theory, if reactions were allowed to reach equilibrium, then the ideal or *normative* composition[1] of mineral phases can be computed (Keesmann et al. 1982: Kresten 1984: McDonnell 1986). Under the microscope (where the mineral assemblage can be quite different from that predicted), the most prominent minerals usually found in slags are FeO as *wüstite* and the olivine *fayalite*, an iron silicate $2FeO.SiO_2$. These are commonly observed embedded in a glassy groundmass which can have a highly variable composition (in the past, it has been described generally as having a composition equivalent to that of the plagioclase feldspar *anorthite*, a calcium-aluminium silicate $CaAl_2Si_2O_8$). In addition, metallic iron and magnetite may be observed. For the sake of sake of simplicity, the proportions of minerals computed from analysis may be grouped into silicates, glass and iron oxide[2] and plotted onto a ternary diagram for purposes of comparison.

A basic measure of the efficiency or otherwise of the smelting process is that of the amount of iron oxides removed from the furnace in the slag. Fluxing of the gangue minerals will produce silicates and, if the process was to be efficient, no more iron oxide should have been removed. However, this was seldom the case. At its simplest, we can say that the smaller the amount of iron oxides appearing in the slag, the more would have been available for reduction to metallic iron in the furnace, and hence the greater the efficiency of the process. Conversely, the more iron oxides in the slag, the less productive the smelt. But other factors have to be taken into account when considering 'efficiency'. Since iron oxide acts as a flux to lower the slag melting point, higher contents will have made separation of metal and slag easier than for lower contents where the melting points of slags formed will be higher. Higher furnace temperatures in turn mean greater difficulty in maintaining the necessary reducing conditions, with the consequent need for higher fuel to ore ratios. The early smelter therefore may well have traded off a lower yield in terms of iron produced against ease of working.

Turning now to look at slags from Irish sites, we may make the general observation that reported finds comprise FBs, SHB's and assorted debris. While some excavators have given the total weight of slags found, they are exceptional. Also, descriptions of

TABLE 6.4.1

Analyses of Irish iron 'slags'

%	1	2	3	4	5	6	7	8
Fe_2O_3	12.62	7.70	12.11	4.02	38.23	9.66	12.61	14.57
FeO	49.56	63.23	63.89	58.70	17.88	53.33	61.54	56.07
SiO_2	22.54	17.34	10.24	26.90	26.90	27.32	15.80	18.63
Al_2O_3	4.58	3.86	3.61	2.32	4.81	2.67	2.83	2.64
MgO	1.85	0.29	0.31	0.84	0.80	0.38	0.57	0.39
CaO	2.62	2.34	2.84	1.76	5.15	2.77	0.58	0.51
MnO	0.18	2.68	2.80	1.79	0.26	1.10	0.98	0.37
TiO_2	0.62	0.13	0.15	0.27	0.44	0.19	0.36	0.32
P_2O_5	0.23	0.10	0.06	0.48	0.15	0.99	0.16	0.13
Na_2O	2.70	0.69	0.76	1.12	0.63	0.78	1.65	3.21
K_2O	0.44	0.87	2.04	0.53	1.86	0.70	0.40	0.25
ppm								
Cr	772	431	336	nd	nd	199	141	857
Cu	382	121	554	801	645	361	277	226
Ni	558	238	178	222	291	670	328	422
V	nd	nd	nd	103	155	nd	nd	nd
Total	97.94	99.23	98.81	98.73	97.11	99.89	97.48	97.09

nd = element not detected

1 = Ballyhenry, Co. Antrim
2 and 3 = Clogher, Co. Tyrone smelting area
4 and 5 = Lisleagh, Co. Cork, slag dump in ditch
6 and 7 = Kilpatrick, Co. Westmeath, smelting area
8 = Altanagh, Co. Tyrone

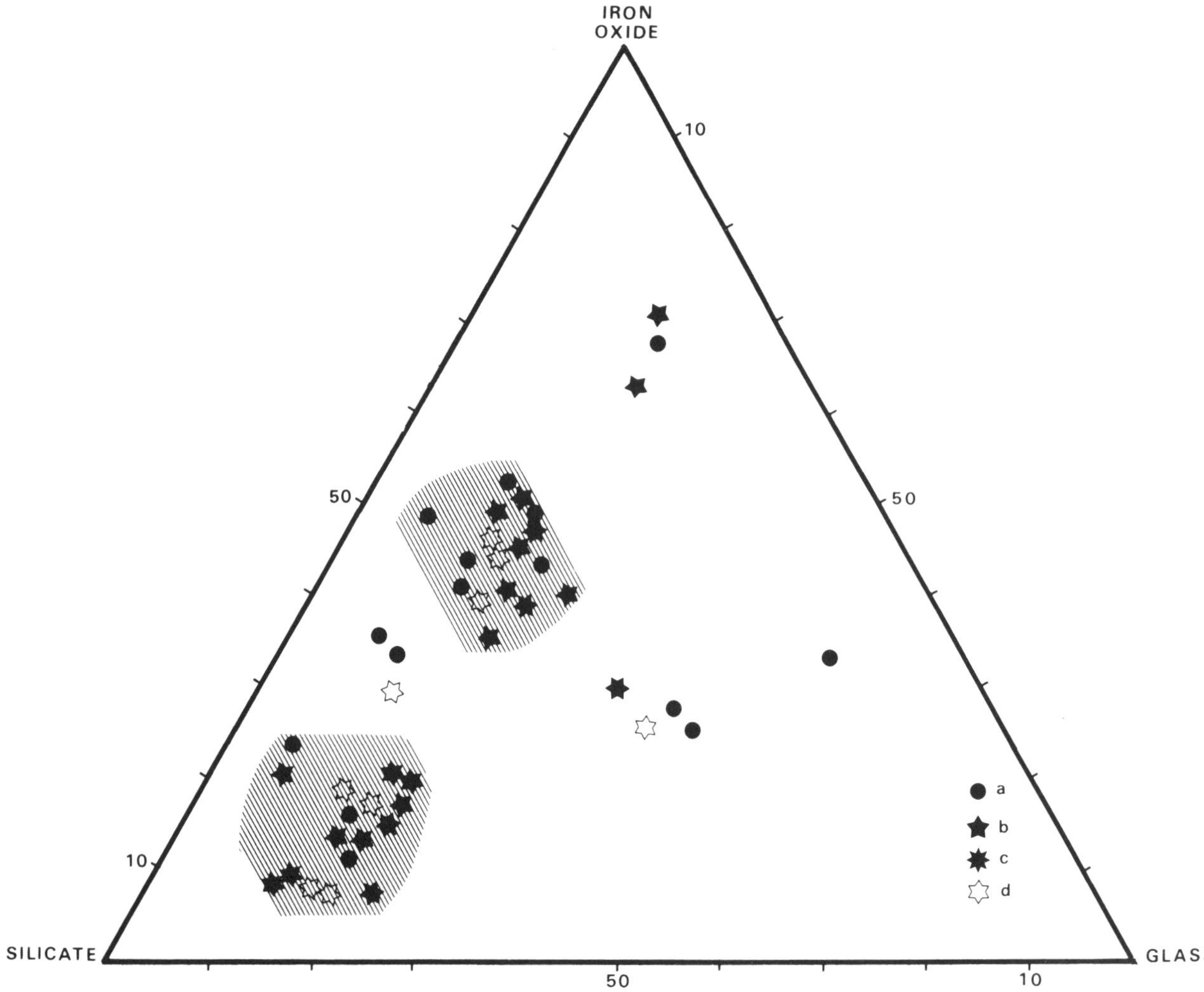

Fig. 6.4.1 Ternary plot of the normative constituents of slags from Irish sites, grouped as silicates, glass and iron oxides: (a) Lisleagh, Co. Cork, (b) Kilpatrick, Co. Westmeath, (c) Clogher, Co. Tyrone, (d) Altanagh, Co. Tyrone.

individual slag pieces are usually sketchy. At Ballyvourney, Co. Cork, O'Kelly (1952, 33f) noted 57 complete 'furnace bottoms' and fragments of a further 80. These ranged in size from 9–18cm in diameter, and from 6–11cm in depth. The excavations at Lagore produced 34 plano-convex blocks, ranging from 12–19cm in diameter and 3–8 cm in depth (Hencken 1950, 230f). One of the Lagore blocks was at least partially analysed to give '. . . magnetic iron, ferric oxide and ferrous sulphide . . .'. In neither case is the overall weight given.

Slags from Lisleagh, Co. Cork weighed in total 800kg, and included large FB's, up to 30cm in diameter and 10cm deep. Comparable finds came from Kilpatrick, Co. Westmeath, where FBs of similar size were found with SHBs down to 10cm in diameter by 3cm deep. In the course of this study, samples of slags from a number of sites have been subjected to bulk analysis, and representative results are given in Table 6.4.1. These analyses have been processed to compute their normative compositions. Grouping the calculated norms into silicates, glass and iron oxide, and plotting them onto a ternary diagram shows two bands, (Fig. 6.4.1), one where the iron oxide content lies in the range 5–35%, silicate 60–80%, the other in the range iron oxides 50–75%, silicate 30–50%. The lower iron oxide/higher silicate band probably results from reasonably productive smelting, while the other group represents less efficient recovery of iron. The scatter of points illustrates the problems of trying to classify slags neatly.

The conclusions which can be drawn from available data are meagre, and really cannot be taken further until detailed research on excavated Irish assemblages has been carried out. This will be true both in terms of the possibilities of distinguishing broadly between smelting and smithing slags, and hence gauging the proportions of these activities on individual sites, and of linking smelting slags to ore types.

* * *

6.5 – Irish bloomery smelting furnaces and smithing hearths

Although the most certain evidence for iron smelting on a site is the survival of a smelting furnace, until comparatively recently it has been difficult to list sites on which the remains of smelting furnaces had been identified with certainty. Prior to the start of my own work in 1970, finds of the remains of furnaces from the period (see Appendix 1) under review had been claimed for Cleggan, Co. Antrim (Mogey 1946),

Dressogagh, Co. Armagh (Collins 1966), Ballyvourney, Co. Cork (O'Kelly 1952), Garryduff I, Co. Cork (O'Kelly 1964), Glannafeen, Co. Cork (Ó Cuileanáin 1955), Oldcourt, Co. Cork (Murphy 1961), Boho, Co. Fermanagh (Proudfoot 1953), Rathgurreen, Co. Galway (excavated by the late M.V. Duignan: information courtesy of Dr. J. Waddell), Aghnaskea cairn A, Co. Louth (Evans, E.E. 1936), Aghnaskea cairn B, Co. Louth (Evans, E.E. 1940), Church Island, Valencia, Co. Kerry (O'Kelly 1959), Liathmore, Co. Kerry (Leask and Macalister 1943), Inishkea North, Co. Mayo (Henry 1945), Creevykeel, Co. Sligo (Hencken 1939), and Kiltera, Co. Waterford (Macalister 1938). Of this list, however, only Dressogagh and some of the structures from Ballyvourney and Garryduff can be included in a definitive listing of Irish smelting furnaces with any confidence due to poor reporting.

Since 1970, structures used for the smelting of iron have been identified definitely at Lisleagh, Co. Cork (Monk 1988 and forthcoming), Taylorsgrange, Co. Dublin (Keeley 1987), Reask, Co. Kerry (Fanning 1981), Knowth, Co. Louth (Eogan 1972), Altanagh, Co. Tyrone (Williams 1986), Clogher, Co. Tyrone (Warner 1972 and forthcoming), Kilpatrick, Co. Westmeath (Swann 1973 and forthcoming), and Rathgall, Co. Wicklow (Raftery, B. 1976b). Furnaces have also been reported from Inishcealtra, Co. Clare (de Paor 1970) and Coolcran, Co. Fermanagh (Williams 1985b), but the published information is not sufficient to make the identifications certain. McCabe (1968) identified (during field survey) as a furnace a structure in a ringfort at Ballymakellett, Co. Louth. Ivens (1984a, 77) has suggested tentatively that a feature dug into natural at Movilla Abbey, Co. Down might be the remains of a furnace.

All of the archaeological evidence to date indicates that during the period under review, iron was smelted in Ireland only in what has been called the *bowl furnace*. This type of furnace can be characterised as comprising a roughly hemispherical depression in the ground, often (but not always) clay lined, and with no provision for the removal, or *tapping*, of slag during the smelting process[3]. In the broader theatre of European archaeology, debate has centred on whether such furnaces would have had any form of low superstructure, rather than just simply being holes in the ground closed over, say with sods or clay (e.g. Cleere 1972; Tylecote 1986, 132ff). Unfortunately, none of the furnaces so far excavated on Irish sites has provided evidence one way or the other. Two stone-built, clay lined structures, one at Garryduff I, Co. Cork (O'Kelly 1964, 99 and Plate XVI), the other at Clogher, Co. Tyrone are best interpreted as smithing hearths (see below). Also, it is not impossible that the structures reported from Aghnaskea cairns A and B, and at Glannafeen and Ballymakellett are the remains of smithies as opposed to smelting furnaces.

Experimental work by Wynne and Tylecote (1958), by Tholander (1987) and by the author (Scott forthcoming) demonstrates that smelting in a bowl furnace without a superstructure is a very difficult process to work. O'Kelly (1961a) conducted his experiments in small bowl furnaces[4] some 30cm in diameter and 20cm deep, in some cases adding a low superstructure about 30cm in height. Given the different densities and packing factors of ore and charcoal, it will be obvious that a hemispherical depression 30cm, say, in diameter, will only be able to hold a very small amount of iron ore. Furthermore, in such a furnace, the burning of the charcoal fuel, which also acts as a support for the charge of ore, is often not of sufficient length to allow either significant reduction of oxide to metal or adequate slagging before the heavier material slips to the bottom. This would be the case particularly if softwood or peat charcoals were to be used. Using a low superstructure, so that the ratio of height to diameter exceeds 1:1, a greater capacity can be achieved, along with a longer process in which the stay for the ore in reduc-

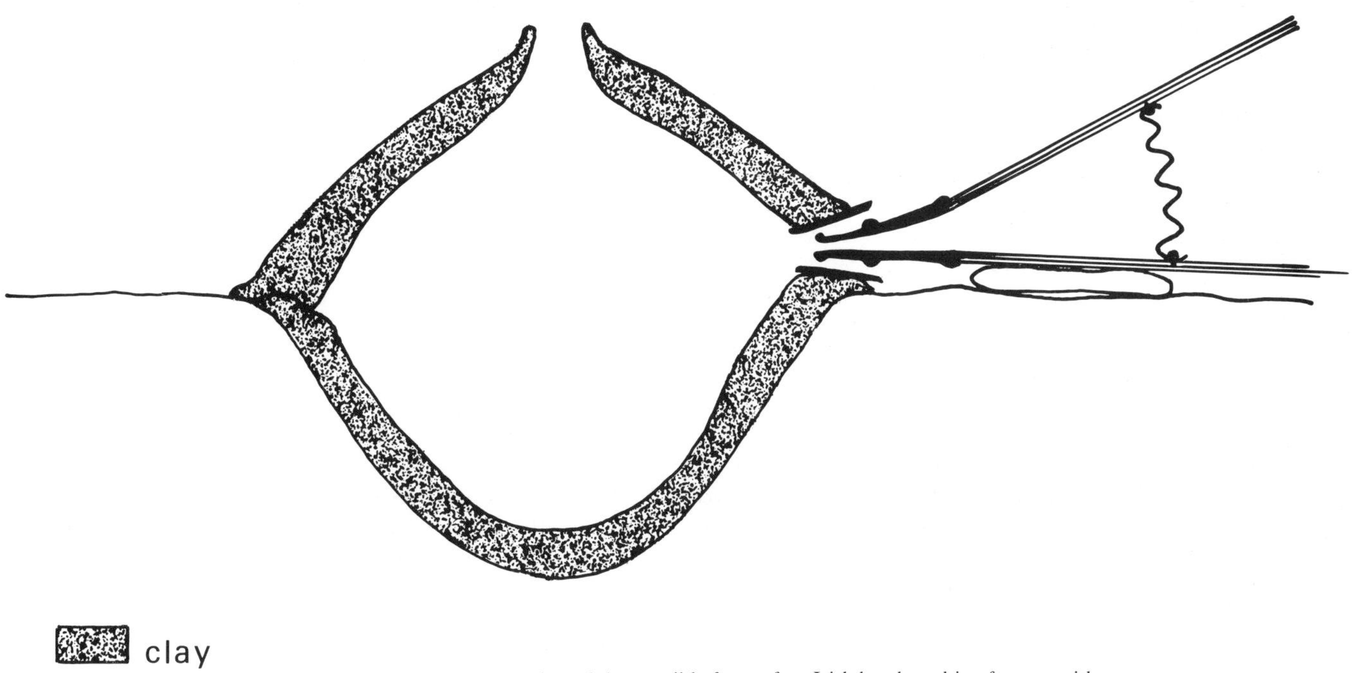

Fig. 6.5.1 Conjectural reconstruction of the possible form of an Irish bowl smelting furnace with low superstructure (not to scale).

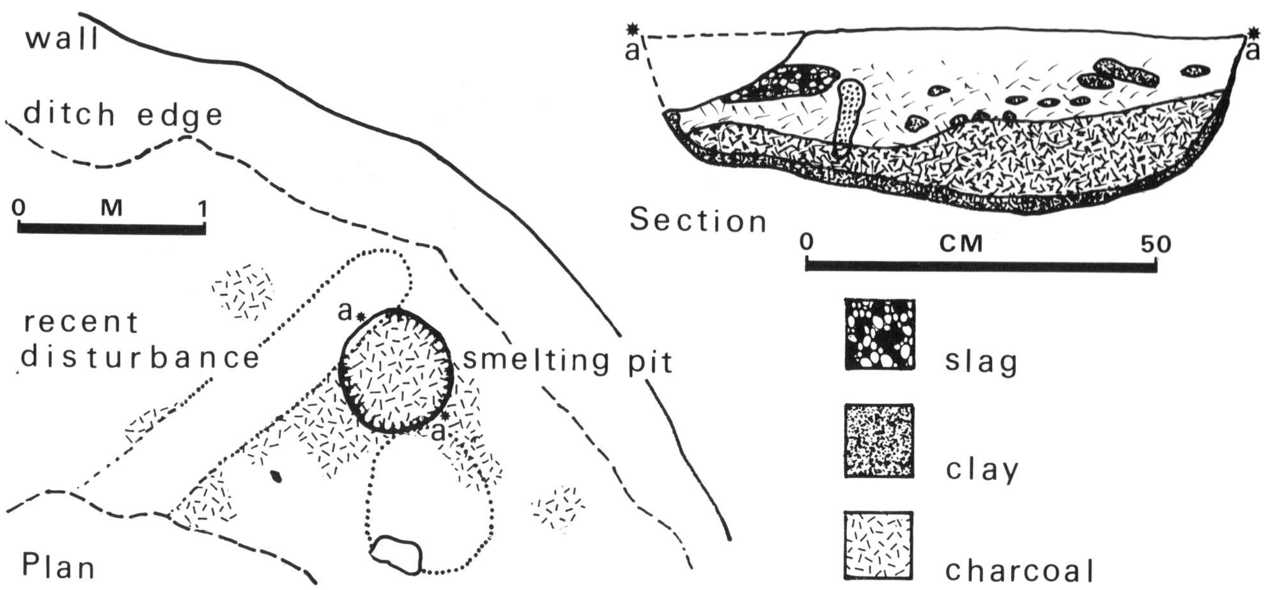

Fig. 6.5.2 Smelting remains at Rathgall, Co. Wicklow (Courtesy of Dr. B. Raftery)

ing conditions is extended. However, a corollary of this is that the superstructure must be of sufficient wall thickness (probably greater than 8–10cm) to provide insulation against heat loss. Alternatively, the insulation provided by thinner walls could have been greatly enhanced by banking earth or sand around the superstructure: this would leave no trace archaeologically. While one would imagine that thick pieces of heavily fired clay would be recognisable on site, nothing resembling superstructure remains has been reported from an Irish excavation. Furnaces undoubtedly were driven by bellows blast through clay nozzles or *tuyères*, and by contrast, examples of different forms have survived (see below).

Almost certainly, the difficulty of identifying and describing the precise form of the 'bowl furnace' in Ireland (in other regions as well, one suspects) is that not only could it be used, closed, to provide the reducing conditions necessary for smelting, but also could be used afterwards, open, as a hearth for the reheating of the spongy mass of iron and slag prior to, and during its consolidation. We might draw an analogy here with the re-use of the superstructure of slag-pit shaft furnaces (e.g. Bielenin 1987). Here, once the smelting process was complete, with the basal pit filled with slag, the shaft was lifted, complete, and placed over an empty pit ready for the next smelt. It could be suggested that a superstructure over a bowl furnace was lifted off after the smelt, and the bowl then re-used for bloomsmithing, and even as a forging hearth for the production of finished artefacts. Thus, while the superstructure undoubtedly would been damaged to some degree during a smelt, it could have been readily patched and re-used on a number of occasions, only being discarded when it fell apart. Alternatively, if a superstructure was constructed for each smelt (a rather wasteful process, but one used by O'Kelly in his experiments), using fresh clay, this could have been broken down for the removal of the bloom, and scattered as unrecognisable fragments around the furnace. Use of a fresh clay covering could explain why superstructures have not survived in recognisable form, while tuyères (which must have been fired before use) have. While purely conjectural, the type of furnace structure which we might reconstruct from the fairly sparse evidence is shown in Fig. 6.5.1.

Among the earliest Irish iron smelting furnaces excavated to date are those associated with the EIA B/LIA A occupations of the hillforts at Rathgall, Co. Wicklow and Clogher, Co. Tyrone. At Rathgall, a small bowl furnace, roughly oval in shape, 50cm by 40cm, and 25 cm deep, had been dug into the top of the silted up Later Bronze Age ditch (Fig. 6.5.2). A date of AD 265 ± 70 (Raftery, B. 1976b, 347), calibrated to give a range of AD 180–540, has been obtained for charcoal from the pit. At Clogher, excavation of an area of the Earlier Iron Age ditch revealed a smelting and smithing complex (Fig. 6.5.3) for which a radiocarbon date of AD 395 ± 45, calibrated to give a range of AD 390–620, was obtained (Warner pers. comm.). Two small, hemispherical bowl furnaces (possibly the remains of a third, but badly disturbed) had been dug into the silt of the ditch. What survived of furnace 1 measured 40cm in diameter by 15cm deep, while the remains of furnace 2 measured 40cm in diameter by 15cm deep: both were lined with clay which showed signs of intense heat, while furnace 1 contained large fragments of slag. Elsewhere, on the main ringfort (built in LIA A times inside the EIA B settlement) although ironworking debris was widely scattered along with the debris of other industrial operations (non-ferrous metalworking, enamel and glass making) no furnace remains were found (Warner pers. comm.). The badly disturbed remains of at least three bowl furnaces, unlined, with a calibrated date in the range AD 430–660, were uncovered during excavation of the multi-period site at Altanagh, Co. Tyrone (Williams 1986, 51ff).

Excavations in a ringfort at Lisleagh, Co. Cork, revealed two foci of ironworking and debris. In the upper ditch fill, an extensive spread of slag blocks (Plate 6.5.1) produced some 800kg of material, as well as a large amount of highly burnt clay fragments. A number of what, by Irish standards, are exceptionally large plano-convex slag blocks (up to 35cm in diameter and 10cm in depth) were included in what was probably a dump of material built up over a

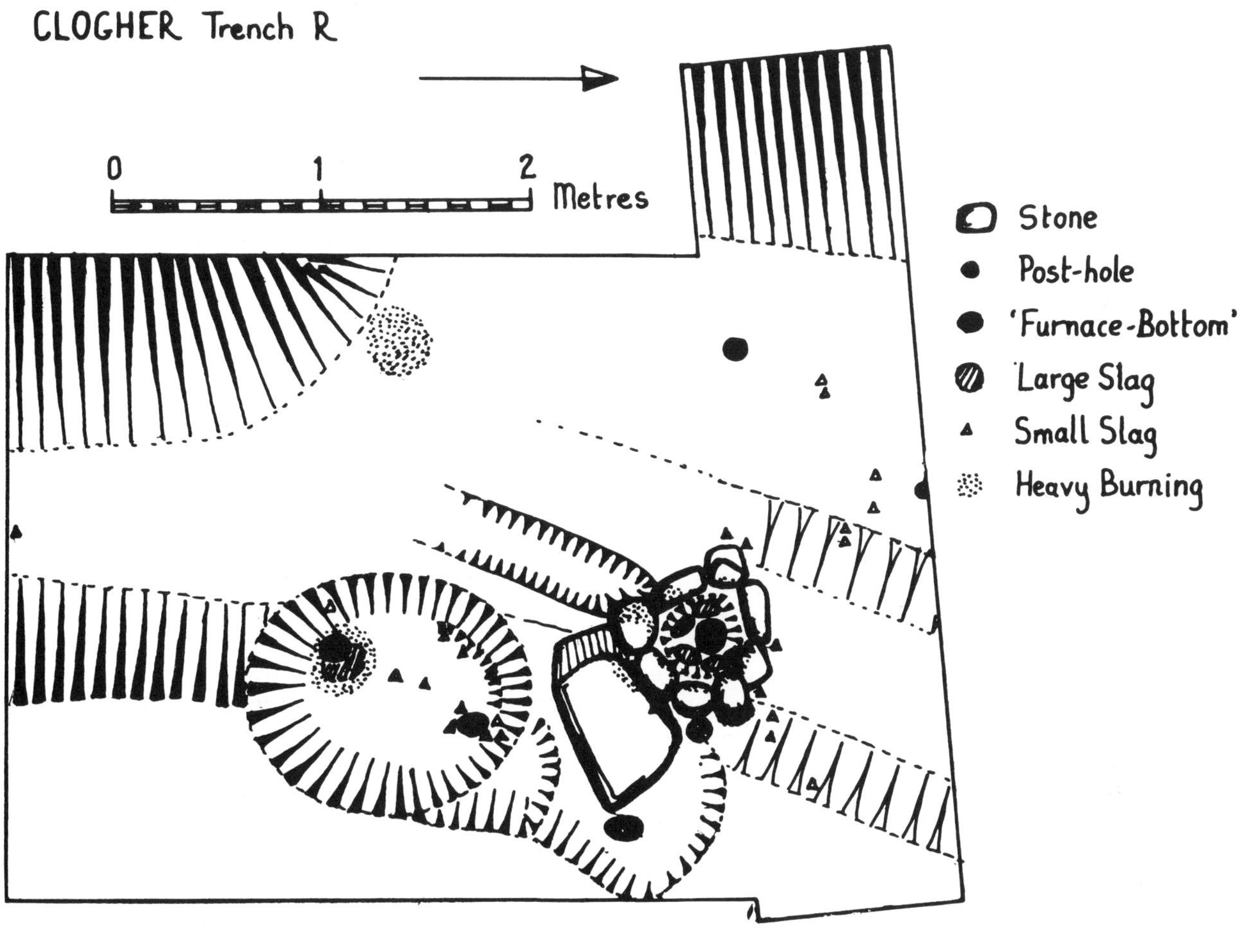

Fig. 6.5.3 Smelting and smithing area at Clogher, Co. Tyrone (Courtesy of R.B. Warner).

period of time from operations in the interior. In the ringfort interior, the remains of a small bowl furnace, some 50cm in diameter and 20cm deep (Monk pers. comm.), cut by a later feature, were found (Fig. 6.5.4). This was not of a size to account for the mass of debris in the ditch. Uncalibrated C-14 dates for this activity centre around AD 665–840.

Inside the monastic enclosure at Kilpatrick, Co. Westmeath whose main period of activity fell in the period 7th-8th centuries AD (Swann 1973), the excavator uncovered a concentration of ironworking debris associated with the remains of a bowl furnace which measured 45cm in diameter by 20cm deep. This had a thick clay lining daubed over a packing of stones (Plate 6.5.2). There were a number of large plano-convex slag blocks up to 25 cm in diameter and 10cm in depth.

At the monastic settlement at Reask in Co. Kerry, founded in the 5th century and continuing in use well into LIA A times, the excavator (Fanning 1981) uncovered the remains of ironworking inside the foundations of two of the small stone cells. Inside structure D, a group of three interconnecting pits, highly burnt, contained slag and charcoal, while a further pit, again showing evidence of intense heat, also contained slag and charcoal (Fig. 6.5.5). In structure G, the badly disturbed remains of pits containing charcoal, slag, and possible fragments of tuyères, probably associated with smelting or smithing were uncovered. The group of three pits in structure D (labelled A-C on Fig. 6.5.5) represents the remains of three bowl furnaces. Furnace A was 45cm in diameter by 10cm deep, furnace C was 30cm in diameter by 13cm deep, and these were partly cut by furnace B measuring 42cm in diameter by 10cm in depth. Pit D, set away from the others was irregular and larger (maximum diameter 1m by 22cm deep), although showing intense burning and a fill containing charcoal and slag. This may well have been used for bloomsmithing, The remains in structure G are less certain due to disturbance. However, in addition to ironworking debris, crucible fragments were found, suggesting that this structure was a general workshop rather than simply a forge. Outside structure D were the remains of a peat stack beside an area of heavy burning where peat charcoal was found, and a deposit of slag (Fig. 6.5.6), and this, in part, may be interpreted as a dump area associated with the ironworking and other activity (see below).

At Knowth, Co. Louth (Eogan 1972), two areas of iron production pre-dating the Anglo-Norman re-use of the megalithic mound as a motte were uncovered. The remains of a small bowl furnace, probably around 30in diameter, were found dug into the fill of the ditch which had been dug round the mound in Iron Age times, while on the mound itself, two small clay-lined bowl furnaces were uncovered (Plate 6.5.3). This activity most likely belongs to LIA A.

A problematic site is St. Gobnet's House, Ballyvourney, Co. Cork (O'Kelly 1952). Here the excavator uncovered in a round house a number of pits containing slag, 'furnace bottoms' and burnt clay. These he interpreted as furnaces. A limonitic ore was locally available, and O'Kelly suggested that this

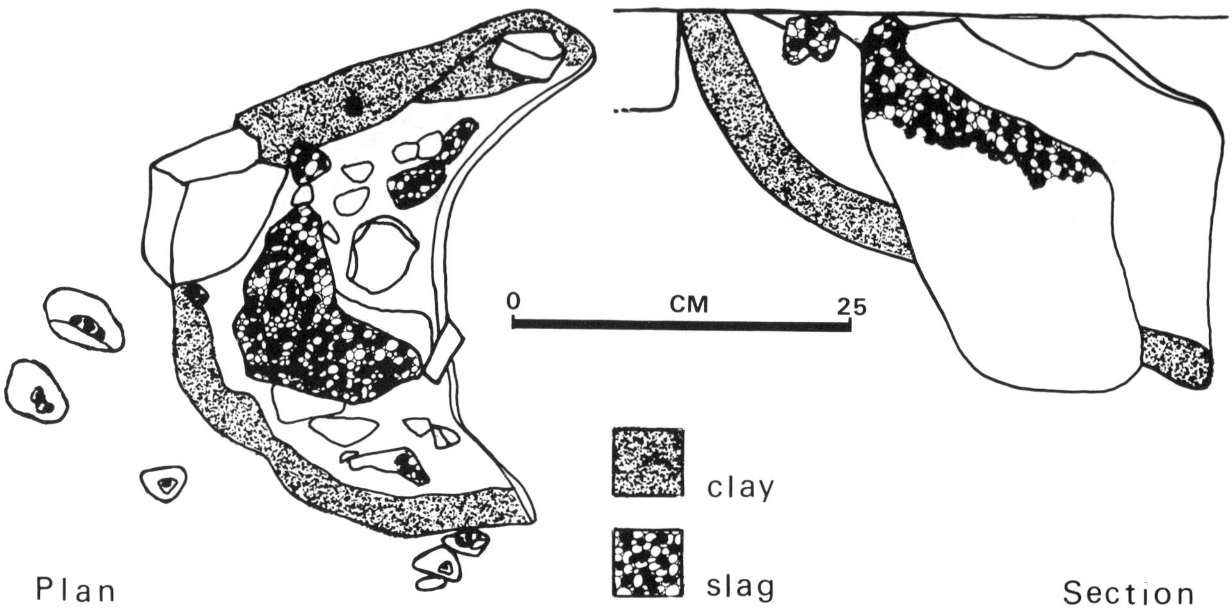

Fig. 6.5.4 Smelting furnace at Lisleagh, Co. Cork (Courtesy of Dr. M. Monk).

accounted for a high level of smelting activity on the site. But the maximum size among the 57 complete 'furnace bottoms', whose median size was 11–14cm, was 18cm in diameter, while the smallest were 9cm in diameter, the thickness varying from 6–11cm. It seems likely, in view of the remains of bellows protectors (see below), that a major part of the activity here was smithing. Equally, at Garannes, where only two small SHB(?)s were found, and where blacksmiths' tongs occurred, activity is more likely to have been smithing rather than smelting (although tongs would also have been needed by smelters).

The later period of occupation at Garryduff I, a ringfort in Co. Cork, produced what the excavator identified as six small bowl furnaces ranging from 30–45cm in diameter and from 23–30cm in depth (O'Kelly 1964, 101). These were accompanied by lumps of slag and of burnt and vitrified clay, and by tuyère fragments. At this site also, lumps of local limonite were found throughout the occupation layers. By experiment, O'Kelly showed that some had been roasted preparatory to smelting (O'Kelly 1964, 103). The site also produced at least 44 'furnace bottoms' in addition to the other slag lumps. Like Ballyvourney, it is difficult to be certain of the nature of activity represented, although a distinctive structure, tentatively identified as as smithing hearth, suggests that here again we have mixed smelting and smithing.

Although no furnace remains were recovered, excavations at Ballyhenry, Co. Antrim (Lynn 1983b), the site produced slag lumps amongst which were found two fragments of a small, unconsolidated iron bloom (total weight 250gm: Plate 6.5.4) and ore, which constitute *prima facie* evidence of smelting on the site. This was discovered during screening of a random selection of slag pieces from various sites in Northern Ireland by X-radiography. This technique is useful in a variety of ways in the study of archaeological assemblages involving iron and ironworking (Scott 1976b), and it likely that wider screening (unfortunately not possible in the course of this present study) would produce further examples of raw blooms. Consolidated blooms have been found at Lough Faughan crannog, Co. Down (Collins 1955) and at Carrigmuirish (Tylecote 1986, 192, Fig. 133) and the Brothers' Cave, both in Co. Waterford (Plate 6.5.5), but they do not constitute good evidence for smelting on those sites, since it seems likely that iron could have been distributed in this form for working up at smithies well removed from the smelting site. Actual ingots have been found at Lissue, Co. Antrim (Bersu 1947; Warner pers. comm.), and at Garannes, Co. Cork (Ó Ríordáin 1942, 103, Fig. 7 nos 441 and 446).

Aside from furnace structures, blooms and hearths, another survival strongly suggestive of iron smelting and smithing is the tuyère through which the furnace was blown. These have been found in three main forms on Irish sites. The first is comparatively small, consisting of a clay tube, with an aperture from 1.0–1.5cm in diameter, and a wall thickness from 2–3cm (Fig. 6.5.7a–d). No complete examples have been found, although tuyères from Arnbathie in Scotland and Glastonbury Lake Village in England (Tylecote 1986, 143, Fig. 87) indicate a range in length between 9 and 15cm. Examples of this type have been found at Cathedral Hill, Armagh (Brown and Harper 1984, 149–150, Fig. 20), Garannes, Co. Cork (Ó Ríordaín 1942, 138, Fig. 25) and Lagore, Co. Westmeath (Hencken 1950, 127, Fig. 60).

The second type of tuyère to be found on Irish sites may be considered to be a much more massive version of the first (Fig. 6.5.7e–g). with diameters ranging from 13–15cm, wall thicknesses from 3–5cm, and apertures around 3cm in diameter. The examples so far noted come from Ballyvollen, Co. Antrim (Williams 1985a, 99), Cathedral Hill, Armagh (Brown and Harper 1984, 150, Fig. 20), Garryduff I (O'Kelly 1964, 100, Fig. 22), Lisleagh, Co. Cork (M. Monk pers. comm.), and Carrigmuirish, Co. Waterford (unpublished: referred to in O'Kelly 1964, 101, note 107, and 100, Fig. 22A). The most complete specimen comes from Ballyvollen and is just over 12cm in length.

The third type of object normally described as a

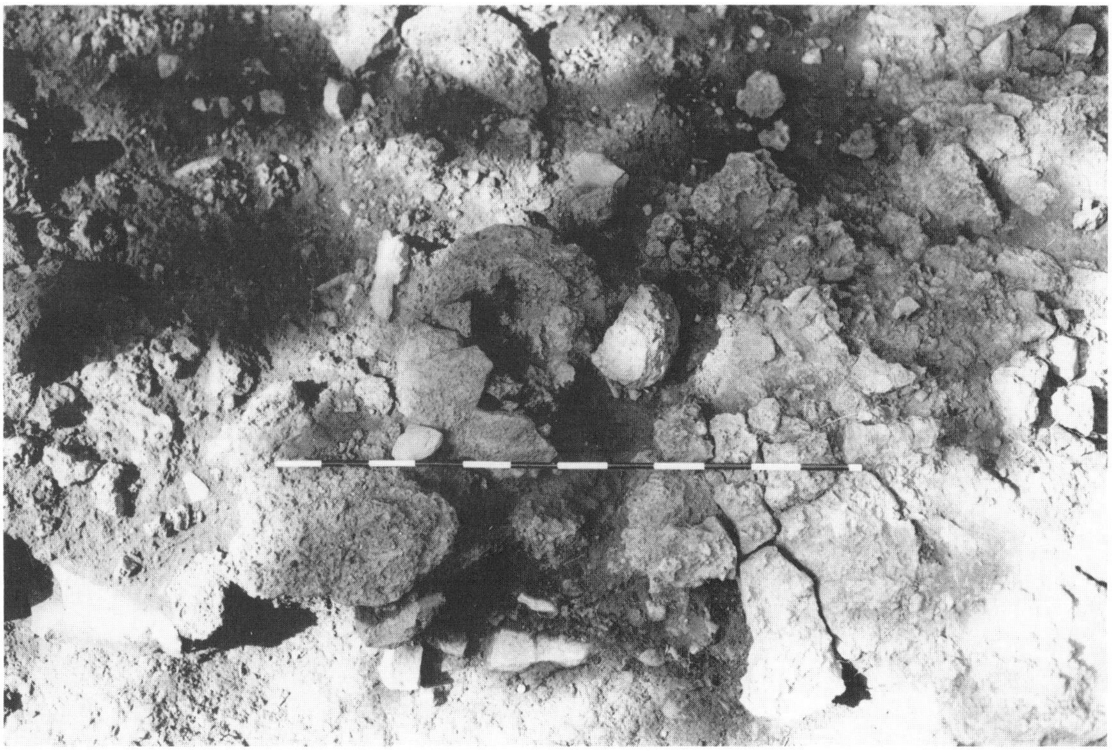

Plate 6.5.1 Dump of slag in the ditch of the ringfort at Lisleagh, Co. Cork (*Courtesy of Dr. M. Monk*).

tuyère[5] to occur in Ireland is a perforated clay disc, either flat or slightly convex in section, from 12–16cm in diameter, 1–4cm in thickness, and with an aperture from 1–3cm in diameter (Fig. 6.5.7h-j). Examples have been recovered from Ballyvourney (O'Kelly 1952, 33, Fig. 3 nos 1–3), Garryduff I (O'Kelly 1964, 100, Fig. 22), and Lisleagh (Monk pers. comm.), all in Co. Cork, from Cathedral Hill, Armagh (Brown and Harper 1984, 150, Fig. 20) , and from Reask, Co. Kerry (Fanning 1981, 116, Fig. 18).

The purpose of a tuyère was to act as a protective nozzle for the air tube from the bellows at the point where it entered the furnace or hearth. The two tubular types of tuyère thus present no problems of identification. The use of the perforated clay discs is less immediately explained, at least in terms of smel-

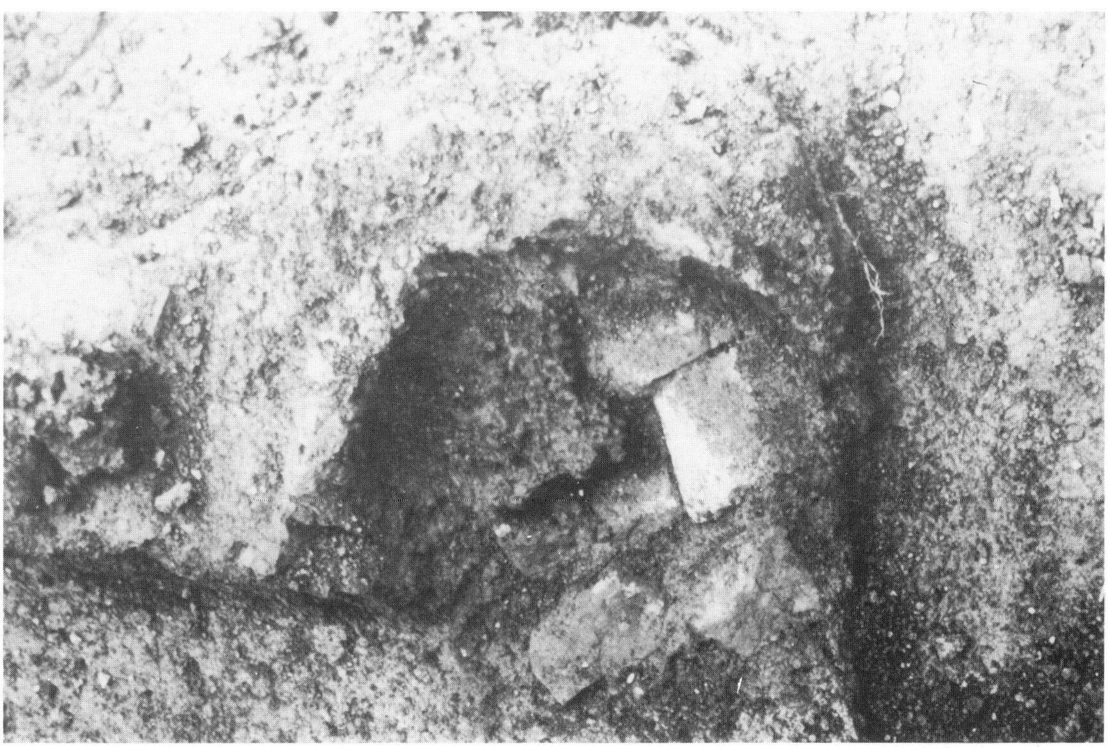

Plate 6.5.2 Smelting furnace at Kilpatrick, Co. Westmeath during excavation (*Courtesy of D.L. Swann*).

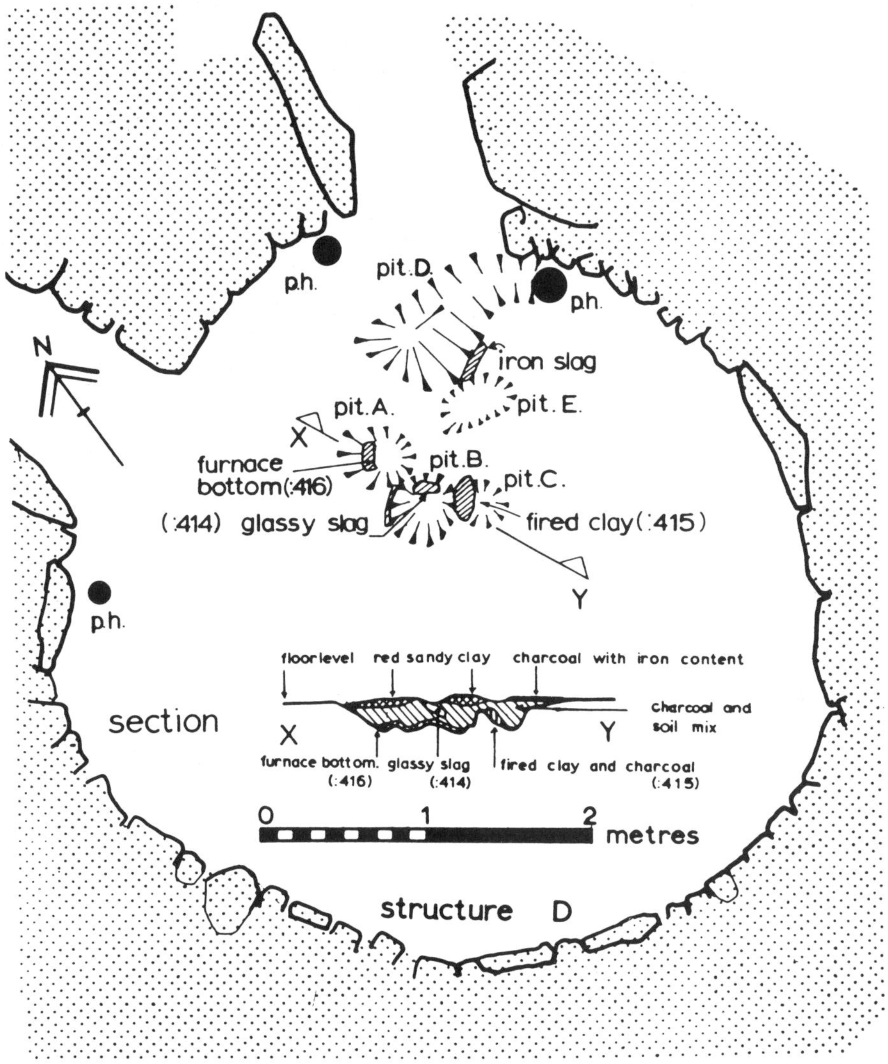

Fig. 6.5.5 Detail of industrial remains in the interior of structure D at Reask, Co. Kerry (after Fanning 1981).

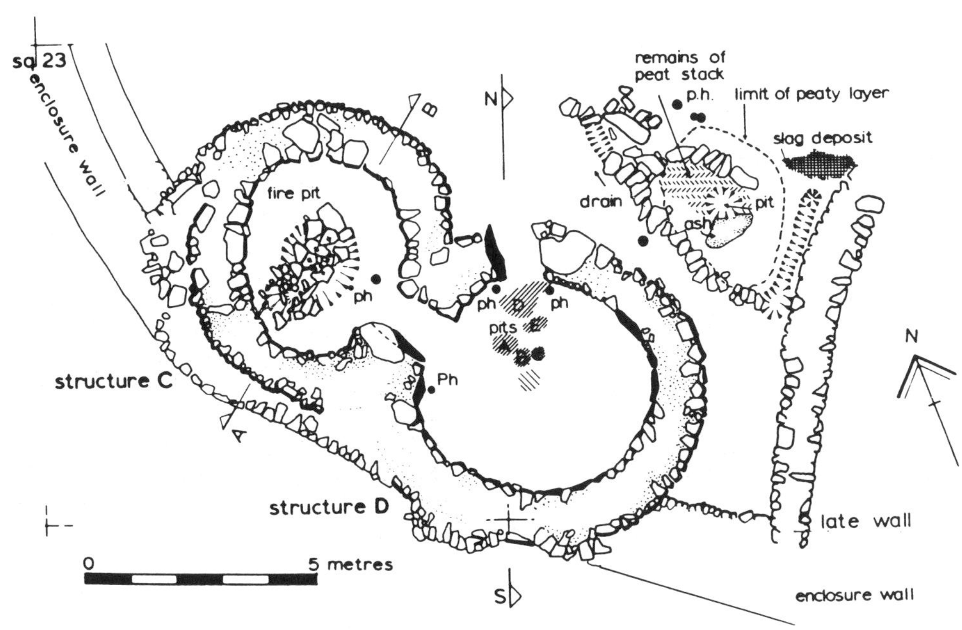

Fig. 6.5.6 Structures C and D at Reask, Co. Kerry, and the peat stack outside (after Fanning 1981).

Plate 6.5.3 Smelting furnace before excavation at Knowth, Co. Louth (*Courtesy of Professor G. Eogan*).

a

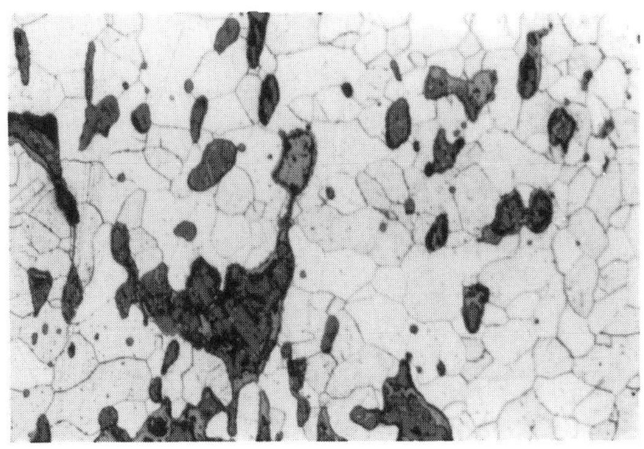

b

Plate 6.5.4 (a) section through the bloom from Ballyhenry, Co. Antrim, (b) microstructure of the bloom showing ferrite and slag x100 (*Courtesy of Dr. C.J. Lynn*).

a

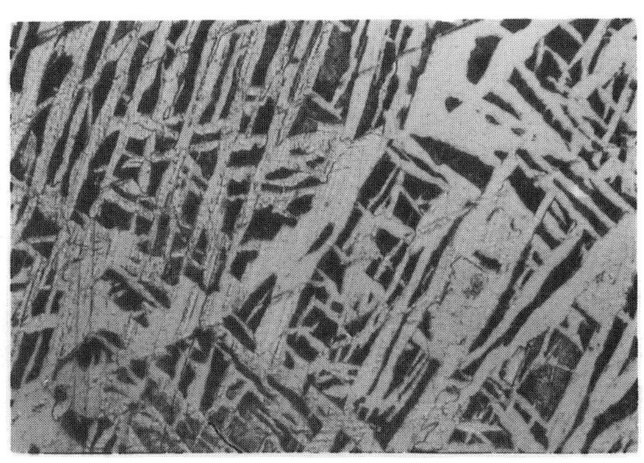

b

Plate 6.5.5 (a) section through the bloom from the Brothers' Cave, Co. Waterford, (b) microstructure of the bloom showing Widmanstätten ferrite and pearlite x100.

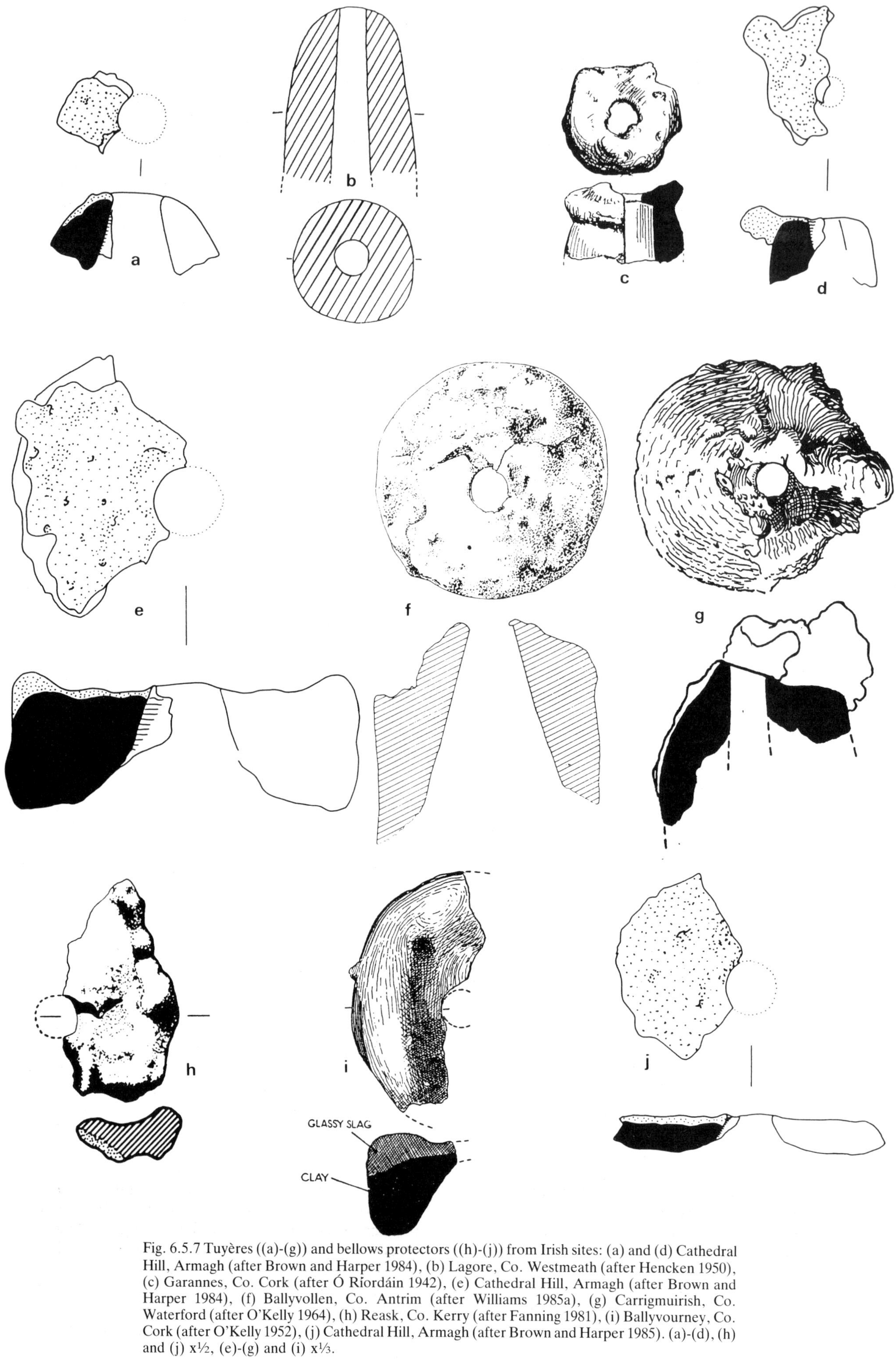

Fig. 6.5.7 Tuyères ((a)-(g)) and bellows protectors ((h)-(j)) from Irish sites: (a) and (d) Cathedral Hill, Armagh (after Brown and Harper 1984), (b) Lagore, Co. Westmeath (after Hencken 1950), (c) Garannes, Co. Cork (after Ó Ríordáin 1942), (e) Cathedral Hill, Armagh (after Brown and Harper 1984), (f) Ballyvollen, Co. Antrim (after Williams 1985a), (g) Carrigmuirish, Co. Waterford (after O'Kelly 1964), (h) Reask, Co. Kerry (after Fanning 1981), (i) Ballyvourney, Co. Cork (after O'Kelly 1952), (j) Cathedral Hill, Armagh (after Brown and Harper 1985). (a)-(d), (h) and (j) x½, (e)-(g) and (i) x⅓.

ting, even though all of the known examples obviously have been subjected to intense heat and are always associated with ironworking remains. The best interpretation is that while they were used, like the tubular tuyères, to protect the bellows tube from intense heat, it was in the open hearth used for bloomsmithing and forging, and not in a smelting furnace. There is a good (if late) illustration of how this would have worked 12th century AD carvings on the stave church at Hyllestadt in Norway which depicts scenes from the story of the hero Sigurd. One of the scenes (Plate 6.5.6) shows the reforging by the smith Regin of the broken sword of Sigurd's father, and in front of the bellows, protecting them from the heat, can be seen a plano-convex block. This is a depiction of the type of stone bellows protectors found at Snaptun in Denmark (Fig. 6.5.8), and at Haithabu (Thomsen 1971, 101ff). Similar finds are reported from La Tène and Roman Europe (Pleiner 1958, 129, Fig. 27). Thus, it seems that the best interpretation for these discs is as part of the equipment of the forge hearth, used to protect the bellows, and not as part of the smelting furnace. It may well be that we can identify these with the *derc* which occurs in texts referring to smithing (Chapter 7.5). A suggested reconstruction is shown in Fig. 6.5.9.

At Clogher, in addition to the smelting furnaces, there was another structure, stone built and clay-lined, and sited beside a massive, flat-topped limestone boulder (Plate 6.5.7). It can be seen that the clay lining runs out against the top course of stones, thus that the entire height of the structure, approximately 50cm, has been preserved (Plate 6.5.8). All around was scattered small slag fragments, and its association with such a distinctly shaped dolerite boulder, whose flat surface would make it a natural anvil, invites the interpretation as a smithing hearth. The quite natural association of smelting and bloomsmithing has been noted often, as for example at Bryn y Castell in north Wales (Crew 1987), and at Skoszyn site 3 in the great iron producing complex in the Holy Cross Mountains of Poland (Bielenin 1987, 46). At Clogher, it seems likely that the small bowl furnaces were used for the smelting of iron, while the larger structure was used both for heating up the bloom for consolidation on the flat surface of the anvil stone, and as a smithing hearth for the forging of artefacts. Despite the surviving differences in form, the bowl smelting furnace and hearth are still, essentially, clay-lined, roughly hemispherical depressions in which heating operations were carried out. A structure seemingly similar to the Clogher smithing hearth was reported by O'Kelly at Garryduff I (O'Kelly 1964, 99f), although not illustrated other than by an uniformative plate. However, its description suggests that it was a smithing hearth too. At both Clogher and Garryduff there thus appears to be differentiation between smelting furnaces and (bloom)smithing hearths.

At Reask, (bloom)smithing and smelting appear again to have been carried on side by side in both structures D and G. It could be suggested that in structure D, smelting was carried on in the three central pits, while that nearest the entrance was used for bloomsmithing. From the way in which structures C and D were joined together, one might be reminded of the 'snail' structure at Bryn y Castell which Crew (1987, 93, and 95, Plate I) interprets as a smithy. However, although there was a large hearth inside structure C at Reask, none of the debris of ironworking, so prominent in conjoined structure D, was found. The peat dump and pit could also have been associated with the preparation of charcoal and the roasting of ore. The bowl furnaces in structure D produced charcoal from oak, hazel, alder, birch and holly, while in structure G, charcoal of oak, birch, willow, holly, ivy and hawthorne was found (Fanning 1981, 164–165). In addition, radiocarbon dates for charcoal taken from structure G gave dates trending towards the 2nd century BC to 3rd century AD, indicating that it derived from peat (Fanning 1981, 164). Peat charcoal is known to have been used for ironworking (cf. the evidence from Wiltrow in Shetland – Tylecote 1986, 140). Given the nature of peat charcoal, and the fact that hardwood charcoals were found in association with ironworking features, there is a strong possibility that different charcoals were being used for different purposes – hardwood charcoal for smelting, peat charcoal for bloomsmithing and forging. At Ballyvourney, O'Kelly found the charcoal of hazel, poplar and willow in association with ironworking debris (O'Kelly 1952, 35).

In addition to Reask and Kilpatrick, other early monastic establishments have produced evidence of ironworking. In the remains of a thatched(?) stone round house on Church Island, Co. Kerry, O'Kelly uncovered a pit, 50cm in diameter and 10cm deep, which he interpreted as the remains of a bowl furnace (O'Kelly 1959, 69f). It contained lumps of slag, and fine charcoal[6]. It seems more likely that this represents smithing activity rather than smelting, as do the remains on Beginish (O'Kelly 1956, 152), Inishkea North (Henry 1945, 141) and the slag found on Skellig Michael (O'Kelly 1973, 3), all in Co. Clare[7]. There certainly need be no argument against smelting on these isolated islands on the grounds of difficulties in transporting ore, since at Church Island, a considerable quantity of slate slabs was brought for the building of the oratory (O'Kelly 1959, 118). Without detailed analytical studies of the slag remains (sect. 6.4), it is impossible to offer a definite opinion, but the scale of activity represented on these sites is more in accord with occasional smithing, rather than smelting.

In view of the literary evidence presented in Chapter 7.6, it is interesting to note that while in the great majority of cases, no special house structures which could be interpreted as forges have been found, ironworking was carried out in substantial structures on a number of sites. Thus, there is an interesting contrast between Ballyvourney, Reask and Church Island on the one hand, and Clogher on the other. This differentiation might just indicate the contrast between specialist establishments and sites where ironworking was carried out as part of a wider domestic economy (cf. Chapter 5.1).

The principal difficulty in interpreting ironworking structures on Irish sites is the probability that a hemispherical hole in the ground showing evidence of intense heat is as likely to have resulted from smithing as from smelting, or from a combination of both. The problems here are perhaps more acute than in other regions where more sophisticated types of smelting furnace were adopted, thus making differentiation between smelting and smithing more clear cut. And

Plate 6.5.6 Detail of the forge of Regin from the carved wooden panel at Hyllestadt, Norway (after Thomsen 1971).

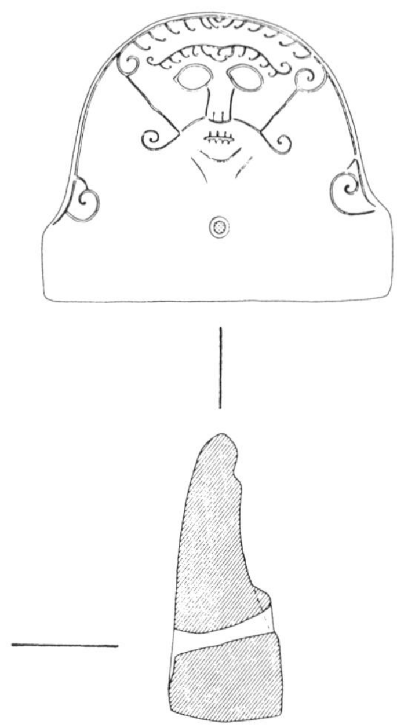

Fig. 6.5.8 Decorated stone 'bellows protector' from Snaptun, Denmark (after Thomsen 1971).

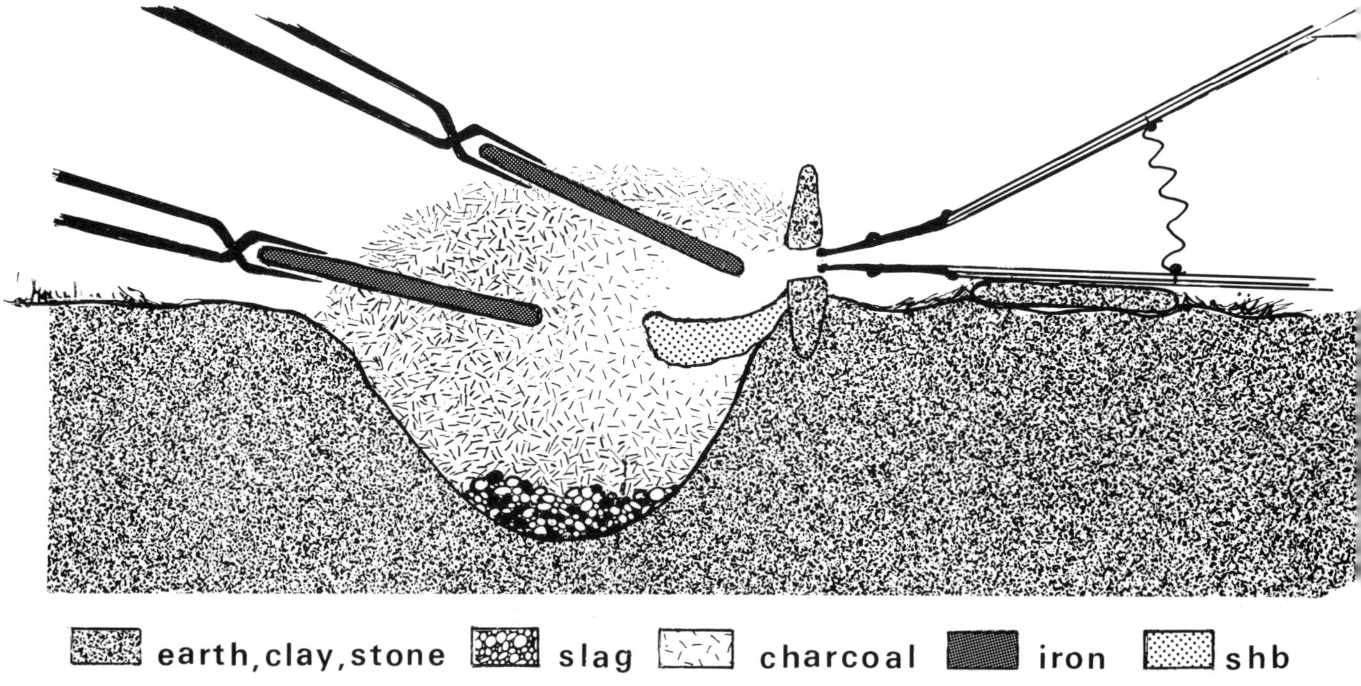

Fig. 6.5.9 Reconstruction of an Irish smithing hearth using a perforated clay disc as a 'bellows protector' (not to scale).

Plate 6.5.7 The Clogher, Co. Tyrone smelting and smithing area, showing the smithing hearth and anvil stone during excavation (*Courtesy of R.B. Warner*).

Plate 6.5.8 The Clogher smithing hearth in section (*Courtesy of R.B. Warner*).

the fact that smelting and smithing slags form a continuum will not make interpretation any easier in the future.

* * *

Notes to Chapter 6

1 I am indebted to Dr. Peter Kresten for providing me with a copy of his SLAGNORM program for computing normative compositions, and for discussing results with me.

2 The silicate group comprises minerals like forsterite, fayalite, tephroite, enstatite, rhodonite and hercynite, while the glass group includes apatite, kalsilite, vivianite, anorthite, mullite, gehlenite. The iron oxide phase is calculated as wüstite.

3 Brown and Harper (1984, 151) quite incorrectly use the term 'tap slag' to describe slags from Cathedral Hill, Armagh, and there appears to be no evidence for flow-patterned slag anywhere on the site.

4 The dimensions and structure given here for O'Kelly's furnaces from his unpublished experiments derive from conversations with the author and from a brief film record of the work which augments the short (1961a) published report. I am grateful to Dr. M. Monk for supplying a copy of the film of O'Kelly's experiments.

5 Tylecote (1987, 117) suggested that such discs are the broken off ends of the more massive tubular tuyères. However, hand examination of the Irish examples shows that this is not the case, at least in Ireland.

6 *Pace* Tylecote (1986, 187), O'Kelly did not find 250kg charcoal at Church Island, but merely conjectured that about this amount had been consumed to account for the amount of 'slag' found on the site (O'Kelly 1959, 116).

7 At Nendrum, Co. Down (Lawlor 1925) ironworking may have taken place in conjunction with the other crafts practised there (non-ferrous metalworking, stone carving), but the evidence is too poorly reported.

Note added in press

In his discussion of the perforated clay discs from Ballyvourney (here described as 'bellows protectors'), O'Kelly suggested that they were the covers for furnaces. Although comparison with the Hyllestadt illustration and Snaptun find indicates that these discs were indeed used in the vertical to protect bellows from the forge fire, recent experimental work by the author suggests that they could also have been used with a bowl furnace to cover the aperture at the top of a low domed superstructure (of the form illustrated in Fig. 6.5.1). One problem with the bowl furnace is the creation of a 'dead zone' opposite the air inlet. By improving the insulation of the furnace (in this case by covering all but the top in sand), and by placing a removable cover over the mouth of the superstructure, the size of this 'dead zone' can be significantly diminished, thus increasing the volume in which the ore can be held at heat under reducing conditions.

A further point to note is that if the wall of the superstructure is in the region of 8–10 cm in thickness, then the hole through it, made as the air inlet, is long enough to remove the need for a separate tubular tuyère provided a metal nozzle is used on the bellows.

In terms of the potential survival of superstructure remains, it is interesting to note that after five smelts, the outer thickness of the wall of the experimental furnace dome was not sufficiently well fired to resist erosion by rain. This suggests that only the inner 2–3 cm of wall would resist weathering, and thus would occur only in very fragmentary condition, making identification on site very difficult.

While only indicative at present, this work (see also p. 212 below) suggests that, as with the bowl furnace itself, perforated clay discs probably served more than one function. It also suggests that the absence of tubular tuyères from a smelting site should not be surprising, and that greater effort should be made to locate clay fragments that might belong to a superstructure, particularly those from the air-inlet area.

CHAPTER 7

The Evidence of Language and Literature

nach indéuin benar *for-cain in cách no-da-ben* *nach sí for-canar*	Any anvil which is struck teaches he who strikes it; it is not taught.

(Old Irish, late 8th century: Carney 1958, 19)

Bristear ór le iarann	Gold is broken with iron
A bheith a' gabháil d'ord ar iarann fuar, a bheith a' tabhairt comhairle dho mhnaoi bhuirb.	To be giving advice to a headstrong woman is the same as hammering on cold iron.
Buail an t-iarann fhad 's tá sé te.	Strike the iron while it is hot.

(Modern Irish: Ó Máille 1952)

7.1 – Introduction

Although we possess a wealth of references, usually oblique, to metals and metalworking in early Irish texts, no writer appears ever to have sat down to describe the ways in which he or his contemporaries worked metal. Sadly, no Irish Theophilus was moved to record his arts for the instruction of future generations and to the glory of God. One can only imagine by how much richer our understanding of the exploitation of metals in Ireland would be had, say, the makers of the Ardagh chalice chosen to write down their methods, in the same way that Theophilus (Book III.xxiv: Dodwell 1961, 76f) described his. Likewise, no Cormac produced a glossary of the names of metals or metalworking processes in his own tongue. All we possess of this type of text consists of derivative fragments such as an uncomprehending early modern Irish *précis* of some metallurgical sections of the *De Mineralia* of Albertus Magnus and the marginal glosses to the early 17th century *Zoilomastix* of Philip O'Sullivan Beare[1], neither of which are of any real value to the student of early Irish metallurgical terminology. We are thus forced to examine every term not only in individual textual contexts, but also with an eye to the general level of technical comprehension displayed by early scribes, in order to determine general semantic trends.

In the law tract *Lebor Aicle*, early jurists have left us with two sections, *Blaí Ord Indeóin* in whose composition we may see the work of a man very much in touch with the working of the Irish blacksmith's forge, and *Bla Miand Midchlais* which is a scrappy and rather opaque piece on mining. These texts are discussed in detail below and, while that on smithing is of some assistance with terminology, it is of most importance when describing (albeit obliquely) manning and working practices. As we shall see later, it also provides corroboration of deductions made from examination of the evidence of archaeological excavation. Against this, while the passage on mining helps somewhat with terminology, it is of scant assistance with the little-known archaeology of Irish metal mining.

It is not my intention to attempt to present here a comprehensive and chronologically-ordered catalogue of references to iron and ironworking, but rather to use selected examples to indicate the immense potential of the sources. Indeed, the reader will observe that I have strayed well outside the chronological limits set for the survey of the archaeological material, by including extracts from texts dating up to the Early Modern Irish period (ca AD 1400). Similarly, the glossary presented as part of the chapter is to be treated only as a framework for future research and is not claimed to be definitive. I have attempted to follow the long-neglected, but useful method adopted by Sullivan in his introduction to O'Curry's *Manners and Customs of the Ancient Irish*, namely that of comparing the semantic fields of various terms with metallurgical theory and the practicalities of the blacksmith's art. An obvious problem of such an approach, but one to which, however, I can do no more than draw attention, is that of the complex history of individual terms. It is evident that the words for 'iron', 'bellows', 'smith', 'a forge' and 'anvil' all belong to the earliest stratum of Goidelic ironworking terminology. But charting the further expansion of the vocabulary is a task for others. While treating all of the recognisable words together, I am fully aware that they do not form a chronologically coherent group, and that various sections occupy different facies of the linguistic stratigraphy. Nevertheless, a start has to be made somewhere. It is also necessary to enter a *caveat* on my own interpretations of textual examples and adjustments to the translations offered in *DIL* and other sources which have been made from a metallurgical standpoint. While it will always be essential to avoid contravention of the laws of science in deriving meaning (and thus avoiding such unfortunate renderings as 're-*smelting*'), it will be equally necessary to underscore any contradictions between linguistic and metallurgical accuracy, and to accept that in the context of early

Irish literature, the two cannot always be satisfactorily reconciled (e.g. Scott 1982, 1983a, 1989b).

One major difficulty in the reconstruction of the early Irish vocabulary of metals and metalworking is our uncertainty as to the place of individual early Irish metallurgical terms in relation to general language, and the degree of understanding with which they were used, indeed of the level of interest in terminological accuracy of the scribes who wrote our texts. While it may be readily demonstrated that sections of the Laws dealing with such activities as smithing (Scott 1983b: sect. 7.5), milling (Mac Eoin 1982), brewing (Binchy 1982), and with household possessions (Lynn pers. comm.) are soundly based on observation, we elsewhere face serious confusion throughout the literature, even within a few lines of one text. To take but one example, the Irish redactor of Lucan's *Pharsalia* (*CCath*) gets himself into a terrible tangle over the composition of electrum. First, on line 4669 he describes it as:

> . . . *electair .i. do thinni commascda airgit ocus credhuma* . . .
>
> . . . of electrum, that is, a mixture of silver and *créduma* . . . (a copper-alloy: Scott 1981b),

but eleven lines later (line 4680) his formula is correct:

> . . . *n-electarda do commusc oir 7 airgitt óengil*. . .
>
> . . . of electrum, a mixture of gold and pale silver . . .

Although late for the purposes of our survey, this text is fairly typical in its use of metallurgical terminology in that it leaves us quite unsure as to whether the redactor had a proper understanding of at least some terms, or achieved accuracy through the working of the law of averages.

A further problem is that a text, as we have it now, may have passed through a number of hands. Thus, while one scribe may have been thoroughly competent in his knowledge of metalworking, we are, as Greene (1954, 26) noted '. . . at the mercy of scribes who might be careless, or given to archaisms, or to wholesale re-editing of their material.'. With regard to this last point, we must remember that words changed meanings through time. Early Irish *findbruinne* probably referred originally to a silver-coloured surface (Scott 1981b, 156) but by the early 17th century referred to brass. We must also bear in mind that the majority of storytellers were primarily interested in metal artefacts as props on the sets on which their characters played out their allotted roles. Weapons and ornaments were usually described in a way that let the audience understand that their possession was proof of status, but told next to nothing of their manufacture. Thus, the poets and the compilers of heroic tales deployed lavish vocabulary in describing the beauty of a weapon and the skills with which it was wielded, but glossed over the detail of manufacture. For example, in the 21 stanzas of the poem *Mo chen a chlaidib Cherbaill* ('The song of the sword of Cerball': *LL* 6793ff), the poet tells of the value of the sword, praises the ease with which it removes heads, elaborates on the number of kings it has been used to kill, marvels at how frequently it has been used in battle and the length of its service, and reaches a crescendo with its role in the battle of Belach Mughna. But nowhere does he so much as hint at the the material from which this obviously splendid weapon was made! Equally, the lay of the sword of Oscar (*Duan. Finn* I.49ff) extols the marvellous history of this weapon, reputedly far-travelled from Troy to Rome to Ireland, again without any reference to the history of its manufacture[2].

Our knowledge of the early Irish terminology of metals comes from writers who, by and large, evidently had no special interest in recording subtleties of metallurgical meaning. The loss can be illustrated from a modern example, that of the Greenland Eskimos who use many different words for snow – a material in whose state and properties they have much interest (Service 1962, 15f). Similarly, the modern farmer not only distinguishes between numerous breeds of cattle, but also between the sexes, meat and milk potential, and age and breeding characteristics. As for the blacksmith, he distinguishes, for example, between a wide range of heating colours on which he makes judgements on the progress of his work. To the layman, snow is white and cold, cows give milk and meat and glowing iron is red- or white-hot. There undoubtedly was a jargon of metalworking by which the details of technique were distinguished from one another between workers, but this proletarian element is not often evident in the vocabulary of the *literati*. Nevertheless, we must also be aware that where the import of a passage is obscure to us, it could well have been perfectly comprehensible to its writer's contemporaries, and that it is our comprehension that is inadequate.

As will become apparent in this chapter, the metallographic assessment of Irish ironworking skills (Chapters 3–5) is often in marked contrast to the technical mastery proclaimed by so many early texts. The early native smiths were undoubtedly competent *shapers* of ferrous metal, but showed limited ability in the period under review either to select stock or to manipulate its properties. It well may be that these skills were guarded by the master-smiths who serviced a primarily aristocratic market, thus limiting the currency of some jargon terms. Those lower down the order, while producing usable artefacts, nevertheless were either unable to control the smelting operation or to apply treatment to modify the properties of their stock. Such an interpretation accords to some degree with the stratification of expertise suggested by the archaeological and metallurgical evidence (Chapter 5.2) Despite this, a not inconsiderable body of terminology may be identified. Furthermore, virtually all of that connected with ironworking apears to be Celtic in origin and form. Given that Irish scribes developed the habit of equipping their heroes with weapons modelled on more prestigious Scandinavian models (cf. Mallory 1982), it might be suggested that their references to sophisticated techniques such as quenching and tempering were the result of observing Viking smiths who undoubtedly worked here from the later 10th century AD onwards.

As with the borrowed terminology of aspects of

shipping (another field of notable Scandinavian superiority over the Irish: e.g. Sommerfelt 1975), we might expect to find some metallurgical loans or calques. But apart from de Vries' unconvincing offering of a Mid. Ir. form *staílinn* from O.Norse *stál* ('stahl, waffe': de Vries 1962, 542: seemingly unattested elsewhere), none are forthcoming. In terms of tools, the verb *slípaid* 'smoothes, polishes, whets' (sect. 7.7 under *límaid* and forms from it), is an O. Norse borrowing (*Vendr. Lex.* S-135). Scandinavian smiths in an urban industrial setting were in constant contact with the mainstream of European metal technology as may be seen from the quality of their edged tools and weapons. In contrast, the majority of Irish smiths were apparently cut off from this and locked into the relative isolation of rural life. The opportunities for observing, questioning and, above all, practising under the eye of an adept would have been scarce. Thus, while a few master-craftsmen could produce weapons which gave rise to, or enhanced, genuinely native traditions of superior weaponry, and which incorporated technology imperfectly recorded in native tradition (such as Celtchair's spear and Cú Chulainn's sword – see sect. 7.6), they were in a minority. In other words, the few well-made weapons were far outnumbered by a mass of artefacts made to less exacting standards. Their frequency in the archaeological record relative to that of their technological inferiors may have been further reduced not only by the unfortunate ease with which iron corrodes, but also by their status value which ensured a long stay in the realms of the living. Some evidence of the preservation of treasured weapons as heirlooms comes in the tale *Ceart Claidib Cormaic*, where Socht possessed what he claimed to be the sword of Cú Chulainn, of which it was said:

> *Audacht ceneoil 7 aitri 7 seanaitri leosom in claideb sin.*
>
> They held this sword to be a tribal heirloom both of fathers and grandfathers.
> Meyer 1891, 199.59

Thus, while the terminology of excellence might be known and partially understood by laymen, access to the technology was by no means so easy. A body of vocabulary with strong metallurgical connections did exist in general and fairly superficial usage, and we must now look to its possible origins in the jargon of early Irish metalworkers.

It is instructive to compare the list of early Irish metallurgical terms (sect. 7.7) with those to be found in a modern glossary of metallurgy. For example, the majority of terms which are listed in *A Dictionary of Physical Metallurgy* (Tyrkiel 1977) would probably be quite opaque to the layman: they belong to the field of *specialised metallurgical language* (cf. Baldinger 1980, 120ff). Terms like *lattice*, *creep*, *solid solution* will, however, be familiar to other scientists since the phenomena they define are not confined to metallurgy. Specialised metallurgical language is thus a sub-set of a wider *specialised scientific language* which itself is a dialect of *general language*. Languages invariably contain specialist components comprehensible only to initiates and this is by no means confined to the sciences or humanities. The specialist languages of the archaeologist or computer scientist will be as difficult for the layman to follow as will be those of the motor mechanic or baker. While individual words may pass from specialist dialects into general speech, their specialist meanings will not necessarily be completely carried over. One result of the increasing modern sophistication of the mass media in the developed world (cf. Toffler 1981) is that the layman is exposed to scientific terminology at a relatively superficial level and becomes accustomed to use terms like *electricity*, *television*, *computer*, *X-ray*, *antibiotic*, in everyday speech. It is, however, usage conditioned by observation of effects rather than any deep understanding of mechanism. The person who switches on the television is impressed by the appearance of image and of sound, not directly by the antics of the electron. It is this familiarity with *effect* and not *cause* which allows the superficial use of specialist terminology, but would prevent comprehension of an exchange couched in the specialist language from which it has been borrowed. Similarly, we might, at one level, view the early Irish scribes as akin to the televiewer in their knowledge of smithcraft. They might well have learned that an iron sword somehow improved in property if heated and rapidly cooled, but not have any clear idea of precisely how this was achieved. Familiarity with the *effect* – a superior cutting edge – for them would not be familiarity with the *cause*, leading to an uncritical use of terminology (Scott 1983a, 20f).

The modern metallurgist can understand and describe the changes that occur in his materials through modern instrumentation and theory, at the atomic level. The blacksmith of the 1st millennium AD did not have access to this level of knowledge, but worked purely empirically. Even where explanation of behaviour was attempted outside the classical world, it was usually paralleled by religious mystique (e.g. Eliade 1978, 43ff; Forbes 1964, 52f). In trying to reconstruct early metallurgical terminology and recapture the sense of its components, we run the risk of expecting a degree of interpretative sophistication that simply could not have existed. Of the fifty-eight headwords, plus the associated terms given in the Glossary (this does not include those words whose usage seems to be restricted to non-ferrous metals and their working – for some of these, see Scott 1981b,c), only some twenty appear to have been restricted in usage to craftwork in general, and ironworking in particular. Of these, only *gobae* 'blacksmith' and *íarn* 'iron' seem to be solely the property of the ironworker. In fact, most of the rest of the words listed have such wide semantic fields that it is difficult properly to describe them as belonging to specialist language at all. The prominent feature of many of the words included in the early Irish vocabulary of ironworking is their wide semantic fields. In contrast to this, one would be surprised if more than a few of the nearly 2300 terms listed in *A Dictionary of Physical Metallurgy* could be properly defined as belonging to general language.

Irish *teine* encompasses not only 'fire' and 'fireplace', but also 'forge fire', 'hearth' (as in 'hearth and home'), and 'dwelling'. Similarly, *bolg* includes in its semantic field 'bellows', 'bag, satchell, sack', 'belly', 'bubble', 'blister', 'berry, bud', and 'ball'. The first represents the development from the initial meaning 'fire' to take in not only 'flames' but also the structure on which they played and, by further extension, the domicile of which it was in turn the focus. Again, the word for 'bellows' seems to centre its meaning on 'a closed container'. The semantic stratigraphy of these and other words shows that their ferrous metallurgical meanings are secondary extensions of their semantic fields. The concept of fire as an agent of transmutation is at least as old as the culinary arts, and the inclusion of the metallurgical meaning 'smelts, refines(?)' in the semantic field of *berbaid* comes as no surprise. But this development is certainly to be seen as pre-dating the beginnings of iron-use among speakers of the Goidelic dialect.

Fire was the agent which transformed soft, sticky clay into hard ceramic, baked dough into bread, fused sand into glass and enamel, melted wax and hardened wood. Above all, it performed the transmutation of solid, friable 'stone' into malleable metal and then softened and melted that metal. It is easy then to see how the words used to denote non-ferrous metallurgical smelting and melting processes were borrowed from the wider vocabulary of cooking. The implication of the molten state carried through into ferrous metallurgy, although the original users would have had no difficulty in distinguishing, through context, between *berbaid* 'cooks, boils', *berbaid* 'smelts/melts copper' and *berbaid* 'smelts iron'. Just as the goldworker and copperworker took existing words which expressed ideas predating metal use to encompass specialised aspects of their crafts, so the later-emerging ironworkers simply used their terminology to describe common techniques, only having to look elsewhere for terms to describe new processes.

The evolution of pre-ferrous metallurgical vocabulary involved the shift of familiar terms into the language of the specialist (Scott 1983, 22f). This process may be observed in the uptake of words for 'bellows' in both IE and non-IE languages. Irish *bolg* 'bellows (bag, belly etc)' corresponds semantically to Lat. *follis* 'bellows (animal skin, skin container)', Gk. ἀσκός 'bellows (wine skin)', Indo-Aryan *dṛ́ti* and *bhástrā* 'bellows (leather bag, hide)'. The Chinese word for 'bellows' was *tho*, whose character is noted by Needham as being closely related to that for *nang* 'skin bag' (1970, 157 and n. 9). Similarly, we might note the analogous terms for 'smelter' in Skt. *dhmātṛ́* and O.Icel. *bráðna* (cf. Eng. *breath*), both of which are cognate with verbs which in other languages mean 'breathe, blow etc.'. These semantic parallels do not in any way imply a common source for the technology they denote, but rather a common approach to the solution of problems. The primary purpose of the bellows is to blow air, and what better way to relieve the lungs of blowing the fire than to force the air from a flexible container, duplicating the actions of the diaphragm? The borrowing of existing tools for metallurgical use is surely parallelled by the borrowing of their names into specialised language. The words will have remained the same in form, but their semantic fields will have been widened in consequence. In general speech, they would have been quite comprehensible, but in the specialist language of the workshop their new meanings presumably were opaque to the uninitiated.

Unlike the modern situation where specialist terminology (containing a significant proportion of contrived words) diffuses out from specialist language into general speech with little traffic in the opposite direction, much of the early specialist terminology arose from the appropriation of ordinary words for specialist usage. In the Goidelic dialect, it is only the names of metals and the words for smith and artificer that have restricted semantic fields in comparison with the words denoting processes[3]. Some of the concepts embraced by early metallurgical terms are older than metallurgy itself – especially the role of fire as transmuter. And it is thus quite understandable that terms originally employed in cooking came to be used to describe metallurgical processes. Thus, at least some problems of interpretation are undoubtedly caused through semantic shift and the widening of the semantic field – the use of terminology changing and expanding with circumstances. This undoubtedly has led to words used for one particular metal or process in one period to come to mean something different at a later date. We may consider here the original sense behind Greek μέταλλον 'searching/prospecting' which ultimately came to mean 'metal' (cf. Lat. *metallum*), having first narrowed down on 'a mine' (Halleux 1974, 19f). Similarly, Gk. χαλκεῖς developed in meaning from 'metalworker, copperworker' (χαλκός 'copper') to 'blacksmith' (*LS* 1973). Other terms specifically relating to ferrous metallurgical processes, such as those relating to manipulation of hardness, need not necessarily have their origins in ferrous metallurgical periods, since fire was known both to harden wood and ceramics and to soften chert and copper alloy, making them easier to work.

I have proposed elsewhere (Scott 1983a) a schematic development of metallurgical vocabulary from pre-metallurgical times, summarised in Tables 7.1.1–7.1.4. We may envisage a group moving from stone-use, through no-ferrous metal use, to ferrous metallurgy. In the transition from stone- to metal-use, while there would be a need to have words for the new materials, most of the words for tool types and pro-

Table 7.1.1

Implement	general purpose definitions
Anvil[1]	a flat and unyielding working surface on which operations involving hammering and grinding can take place
hammer	a tool for percussive shaping
bellows[2]	a device including a flexible air-bag for blowing a jet of air into a fire to make it hotter
furnace	a closed oven with an air inlet, used for baking bread and firing pots
(crucible) pot	a fired clay container for liquids
tongs	a hand-held gripping and lifting device
charcoal[3]	a fuel made by "baking" wood

[1] Old Irish *indeoin* = "anvil" also refers to the lower millstone in a rotary mill: cf. Armenian *sal, i, iv* = "flint, flagstone" (modern Armenian = "anvil"); also the similarity of meaning between Skt. *dṛsad*, "a stone on which grain was pounded", and its later meaning of "lower millstone".

[2] Cf. the list of the semantic parallels given above; also, in passing, Skt. *dhāmtṛ́* "smelter" (lit. "blower").

[3] It would be of great interest to see whether charcoal-burning could be shown to be a pre- rather than post-metallurgical development.

Table 7.1.2

material/ process	empiric definition
native metal	a shiny type of stone which does not chip when hammered, but can be "moulded" gradually like clay: melts and flows like water when heated in a hot fire
metal ore	a special type of stone, often brightly coloured; breaks up when hammered; can be converted into the shiny, new type of stone
smelting	converting "ore" into "metal" by strong heating; involves the production of liquids, one of which solidifies like ice as "metal", the other as "stone"
melting	using heat to turn "metal" into a liquid like water which can then be poured into containers
refining[1]	melting metal by strong heating so that impurities float to the surface and can be removed. This concentrates its "desirable properties"
casting	pouring liquid "metal" into specially shaped containers and allowing it to solidify into the shape of the container
smelting furnace	a closed hearth like a cooking or pottery oven using charcoal as fuel and powered by bellows
shaping	hammering a piece of "metal" to change its shape without chipping pieces off

[1] An obvious and pre-metallurgical example is the production of salt by evaporation of seawater (cf. Old Irish *murluaithé* = "sea ashes" [lit.] = "salt").

cesses would already exist (Table 7.1.1). As our hypothetical group made more progress in their metallurgy, some new words would have been required, but again, many could have been modified from existing vocabulary (Table 7.1.2). An area in which new words would obviously have come into use would be specific metal names (Table 7.1.3). By the time the group discovered iron metallurgy, virtually all of the terminology required to communicate metallurgical information would have been acquired (Table 7.1.4). While this is a simplistic model, which pays no attention to the fact that languages are never static, it nevertheless illustrates how, in a pre-scientific context, specialist jargon could quite easily have been perfectly useful as a tool of the metalworker, while appearing diffuse and confusing to the modern metallurgist. The development of metallurgical vocabularies is, like the development of technology itself, a *cumulative* process whose rate of progress will have been determined by rate of technological development among different language groups. Certainly, the development of metalworking vocabulary in Goidelic (and, one suspects, in other languages: cf. Pleiner 1971b on Vedic terminology and Bailey 1977 on Khotan Saka terminology) consists principally of the imposition of specialist roles onto parts of the semantic fields of terms from general language, and of the widening of semantic fields to accommodate need.

Table 7.1.3

material/process	empiric definition
arsenic	a white smoke which boils off like steam from some copper when it is melted; copper which gives off this white smoke is harder than other coppers. Copper can be softened by its loss
silver	a white, shiny metal, not useful for making tools, but ornaments
tin	a white metal that is easier to melt than copper; it is not useful for making tools, but hardens copper when mixed with it
lead	a heavy, soft, silvery-grey metal, which melts like tin; it is not useful for making tools
alloy	a mixing together of two or more molten metals which produces a new metal, often of superior qualities
annealing	softening metal by heating[1]

[1] Cf. the use of heat to make chert easier to work.

Table 7.1.4

material/process	empiric definition
iron	a hard, silvery metal that cannot be melted in the fire
iron smelting	the conversion of "iron stone" to metal
bloom	the mass of impurity left by the smelting process; if this is hammered at heat, the dross gradually breaks away to leave a mass of iron
welding	"joining" two pieces of iron by hammering them together at white heat
carburising	long heating of iron in the forge fire[1] to increase purity and thus make the metal tougher
steel	the purified, toughened iron produced by the above process; it holds a cutting-edge better than "impure iron"
quenching	making "steel" exceptionally hard by immersing it, red hot, in liquid
tempering	making quenched steel more flexible by "annealing"

[1] Heating to harden would also have been known from the firing of ceramics and "hardening" of wood.'

Whereas today's technocracy feels unconstrained in the creation of specialist vocabulary (and we might take as but one example the continuing systematisation of chemical nomenclature[4]), early workers adapted existing terms from general speech for their particular needs. It is this evolution, coupled with the necessarily empirical approach to explanation of mechanism which makes the terminology so difficult for us to deal with today. There is now little excuse for the scientist who fails to make proper use of terminology to describe well-observed phenomena. But when terminology was at best often ambiguous in the absence of scientific theory and method, and where writer and worker were in much closer communion, we must be prepared to accept that the *literati* might often have understood the language of the crafts better than it would seem, their treatment being the result of interest in effect rather than accuracy in their presentations. In this light, it is necessary to work, not to bring their terminology into line with modern perception, but rather to find ways of sharing their understanding.

* * *

7.2 – Mining and smelting

Although the ubiquity of iron artefacts and the debris of iron smelting and smithing operations demonstrates the antiquity of iron use in Ireland, our knowledge of the winning of iron ore is scant indeed (Chapter 6.3). The bulk of our knowledge of mining (and then it is principally of localities) in Ireland as a whole, comes from two periods (both outside the chronological scope of this survey), that of the Elizabethan entrepreneurs in the earlier seventeenth century (Scott 1985), and the later 18th and earlier 19th century industrial exploitation. In the 19th and early 20th centuries, the interbasaltic laterites of Co. Antrim were extensively mined for export, and a wide network of workings with traces of their associated transport system are still visible (Wilson 1964). Although now much decayed and dangerous, they probably represent the most extensive traces of iron mining anywhere in the country. Some traces of the 18th and 19th century working survives in Wicklow, but of the period before the advent of the English interest displayed by Boyle, Petty and their contemporaries, we know virtually nothing. Textual evidence is scant and often obscure, and those passages that do refer to mining are mostly of a general nature, and not confined to the winning of iron ore.

The main resource areas that would have been available to the early Irish have already been discussed in Chapter 6 and the literature has little to add in this respect. As we have already noted, the paucity of physical evidence for prehistoric and early mining is unremarkable in view of the intensity and spread of later activity. The literature, however, adds little; indeed, the one surviving section on mining in the laws probably gives rise to more problems than it solves! One may understand the superficial prominence given to at least some products of the blacksmith's art, but equally in the absence (noted by Greene – 1954, 26) of proletarian literature one must appreciate that the unglamorous siliceous ancestry of a hero's sword would be of little interest to the compilers of tales.

The 11th century *Lebor na Cert* ('Book of Rights': *BR*^{2}) purports to list the tributes due to the Connacht dynasty, and the stipends which they in turn would disburse. As a historical record, it is of little value, but embedded in the extensive lists is the tribute supposedly expected from an obscure group known as *Na Cuirc* which stands out from all others. Instead of giving to their overlord fine weapons, garments or jewellery, the *Cuirc* were to give iron blooms. In fact, the *Cuirc* are the only group to have been noted down for tribute in other than finished artefacts. The prose section reads:

> *Ceathracha ar chéd bó 7 seacht caecait caer iaraind, caeca ar trí cét muc 7 caeca ar trí céd dam óna Corcaib indsin . . .*
> *BR*^{2} 48.681f

> A hundred and forty cows and seven times fifty blooms of iron, three hundred and fifty swine and three hundred and fifty oxen, that is from the Cuirc . . .

while the equivalent verses are:

> *Airdchís na Corc cen chruaidi*
> *do thobairt cach aen-uairi*
> *do ríg Maigi hAe na n-each*
> *seacht fichit bó, ní hainbreath.*
>
> *Seacht caeca do chaeraib iairn*
> *seacht caeca muc co mór-gliaid*
> *seacht coeca dam, dílsi smacht*
> *do-bearar do ríg Condacht.*
> *BR*^{2} 52.742ff

> The rent of the Corcraige without hardship to be given each time to the king of Mag nAí of the horses is seven score cows, no unjust judgement.
>
> Three hundred and fifty iron blooms,three hundred and fifty noisy swine three hundred and fifty oxen, appropriate rule, are given to the king of Connacht.

We have seen in Chapters 3–5 that the comparatively meagre archaeological evidence for the later 1st millennium AD points to regular production centering on 1.5–2.0 kilos of iron from one smelting operation. Taking the term *cáer iarann* as referring to the consolidated bloom (below: cf. sect. 7.7 under *cáer*), this tribute would have amounted (in theory at least) to around 450–600 kilos of iron, a substantial amount in the context of early Irish production. Following this line of thought a little further, it is possible to extrapolate from the experimental work of Pleiner (1969a) and Tylecote et al. (1971). Using their figures to estimate the average yield of smelting at about 25% efficiency, the *Cuirc* tribute would have involved the mining of some 2000 kilos of ore. This does not allow for the losses by oxidation during the consolidation of the bloom, and we can probably raise this figure to over 3000 kilos. Using fuel to ore ratios in the order of 3:1, this, in turn would have required in the region of 10 tonnes of charcoal. Experimental charcoal burning conducted by Pleiner (1969a, 469f) showed a yield in the order some 17.5% by weight produced of charcoal from pine. Thus, to get 10 tonnes of charcoal would have required the processing of nearly 60 tonnes of wood[5]. Bloomsmithing uses as much charcoal as smelting, and so we can raise the requirement for wood to at least 120 tonnes. While these figures are, of course, purely hypothetical, the organisation and effort to mine and process the ore (let alone the smelting) which this tribute listing represents is large. While Dillon's verdict on *Lebor na Cert* as a propagandist piece, with no real pretension to quantitative accuracy must stand, the fact that it was felt that consolidated blooms were a worthy tribute may be indicative of continuing small-scale production which is suggested in Chapters 3–5 as the norm for the period under review and for native production in LIA B also.

Literary references to mines and mining are comparatively scarce (Scott 1988), although it is possible to retrieve the main elements of vocabulary. For iron, the most obvious correlation between language and archaeology comes from the place-name *Sliabh an Iarainn* (lit. 'Mountain of Iron') for the siderite deposits around Lough Allen which we know were exploited in Elizabethan times. But this is exceptional! The Irish words *mianach* 'a mine' and *mianaige* 'miner', derive from *méin* 'ore, mineral (metal?)' (*DIL* M:83.17f, 127.46ff; *Vendr. Lex.* M:29). Closely associated with these terms are derivatives of *claidid* 'digs, excavates', including *clas* 'trench, pit etc.', *clasaige* 'miner' and *tochailt* (vn of **do-claid*) 'act of digging up, excavating, mining'.

Of the literary references, none appears to be specifically directed at the mining of iron ore. Thus, the interesting verse in the *Dindsenchas* of *Ailén Cobthaig* quoted by Ó Corráin (1974, 67) concerning the mining activities of the *Síl ṁBuinne* in the Bantry area of Co. Cork cannot with certainty be associated with a particular metal:

> *Regthair etarru ocus tír*
> *oc buain míanna, mór in bríg;*
> *it é dogénat in ṁbrú*
> *síl ṁBuinne do Bentraigiu.*
> *Met. Dinds.* IV, 308.45f

> They shall come and go between the Rock and the land, quarrying ore with great toil; they that do the crushing(?) shall be the Síl mBuinne of Bantry.

The expression *oc buain míanna* is best translated as 'quarrying ore', since *búain* (vn of *boingid*) includes '(act of) cutting, striking, reaping, harvesting' in its semantic field, and its use here might be taken as suggesting opencast or drift mining, rather than deep underground working: equally, however, we might take 'reaping' simply as harvesting metallurgical fruits in a general sense. The other point of note in this passage is the slight hint of the transporting of ore away from the mine operations in the phrase *Regthair etarru ocus tír*, a correspondence with some archaeological evidence (such as the finds of iron slag on Church Island, Co. Kerry – Chapter 6.5). In this case we might here see a dim reflection of ore being transported from an island (the 'Rock', *Ailén Cobthaig*, tentatively identified as Whiddy Island, Co. Cork, from which general area iron ore was recorded by Kinahan) to the mainland. Ó Corráin (1974, 68) also draws attention to the circumstances surrounding the death of Conall Cernach whose escape from his enemies was foiled by ore-processors whose activities so muddied the waters of a river that he was unable to find the ford and as a result was caught and killed:

> *Batar mianaigi ac nighe mianaigh isin abaind leth súas de 7 noticedh a buadred sin cuici.*

which he translates usefully (cf. *búaidrid* 'troubles, disturbs etc.': Chapter 7.7) as:

> There were miners upstream washing ore, and the effluent came to him.

Interestingly, this tale is set in the area of *Sliabh an Iarann*, where siderite nodules occur in stream beds (Chapter 6.2). The crushing of ore mentioned in the first passage, and its washing to concentrate the heavier fraction rich in metallic minerals, graphically described for 16th century Europe in Book VIII of the *De Re Metallica* of Agricola (Hoover and Hoover 1912, 267ff, esp. note 8, p. 279f), are processes of great antiquity. The term used for 'crushing' is from *bruïd* 'breaks in pieces, smashes, crushes', a rather apt term.

Given that the winning and preparation of ore must have been fairly prominent activities, it is somewhat surprising that treatment in the laws seems scant. In the 8th century *Tír Cumaile* ('Divisions of Land': Mac Niocaill 1971, 82f = *CIH* II:676.6) we find that one of the occurrences which enhances the value of land is a mine:

> *Cis lir do-formaiget loigi forsna tiraib seo? . . . airgetlach . . . Mad airgetlach do mein uma no iarainn, do formaig .u. seotu.*

> How many things increase the values of these lands? . . . a mine . . . If it be a mine of copper or iron [ore] it adds 5 *séts*.

The word *airgetlach* is difficult since it ostensibly derives from *argat* 'silver'. It is explained in a Commentary as *.i. loc i mbi mein airgit .i. mein in airgit no in uma no in iaraind (CIH* II, 384.34) 'that is a place in which there is a mine of silver, i.e. a mine of silver or of copper or of iron'. It is a rare term, there seeming to be only two occurrences, both in *Laws* (VI.38), and it is difficult to see how a term for a silver mine would come to be used as a general term for all types of mine. While it could be an archaic term, or a scribal mutilation, one reasonable explanation is that the passage above should be emended to read . . . *Mad airgetlach* no *mein uma* . . ., with *airgetlach* simply being translated as 'silver mine', thus altering the translation to the more comprehensible

> How many things increase the values of these lands? . . . a *silver mine* . . . If it be a *silver mine* or a mine of copper or iron ore it adds 5 *séts*.

With their sense of propriety, it is not surprising to find that the jurists were concerned by the possibility of 'claim-jumping'. There are several passages which include provision for gaining redress through distraint against persons removing ore without permission:

> *Athgabail trise . . . i fothla t'airgetlaig . . .*
> *CIH* II, 384.15f

> Distress of three days . . . for stealing from your silver mine . . .

and again,

> *Im claide alla for rud for umad . . .* glossed *.i. in rudh no int uma fora claiter in aill .i. coitchend do tuaith i ssund, is aire rosaig .u.e. .i. int iarann ina caeraib 7 int uma ina thinnib; no int uma ina cloithib ar .u.thi, 'na tanalaig ar .iii., 'na naicdib urluma ar aine In ru 'na trillsenaib ar .u.thi, 'na screplaib ar .iii., 'na mein no 'na dath urlam ar aine.*
> *CIH* II:390.8, 390.34ff

> (Distress of five days) For digging *on* a cliff for *rú*

> or for copper [ore] i.e. the *rú* or the copper for which the cliff is dug i.e. this is common to the people, wherefore it reaches five days i.e. the iron as fettled blooms, the copper as *tinni* or as *cloi* five days, in preparation three days, as finished goods one day. The *rú* in sheaves five days, in *screpuil* three days, as finished metalwork or dyestuff one day.

The idea is to distinguish between valuable materials unlawfully appropriated from communal land, and between the length of distraint applicable depending on the degree of processing which raw materials had received. The editors of *Laws* took *rú* in this difficult passage to refer to iron ore, but it more likely refers to a dye-plant, possibly madder[6]. In this case, we can perhaps more freely render this passage as:

> For digging madder or copper ore from a [sea] cliff, there is a stay of five days because this is communal land. Smelted iron as consolidated blooms, or smelted copper as ingots of impure metal or as cast ingots of refined copper attract a stay of three days. The freshly-picked madder in sheaves has a stay of five days, in *screpuil* a stay of three days, while the dye prepared from it attracts a stay of one day.

This gloss echoes part of an obscurely-worded passage in the laws entitled *Bla Miand Midchlais*' ('The Exemption of ore is a mine') in which restitution for illegal excavation and injuries during mining operations are briefly set out. The impression gained is that, unlike the passage on smithing discussed in sect. 7.5, this was drawn up with little interest in, or knowledge of mining. Three main versions, of which the first is probably the earliest, are given in *CIH* (I, 270.15f; III, 940.11f; IV, 1256.3f). The translation set out below was kindly provided by Prof. G. Mac Eoin from *CIH* I, 270.15f and III, 940.15f[7]:

Bla Miand Midchlais

> The exemption of minerals is a mine. He who digs ore from a mine is exempt from liability, provided that the ore was not privately owned, or if it was, provided permission has been given by the owner. Exemption exists provided there was not excessive destruction, for if there was, it is to be treated as an unlawful excavation. The exemption extends to idlers and unprofitable workers, provided that the work is being carried out with proper entitlement; there is one third restitution for all damage caused while moving it [the ore] up or down, this being [lit. and] a reduction from half honourprice to one third restitution. [One third of restitution for collaborators, for every profitable person, and for every animal: *CIH* III, 940.15] If the ore is in private ownership and the owner has not given permission, there is a fine of five *séts* for it and the restoration of the ore in the form in which it is found, whether it be in ingots or semi-products or as finished artefacts. And there is full liability for all damage caused in transporting it, as well as for the mine [itself] until the owner has acknowledged [its repossession] some time after it has been restored to its former state, and he has the option of not restoring it. In this case full payment is due as for the first slipping of the sledgehammer without foreknowledge of a fault [see *Blaí Ord Indeóin* – sect. 7.5].

The term *midchlais* (lit.) 'middle/central trench/digging', here simply translated as 'mine', seems to make little sense in the passage *as we have it now*. It could, for example, be argued quite reasonably that the prefix *mid-* was added to provide alliteration. If, however, we look at early Greek mining regulations, we find the terms μεσόκρῖνεῖς and ὅρμοι, which have been interpreted as referring to the dividing markers between adjoining properties, or to the pillars left in an ore seam to support the overburden (Healy 1978, 81f). Gk. μεσόκρῖνής translates literally as 'parting in the middle' (*LS* 1106), being a compound of μεσό 'middle etc.', with κρίνω whose meanings include 'separate, put asunder, distinguish' (*LS* 996). One possible translation of *midchlais* is thus as the baulk left between adjoining workings. Translation of *midchlais* as 'support pillar' would also seem to make sense, in terms of protecting mines from their removal which would have left adits in danger of collapse, although such robbery would have been likely to bring about its own instant retribution! Some stipulations for compensation for injury could stem from knowledge of the condition of a mine after supporting pillars had been removed (cf. Hoover and Hoover 1912, 82 note 6).

The section which begins .... *slan s̄ ma roib forsgris foithe* ... '... exemption exists provided there was not excessive destruction ...' could refer to old European tradition that mining rights applied *below* ground, not usually extending to the land surface which could be under different ownership[8]. Against this is Davies' suggestion that '... distinction between ownership of the soil and of underlying minerals can only arise with a developed mining industry.' (1935, 2), something not really in evidence in Ireland until the end of the 16th century. The concept, first codified in the Greek and Roman law which influenced Irish jurists (Ó Corráin et al. 1984), was perpetuated in various mediaeval legal codes. And, as we have seen earlier, the presence of minerals increased the value of land. Nevertheless, it seems possible that we may have a garbled attempt to differentiate between the land surface (which could have value purely for agricultural purposes) and minerals which could be extracted from below it without disturbance.

It is a common, but frequently unjustified assumption among archaeologists in Ireland that, if accessible outcrops of iron ore are not to hand to explain 'slags' on their 1st millennium AD sites, that the source was bog ore (cf. Chapter 6.2). Examination of the distribution of iron ores in Ireland (Fig. 6.2.1) shows clearly that early ironworkers were never too far away from sources of bedded ores. Furthermore, one would be hard pressed to argue that opencast mining alone provided not only the iron ores required in the period of iron-use, but also the copper of the Bronze Age, particularly in the light of the work of Jackson (Jackson, J. 1978: but cf. Briggs 1984). It seems inevitable that miners had to follow usable ore seams underground, and this is certainly the strong implication of *Bla Miand Midchlais*. We may interpret the references to compensation for injury to man and beast during the movement of ore, as covering the haulage of ore from underground galleries to the surface. Again, while the reference to *rob* 'animal' may simply be formulaic, it is interesting to note the suggestion of the use of animals in such work.

After ore was mined, dressed, and transported, the next stage was to smelt it. (cf. Chapter 6.1). Three verbs, *berbaid* 'boils, cooks, melts, dissolves, smelts, etc.' (*DIL* B:82.22f), *bruinnid* 'smelts, refines' (*DIL* B:214.51f), and *legaid* 'melts, dissolves, smelts etc. (*DIL* L:72.34f), and forms from them, denote the smelting of iron and copper ores, as well as other metallurgical operations (Scott 1981b, 1983a) and their products. There appears no consistency of use with iron, although a purely qualitative judgement would give *berbaid* the most frequent occurrence. The most immediately noticeable feature is the consistent implication of the molten state, which has misled many scholars into believing that 'molten iron' would be an acceptable translation from texts predating the 16th century introduction of blast-furnaces to Ireland. The explanation is to be found in their use with the non-ferrous metals, in whose working the molten state was a regular occurrence. As suggested above (sect. 7.1), the vocabulary of ironworking was an expansion of that earlier used by workers in non-ferrous metals, and it is not difficult to imagine that metalworkers of the 1st millennium AD easily distinguished albeit empirically, between *oc berbaid méin iarainn* 'smelting iron ore' as a solid-state reduction process where only the slag was molten, and *oc berbaid méin umae* 'smelting copper ore' in which both slag and metal were molten. The *process* – conversion of friable stone to malleable metal – remained the same, even if the forms of the end-products were distinctly different.

Although there is an increasing body of material evidence for the form of the smelting furnace in use in early Ireland, the literature is unhelpful. We do at least know that the furnace/hearth was blown by bellows, *bolg*, which in some cases were made of leather, through a tuyère (sect. 7.5). All but one of the potential candidates for 'smelting furnace' is either too general in meaning (*teine* 'fire, hearth, forge-fire'), or like *sornn* 'furnace, kiln, oven', a Latin loan from *fornus*, without (as far I can determine) any metallurgical uses. In the *Triads*, to the blacksmith is attributed . . . *fulacht na Morrígna* . . . '. . . the cooking-pit of the Morrigan . . .' (*Triads* 120). On the face of it, this makes little sense, unless *fulacht* (*DIL* F:485.23ff) here referred to the sort of pit known to the archaeologist as *fulacht fiadh*, in other words, a forging hearth of gargantuan proportions. There is a seemingly obscure word *nes*(*s*), always associated with clay and/or clay vessels, and confusingly translated as 'a wooden mould for shaping a smith's clay furnace, the clay furnace itself'. In the *Triads*, the blacksmith's *nes*(*s*) is listed as one of the three 'renovators of the world':

> *Trí aithgine in domuin: brú, uth bó, ness gobann.*
> *Triads* 148

Three renovators of the world: the womb of a woman, the udder of a cow, the *nes*(*s*) of a blacksmith.

In the elegy to a dead smith (sect. 7.4), *nes*(*s*) certainly refers to the forge hearth. Before shaping under the hammer, iron is heated in the forge fire, and so the reference to renovation/reformation here is clear. And from the archaeological evidence (Chapter 6.5), we can suggest that forge hearth and smelting furnace were often the same structure, hence reinforcing the imagery. What is less clear is how the connection with a 'wooden former' arose. Since we are told in Cormac's Glossary (*Thr. Ir. Gl.* 32, under *Nescoit*) that Gobniu was wont to swat intruders with his *nes*(*s*), this suggests a piece of wood rather like a riot-stick. In fact, this makes sense if we remember the shape of a tuyère (cf. Fig. 6.5.7a–g), and consider that one easy way to form one of the larger tuyères would be to mould the clay round a short pole. Thus we can translate:

> *ness . . . is uime dognī ind aurnaisi criadh . . .*
> (*Anecd.* iv 83.24)

as:

> *nes*(*s*). . . around it the earthen utensils are moulded . . .

leaving Gobniu with his forge hearth and truncheon, and ourselves with a puzzle that admits of no easy solution!

The end-product of the smelting process was the bloom, the spongy mass of iron particles and slag. The terms *cáer*, *bruth* and *bruithne* occur regularly in contexts which suggest that they can mean 'bloom' as well as consolidated ingot. It is interesting to note that among its alternative meanings, *cáer* includes 'berry or cluster of fruit' (*DIL* C:19.28f), since the botyroidal shape implied is precisely that which can be observed in furnace slag flow patterns. The second term, *bruth* is invariably used in contexts of hot forging, suggesting that it could be used to denote the hot bloom undergoing consolidation, or a hot ingot being forged. In the two versions of the motif of the 'Isle of the Demon Smiths' (sect. 7.5), both *bruth* and *cáer* are used to denote the glowing hot mass of metal hurled at the travellers' boat. In his discussion of the obscure term *findbruinne*, M.A. O'Brien (1932) suggested that the second element of the word was *-bruithne* (part. of *berbaid*), referring to a glowing charge in the furnace, suggesting that its primary use was in referring to hot metal. The term *bruthdamnae* (cpd. of *bruth* + *damnae* 'the material out of which things are made – *DIL* D:64.63) suggests composite stock – i.e. an ingot made up by forging a number of blooms together, as in:

> [130] . . . *secht cendphartaib iarndaib a cind cacha slabraidé trummither tinni deich [m]brudamna cach n-ae.*
> *BDD* p. 123

> [130] . . . seven iron knobs at the end of each chain, each of them as heavy as an ingot made of the blooms from ten smelts.

The *Na Cuirc* passages indicate that *cáer*, like *bruthdamnae*, was used to describe consolidated stock. While *cáer* and the derivatives of *berbaid* are used in connection with iron, *tinne* 'bar, rod, ingot' is used with both ferrous and non-ferrous metals. The enigmatic *tanalaig* could refer to almost any form of metal, although it may just refer to stock that has been forged into sheets preparatory to forging up into a finished artefact.

* * *

7.3 – The nomenclature of iron and steel

The work of the 19th century metallurgists (e.g. C.S. Smith, 1960) laid firm foundations for the development of modern metallurgy. The fruits of their labours, the explanation of the structure of metals, has also led to the emergence of a systematic nomenclature. The modern metallurgist, knowing the composition and structure of his material, can also name it with precision. In earlier times, nomenclature was a much less certain subject. While metalworkers could broadly distinguish between 'irons' and 'steels' on the difference in physical properties, the lack of the tools of analysis and the underlying theory meant that the *reasons* for these differences were mainly beyond their reach. From classical times, however, ironworkers were able to distinguish between metals smelted from the ores of different regions and value the products accordingly. Thus, *ferrum noricum* possessed qualities against which the iron of other production centres of the Roman Empire was measured. Piaskowski (1982a) has presented detailed arguments for the identification of some high-nickel irons with *Chalybean* steel (cf. Photos 1989b), a product which would be have been identified as coming from special ores, rather than resulting from secondary enhancement of properties through a process distinct from smelting[9]. While being unaware of the complex processes which led the iron ores of the province of Noricum to produce a superior metal, classical ironworkers marked it out for its physical properties. This empirical approach persisted for centuries, so that, for example, the 18th century writer Ole Evenstadt could confidently predict the properties of the iron that could be smelted from bog-ores (Jensen 1968).

(a) Iron

The early Irish word for 'iron' is *íarn* (Mod. Ir. *iarann*), and it is cognate with the words for iron in the other Celtic dialects and with those in the Germanic dialects. Both groups go back to the closely related forms Celt. **ī̆sarno-*, Germ. **īsarna-*, and provide, throughout the regions of their currency, a wealth of place- and personal-names (e.g. Jackson, K. 1953; Evans, D.E. 1967; Birkhan 1970). In Ireland, the best known of these is *Sliabh an Iarainn* ('Mountain of Iron') given to the area around Lough Allen in Co. Leitrim which is rich in sideritic iron ore. Although many attempts have been made to connect the Celtic word for iron with a common ancestor (IE **ai̯os*, Skt. *ayas*, Lat. *aes*: e.g. Pokorny 1914, 1920; O'Rahilly 1942; Benveniste 1956: but cf. Pleiner 1971b), such relationship is not readily established. The wide range of dissimilar names for 'iron' in the IE language family has often been often been remarked upon: thus, Celt. **ī̆sarno-* and Germ. **īsarna-* stand, for example, in contrast to Lat. *ferrum*, Gk. σίδηρος, O. Ch. Slav. *želězo*.

The disparate origins of words for 'iron' may have a hitherto underestimated significance for our understanding of the passage of the knowledge of iron and its technology into Europe along different routes (Figs. 2.2.1 and 2.2.2) from a single, east Mediterranean technological source. Although detailed discussion is beyond the scope of this work, it may be remarked nevertheless that iron technology is certainly younger than the dispersal of the Indo-European language family. Thus, the number of unrelated words for iron in areas which were comparatively early exploiters of the new technology might well indicate spread from several quite distinct centres. In comparison, we may note in passing that the common Celtic-Germanic terminology of 'iron' contrasts with the situation in which the words both for 'gold' (Ir. *ór*, O.H. Germ. *gold*: Scott 1982, 117f) and 'silver' (Ir. *argat*, Germ. **silubra-*) have quite different origins in each group.

(b) Steel

Early distinction between iron and 'steel' in the Old World was usually made, in practical terms, on the ability to take and to retain a superior cutting-edge. It is thus no surprise to find Latin *acies* 'an edge, sharp edge' (cf. Pliny XXXIV:xli.143 *stringere aciem* 'to draw out a sharp edge'; similarly in Gk. στόμωμα, *LS* 1649), later developing (through *aciarum*) into French *acier* and Italian *acciajo*. Another property of steel, especially when quenched, is its hardness, which led to its naming from words for 'hard'. Welsh *dur* 'steel' is a borrowing from Latin *durus* 'harsh, hard etc' (although, *pace* Jackson, K. 1953, 78, the Latin was not always used on its own to signify steel, but frequently in the expression *ferrum durum*). Early Irish has no unambiguous term for 'steel' to parallel either the Romance derivatives of Latin *acies* or those from Germanic **stahl(i̯)a-*. This means that as with the words for copper and its alloys, our interpretations of words for ferrous metals will be less precise than would be desirable. In addition, there are resulting difficulties in the identification of words specifically referring to the various forms of heat-treatment to which iron-carbon alloys are susceptible. And, given that many laymen today could not properly distinguish between 'iron' and 'steel' (cf. sect. 7.1) it seems probable that ferrous metals would have been lumped together under a heading 'iron' by many scribes.

References to judging the quality of iron (as opposed to the relative wealth of references to assaying precious metals – Scott 1981c), are scant, and both of those known to me are obscure. The first, in the 8th century *Audacht Morainn* (Kelly, F. 1976, 12.39), reads:

Ad-mestar iarn asa thoicib túath tacarthaib

Let him estimate iron by its properties at disputes(?) of tribes.

and the second, from the compilation *Bretha Nemed* (Gwynn 1942, 41.23) which reads:

Meisir iarn iarna thomhus . . .

Let you judge iron according to its weight . . .

Both of these occur in sections of tracts offering advice to rulers (a sort of 8th century business management manual). The *Audacht Morainn* admonition is quite opaque, unless it implies that the material from which a weapon is made should be judged by the damage done to an enemy. In each case, the comments on iron are included in lists of similar advice on judging the quality of silver, gold and copper. The use of *mesir* (<*midithir* 'judges') is reminiscent of the gloss *mese no brathnigthe i forloiscthe* (*Thes.* I, 71.28) against *argentum igni examinatum* (Scott 1981b). The purpose of the admonition to judge (assay?) iron by its mass (*tomus* 'act of weighing etc': *DIL* T:245.7f) is unclear. One possible interpretation is that we have here a reference to the sort of weight loss of which Aristotle wrote in the *Meteorologica* (see below). It is, however, difficult to see how distinction could be made between 'iron' and 'steel' in this fashion.

Early Irish would appear to lack an unambiguous equivalent to Germ. **stahl(i̯)a-* which Birkhan (1970, 125f: cf. *IEW* 1011) derives from a form **stok-l(i̯)o-*, proposing its earlier use to denote a non-ferrous metal (. . . *stokli̯o- *ursprünglich einen anderen Gegenstand oder ein anderes Metall – etwa gehärtete Bronze?*). While it seems almost axiomatic that early terms for 'steel' should develop from early forms whose primary sense was either 'hard' or 'sharp (of a cutting-edge)', for Irish, it is difficult unambiguously to demonstrate the former, while the latter is patently not the case. Indeed, we may note that both Lat. *acer* and *durus*, whose progeny in other languages refer to 'steel'(e.g. Mod. French *acier*, Welsh *dur*) were borrowed into Irish (as *aicher* and *dúr*), but made no perceptible impact on the early Irish metallurgical vocabulary.

Ironworkers in the Celtic-speaking regions were familiar with the production of often homogeneously carburised iron, and with the technique of *fold-welding* (cf. Chapter 1.4) which could produce stock closely approximating to homogeneous mild steel, as may be seen in Britain, for example from the remains of chariot-tyres and sword-blades from the deposit at Llyn Cerrig Bach, Anglesey (Fox 1946, 75–76; McGrath 1968: cf. Tylecote 1986, 149f). At least some Irish smiths of the Earlier Iron Age had learned the merits of this technique, as seen from the axes of the Kilbeg hoard (Chapter 4.2) which parallel pieces such as the Wietzrno-Bóbrka and Eskilstuna axes (Chapter 2.2). There is also evidence of knowledge of the technique in the description in *Táin Bó Cúalnge* of a sword described as . . . *claideb secht ṁbrattomon (ṁbrotha)* . . . '. . . a sword [made] of seven ingots . . .' (*Táin* 4486–4487). Furthermore, smiths of the European Iron Age had been exploiting the ability of carburised iron to form exceedingly hard structures when cooled rapidly from high-temperatures (Chapter 1.4) and this could not but have been reflected in their metallurgical vocabularies. While the metallographic evidence for early Irish steel production is not particularly satisfactory, there are at least some grounds for believing in some familiarity with the technology, and it seems quite improbable that this was not similarly reflected in Irish metallurgical vocabulary.

Although the word *crúaid* is usually translated uncritically as 'steel', presumably on the basis of Mod. Ir. *cruach* (Scots. *cruaidh*) 'steel', there is scarcely an instance of its early use where this rendering is not open to doubt. Its wide semantic field, including 'hard, harsh, severe, rough etc' means that descriptions such as *crúaidiarann* lit. 'hard iron' are by no means uncommon, and there is a temptation to simply dismiss it as a calque on Lat. *durus*. It is likely also that the form *crúadán* listed in *DIL* (C:556.1) stands for *crúan*[10] '(red) enamel'. The verb *crúad(aig)id* (<*crúaid*) 'hardens, stiffens' (*DIL* C:555.57f) appears to have no extant metallurgical connections whatsoever, except in the passage in *CCath* referring to Caesar's *lorica* (see section 7.5): it normally describes operations involving organic materials (e.g. corn drying).

Examples of the use of *crúaid* in contexts suggesting that 'steel' might be the intended meaning (and it might be noted that the *DIL* citations (C:556.29f) are less than helpful) include:

cnedh ó crúaid gurm
a wound from *blue steel*
IGT Decl. ex. 1129

O tharraidh Lughaine lonn caladhcholg crúaidh
When fierce Lugh obtained the hard steel(?) *colg*
Duan. Finn XX.91

. . . *in claidem on frisna geibid iarn na cruaidh do cruaidiph na cruinne* . . .
CCath. 5247–5248
. . . that sword against which neither iron nor any steel in the world prevails . . .

. . . *7 de chirclaib cruaidíairn* . . .
. . . and two rings of hard-iron/steel(?)
LL 218b.30953

. . . *cloidem cru(a)idgher curata* . . .
. . . a sword hard and keen, fit for a hero . . .
Aen. 1954

One feature of descriptions of (presumably) iron weapons is the frequent attribution of colours. We find the adjectives *dub* 'dark, swarthy, black' (*DIL* D:425.38ff), *glas* 'green, bluish green, blue, greyish blue' (*DIL* G:95.13ff) and *gorm* 'blue, deep blue, green, dark, swarthy, black' (*DIL* G:37.73ff) associated with weapons. While it is tempting to adopt a general rendering of 'metallic, iron-grey' when these adjectives are used with *íarn*, it is also possible that *gorm* in particular refers to the blue of tempered steel: thus, *crúaid gurm* conceivably could be translated as 'blued, tempered steel' (cf. sect 7.6). This, however, must be treated with great reservation when trying to use colour adjectives to demonstrate either the early production of steel or use of heat-treatments.

Detailed reconstruction of the stratigraphy of metallurgical terminology is beyond the scope of this work. It does, however, seem worth considering whether the meaning 'steel' as part of the semantic field of *crúaid* is not a comparatively late development, replacing an older term, in the same way that the Latin loanword *stán* replaced an earlier Goidelic *créd* as the term for tin (Scott 1981b, 156–157).

In *CCath*, Stokes read the word *úriarann* which he translated as 'steel'. The word *úr* (*DIL* U:87.69ff) literally translates as 'new, raw, fresh, uncooked etc', and its use with *íarn* is difficult to explain: there seem to be no other instances outside *CCath.* Potentially the most interesting occurrence is in the description of Caesar's *lorica* (detailed in sect. 7.5). If the word *íarn* did not appear, we might be justified as taking this to be a description of the preparation of leather body armour. However, it seems most likely that Stokes has misread for ***dúr****iarainn* 'of hard iron' (genitive instead of *de* + dative, thus equating with Lat. *ferrum durum*), and that in fact *úriarann* is a ghost term which may be suitably emended in *CCath.* Such emendation also has the advantage of preserving the alliteration with initial *d-* in the preceding five words.

There does occur one comparatively common term, *íarn aithlegtha,* in early Irish texts with strong claims to be translated 'steel', although it has formerly been translated by the unfortunate phrase 'remelted iron' which, in the context of pre-16th century Irish texts is unhelpful (the other common rendering, 're-*smelted* iron'[11], is a technological nonsense – Scott 1981b, 154–156). Now 'refining' means the removal of impurities, and early Irish *ór aithlegtha* for refined gold (Scott 1981c, 247f), as in . . . *denum d'ór aithlegtha* . . . '. . . made of refined gold . . .' (*St. Ercuil* 76.1444) excites no comment. But *íarn aithlegtha* lit. 're-*melted* iron' cannot be so translated in texts which predate the advent of the blast-furnace, and the examples cited below, representative of a large number of such descriptions, suggest that the original sense should be 'steel', i.e. a metal figuratively refined in order to concentrate its best properties. It is possible, of course, that the term could refer to iron smelted from hammer-scale. As an almost pure oxide, this would produce little slag, thus allowing for a higher than usual uptake of carbon by the reduced iron. As noted in Chapter 1.4, there is a considerable loss by oxidation during forging, and the floor of the smithy could have provided a source of very high grade 'ore' from which a steel could be be produced by primary carburisation. However, we have no archaeological evidence for this process.

The empirical view of steel as purified iron would not be a new idea in view of the much-debated 'cast iron' passage in the *Meteorologica* of Aristotle (Book IV.VI, 383.3f: Lee 1962, 322ff):

> ταῦτα δὲ ἄλυτα μὴ ὑπερβαλλούσῃ θερμότητι, ἀλλὰ μαλάττεται, οἷον σίδηρος καὶ κέρας. τήκεται δὲ καὶ ὁ εἰργασμένος σίδηρος, ὥστε ὑγρὸς γίγνεσθαι καὶ πάλιν πήγνυσθαι. καὶ τὰ στομώματα ποιοῦσιν οὕτως · ὑφίσταται γὰρ καὶ ἀποκαθαίρεται κάτω ἡ σκωρία · ὅταν δὲ πολλάκις πάθῃ καὶ καθαρὸς γένηται, τοῦτο στόμωμα γίγνεται. οὐ ποιοῦσι δὲ πολλάκις αὐτὸ διὰ τὸ ἀπουσίαν γίγνεσθαι πολλὴν καὶ τὸν σταθμὸν ἐλαττω ἀποκαθαιρομένου. ἔστι δ΄ἀμείνων σίδηρος ὁ ἐλάττω ἔχων ἀποκάθαρσιν.

> . . . but those that solidify because of cold and the evaporation of all their heat are indissoluble save by excessive heat, but can be softened like iron and horn. Wrought iron will indeed melt and grow soft, and then solidify again. And this is the way in which steel is made. For the dross sinks to the bottom and is removed from below, and by repeated subjection to this treatment *the metal is purified and steel produced* (my italics). They do not produce the process often, however, because of the great wastage and loss of weight in the iron that is purified. But the better the quality of the iron, the smaller the amount of impurity . . .

Discussion of this passage (cf. Pleiner 1969b, 25f) has centred on the use of the verb τήκω ('melt, melt down': *LS* 1786), although as Lee (1962, 327) points out, of the other terms with which it is used, μαλάττεται implies *softening* rather than liquefaction, while ὑγρός can be used both of softened and molten materials. Olshausen (1915) suggested that the passage was intended to refer to the reforging of a bloom. By this explanation it would be possible to reconcile both the suggestion of 'softening' (the iron) and 'liquation' (the slag). This does not, however, take into account the claim that '. . . the metal is purified and steel produced'. Another example of the use of τήκω to denote the melting of metals, tin and iron, is provided by Hesiod, who wrote (*Theogony* 861–866: text from West 1966, 143, translation from West 1988, 28):

> πολλὴ δὲ πελώρη καίετο γαῖα αὐτμῇ θεσπεσίῃ, καὶ ἐτήκετο κασσίτερος ὣς τέχνῃ ὕπ' αἰζηῶν ἐν ἐυτρήτοις χοάνοισι θαλφθείς, ἠὲ σίδηρος, ὅ περ κρατερώτατός ἐστιν, οὔρεος ἐν βήσσῃσι δαμαζόμενος πυρὶ κηλέῳ τήκεται ἐν χθονὶ δίῃ ὑφ' 'Ηφαίστου παλάμῃσν· ὣς ἄρα τήκετο γαῖα σέλαι πυρὸς αἰθομένοιο.

> The huge earth burned far and wide with unbelievable heat, melting like tin that is heated by the skill of craftsmen in crucibles, or as iron, which is the strongest substance, when it is overpowered by burning fire in mountain glens, melts in the divine ground by Hephaistos' craft: even so was the earth melting in the glare of the conflagration.

It seems clear from Aristotle's use of ἀποκαθαίρω ('clean, cleanse, refine': *LS* 200) that he regarded 'steel' (στόμωμα) as a 'purified' form of wrought iron (εἰργασμένος σίδηρος). Halleux (1974, 189ff) discussed in general the classical literature containing potential allusions to 'cast iron', and concluded that while cast iron was produced, it was occasional and accidental. Lee (1962, 328) suggested that liquation applies to slag, and softening to iron which is taking up carbon from the forge fire and being carburised to steel. However recent work by Photos (1988, 1989a) has shown quite conclusively that the furnaces depicted on 6th/5th century BC Greek ceramics could have been used for the decarburisation of cast iron. Given that cast iron would have been of little immediate use due to its brittleness, its conversion to metal of superior properties could have been easily viewed as purification. Photos draws particular attention to the vessels placed over the tops of these structures, and notes that Greek στόμα also refers to 'mouth' (cf. *LS* 1648). There seems little doubt that Aristotle was describing the decarburisation of cast iron to steel, although it is likely that this was a process which was going out of use (cf. Chapter 3.1). Attractive as it is to interpret *íarn aithlegtha* as steel produced by such a process, there is no evidence forthcoming for its application in Ireland.

We may contrast the passages above with that cited by Halleux (1974, 190) in which the smelting of iron is clearly separated into the liquation of slag and the agglomeration of the iron into the bloom. The reference there to 'melting' is very similar to the use of Lat. *liquo* in a passage in the *Historia Naturalis* of Pliny (Book XXXIV.146: Rackham 1952, 235), which reads:

> *Mirumque, cum excoquatur vena, aquae modo liquari ferrum, postea in spongeas frangi.*

> And it is remarkable that when a vein (*of iron ore* – my insert) is smelted (*Rackham translates* excoquatur *as 'fused'*), the iron becomes liquid like water and afterwards acquires a spongy and brittle texture.

Here again, the reference is likely to be to the melting of the slag.

The same *liquo* occurs in the writings of Adomnan, when describing one of the many miracles ascribed to Columba (*Life of Columba* Book II, 29.76a: Anderson and Anderson 1961, 391):

> *Quod monaci scientes experti ejusdem pugionis ferrum ignis resolutum calore per omnia monasterii ferramenta liquefactum diviserunt inlinitum.*

Anderson and Anderson translate this as:

> When they knew of this, skilled monks softened the metal of that dagger in a hot fire, and distributed it, melted, by overlaying it upon all the iron tools of the monastery.

While it was obviously not beyond the saint's influence to bring about a lowering of the melting point of iron, it is equally likely that Adomnan was translating directly from his native Irish *legaid* and using it in the same sense as his fellow countrymen, that of heating up iron as a preliminary to forging operations (as opposed to its use to signify melting in non-ferrous metallurgical operations). We can thus better render the sense of this passage as:

> When they knew of this (*the miraculous powers of a kitchen knife* – my insert), skilled monks softened the metal of that dagger in a hot fire and distributed it *by welding it onto* all the iron tools of the monastery.

The dagger was cut up into pieces, which were heated prior to their being welded to other iron artefacts, so that the blessing bestowed by Columba on one tool went far as possible. Similar usage may be observed in Book I of the *Periphyseon* of the Irish philosopher John Scottus Eriugena (Sheldon-Williams 1968, 202) where we find:

> . . . *sicut aer per totum illuminatur lumine et igne ferrum totum toto liquefactum* . . .

> . . . as air is totally illuminated by light and the whole lump of iron is *liquefied* (my italics) by the whole of the fire . . .

In his notes to the translation, Sheldon-Williams (1968, 243, note 250) equates Eriugena's use of *liquefactum* with Gk. πῦρακτέω which has the intransitive meaning 'to be inflamed', with no implication of melting (*LS* 1555). Thus, in the context of the discussion, this particular passage could, perhaps, be better rendered:

> . . . as air is totally illuminated by light and the whole lump of iron is *suffused with light* by the whole of the fire . . .

As with Adomnan's telling of the story of Molua's dagger in Latin, so we can see that his compatriot Eriugena translated his quote from the Greek using a Latin word which corresponded to the shades of meaning of his native tongue.

It seems clear that *legaid* was used in ferrous metallurgical contexts to denote heating of metal. Were it not for the consistent references to the superior properties of *íarn aithlegtha*, one might, as a result, be tempted to see *aithlegtha* as referring to annealing. However, this reheating process quite evidently was seen as significantly improving the physical properties of the metal, and, in 1st millennium AD Ireland, can surely only have been *carburisation* – i.e. steelmaking. Interpreting *íarn aithlegtha* as 'carburised iron' – *steel* – we can begin to make proper metallurgical sense of passages in early Irish texts, such as:

> . . . *claideb secht ṁbrattomon* (*ṁbrotha TBC*[2] 3287) *do iurn aithlegtha* . . .
> *Táin* 4486–4487
> . . . a sword (made) of seven billets of carburised iron . . .

> . . . *da cáeir d'iarann aithlegtha, 7 iad ar dercclasadh* . . .
> *BNnÉ* 72.247a
> . . . two flaming-red ingots of carburised iron . . .

> . . . *lūrig* . . . *don fúairíarann athlegtha* . . .
> *Fianaig*. 84, 31.28
> . . . a corselet . . . of cold, carburised iron . . .

In the light of the metallographic evidence derived from Irish iron artefacts presented above, as well as from foreign comparanda, perhaps the most telling reference is that from *Táin* which speaks of a sword being made from a number of peices of *íarn aithlegtha*. The terms *bruth* (*DIL* B:217.20f) and *bruthdamnae* (for *brattomon*: *DIL* B:217.59: see under *berbaid* sect. 7.7) should refer to metal under heat, and given that *bruth* is a noun derived from *berbaid* 'smelts', we might also consider that there is a further implication of heating under reducing conditions. Thus, *bruth* could be taken as referring to iron undergoing carburisation to produce *íarn aithlegtha*, i.e. steel[12]. Thus, it is most difficult not to interpret the description as being of a fold-welded blade, made by welding and forging together of pieces of steel to produce a single ingot whose overall structure would approximate to a homogeneous carbon steel (Chapter 1.4).

Another interesting use of *legaid* in connection with iron (or rather in connection with artefacts of types which should at the date of the text have been made of iron) comes from one of the Commentaries on the law tract *Breatha Comaithchesa* (*Laws* IV, 130: *CIH* I, 199.7f; II, 579.25f). In discussing items which a landowner had to possess in order to erect different types of fencing (Thurneysen 1925, 314f), a commentator has added the following:

> . . . *rama fri clais .i. a tir buic, screpal a fiu, 7 a leagadh ænfecht; soc fri coraigh, screpal a fiu, 7 a leaghadh fa tri; biail fri duirimme .i. a leagadh fa di gora fiu screpal; fighba fri felmag no frisin naile; screpal a fiu, 7 a leagad fa di no sithaile no beoleagadh;* . . .
> *CIH* I:199.11f

DIL (L:66.57f) suggests that *leagad* should be read as *lecad* (vn of *lecaid*) 'act of placing setting, positioning', citing the above passage as an obscure example. However, one may be equally justified in retaining the reading of the word as *legad* and in interpreting the passage as referring to surety in terms of iron artefacts of varying quality. Associating *leaghadh fa tri* with a composite artefact forged up from pieces of carburised stock, we might translate this as:

> . . . a spade for a trench i.e. in soft land, a trench, a *screpall* is its worth, and it is to be *carburised* once; a ploughshare for a stone wall, a *screpall* is its worth, and it is to be made of three pieces of *carburised* metal; a bill-hook for an oak fence, i.e. it is to be made of two pieces of *carburised* metal so that it is worth a *screpall*; a wood-axe for a *felma* fence, or for a palisade, its worth is a *screpall*, and it is to be made to two pieces of *carburised* metal, or it is to be *carburised* for a long time (*i.e. to ensure deep carburisation* – my insert).

Carburisation and the forging of composite pieces is not uncommon, as we have seen from the metallographic chapters. However, this passage, like so many of the commentaries, must be treated with great caution.

Even without the ambiguity surrounding the textual use of *crúaid*, *íarn aithlegtha* can surely only be seen as the early Irish equivalent to Germ. derivatives of **stahl(i̯)a-*, only later supplanted by *crúaid* from which developed Modern Irish *cruach*.

* * *

7.4 – The Blacksmith

The early Irish word for 'smith' is *gobae* (masc. n-stem), Mod. Ir. *gabha*. The modern word is usually applied to a worker in ferrous metals, but is also used with qualifiers in other connections, *gabha dubh* 'blacksmith', *gabha geal* 'silversmith' and *gabha buí* 'goldsmith' (*Ó Dónaill* 598). These are not found in earlier sources, and may result from the influence of the English terms. The older use invariably signifies a worker invariably associated with ferrous metal to the exclusion of all others. The other Celtic languages possess cognate words for 'smith': Welsh, Cornish, Breton *gof*, Scots Gaelic *gobhainn*. The root occurs also in the Gaulish personal name as *Gobannitio*, borne by an associate of Vercingatorix (Evans, D.E. 1967, 350–351). The etymology of *gobae* is quite obscure (Birkhan 1970, 137f). Superficial similarities with Balto-Slavonic words for 'smith' (e.g. Pol. *kowal*) are not due to etymological relationship[13].

Celtic **gobann-* has a widespread occurence in the guise of cognates forming elements of personal and place-names (Evans 1967, 350–351). Irish place-names based on *gobae* (e.g. *Ballygowan* = *Baile an Gobhainn*, 'town of the smith', *Brigown* = *Brí gobhann*, 'hill of the smith', *Ardaguivna*, *Carrowna-guivna* etc.) and on the word for the forge, *cerdcha* (e.g. *Tullycardy* = *Tullach na cerdcha*) reflect the local importance attached to the blacksmith and his forge[14]. The word is also well known from the name of the Smith-Deity of the *Túatha dé Danann*, *Gobniu*, who furnished their weapons, as well as in personal names: for example, one of the inmates of Da Derga's hostel was *Goibnenn* mac Luirgnig (*LU* 7712), while one of the kings of Uí Fhiachrach Aidne is named in the Annals for AD 538 as *Gobniu* mac Conaill (Kelleher 1971, 64–65: for other exx. see *DIL* G:129.3f).

The metallurgical counterpart of *gobae* is the artificer *cerd* who is described as principally working in precious metals and copper alloy, but who also worked in other media (such as ceramics: *roth cruind forsandenat nacerda innalestrai*. . . 'a round wheel on which the potters make the vessels . . .', *Thes.* I, 23.20). The word is cognate with Welsh *cerdd* 'art, poetry', Gk. κέρδος 'gain, profit, craft', Lat. *cerdō* 'labourer, craftsman', (and also appears in the name of a Ligurian group, *Cerd-iciates*, referred to by Livy: Holder 1896, I, 991). The spheres of activity of *gobae* and *cerd* invariably seem to be treated in the literature as mutually exclusive[14]. This apparent demarcation occurs in *Blaí Ord Indeóin* (see sect. 7.5) where we find:

> *IS diles int or 7 int airget 7 int uma a cerdcha in goband, 7 noco dilus i cerdcha in cerda, uair aithne indeithbir he i cerdcha in gobann 7 nocon ed a cerdcha in cerda.*
> *CIH* I, 268.10f

Gold, silver and copper-alloy are unindemnified in the forge of a smith, but not in the workshop of a *cerd*, since they are inappropriate to the blacksmith's forge but not to the workshop of a *cerd*.

In another passage we find the division of labour spelt out even further as:

> . . . *GOBAIND .i. doniat in goibneacht 7 UMAIDE .i. imrit int uma 7 CEARDA .i. imrit int or 7 int airgus* (leg. *airget*) . . .
> *CIH* VI, 2277.36

> . . . blacksmiths, i.e. they practise the art of blacksmithing, and *umaide*, i.e. they work in *uma* (copper or copper alloy: Scott 1981b, 156), and *cerda*, i.e. they work in gold and silver . . .

In *CMT*[2] (§54f) a warrior seeking to enter Tara is challenged by the doorkeeper to declare a craft not already possessed by one of the court. He is only admitted having claimed skill in some ten crafts, including blacksmithing and non-ferrous metalworking. The contrast portrayed between the craftsmen who practised only one craft, and the newcomer claiming mastery in all underscores the demarcation presented by legal texts[16]. There is, however, a somewhat ambiguous stipulation in the 8th century tract *Críth Gablach* (*Críth Gabl.* §15.191) which hints at a wider manufacture of ironwork than might otherwise be assumed. Among the possessions of a *mruigfher* (one of the grades of farmer) is supposed to be . . . *míach gúaile fri herna.* '. . . a sack of charcoal for (*working with*) iron.' The archaeological evidence shows that iron was forged and/or smelted on virtually every occupation site of Later Iron Age date, and it could thus be that the charcoal referred to in this passage was destined for the smelting furnace or smithing hearth[17]. We know that the early jurists had a passion for the construction of elaborate schemes encompassing all aspects of life, and that there are inevitably areas where legal stipulation entered the realms of fiction. Nevertheless, the distinction between gobae and *cerd* seems to be consistent, even though it is likely that farmers of the small, mixed economy Later Iron Age farmsteads (just like some of their modern small-farmer counterparts), were accustomed to carry out some basic ironworking on a seasonal basis as a part of a range of industrial activities (such as weaving, wood- and bone-working, pottery making: cf. Chapter 5.1).

Given the ostensibly rigid demarcation between *gobae* and *cerd* in law tracts, it is thus most interesting to find that in the Goidelic dialects the word for 'forge, smithy', *cerdcha* (Mod. Ir. *ceárta*), derives from *cerd*, while in the Brittonic dialects the word for 'a forge' is always derived from the word for 'smith': thus, for example, Welsh *gefail*, Bret. *govel*). And Goidelic also differs in this respect from Indo-European (O.H.Germ. *smith*, *smitta*: Lith. *kalvius*, *kalvis*: Alb. *kovac*, *kovacane*), and other language groups (e.g. Hung. *vasverö*, *vasveröház*). Another area in which Goidelic *gobae* differs from its analogues is in its virtual non-appearance in compounds. Irish has nothing, for example, to parallel the use of Germ. *smid* as an agential suffix (e.g. O. Icel. *skósmiðr* 'shoesmith', *ljódasmiðr*, 'song-smith (i.e. poet)', O. Swed. *eersmidhe* 'bronzesmith', Calissendorf 1979, 157f). The only instances of *gobae* forming a compound before the Modern Irish period appears to be in the personal names *Dubgoband* (gen. sg. of *Dubgobae*: *DIL* D:429.76) and *Engoba* (*Aengoba*: *Leb. Gab.* IV, §368), and in the term *prímgoba* 'chief/master smith' used to describe the highest grade of the profession (cf. *DIL* N-O-P:203.54f; *BNnÉ* I, 11.7).

Without wishing to become enmeshed in the treacherous web of speculation as to the dating of the implantation of the Goidelic dialect in Ireland (for a summary of arguments, see O'Rahilly 1946, 193ff; Dillon and Chadwick 1972, 4f; Caulfield 1981, 213f; Mallory 1984; Mac Eoin 1986: cf. Chapter 3.1), the different origins of Irish *gobae* and *cerdcha* gives rise to an interesting line of thought. On the basis of interpretation of the archaeological evidence for the nature of the technological transition from non-ferrous to ferrous metal dependence, it is likely that the initial introduction of iron-use was effected by small groups only, the process beginning sometime around the 7th century BC. Assimilating into the Later Bronze Age population, the only major changes that they would have been likely to have brought about in the metallurgical vocabulary of a region with such a rich fund of metallurgical expertise would have been the *addition* of words for which no necessity had previously existed, i.e. *gobae* and *íarn* (possibly also *argat* 'silver'): the native formation for a metalworker's shop would have existed already. If those who brought iron technology to Ireland were also the first to introduce the Goidelic dialect, would not derivation of the word for 'a forge' have been the same as in other regions where Celtic dialects were spoken? We might thus infer that the first establishment of Goidelic speakers in Ireland *preceded* the implantation of iron metallurgy (cf. Chapter 3.1). While the date for this is given here as ca 7th century BC, arguments for a later introduction of iron metallurgy do not affect the basic premise of the prior arrival of Goidelic.

Although usually described as *saer* 'wright, carpenter', Joseph, father of Jesus, is called *Iósep gaba* 'Joseph the blacksmith' in one passage in *PH* (113.2585), suggesting that the blacksmith's calling was not considered inappropriate (if in contradiction to tradition) to the Holy Family. In his discussion of early Irish *Imrama*, Oskamp (*ICM*[2] 31f) draws attention to the episode of the death of a smith in the second voyage episode of *Betha Brennain Clúana Ferta*. Following the death of his smith and, finding his craft without an anchor, Brendan blesses the hands of one of his companions, thus bestowing on him the skills of the forge (*BNnÉ* I, 65, §96). Oskamp suggests that in an earlier tradition, the character who took on the role of the smith was a wright (*saer*) who otherwise plays no part in the tale as we see it now. He goes on:

> It seems therefore very likely that at an earlier stage of the tradition the miracle was that the wright took over the smith's work, perhaps thanks to magical incantations – the priest and the blessing being the Christian adaptation of the motif.

This focusses nicely on the high status afforded the blacksmith in Irish literature, underscoring not only the secular demarcation between workers in iron and workers in other materials, but also the traces of the magico-religious status of the smith which survived the coming of Christianity, and lasted through to recent times, to be preserved in the folk-tales recorded by, amongst others, Lady Gregory (1970, 274–275) and Danaher (1962). Sithcenn, the smith who practised his skills at Tara is described as . . . *Sithcheann cosin ngabaind bai i Temraig; ar ba fisid side 7 fhaidh amra* . . . '. . . Sithcenn the Smith of Tara who was a man learned in occult arts and a prophet of note. . .' (Stokes 1903, 194). Although far less emphasis was given to his supernatural powers than is the case, for example, in Norse literature, even the 7th century prayer ascribed to Patrick and known as *Faed Fiada* included an invocation against the spells of women, druids and smiths:

Tochuiriur etrum indíu inna hule neurtasa. . .
fri brichta ban ocus goband ocus druád . . .

I summon today all these virtues . . .
against spells of women, smiths and druids . . .
Trip. I, 50.19ff.

We get a further glimpse of the supernatural power ascribed to the pre-Christian blacksmith in *Scéla Eogain* (*CMM* 68.11). In the account of the birth of Cormac mac Art to Achtan, the daughter of Olc Aiche, a man described both as *gobae* and *druí* (*SG* I.235; Ó Cathasaigh 1977, 49), we are told:

In tan ro génair Cormac fo-caird in druígoba Huilc Haiche cóic cresa imdegla fair, ar guin, ar báduth, ar thein, ar adgaire, ar chonaib [.i.] ar cach holc.

When Cormac was born the *druid-smith* Olc Aiche put five protective [magic] circles about him, against wounding, against drowning, against fire, against enchantment(?), against wolves, that is to say, against every evil.

While various professions are characterised by skills or associated tools in *Triads*, the supernatural is pressed on the blacksmith:

Tréde neimthigedar gobainn: bir Nethain, fulacht na Morrígna, inneóin in Dagda.

Three things that constitute a blacksmith: Nethin's punch, the fire-pit of the Morrigan, the Dagda's anvil
Triads 120

It is also worth remembering in this context the association of ironworking with Neolithic tombs (cf. Chapter 6.5), which provides archaeological support for the link between the smith and the supernatural.

The most prominent role of the blacksmith in the heroic literature of other early societies, that of armourer to the gods, is only occasionally permitted to show through in the person of divine *Gobniu* (for detailed discussion, see O'Rahilly 1971, 308–317, esp. 314f; *CMT*[2] p. 125). We may see Gobniu as the pre-Christian Irish equivalent of Greek *Hephaistos*, Roman *Vulcan*[18], and Germanic *Weland*, and of the veritable pantheon of divine smiths to be found in the mythologies of iron-using peoples past and present (Eliade 1978, 97ff). He gave his name to a (now lost) law-text on blacksmithing (**Bretha Goibnenn*) one of a group including *Bretha Déin Chécht*, assigned by Binchy to his *Nemed* collection of tracts (Binchy 1955, 1966, 2: Kelly, F. 1988, 62). It is referred to in a commentary from *Uraicecht Becc*:

In glasaigne is ferr .i. eolus aicgi i mbrethaib crethine 7 luctine 7 dianecht 7 goibnend . . .
CIH V:1614.12

The best apprentice advocate, i.e. an advocate with knowledge of the judgements of Crédne and Luchtaine and Dían Cécht and Gobniu . . .

For the battle of *Magh Tuired* fought by the *Túatha Dé Danann* against the *Fomoire*, Gobniu and his fellow craftsmen (*Crédne*, the *cerd*, the bronzesmith, and *Luchtaine* the *saer*, the wright) promised to supply from their forge[19] infallibly lethal weapons:

96 *Rotinálid trá maithe Túath nDé Danann go Luch. Roimcomhoirc a gaboinn .i. Gaibne, cía cumong conánocuir dóib.*

97 *"Ní anse," al sé. "Gé bet fir Érenn isin cath go cenn secht mblíatan, gaí detáet dia crunn ann nó claidem memais ann, tarcéba arm núa úam-sai ina inoth. Nach rind degéno mo lám-so," ol sé, "ní focertar imrold de. Nach cnes i ragae noco blasfe bethaid de íer sin. Níbo gníthe do Dulb gobhae na Fomore annísin. . . ."*

96 Then the men of rank among the Túatha Dé were assembled around Lug. He asked his smith, Gobniu, what power he wielded for them.

97 "Not hard to say", he said. "Even if the men of Ireland continue the battle for seven years, for every spear that separates from its shaft or sword that will break in battle, I will provide a new weapon in its place. No spearpoint which my hand forges will make a missing cast. No skin which it pierces will taste life afterwards. Dolb, the Fomorian smith, cannot do that. . . ."
CMT[2] 50–51

and in the pauses in the fighting, Gobniu and his fellow armourers provided a quite miraculous logistic support for the warriors:

122 *Ba ingnad trá liasna Fomore alaill tárfas dóib isin cath. Botar cloíte a n-airm-sie .i. a ngaoí 7 a cloidme; ocus an romarbad dia feruib-sium ní ticdis íernabháruch. Níba edh immorro de Túathaib Déa: ar cía noclodis a n-airm-sium andíu, atgainidis amuhárach fo bíth roboí Goibnenn goba isin cerdchai ag dénam calc 7 gaí 7 sleg. Ar dogníth-side na harma-sin fria téorai grésai. . . .*

122 One thing which became evident to the Fomoire in the battle seemed remarkable to them. Their weapons, their spears and their swords were blunted; and those of their men who were killed did not come back the next day. That was not the case with the Túatha Dé Danaann: although their weapons were blunted one day, they were restored the next because Gobniu the smith was in the smithy making swords and spears and javelins. He would make these weapons with three strokes. . . .
CMT[2] 54–55

In addition to the supernatural powers ascribed to the smith in European folklore, magical properties also attached to the metal (e.g. Eliade 1978). In *Duanaire Finn*, Fionn and his companions encounter three witches who assault them with magic:

> *Glaisíarna garbh geinntlidhe*
> *do cuireadh leó for crannoibh*
> *gabhair meiscce is meirbhlighe*
> *Fionn is an Fhian da ffaixin.*
>
> *Le draoítheacht go drochdhalach*
> *do shíabhrattar ar fflaith-ne*
> *na senóir críion crothanach*
> *fagbhaid Fionn ara haithle.*
>
> Rough grey iron of wizardry they had mounted on poles: giddiness and faint sickness came over Fionn and the Fian at the sight of them.
>
> With magic of evil-dealing they spell-bound our chief; they left Fionn thereafter a withered quaking ancient.
> *Duan. Finn* I, 92.110f

Iron finds magical use in search of truth, in the tale *Iarnn Luchtna*, a section of of the series of ordeals outlined in *Scél na Fír Flatha* (Stokes 1891, 192.23). Here we find:

> *.i. Luchta drai dochoidh dia fhoglaim in-Letha, con-aca é ní ingnadh occa ic delugud fhirindi 7 breigi .i. iarnn do senadh lía ndruidib, 7 a chor a teinidh iarsin com[b]ad dearg. 7 a tabairt for bois in lit[h]ig. No loiscedh immorro hé dia mbedh cin occa. Ni denad urchoid dho mina bheth cintach. At-bert Luchta iarsin friu "Noricfaidh a leas againdi fir Erenn", for se, "sud do delugud etir firindi 7 breig". Dobretha Luchta a iarnn senta leis iartain, co mbaí ic delugud etir gai 7 fhír conidh de sin leantar iarnn se[n]ta beus ag Gaeidelaib dogrés.*
>
> That is, Luchtna the druid went to study in Brittany, and there he saw a strange thing (used) for discerning truth and falsehood, namely, some iron was hallowed by the druids, and then cast into a fire until it became red, and then it was placed on the palm of the accused. Now if guilt were with him the iron used to burn him. But it did him no harm unless he were guilty. Thereafter Luchta told them that it would be needed "for us, the men of Erin," he said "to distinguish between truth and falsehood." Luchtna afterwards brought with him his hallowed iron, and it was (used) in distinguishing between truth and falsehood. Hence then (the ordeal of) the hallowed iron is still continually practised by the Gael.

It is likely that there are many more references of this nature, and this is a field which would repay further study.

In Christian times, the smith still enjoyed high status, particularly when his work was dedicated to the glory of God, although his tools were also pressed into the service of the Devil as instruments of torture, as in the poem 'The Harrowing of Hell' (Bergin 1910, 112), where we find the verse:

> *Éirgidh, indlidh bar mbulga,*
> *tógbhaidh suas bar seanurda;*
> *ná déntar uabar oili*
> *do bhualadh in brégoiri.*
>
> Arise! make ready your bellows, take up your old sledgehammers; let not the smiting of the deceiver be a second time an idle boast.

He was also circumscribed by Christian observance, as, according to *Cáin Domnaig* ('The Law of Sunday'), smiths, like druids and satirists, were forbidden either to travel or to receive payment on the Sabbath (O'Keeffe 1905, 208.31). A number of prominent figures of church and state boasted of family connections with blacksmithing (cf. Joyce 1903, II, 325f), including St. Bairre of Cork, of whose father, Amairgen, it was said:

> . . . *In tAmhirgin sin dino fa gabhai amhra hé, 7 is hé fá primh-gobha do righ Raithlenn an tan sin* . . .
> *BNnÉ* I, 11 §2
>
> . . . This Amairgen was a notable smith, chief smith to the king of Rathlenn at that time . . .

If the identification of Garannes, Co. Cork (Ó Ríordáin 1942) with *Ráth Raithleann* is correct, then it is a nice thought that there we might have some of the remains of Amairgen's work! Fortchern, called *goba* in *Bethu Phatraic*, one of three smiths listed as members of the saint's retinue (*Trip.* I, 266), was said to be a son of Laegaire, the high king of Ireland.

The master smith, who commanded a high honour-price (Kelly, F. 1988, 62), entertained royalty in the heroic literature, and was seated well above the salt according to the theoretical protocol specified in *Tech Midchuarta* (*LL* II, 116f). This status parallels that of his counterparts in the Germanic world. His forge, recognised as an important focus in the community, was one of the places to which, according to the Laws, notice of a waif and of straying horses had to be sent, amongst other places.

The smith's anvil was likened to the cornerstone support of honour:

> . . . *inneoin fothaighthi onóra a teglaig* . . .
> . . . an anvil for supporting the honour of his household . . .
> *MR* 108.17

and was used in metaphor to epitomise masculinity:

> . . . *a threnníafer talman 7 a inneoin fhothaidh na ferrdachta* . . .
> . . . strong champion of the men of the world and firm anvil of manliness . . .
> *St. Ercuil* 12.184.

In the *Dindsenchas* of *Loch Lein* (*Met. Dinds.* III, 262.28), we find the expression *Indeóin na nDése* 'The Anvil of the Dese', which survives in the townland name *Mullaghnoney*, *Mullach Inneóna*, in Co. Tipperary (*Hog. Onom.* 551). There is a trace of a pre-Christian stratum of the magico-religious significance of the blacksmith's anvil in the tradition of the choice of Níall Nóigíallach for the kingship. In *Echtra Mac nEchdach Mugmedóin* (sect. 7.5), one of the proofs of his fitness to be king over that of his brothers was that he rescued the anvil and block from a burning forge[20], while they carried out other tools and artefacts. The poet attributes to his father, Eochaid, the following (Joynt 1910, 110):

> *Inneóin Temrach tair co trom[m] bud rí Níall menmach macdond.*
> Tara's mighty anvil in the east; bold Niall, the boy-chief, shall be king.

The importance of the anvil as the focus of the blacksmith's work may be judged from the legal stipulation that notice of distraint against a smith could be given by tying a withe around his anvil, as in:

> *Caidi athgabál cach aes[o] dana? ní, athgabál goband cétamus: gat im inde(o)in 7 apud nad derna atmat fuiri co nderna cert frit do neoch adgair fair; a chomut cach oen oca mbi inde(oi)n.*
>
> How is each member of the professional classes distrained? Answer: first, to distrain a smith: tie a withe about his anvil and give notice that he is not to work any material on it until he shall have done right by you in respect of what you are claiming from him: the same [applies] to everyone who has an anvil [for his work].
> Binchy 1973b, 78–79

Eliade (1978, 60f) draws attention to examples from modern ethnography, where the anvil is considered a focal point in the household of a blacksmith. It is also interesting to note that *O'Reilly* (305) gave the meanings 'navel, centre of a pool', in addition to 'anvil', for *inneóin*.

These few examples highlight the tradition and mystic trappings of the pre-Christian blacksmith and his craft. Analysis of their deeper significance is beyond the scope of this work, although we may remark that is certainly a phenomenon of considerable antiquity, and indeed must stem back to the transition between copper-alloy and iron technologies (cf. Chapter 2.1). By providing edged weapons superior to those previously in existence, the blacksmith must have gained commercial advantage over his competitors, while as the weaponeer of the Gods, his spiritual ascendancy was also assured. The mythology of the Old World frequently shows the smith as the provider of the weapons with which the deities overcome their enemies, a theme which may well be derived in part from (amongst other things) the profound changes to the old technological order.

While the blacksmith is primarily depicted as the *shaper* of iron, there are a number of references whose translations have contained the implication that blacksmiths also conducted the smelting process (sect. 7.2). As we have already seen, what archaeological evidence there is suggests that not infrequently, ore was transported for smelting on occupation sites rather than being fully processed at the mine itself. There appear to be no examples where smelting is attributed to other than *gobae* or *cerd*, apart from the passage in O'Clery's late recension of *Lebor Gabála* where the term *bruithneóir*[21] (<*berbaid*: *DIL* B:216.24f) is used:

> . . . *nir baoth . . . an bruithneoir ag bruith tinneadh 7 caor i cceardcha é* . . .
> he was not a foolish smelter *smelting* bars and ingots in the forge . . .

The literary evidence for the early Irish smith as the sole smelter of iron is, however, not particularly strong, resting on a handfull of often uncertain passages. For example, the passage cited above could well refer to the *carburisation* of stock (for a discussion of this process, see sect. 7.3) given the ambiguity of *berbaid* when referring to ferrous metals.

Perhaps the most straightforward comes from *Togail Troí* and reads:

> *Tuaircnech na ṅgoband ic méinlegad miannaig.*
> *LL* IV, 218b.30950

Túaircnech generally means 'a beating, striking, hammering etc.', and its use suggests the appreciation of the writer of the effort required to sustain the air-blast during smelting. Thus this passage invites translation as 'the pounding of smiths smelting ore'. There is indirect support for this rendering in the passage:

> . . . *sithiger urbuinde gabond a ṡron. mar sigemnaig builg oc berbaid míannaig súgad* . . .
> *Thr. Ir. Gl.* p. 36
> . . . his nose was as long as the tuyère of a smith, and his panting like bellows-blowing in the smelting of ore . . .

From the same source comes a moving verse from an elegy on a dead blacksmith by his wife:

> . . . *ba binde no chandais dord*
> *fri derc a niss a di bolg*
> *Thr. Ir. Gl.* p. 32
>
> . . . sweetly his two bellows chanted a refrain against the *derc* (sect. 7.5) of the forge hearth.

Now this last review of a loved one could be translated either as referring to the dear departed smelting ore, or simply heating stock in the hearth as a preliminary to forging. Given the ubiquity of the debris of iron smelting and forging on Irish sites of the 1st and earlier second millennia AD (Chapter 6.2), it would indeed be surprising to find the imposition of rigid demarcation between *gobae* and *bruithneóir* as is stipulated in the laws between *gobae*, *cerd*, *umaide* and *saer*.

In summary, we may view the blacksmith as a person who specialised in the shaping of iron. Some smiths may also have been involved in the smelting of iron, although the textual evidence for this is somewhat tenuous. Even after the establishment of Christianity, he enjoyed high social status, although still hedged around with taboos which survived from his pre-Christian sanctity, and which, to a certain extent, had been assimilated into christianised Irish society.

* * *

7.5 – The Blacksmith's Forge

As with other aspects of early iron metallurgy, most Irish texts deal only obliquely with the forge, its organisation and operation. But legal texts can augment the references in stories and poetry to provide a clearer picture of the forge and its organisation than is available for any other area of ironworking encountered in the literature. In many instances, descriptions can only have come from a writer's first-hand observation of a local blacksmith at work. The overall picture that emerges of a forge of some thousand years ago is of one in which the rural blacksmith today would probably feel quite at home. While numerous incidental references allow us to piece together a general inventory of the forge furniture and (to a lesser degree) the types of structures we might expect to identify through archaeological excavation, they tell us little about actual organisation and operation. We are thus fortunate to possess a very detailed description, and one which will be shown to be soundly based on practical observation, in the legal compilation *Blaí Ord Indeóin* ('The exemption of the hammer is the anvil': Scott 1983b) which purports to establish criteria for assessing culpability in cases of injury to workers in, and visitors to the smithy.

A very good example of the valuable information provided by incidental references may be found in the prose tale *Echtra Mac nEchach Muigmedóin* ('The adventure of the sons of Eochaid Muigmedón': Stokes 1903, 194) and its counterpart, the poem *Echtra Mac Echdach Mugmedóin*, attributed to Cúán Húa Lothcháin (*ob.* 1024 *AU*: Joynt 1910, 109.70f), in which the testing of Níall Nóigíallach and his brothers for their fitness for the kingship of Ireland is described. In the prose tale, the advice of a druid is sought, and he in turn sends for Sithcenn, a smith from Tara, renowned as a sage and a seer. The sons are taken into a forge to which the smith sets fire in order to see what each of the sons will choose to save.

Echtra Mac nEchach Muigmedóin

[6] *Ro loisc iarsin in goba in cheardcha forro. Doriacht Niall immach 7 in indeoin cona cip lais. Niall confortamlaig, ar in drui, 7 bud indeoin fhothamail he co brath. Doriacht Brian 7 tuc na huird leis. Brian da bur cathraib, ol in drai. Doriacht Fiachra, 7 tuc dearb corma 7 na builg leis. Bar sciam 7 bar n-dán la Fiachra, ol in drai. Doriacht Ailill 7 in comrar a mbadar na hairm les. Ailill do bar ndigail ol in drai. Dorocht Fergus 7 cual crinaig lais 7 crand ibair inti. Fergus crin, ol in drai. Ba fir on, ar ni maith sil Fhergusa cenmotha oen . . .*

[7] *Conid dia forgell sin ro can in seanchaid*:

Coic maic Echach, Niall indeoin oll,
Brian ord fri tuaircain fir,
Ailill comrar gai fri fine
Fiachra sidhi, Fergus crin.

[6] Then the smith set fire to the forge in which the sons were. Niall came out carrying the anvil and its block. "Niall vanquishes" says the druid, "and he will be a solid anvil forever." Brian came (next), bringing the sledgehammers. "Brian to your fighters", says the druid. Then came Fiachra, bringing a pail of beer and the bellows. "our beauty and science with Fiachra", says the druid. Then came Ailill with the chest in which were the weapons. "Ailill to avenge you!" says the druid. Last came Fergus with the bundle of dried wood and a bar of yew therein. "Fergus the withered!" says the druid. That was true, for the seed of Fergus was no good, excepting one . . .

[7] To bear witness of that, the *seanchaí* sang:

"Eochaid's five sons, Niall the great anvil,
Brian the sledge-hammer for true striking,
Ailill the chest of spears against a tribe,
Fiachra the blast, Fergus the withered."

Echtra Mac Echdach Mugmedóin

[70] *Atagar iat 'sin cerdcha con óengobaind na hElgca,*
'musloiscther hí for in claind ropo ní do dáil decm[a]ing.

[70a] *Is and atrubairt riu thaill tessargain cruid in gobaind*
amail dogéntais in mbeirt co mbad foe Hériu ardneirt.

[71] *Ruc Brían na hurddu ria ais, ruc Aillil in gaisced lais,*
ruc Fiachna in taulchuma thair, ruc Fergus in cúal crínaig.

[72] *Na builg, na huird – tenn in treóir – is in cepp i mbái ind inneóin*
is ind indeóin fodéine rostessairg Níall nárf-héile.

[70] (Thereafter the five youths) are led to the forge with Elg's foremost smith; It is set on fire about the sons, somewhat of a hard test that.

[70a] It was then (Eochaid) told them to save the smith's chattels even as they should achieve the task should Erin of high strength be beneath their sway.

[71] Brian carried out the (sledge-?) hammers on his back, Aillil carried out the weaponry, Fiachna carried out the water-trough, Fergus carried out the bundles of well-dried firewood.

[72] The bellows, the hammers - stout the strength – the block on which the anvil stood, aye, and the anvil itself, Níall the noble-modest saved.

The explicit motif of this episode (but cf. note 20) is the way in which the young men reacted and the implements which they chose to save. We are presented with a useful list of the sorts of items expected to be present in a forge of the period. Interestingly, no mention is made of tongs, although most of the rest of the items that would have constituted *airnise gobann* (*airnise* 'implements, tools' *DIL* A:233.82f), the equipment of a smith, described in the laws as . . . *bulgu 7 urdu 7 indeóin cona cip* . . . 'bellows and hammers and the anvil with its block . . .') are included. Níall's choice of the anvil, marking him out as worthy of the kingship, as noted earlier, is an example of the survival of pre-Christian traditions concerning the smith and his tools (see sect. 7.3). While the poem ascribes to Fiachna the carriage of a water-trough, presumably for quenching and cooling of tools and artefacts, the prose writer has taken *taulchoma* in its more specific meaning of 'wine/beer-vat', showing a nice sense of priority, and providing a reminder of the working conditions of the forge and the need to quench and cool the workers!

The episode of the 'Isle of the Demon Smiths' which occurs both in *Betha Brennain Clúana Ferta* (*BNnÉ* I, 71.123ff) and *Imram Curaig Maele Dúin* (*ICM²* 142.7ff), provides further useful information on the smith's forge:

Betha Brennain

[123] *Lá ele do Bhrénnain acc siubal na fairrge, co facaidh oilén hi ccomfhoccus dó, . . . lan do thigibh amhail cherdcadaiph. . . .*

[124] . . . *an uair dochúalattar torann-glés na mbolc acca seidedh, 7 comthuarccain na nord ag bualadh na nindéon. . . . Asahaithle sin iaramh doconncatar dias dubh-granna ar dath an ghuail gaibhnide ag teacht amach asna ceardchadaibh amhail do beittis ag dol do dhenamh saethair éiccin.*

[125] *Ocus mar doconcattar muinnter Dé, do filletar astech irís 'sna cerdcaibh, 7 tuccattar léo da cáeir d'iarann aithlegtha, 7 iad ar dercc-lasadh, i ndá thenchair ina lamhaibh; . . . Ocus do éirget-tar amach asna cerdchaibh cetna drong gránna díáirmidhe diabhlaidhi, 7 caora teinntidhe i tten-cairibh leo da tteilccenn orrasan. Ocus an tionadh i mbendáis na caora sin annsan bfairrge, do biodh fiuchad mór ard uasa, mar do beith as coire no a haighen fo mbeith teine adhbal-mhór.*

[123] Another day, as Brendan was traversing the ocean, he saw an island near him, . . . full of houses like forges . . .

[124] . . . they heard the thunderous working of the bellows being blown, and the pounding of hammers forging (iron) on the anvils . . . After this they saw a couple, hideously black like the colour of smith's charcoal, coming out of the forges, as if they were going to do some work or other.

[125] And when they saw God's people, they turned back again into the forges, and brought out two flaming red ingots of steel from the forge-fire which they held in tongs in their two hands; . . . And out of the same forges their issued an innumerable hideous and diabolic band carrying red-hot ingots in tongs to cast at them. And where these ingots struck the sea it boiled up high above them, like a cauldron or pot over a huge fire.

Immram Curaig Maele Dúin

[21] *A mbatar iarom ciana for imluadh forsna tonda achondcatar foda uaidib indsi. amal ro scuchsad focus di co cualatar fogar na ngoband ic tuargam brotha forsann ineoin co n-ordaib amal tuarcain trír no cethrair . . . Dothaet an gaba andside asan cerdcha immach 7 bruth mor iairnn isan tenchair ina laim 7 focherd an bruth sin a ndegaidh an curaigh isan muir coro fhiuch an muir uile . . .*

[21] When they had been long voyaging on the waves they saw far from them an island. When they approached it they heard the noise of the smiths forging a bloom on the anvil with hammers like the noise of three of four men forging . . . Then the smith came out of the forge with a huge iron bloom in the tongs in his hands, and he threw that bloom after the boat into the sea so that all the sea boiled . . .

In both versions, we are offered the picture of the forge as a hive of activity and noise and heat[22]. Although the emphasis is on the diabolic nature of the smiths, we are not, in fact, too far from an anticipation of Gerard Boate's vivid description of English-owned blast-furnaces working in Ireland in the 17th century (Scott 1985). Common to both is the motif of the smith carrying a hot mass in his tongs and then hurling it after the travellers. There is an interesting contrast between the descriptions *cáeir d'iarann aith-legtha . . . dercc-lasadh*, and *caora teinntidhe* in *Betha Brennain*, and *bruth mor iairnn* in *ICM²*. It may well not be mere coincidence that the writer of this chose to reinforce the impressions of the noise of heavy hammering caused by several smiths working as a team in his choice of *bruth*, since the final stages in the consolidation of a raw bloom and the expulsion of entrapped slag would have required exceedingly vigorous application of the sledge-hammers. The phrase *da cáeir d'iarann aithlegtha* cannot be translated using 'remelted', but rather as 'two ingots of steel (carburised iron: see sect. 7.3). The description *dercc-lasadh* can be easily translated 'flaming, burning red-hot (*lasaid* 'takes fire, blazes' *DIL* L:56.1f). When used with *íarn aithlegtha, cáer* is best translated as 'an ingot of iron being carburised in the forge fire'. Thus, *bruth* and *cáer* should, in connection with iron, be translated as 'a raw bloom' and 'a consolidated ingot' respectively (see sect. 7.7).

Collating the evidence so far, from the images and imagery used, we can reproduce a rough sketch of a rural forge consisting of some kind of structure in which the smith's work of shaping metal focussed on a small anvil supported on, or driven into a substantial wooden block, his rack of tools included tongs and hammers. Central to the structure would have been the charcoal-burning forge hearth blown by bellows, while somewhere nearby, a water-trough stood waiting to reduce hot metal to ambient temperature.

To fill in some more detail we must now turn to the wealth of observation contained in the section from the laws entitled *Blaí Ord Indeóin* ('The exemption of the hammer is the anvil'). While it was presented in the manner of a proto-Health and Safety at Work Order and Industrial Injuries Compensation Act (down to the stress on 'reasonable care'), the legislators were obviously keenly aware of the working of the blacksmith's forge and the range of accidents that could occur there. Binchy (*CIH* I:viif) has erected extensive warning notices around the perimeter of the minefield that is the corpus of early Irish legal material. However, the particular passage to be discussed presents an acceptable level of risk, since the features important to this study may be checked against modern forge practice. It was on the foundations of eminently practical observation of actual work-in-progress that successive legislators transformed a simple but sturdy construction into the delightful baroque folly that has come down to us.

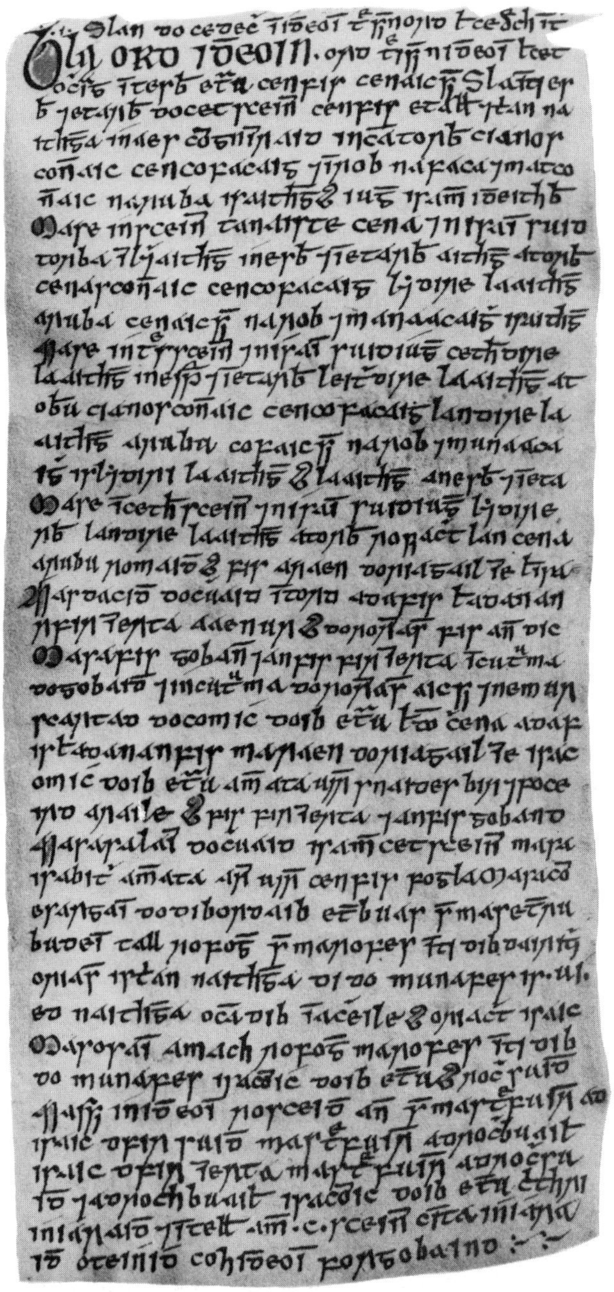

Plate 7.5.1 Opening of the text of *Blaí Ord Indeóin* from Trinity College Dublin manuscript E 3 5 (*Courtesy of the Library of Trinity College Dublin*).

The tedious detail of compensation for injury and loss, left in the translation of *Bla Miand Midchlais* (sect. 7.2) has been omitted to clarify the points which here concerns us. *CIH* gives two main recensions of the text (I:267.2ff; III:936.1ff), both similar in all but a few details, and includes several other fragments (cross-referenced in the margins of the main texts). The translation here is based on that given in *Laws* (III:186ff) with emendations, and consists of abstracts from the texts of the main issues dealt with. Italics within or without round brackets are carried over from the *Laws* translation: italics within square brackets represent my own comments or alternative versions. Plate 7.5.1 reproduces[23] the opening lines of the text found in Trinity College Dublin Manuscript E 3 5, 28b.

Blaí Ord Indeóin 'The exemption of hammers is the anvil'

. . . the person who plies the sledge [*hammer, sledge-hammer*] on the anvil is exempt *(from penalty for injuries arising from the work he is engaged on)*
. . . he is exempt, though the anvil break the sledge [*hammer*], or the sledge [*hammer*] break the anvil, or though an idler has been thrust between them without his knowledge or *his* having seen *him*. He is exempt *from fine* for *injury to* idlers and unprofitable workers in the first slipping [*glancing of the hammer off the anvil*] . . . if he has no knowledge of *any* defect [*in the implements he is using*]; and he pays . . . compensation for *injury to* fellow labourers *and* to all profitable workers, whether he has seen them or not . . . and for *injury to* beasts which he has not seen; but if he has seen the beasts, it (*the fine*) is *full* compensation.
If it be the second slipping [*glancing of the hammer off the anvil*] . . .
If it be the third slipping . . .
If it be the fourth slipping . . .
If it be the head of the [*sledge-*] hammer that has slipped [*from its haft*] . . .
If it be from [*the striker's*] hand it (the sledge-hammer) slipped . . .
If two [*sledge-*] hammers while being wielded have come in collision, and if injury to the persons engaged [*in the forging operation*] only has resulted . . . If it be injury to someone else . . .
If it be the anvil which has slipped off the block . . . and if (*this happened*) in consequence of bad fixing [*of the anvil into the block*] . . . if *it occurred* in consequence of bad striking . . . If it (*the slipping off*) be in consequence of bad fixing and bad striking . . .
The injuries from the sparks of the iron [*hot hammer scale*] . . . and of the hearth . . . The injuries *done* by the iron *in carrying it* from the fire to the anvil . . .
If the anvil has been injured [*damaged*], *and* if *it was* in consequence of bad striking . . . but if *it was* in consequence of *the iron* having been badly held . . .
If it be the sparks from the hearth [*forge fire*] that have caused the injury . . . the *bellows*-blower

alone is to pay for it, unless the smith has urged him *to blow strongly* . . .
If it be the iron that has caused the injury when being struck, it is to be considered whether it was in consequence of bad holding or of bad striking . . .
If it be the sparks of the iron that have caused the injury . . .

'Profitable workers' here are persons who fell asleep before the iron was put in the [*forge*] fire. 'Idlers' here are persons who fell asleep after the iron had been put in the [*forge*] fire.
[*Every inappropriate thing left in a* corn-drying *kiln, a kitchen, a forge or a mill is unindemnified.*]
[*Gold and silver and copper alloy are unindemnified in the forge of a smith, but not in the workshop of a* cerd *since they are inappropriate to a blacksmith's forge but not to the workshop of a* cerd.]
[*when the smith is aware* [of a hazard] *and his apprentices are not* . . .]

A superficial reading of the text presents us with the faintly bizarre vision of industrial chaos in which drowsing customers are incinerated by showers of sparks thrown out from the forge-fire by fanatical bellows-blowers, wandering livestock and stray visitors are felled by hammer-wielding myopics, the air is thick with flying hammer-heads, red-hot stock and even the occasional anvil, while the smith and his assistants are embroiled in permanent litigation! But closer scrutiny shows that the early jurists had analysed keenly the operation of a forge before developing their regulatory scheme. In fact, it seems reasonable to suppose that the original compiler was one with close connections with blacksmithing. The text is of importance in several respects in that not only does it confirm the range of forge-tools and their terminology, but also casts some light on the manning of the working forge. Overall, great stress is laid on the competence of the smith, not only as a worker in iron, but also as a manager. The text distinguishes various types of mishap and sources of hazard, which seem perfectly reasonable, not only in the forge of antiquity but also in the modern blacksmith's forge. Two main sources of accidents, both to workers and visitors to the forge, requiring compensation from culpable parties, were seen as being:

(a) improper preparation and maintenance of tools,
(b) lack of experience and/or carelessness on the part of workmen, and carelessness on the part of visitors to the forge.

Two main types of injury are distinguished:

(i) burns caused by contact with hot stock, hammer-scale or sparks, and
(ii) injuries caused by impact.

Some accidents, such as coming into contact with the hot iron as it was carried to or from the forge fire, require no comment, while the descriptions of others provide a wealth of information. In developing their regulations, early jurists have provided a wealth of incidental detail on the type of layout and working conditions that might have been found in an early Irish forge.

The shocks to which wooden-hafted metal hammer-heads would have been subjected in everyday forgework would have naturally led to the haft breaking from time to time. The careful workman of any period would periodically inspect for signs of impending failure, if for no other reason than to save himself from injury. A skilled blacksmith today will make a number of hammers of different weights, for different jobs, ensuring that the wooden haft is of such a length that its 'balance' or 'weight' is precisely to his taste and that the head is firmly attached to it. Heavy sledges are treated in the same way as the lightest hammers. Given the obvious skill of at least some early Irish smiths in forging iron, it is not unreasonable to assume that they took the same care of their tools as their modern counterparts. There would always have been plenty of scope for the careless or unskilled worker to take inadequate precautions. An improperly secured hammer head is potentially lethal both to the wielder and to those in the vicinity since, should it become detached during the swing, it has built up a fair momentum and will

have much kinetic energy to transfer to whatever it strikes. It is easy to see how early jurists anticipated custom from this source. Similar penalties for the careless use of tools, such as flails used in a corn-drying kiln, are stipulated elsewhere (e.g. *CIH* I:273.29f).

The successful forging of hot iron, indeed of any metal, depends on the smith having mastered completely the art of striking precisely through the point of contact between the piece of stock and the striking platform. When iron is struck in this way, the shock is largely absorbed by the anvil and hammer haft; a good analogy is the lack of shock felt by the cricketer, golfer or hurler who makes a clean strike of the ball. But if the blow is improperly placed or the stock not held in precisely the correct position, then the shock transmitted both through the hammer haft and tongs can be considerable. Anyone who has tried their hand at blacksmithing will ruefully acknowledge the ease with which the novice can sustain a badly sprained wrist! It is not only the person who holds the iron in the tongs in one hand and the hammer in the other who is endangered by inaccurate striking. When the heavy hammer is plied by a second worker, a vigorous blow away from the point of contact between iron and striking platform is capable of transmitting a severe shock and of causing severe injury to the hands and wrists of the person holding the tongs. This certainly underlies the distinction made by the jurists between *drochbualad* 'bad striking' and *droch-chongbáil* 'bad holding'.

Any lingering doubt as to the eminently practical basis of these 'regulations' is surely dispelled by the reference to the dangers of hammers colliding during forging. One of the first lessons which the apprentice blacksmith learns is to guard against letting one hammer get in the way of another. Those who have seen skilled smiths forging hot metal as a team cannot but have been impressed by the co-ordination and split-second timing involved. One smith will hold the iron in the tongs against the anvil, while his colleagues deliver alternate blows as directed. This operation carried out by experienced workers assumes the rhythm of the dance – bars of waltz time when three persons are involved – and is a positive delight to watch. Timing and strict observation of the conductor's directions are essential since a loss of timing will result in a collision of hammers and the risk of injury. It seems inconceivable that this particular stipulation could have been introduced other than as the result of observation, however arcane the legal jargon. This practice was not only noted by the jurists, but also by the writer of *ICM*[2] who described the sounds emanating from a forge as . . . *ic tuargam brotha forsann ineoin co n-ordaib amal tuarcain trír no cethrair* . . . '. . . forging a bloom on the anvil like the noise of three or four forging . . .'.

References to accidents involving the dislodging of the anvil from its supporting block are readily comprehensible when we remember that early Irish *indeóin* did not refer to anvils of today's mass, but rather to artefacts of the size of a small shoe-last. Although for the period under review we have only the Garryduff example (Fig. 1.5.2a), numerous examples occur on the Continent and are frequently depicted in mediaeval manuscripts (e.g. Tylecote 1981, esp. Figs. 39–40). Such anvil and block arrangements have been used in recent times, for example in Iran (Wulff 1966, 50–51, and Fig. 71). Like the Garryduff example, these anvils had a pointed butt which allowed them to be hammered into a large block which acted not only as a firm support but would also have contributed an overall resilience or 'spring' to the whole (cf. . . . *T'indéin do béimim i cepp* . . . '. . . A striking of your anvil into its block....': O Neill, J. 1907, 108). It is useful that this text tends to confirm our understanding of the early arrangement of anvil and block (cf. . . . *in cepp i mbái ind inneóin, is ind indeóin fodéine* . . . '. . . the block on which the anvil stood, (and) the anvil itself . . .': Joynt 1910, 91.72).

The small striking platform offered by such an anvil would not have posed a major problem for the experienced smith, provided it was seated properly in its block. But even if an anvil was firmly driven into the block, an imprecisely placed blow could easily have dislodged it with the painful consequences anticipated. It is also reasonable to suggest that with anvils of the size used in antiquity, where the shock should be absorbed by the wooden block, misplaced hammer blows or improperly held stock would have tended to displace the anvil, with a liklihood of damage both to it and to those in close proximity. Similarly, if the anvil had not been firmly secured to the block, then subsequent hammering would have been likely to dislodge it. This is probably also the primary source of the reference to damage to the anvil, although the use of a large stone block as a striking platform is also known, both from the iron-working site associated with the hillfort at Clogher, Co. Tyrone (Warner 1972) and on the Continent (as at Suderschmedeby-Auburg in Schleswig-Holstein: Hingst 1981, 84 abb. 10). Such anvil-stones would likely have been more commonly used during the vigorous forging necessary to consolidate raw blooms, and thus prone to fracture. Again, the distinction between *drochṡuidiugad* 'bad setting/fixing' and *drochbualad* 'bad striking'.

The reference to the bellows focusses attention on this important piece of equipment, although without providing any clue as to its construction. It is also interesting to note the references to the use of smiths' bellows in what one might term extra-curricular activities. Joyce (1903, II, 305f and 307, Fig. 296) argued that bellows mentioned in the Laws and also in various other early texts were of two chambered construction[24], linked by a pivot, and with simple flap valves which closed on the down stroke. He based this both on the expression *dá bolg* (lit. 'two bags') which occurs in the elegy on the smith in *Sanas Cormaic* (sect. 7.4), and on a passage in *Tochmarc Emire* which describes training in the martial arts undergone by the young Cú Chulainn, and in which the term *cetharbolg* occurs:

> *O ro-ancatar Domnall, for-céta leiss: aill for licc dercáin fó-séted cetharbolcc – no clistis fuiri iarm [co]nápdar duba ná glassa a fond, aill for slig fris-ndringtiss, con-clistis fora rind, [co]ná ferad fora fonnib.*
> Meyer 1890, 444.25–28

When they came to Donall they were instructed by him, at one time on a flagstone with a small hole in it under which he used to blow bellows – they used to dance on it then in such a way that the soles of their feet were not black or green – at another time on a spear on which they used to climb and dance on its point in such a way that it did not harm their soles.

Joyce (1903, II, 307) suggested that the idea of this particular exercise was to strengthen the legs by working a double bellows arrangement with the feet, while providing a continuous air-blast to a furnace. In fact, in view of the feat performed on the point of a spear which follows directly, the exercise was certainly to train in agility. It is this which provides the clue, since it is obvious that the *licc dercáin* must have been in the vertical, not horizontal. At first this passage is puzzling if we translate *derc* as 'tuyère' in the conventional sense. However, the solution becomes glaringly obvious if we identify the *derc* or *licc dercáin* with the bellows' protectors depicted on the Hyllestadt carving (Plate 6.5.6) and represented by the Snaptun find (Figs. 6.5.8 and 6.5.9). Even if the top of the block was squared off (like the majority of those known from Europe), and not chamfered to an edge, dancing on such a small surface directly above, and heated by, a mass of glowing charcoal blown by bellows gives a whole new dimension to the traditional Irish jig!

During the siege so graphically described in *Orgain Denna Ríg*, fire was used as a weapon, and one hundred and fifty smiths' bellows were pressed into action as an aid to speed the progress of the conflagration[25]:

> . . . *7 ro sétea na trí chóicait bolg goband dōib immon tech, 7 cethrur ōclách for cach bulg, combo te dont ṡluag.*
> Greene 1955, 438*ff*

> . . . and the three-times fifty blacksmiths' bellows which they had around the house with four warriors at each bellows, were blown until the house became too hot for the host.

On the basis of the term *cetharbolg* (lit. 'four bags') in the *Tochmarc Emire* passage, Joyce attempted to reconstruct a four-chambered bellows arrangement. While his suggestion that smiths used bellows with more than one chamber seems not unreasonable, his four chambered arrangement (1903, II, 308, Fig. 297) seems both over-elaborate and smacks of the imposition of a modern design onto an ancient instrument. One might point out that the use of one or more drum bellows[26], each of which would qualify for description as *bolg*, could provide a translation which did not require extravagant reconstructions since, for example, *dá bolg* and *cetharbolg* would simply mean 'two/four single-chamber bellows worked in series to provide a continuous air blast'. Even the assertion that four warriors were deployed to work each bellows to fan the flames at Dind Ríg, need require no more explanation than that the musculature of four could create a more concentrated blast than that of one. Indeed, this is certainly the meaning of the phrase in the 11th/12th century *Imtheachta na nOinmhideadh* (*Lib. Hymn.* p. 90):

> *geibidh tennchair caor nduibh-dheirg*
> *ima dregaid alla uird*
> *sceinnidh tuirenn ar gach leith*
> *ima sech seinnid na builg.*

> The tongs hold the black, red metal about which hammers contend yonder, sparks fly on all sides, the bellows are blown alternately.

The degree of heat that such a rig could have put out suggests that Donall would have been made very welcome in the sergeant's mess of the French Foreign Legion!

As we have already seen, the noise of the bellows attracted the attention of early writers who used it as simile for heavy breathing. Thus, in *Táin* the description of the final clash of the bulls contains the sentence:

> *Ra bulgsetar a n-óli 7 a sróna mar bulgu goband i certchai* . . .

> Their cheeks and nostrils swelled like smith's bellows in a forge . . .
> *Táin* 4875

while in *ML* we find a description of two warriors fighting which contains the following detail:

> . . . *agas do líonadar a n-ochta dá n-análaib amhail bolga gabhonn ag gris-shéideadh*

> . . . and their breasts heaved like the bellows of a smith blowing up a heat.
> *ML* 30

Sanas Cormaic contains a delightful description of the 'Spirit of Poetry' which also projects an image of the interior of a working forge:

> . . . *sithiger urbuinde gabond a ṡrón. mar sigemnaig builg oc berbaid míannaig súgad 7 imlécad a anala. ni bendais uird do chǽr dérg ambendais a beoil do thenid.*
> *Thr. Ir. Gl.* 36

> . . . his nose was as long as the tuyère of a smith, his panting like bellows-blowing in the smelting of ore: sledge-hammers would not strike from a glowing ingot such a shower of fire as his lips struck forth.

In modern practice, the blacksmith will restrict the size of the fire in his hearth to that necessary for the work on which he is engaged, occasionally applying extra blast from the bellows as required. It seems unlikely that early smiths would have worked differently, since a primary reason is to minimise oxidation of the stock being heated with the production of hammer-scale. A further important reason is to cut down the formation of clinker which is formed in the modern hearth near the tuyère from the coke or coal fuel. In place of this, the early smith would have had to contend with a mixture of ash and smelting slag when consolidating the bloom after after it had been removed from the furnace. The use of charcoal instead of modern fuels would have made production of showers of sparks on blowing up a greater hazard than today. By comparison with modern wrought iron (produced, it should be remembered in a liquid-state), early bloomery irons contain much slag and would be considered extremely dirty by today's standards. Forging of such stock would have expelled some semi-liquid slag as hot, glowing droplets as well as shedding hammer-scale. Irons in the forge fire pick up clinker as well as developing hammer scale through oxidation, and the smith will usually tap the piece sharply on the anvil before beginning hammering, to avoid scattering this unwanted material as showers of hot sparks. That sparks are given off (quite literally in showers) during forging, being particularly numerous during the forging up of a bloom, is captured in verses from the poet (one who must have seen the working of hot iron) who wrote the *Dindsenchas* of *Loch Lein* in verses which read:

Iar scur cech aidche, gním cert,
dond fhiur foirbthe, nírb amnert,
dia indeóin, focherdded-se
úad co hindeóin na nDése.

Tri fossa focherdded-si
ind indeóin co n-óibligi,
fross usci fedil fedma,
fross do thenid trichemda:

In tress-fhross co nglé-raind glain
do némaind cháim chorcar-glain,
conid iat co ndémna dil
némna locha Léin lind-gil.

After work ceased each night (right deed),
for the accomplished man (no weakling he),
at his anvil he would fling it
from him to the anvil of the Desi.

Three showers would it fling forth -
the anvil with its sparkles:
a shower of water, unfailing, vigorous,
a shower of flaming fire:

The third shower of pure bright quality
was of lovely purple [*jewels*] jewellery
so that these, lovely in purity,
were the jewels of Loch Lein's clear waters.
Met. Dinds. III, 262.25ff

Although 2 *oíbell* 'a spark, flame of fire' (*DIL* N-O-P:115.3f) appears to have little direct metallurgical usage, the image of sparks being struck from hot metal on the anvil is quite arresting. Unsuspecting persons in the vicinity of this material can easily sustain unpleasant burns. Different terms are used in the text for sparks thrown out from the forge fire (*spréd*) and those coming off the hot iron (*crithir*), suggesting a recognition of the different origins of the hot particles thrown out during the working of iron. Anyone who has stood too close to a smith forging hot iron will understand perfectly from the small burn holes in their clothes the hazard that flying hot hammer scale poses. Their use in this text suggests that *spréd* was used to denote the sparks thrown up generally by the fire, while *crithir* (like another term *tuirenn*, elsewhere glossed *scama ferri*) was reserved for 'hammer-scale'.

The incidental references to ironworking in other texts, as we have seen, indicates that the early Irish smith, in addition to shaping metal in his forge, also carried out smelting operations. While our text makes no direct reference to smelting or smelting furnaces (the two terms used are *tellach* and *teine*, denoting 'a hearth', 'a fire'), archaeological excavation strongly suggests that the two operations were usually carried on side by side. Even the simple bowl furnace which is the commonest type to be used here can be a source of showers of sparks, while the more advanced furnaces with provision for slag tapping would have periodically released quantities of molten slag at temperatures in the region of 1150–1300°C, another potential source of serious injury for the unwary or unfortunate in later periods.

The references to burns from the hot stock being forged further illustrate the practical basis for the regulations. As we have already seen, the success of the forging operation depends to a large degree on the holding of the stock being forged and the precise placing of the hammer blows. In addition to actually injuring the person holding the metal, a really ineptly placed blow could easily send the red hot stock flying. Another source of danger comes when hot stock is being cut with a hammer and sett. Unless care is taken over the final blow, the offcut is likely to be propelled with some force along an unpredictable trajectory, with possibly unhappy results.

The literature is invaluable as a source of ideas for the range of tools which should be sought among the archaeological assemblage. While even a brief perusal of the modern smith's forge reveals a wide variety of shapes and sizes, these will be reduced to an amazingly anonymous uniformity by corrosion. Some more substantial pieces, particularly tongs (*tennchor*), have survived, but they are few. We have only found two anvils (*indeóin*), three possible setts (possibly doubling as hammers), and one (dubious) file (*íarnach*). No definite hammers (*ord*), punches or drifts (*tairbir*) have been identified (cf. Chapter 1.5). The file, for example, has left traces on some early metalwork (Chapter 4.1), although outside Viking Dublin the only example from the archaeological record is the now lost piece from Carbury, Co. Kildare (*CIIAA* 589). There is a late reference to *íarnach* 'file':

le cas-chuimilt na gcead chloidheamh ris na h-iarngaib
Ó Dúnlainge 1907, 19

the turning and rubbing of the swords against the files

Where mention is made of the removal of metal by abrasion, it is most usual to find terms used implying grindstones or whetstones, as in:

Baí dano ben and fri bleth arm .i. Crón máthair Fíanluig; is í rus-meil gaí Rúadáin.

Now there was a woman there grinding weapons, Crón the mother of Fíanlach, and she ground Rúadán's spear.
CMT[2] §124.551f

Indeed, there is even a passage in *Lebor Aicill* headed *Bla liac limad* (*CIH* I:290.27f). It may thus be that in this instance, the infrequency of textual references to metal files accurately reflects their lack in the archaeological record.

Blaí Ord Indeóin presents the picture of a lively workplace, manned by a number of workers engaged on various tasks while their customers wait around for their work to be completed. It is interesting to note the demarcation implied between the blacksmith, those who wielded heavy hammers, the bellows blower and the apprentices. In fact, it would seem more likely that these distinctions are more apparent than real, with the smith being less likely to have had an assistant whose sole task was to work the bellows, than either to have worked them himself or asked one of the other workers to blow up the fire while not actively engaged in some other task. As an aside, we might note the modern Irish equivalent to the English 'busman's holiday', *scíth an ordlaí ón inneoin go dtí na boilg*, 'the hammerer's rest from the anvil to the bellows' (i.e. no rest at all!). We might also note the explanation of the word *etarlam* 'opportune(?)' in the glossary compiled by O'Mulconry. (*DIL* E:220.35f) reads *lámdæ dogni an goba cein mbis iarnd i tinid* 'work done by the smith while the iron is in the fire (i.e. when an opportunity presents itself; or, done in the meantime). Of course there would have regularly been occasions (as there are in modern forgework) where it would have been necessary for the smith to call upon one or more others to strike the metal while he held it (hence the concern about colliding hammers). Also, the use of the term *aes urlaide* suggests a specific group whose primary duties were heavy sledging. The aim of legislators was to identify the agent and then to determine whether he bore full responsibility, or if some of the blame could be attached to a second party. The reference to apprentices is unsurprising, since the idea of training to a trade was accepted in other walks of early Irish industry.

Elsewhere in the laws, as well as in literature, we are given a general impression that the blacksmith's forge tended to be a substantial structure. Thus in the voyage tales, the island of the diabolic smiths is . . . *lan do thigibh amhail cherdcadaiph* . . . '. . . full of houses like forges . . .', structures whose arson would cause serious loss of contents. In the legal compilation known as the *Heptads*, we find an interesting exclusion of the forge as a building in which stored goods can be indemnified against loss:

> *Tait .uii. treaba lā ariscuille coir urnadhma aithne ar cia roloiscear ni tuillit aithgin na dire aith cerdchae . . .*
> *CIH* I, 11.26ff

> There are seven houses [structures] in which a valid agreement for custody of articles is prohibited by law, for though they be burnt, they do not deserve compensation, a [corn-drying] kiln, a forge . . .

To this is added the quaint gloss . . . *rofitir inthi rosaithnnister com conuir do aidhe na haithne isna trebaib-seo* . . . '. . . the depositor knows that in these houses is a path to the loss of the deposit . . .'. Similarly, *Blaí Ord Indeóin* states that goods left in a forge, a kitchen and a corn-drying mill cannot be indemnified against loss. However, in that text there is no mention, for example, of setting fire to other than persons as opposed to a building, or compensation that might be demanded as a result of loss or injury.

On the other hand, we do have the superficially bizarre mention of injury to livestock during forging, although this may simply be formulaic (cf. the similar reference in *Bla Miand Midchlais* in sect. 7.2). The word used is *robb*, a difficult term, though one generally referring to domesticated animals (cattle, pigs). The practice of providing horses with iron shoes would appear to be an introduction of the Anglo-Norman period, and evidence for the native shoeing of horses in pre-mediaeval Ireland is sparse. While a horse in a blacksmith's forge would occasion no comment, the common word for 'a horse' *ech* is not used. The language of the passage as we have it now is 11th century, although its redaction from an earlier source is not unlikely. Thus, the archaeological evidence weighs heavily against any interpretation that the passage refers to farriery. It is easy to envisage a dog or a cat as visitors to the forge, but not a pig or a cow. And while one would not be surprised at a dog or cat accidentally getting itself unnoticed into a precarious position, inability to notice a cow or pig in an enclosed space would surely have barred a person from the blacksmith's calling. For it to have been possible for such beasts to have strayed into a danger zone would have required at least some forges to have not been a substantial enclosed structure, but open. The evidence of archaeological excavation to date suggests that in Ireland in the period under review, forging and smelting were very often carried out at best under flimsy structures designed to give protection only against the worst of the weather. While not contradicting either the image of the forge in the poem and prose versions of *Echtra Mac nEchach Muigmedóin*, or the evidence for substantial forge structures from sites like Reask, Church Island and Ballyvourney (Chapter 6.5), this sort of arrangement would explain the legislators' concerns over the safety of livestock in the vicinity of the forge. Cattle in particular are inquisitive animals quite capable of ambling in to inspect goings-on in what to them appear familiar and safe surroundings. Unless firmly shooed away, they could have put themselves in danger in a working environment like this.

While our text is couched in the stilted jargon of the early jurists, it is nevertheless still possible to see in its composition the work of one familiar with the blacksmith's craft. The perhaps irritating attention to fine detail of compensation does not obscure the compiler's understanding of his subject matter. Although it is usually understood that much of the early legal material was compiled by clerics, the provenance of this text, whether ecclesiastical or secular is of little importance to its industrial content. The evidence of literature and archaeological excavation is in good agreement that both churchman and layman were regularly involved in the production and working of iron at the time of its composition. Although one has no evidence whatsoever, it is very tempting to envisage the compiler as one who at some time in his life had worked with hot metal and who had both a feeling for good workshop practice and a sympathy with those who plied the blacksmith's trade. Is it even possible that in some form it was a part of, or related to **Bretha Goibnenn*?

In conclusion, it is clear that while *Blaí Ord Indeóin* may hold little of interest to the legal historian, for the student of early Irish ironworking it paints a vivid picture of a working forge of the 1st millennium AD and one that is in close agreement with the knowledge gained from archaeological excavation and metallurgical and petrological examination.

* * *

7.6 – Heat treatments and jointing

As we have seen from the metallographic examinations of native-produced EIA/LIA A iron artefacts in Chapters 3–5 that early Irish smiths were less competent in the manipulation of the properties of ferrous metals than in shaping them. This is something of a disappointment, since we would expect to find a much higher degree of technical competence to complement the descriptions in the texts. Yet despite the identification of *íarn aithlegtha* as 'carburised iron', i.e. 'steel' as a relatively common term, the number of examples of deliberate secondary carburisation of early blooms or artefacts is surprisingly low. As for heat-treatments, there is little at present to suggest that Irish smiths were as capable as their contemporaries of exploiting the potential to modify the properties of their stock. It may well be that the small sample so far examined is unrepresentative of the level of expertise of the early Irish blacksmith, but until the microscope reveals evidence to the contrary, we must treat textual references to sophisticated manipulation of physical properties with great reserve.

Quenching of steel has been identified in artefacts dating back to the earliest period of iron-use in Europe (see Chapter 2.2), and the products of tempering also appear at an early date. One of the most graphic accounts comes in the episode of the blinding of Polyphemus in Book IX of Homer's *Odyssey* where the noise of the glowing stake as it put out his eye is likened to that made when a blacksmith quenches a red-hot axe or adze. The two techniques underpinned the adoption of iron as the primary utility metal since they gave metal a degree of strength and flexibility superior to that of even the best bronzes. The speed of the quench, dictated by the quenching medium, was early understood, as may be seen from Pliny (*Historia Naturalis* Book XXXIV, xli.144):

> *Tenuiora ferramenta oleo restingui mos est, ne aqua in fragilitatem durentur.*
>
> It is the custom to quench smaller irons (forgings) with oil in case water might make them brittle-hard.

The speed of quench is today controlled by using a wide variety of different media, while in antiquity, brine (still used today), blood and urine were all employed. Although Theophilus' assertion that steel quenched in the urine of a goat fed on a diet of ferns was best for the cutting of gemstones (*De Diversis Artibus* Book III.21) might seem bizarre, it serves to emphasise that this genius of mediaeval craftsmanship was probably well aware of the end properties of steels quenched in different media.

There are several passages in early texts which strongly suggest that the writer was referring to a weapon made of quenched and tempered steel, since the properties of a razor-sharp cutting edge coupled with great flexibility are stressed. In *Ceart Claideb Cormaic*, there is a description of a sword whose ownership is the subject of dispute. Its nickname *Crúaidín* (dim. of *crúaid*) has been taken to imply a steel blade, and the tale relates:

> *Bai claideb amra la Socht... Da fillti a rind conici a dhornchar dosineadh amal cholg. Doletradh finda for usci. No athcumadh finda for ceand, 7 ni fhaiscead in toind, Do[g]níd da leth don duine 7 ní forcluinedh 7 ní fhairigidh co hetircin in leth araile dhe. Atbert-som ba hesin in Cruaidin Coiditcheann .i. claideb Conculainn.*
>
> Socht had a wonderful sword... If its point were bent back to its hilt it would straighten (again) like a *colg*. It would cut a hair floating on water. It would cut off a hair on a (man's) head, and without touching the skin. It would cut a man in two and for a long time one half would not hear and not notice that the other half was missing. Socht said that it was 'Hard-headed' *Crúaidın*, the sword of Cú Chulainn.
> Stokes 1891, 199.59

With the same sword, Cú Chulainn is supposed, with two strokes, to have shaved bare the head of Etarcumul and split him from the crown of his head to his navel (*Táin* 1658): he and other heroes are credited with similar slayings. The motif of the super-flexible blade is also found in *CCath* where one of Caesar's swords is decribed thus:

> *No stuaglubadh Cesair in claidem sin inn aimsir ainesa co tibead a inn re urdurnd. No sceinnedhlingedh in claidem ina coilgdírgi... acht cena ba iar cein máir no tairised don cressnugadh ocus don crithnugudh no gabad fair ic ascnam ina dirghi fadein.*
>
> In a time of pleasure, Caesar used to bend that sword like a bow so that its point touched its hilt. ... it would spring and leap to be again as straight[27] as a *colg*. But it was along time after returning to its straightness before it ceased from shaking and trembling.
> *CCath* 5250f

Again, in *BDD*[2] we are told of the spear of Mac Cecht:

> 7... *cresaigthiseom in gaí sin co comraiged a dá n-ind...*
> *BDD*[2] 849
>
> ... he shakes that spear until its two ends meet ...

The recurrent hyperbole does not mask the underlying recognition that such flexibility was coupled with an especially good cutting-edge. How far the writers who referred to such weapons were aware of how these desirable properties might be achieved is open to question, although there is a potentially illuminating passage in *CCath* apparently concerning the heat-treatment of body armour.

> . . . *.i. cathluirech . . . [d]úriarann . . . 7 a fobairt co frichnamhach firglic a llendaib inmara olarda itharda ola indara fecht ann do boccadh 7 do righniugadh do suntugadh ocus do sofilltiugud a adbair. In fecht aile immorro no foibrigte do crúadugadh 7 do chaitiugud a drol a grethaib gleorda glanuisci . . .*
>
> . . . i.e. (his) battle corselet of hard iron (*steel*?) . . . and (it had been) at one time dipped with cunning care in juicy, oily fatty liquids to soften and to toughen, to give strength (?) and to (make) supple its links. At another time, however, it was dipped to harden and to stiffen its links in bright baths of clean water . . .
> *CCath* 5211f

The two terms used, *fobairt* and *foibrigte* (see under *fo-beir* in sect. 7.7) both derive from *fo-beir* (lit. 'brings under'), in other words 'dips, submerges, plunges'. Reading *dúriarann* as referring to steel (sect. 7.3), then one cannot help but be drawn to the contrast between the media in which the corselet is described as having been treated, and see indications of the distinction between quenching in water and quenching in oil.

There is another recurring motif in early heroic tales, that of the spear associated with the warriors Celtchar Mac Uithechair and Dubthach Dael Ulad, which required to be immersed in liquid to cool its appetite for slaughter. When sensing the imminence of opportunities for the shedding of blood, the spear was reputed to have become so inflamed as to necessitate its being plunged into a cauldron of liquid. The best account is preserved in *Togail Bruidne Da Derga*:

> *Lágen mór il-laim . . . Lónchore mór ara mbélaib méit chore colbthaige. Dublind úathmar and: mescthus beós isin duiblinn isin. mád chían co tairi a fobdud. lassaid fora crand. . . . is écen core co neim día fábdud in tan frisáilter gnim gona duine dí.*
>
> A great spear in his hand . . . A vessel as large as a meat-boiler is placed in front of them: in it is a hideous black liquid. It is quenched in that black liquid. If a long time elapses before it is immersed, a blaze shines from its shaft . . . it must be quenched in a cauldron of poison to check it when the killing of a man is expected from it.
> *BDD* 118 §128–129

In *Mesca Ulad*, the spear is described as being cooled in:

> . . . *Cairi dubfhola da lind aduathmar aidchi remi arna denam tria druidecht da folaib con 7 catt 7 druad cu fobairthea cend na slegisin issind lind nemi sin. In trath na thiced a grith slegi.*
> *LL* 35100f
>
> . . . A blood-black cauldron of horrid noxious liquid before him, composed through sorcery of the blood of dogs and cats and druids. And the head of that spear was quenched in that poisonous liquid when its spear-ardour came (upon it).

while in the early Modern Irish version of *Cath Rois na Ríogh* (*CRR* 78.26), we are told that the spear was quenched every hour in its venomous bath. This motif is similar to that of the berserker's frenzy of Cú Chulainn who, when seized of it at an inopportune moment, had to be plunged into cold water to restore his sense of proportion and prevent him from slaughtering friend as well as foe. The emphasis placed on the cooling medium seems sufficiently strong to suggest a metallurgical basis. (One can only conjecture that Cú Chulainn would probably have felt quite justified in slaughtering friends who ducked him in the sort of brew described in *Mesca Ulad*!). It does not seem unreasonable to suggest that the *BDD* version is what might be termed a stylised description of the preparation of a blade before battle. I would further suggest that the *BDD* version *as we have it now* shows elements of a metallurgical vocabulary in Iron Age Ireland relating to heat-treatments and of a tradition that such manipulation produced weapons with special qualities. What *BDD* offers is technology, forged with the particular genius of the Irish tale-smith, into a mere embellishment. In other words, the story-teller has emulated the Celtic artists who rearranged the patterns of the Classical world into their own perception of shape and movement, and produced his own 'illusive image' of a metallurgical detail[28].

We can find early Irish parallels for accounts by classical writers of repairs required as a result of damage to weapons during battle. It is interesting to note that while Gobniu is attributed with supplying infallibly lethal weapons, he was either unwilling, or unable, to guarantee a similar degree of resistance to wear and tear! His field forge (sect. 7.4) was a hive of activity, preparing weapons for the following day's fray. And in the story of the First Battle of Magh Tuired, there seem to have been as many casualties amongst weapons as humans:

> *Rosníid a slegha re sirthuargain. Rochruaidbrisit a cloidme a cnamchomach a comlannaib in catha sin.*
>
> Their spears were twisted by the continual smiting; in the hand-to-hand combats the swords broke on splintered bones.
> *CMT* §39

We may also note the recognition of the ease with which even a good cutting-edge could be damaged from the passage in the laws entitled *Bla Iarand Airlech* which reads:

> *Bla iarand airlech .i. Slan donti berus int iarunn*

d'airlech na mart no na daine marb; slan ce uachtnaigit risin niarunn s̄ na tuca ar cloich no ar fiacail he, 7 da tuca is fiach fon fath.
CIH I:288.15f

The exemption is the iron in slaughtering, i.e. he is exempt who takes the iron to butcher beeves or [the corpses of] dead people; he is exempt though he injures the iron, but not if he strikes it against a stone or a tooth, and if he does there is a penalty for this reason.

Four verbs contribute to to the terminology of heat-treatments, although their translations present difficulties. *DIL* ascribes to the group of terms derived from the verbs *fo-bádi*, *fo-beir*, *fo-geir* and *fothraicid* the meanings 'quenched, tempered, annealed', in what seems a somewhat arbitrary manner. While the nature of the changes in crystal structure only began to be unravelled in the 19th century, some smiths of the Irish, British and Continental Iron Ages at least had a good empirical understanding of the role of heat and rate of cooling in determining the properties of ferrous artefacts. The *DIL* citations noted in section 7.7 should, in the majority of cases relating to metals, be treated with the greatest circumspection. One can sympathise with the difficulties facing the lexicographers and suggest that when, in the future, any revisions are planned, those metallurgical references involving *fo-bádi*, *fo-beir*, *fo-geir* and *fothraicid* should be translated in their individual textual contexts.

In sect. 7.7, under *fo-beir*, the general meaning 'dips, submerges' would be more in keeping with quenching than tempering. The *DIL* citation, from *CMT* reads:

Rofoibred na fiarlanna isna lennaib fichecha a slessaib na saerclann
CMT §44

which is translated as:

The curved blades were *tempered* in boiling pools of blood in the thighs of warriors.

The passage from which this is taken, however, is a description of a furious battle, and I would venture to suggest that in this context, *quenched* is a more reasonable translation. The *DIL* offerings of 'tempering by immersion' (F:182.73) and 'annealing' (F:240.85) as meanings for *fobairt* are less than helpful. In the section of Trinity College Dublin manuscript H3.18:1337 on the form and value of the *somaíne*, we find:

Is e in biail innraic: airtem fod a genned, .iii. mer lethed a foebuir, tri mer ina maisil, tri lethad a cuil, ordlach a leth isa maisil frit, da mér a llethud anall. A deiccsi fo tri gortaib .i. donnghorad 7 banghorad 7 a gorad foberta; is he ni clui ibar na daurratus fora iarn; . . .
CIH III:921.28f

A proper axe is [as follows]: its head [lit. wedge] is an *airtem*[29] in length; three fingers in its *maisil*; three fingers in the width of its back; an inch its width in the *maisil* facing you, two fingers its width on the other side. It is examined in three heatings, i.e. dark red heat, white heat and a quenching heat. Its iron is not marked by yew or oak; . . .

The use of the adjective *gorm* 'blue (various shades)' in reference to weapons[30], is quite common, as in:

Rotogbadur . . . cloidme gera gormfaebracha.
They took up . . . [their] sharp, blue-bladed swords.
CMT §58

It is also found in conjunction with with *íarn* and *crúaid* (sect. 7.3), and could be taken as indicative of knowledge of tempering. After quenching steel, the smith will reheat the hard, brittle metal, judging its progress by the succession of oxidation colours on the surface. As the temperature rises, colours change from a straw colour at about 235°C, through purple at about 275°C, to blue at around 300°C. A good steel, properly tempered to blue (which, as an oxide coating would be fugitive) would have the right balance of hardness and elasticity. *IEW* (493–494) notes a connection between *gorm*, *fo-geir* (*DIL* F:230.64f) 'heats, inflames etc.', and *guirid* 'warms, burns' (*DIL* G:174.79f) and its verbal noun *gorad* 'heating, warming, burning', from a form *g^uher* 'heiss, warm', offering an interesting technological strengthening of an etymological link.

The two other techniques of the early blacksmith should be discussed here are *carburisation* and *welding*. Of these, carburisation has been largely covered in sect. 7.2 in the discussion of steel. This techique, which today is known as 'case hardening' and used to produce steel 'skins' on steel tool-parts, is the introduction of carbon atoms into the iron lattice by prolonged heating, under reducing conditions, in contact with charcoal (see Chapter 1.4) The term *aithlegtha* may, in ferrous metallurgical contexts be translated as 'reheated' (cf. note 11), a not unreasonable description of the basics of carburisation, hence the identification of *íarn aithlegtha* as the early Irish term for 'steel'. But there is another technique which involves the absorbtion by iron of an element which increases hardness of the stock – *nitriding* (Chapter 1.4) – in which nitrogen is introduced by replacing the charcoal of carburisation with such materials as dung. Now acicular precipitates have been observed during metallographic examination of early Irish iron artefacts and, since this might also be classed as a 'reheating' process, it is not impossible that *íarn aithlegtha* could also refer to nitrided metal although there seems no way, at present, of demonstrating this (cf. note 28).

Welding is a technique represented among the native-produced artefacts submitted to metallographic examination in the course of this study, although there are few native-produced examples of welding to rival, for example, the sandwich-constructed knives being used in Scandinavian Dublin. In particular, the welding together of iron and steel, something at which the Scandinavian smiths were accomplished, was not, apparently, widely practised by Irish smiths. It is a difficult process to master since each material has its own optimum welding heat. The terminology of welding is difficult, since the verbs *dlúthaid*, *táthaid* and *úaigid*, translated as including 'welds', have wide semantic fields, including 'joins, sews, compresses, brings together, unites, cements' (see sect. 7.7). The *DIL* examples for *dlúthaid*, *táthaid* and *úaigid* = 'welds iron' are unconvincing: it is *táth* that survives in Modern Irish as the term to cover this process. It may well be that *dlúthaid* and *úaigid* were merely figurative alternatives occasionally pressed into use. A more likely interpretation of these two would be 'rivetted' when referring to the jointing of metals, although there is a much more specific term in *seim* and its derivatives (*DIL* S:145.83f). None of these terms seem to include the progression from 'joins' to 'adheres to' to 'welds iron' as, for example, does Greek κολλάω (*LS* 972).

The few examples where it seems clear that a writer was implying welding as the technique for the joining together of components contain no special term. Thus, in the passage from *Táin* (4486–4487) cited previously we find the description of a sword as:

> . . . *claideb secht ṁbrattomon do iurn aithlegtha* . . .
>
> . . . a sword (made) of seven billets of carburised iron . . .

obviously referring to a blade made by fold-welding together pieces of carburised metal to give a blade with a relatively even distribution of carbon, capable of giving a uniform response to heat treatment.

* * *

7.7 – A Glossary of Early Irish Ironworking Terms

This glossary covers those terms used in connection with the mining of iron ore, its smelting, and the smithing of iron artefacts, and also highlights terms which have no genuine connection with iron. It has been compiled using the entries in the Royal Irish Academy *Dictionary of the Irish Language* (referred to as *DIL*) as the principal source. While the great majority of their translations cause no metallurgical problems, the compilers have included, due to the wide semantic fields of the various terms, some difficult, misleading and incorrect translations, particularly in terms of smelting and heat-treatments. Frequently they have taken over a translation uncritically from the publication in which it appears, perpetuating previous errors, notably referring to early ability to melt iron. Summaries are given of the various problems and difficulties discussed in earlier sections, and attention is also drawn to *DIL* entries where attribution of a ferrous metallurgical connection seems unfounded. Cognates and Modern Irish equivalents are noted where relevant. Terms which appear to the writer only to have been employed in non-ferrous metallurgical contexts are omitted unless of direct relevance to discussions. Entries are laid out alphabetically, as follows:

headword '*DIL* translation' [***F***ascicle:***C***olumn.***l***ine(s)], ***preferred translation (if any)***: *allied forms* '*DIL* translation' [*F*:*C*.*l*], *preferred trans.* (*if any*)
Words from the same semantic range (laid out as for *headword*)
Comments
Cognates (where appropriate)
Modern Irish equivalents (where appropriate)
Cross references.

Verbs are cited as 3rd pers. sing., pres. ind., nouns in the nominative. A summary of terms grouped by function is given as Table 7.7.1 at the end of this section.

* * *

A

airbuinne cpd.n. 'front spout (of bellows)' [A:170.69f], = ***tuyère***.

Joyce's translation (1903, II, 303) of *urbuinde* (leg. *airbuinne*: *Thr. Ir. Gl.* 36.29) as 'anvil snout' (i.e. 'bick iron') is untenable (cf. *soc*).
See also ***bolg***, ***derc***, ***nes***(***s***), ***séitid***

airgetlach n. 'mine' [A:209.4*f*], = ***silver mine***

A rare term, seemingly with only two occurrences, both in *Laws*. Although translated there as 'mine' (including of iron ore), the emendation proposed in sect. 7.2 makes 'silver mine' a much better translation.
See also ***claidid***, ***clasaige***, ***méin***

áith adj. 'sharp, keen' [A:252.36f]: see also *áithigid* vb. 'sharpens [A:265.23f]; *áithe* n. 'sharpness' [AI:256.71].

cf. *acarb* adj.'very rough, fierce' [A:14.5]; *aicher* adj. (< Lat. *acer*) 'sharp, keen, fierce' [A:95.73]; ? *airlíachtad* n. 'whetting (of a sword edge)' [A:225.26]; *fáebar* n. 'sharp edge (of a blade)' [F:10.5f], commonly used of sword-edges (Mallory 1982, 110f); *fáebarda* adj. 'sharp-edged, keen' [F:11.5f]; *fáebrach* (*fáebar*) adj. 'sharp edged' [F:11.11], used of iron weapons, as in . . . *.i. Slan dona feraib bis acomimfulang a feirgi deithbiri ce fuachtnaighidh iarann faebrach caich dib re cheile* . . . (*CIH* I:277.17f) 'i.e. the men who are sustaining their lawful anger are exempt though the *sharp* (= edged) iron of each of them injure the other . . . '; *fáebraige* n.'keenness, sharpness' [F:11.28]; Mod. Ir. *faobhraig* 'sharpen, whet'; *gér* adj. 'sharp, keen' [G:71.84f], Mod. Ir. *géar*; *meilid* (*melaid*) vb. 'grinds, crushes, sharpens' [M:82.21f], *mleith* vn. 'act of grinding' [M:151.12]; *meilteóir* n.'a grinder' [M:83.16]; *con-meil* vb. 'rubs (together), grinds; attacks, consumes' [C:449.20f]; *imbleith* vn. 'the act of grinding (on a whetstone)' [I:70.48]. The root of Ir. *meilid* and its derivatives are very widely distributed in the IE languages, e.g. O. Icel. *malmr*, Mod. Swed. *malm* 'ore'.

With the *comparanda*, this forms a general set of terms referring not only to the sharpness particularly of the cutting-edges of weapons, but also in the wider transferred sense to behaviour and personality.
See also ***bró***, ***límaid***

aithlegad cpd. (*aith-* + *legad*, vn. of *legaid*) 're-smelting, refining' [A:268.1f], = ***carburising (of iron), refining (non-ferrous metals in general, e.g. gold)***: see also *aithlegtha* part. 'resmelted, refined' [A:268.7f]; = ***'carburised (iron), refined (of non-ferrous metals)'***.

The word *aithlegad* translates literally as 'remelted'. While the verb *legaid* certainly does refer both to the smelting and refining of non-ferrous metals (Scott 1981a, b), the translation of *aithlegad* 're-*smelted*' is a nonsense, unless it referred specifically to the smelting of hammer-scale (pure iron oxide), a process for which we have no evidence in early Ireland. To translate as 'refining' in the context of early bloomery iron technology is not directly acceptable, although it would be unremarkable with non-ferrous metals. As with the other terms for high-temperature metallurgical operations, the implication of a molten state belongs to the original use in a non-ferrous metallurgical context. If *aithlegad* is used in connection with iron, then it may be understood to refer to a re-heating process (cf. the connection with *bruth* – under *berbaid* below) and 'carburise' is thus the most acceptable translation, with the term *íarn aithlegtha* referring to 'carburised iron, steel' (sect. 7.3). And if the early Irish regarded steel as a 'purified' form of iron, then the meaning 'refined' could be inferred. In later periods, when iron was produced by the indirect process, the term would be appropriately translated as 'remelted iron' i.e. cast iron which had then been decarburised under oxidising conditions: but this usage would scarcely pre-date the late 16th/early 17th century introduction of blast furnace technology to Ireland.
See also ***berbaid***, ***bruinnid***, ***crúaid***, ***legaid***

B

benaid vb. 'beats, strikes' [B:71.83ff].

Also 'hammers' as in *nach indéuin benar/ for-cain in cách no-da-ben* 'any anvil that is struck teaches him who strikes it' (Carney 1958, 19).
See also ***búailid***, ***túarcon***, ***urlaide***

berbaid vb. 'boils, cooks; melts, dissolves' [B:82.39f], = ***'smelts (of metals)'***: see also *bruith* vn. 'act of smelting, forging, refining, boiling, seething, cooking' [B:215.35f]; *bruithe* part. 'smelted, refined' [B:215.56], = ***refined of non-ferrous metals?***; *bruth* n. '(glowing) mass, lump; charge (of metal)' [B:217.20f], = ***iron bloom undergoing forging***; *bruithne* n. '(*furnace*) charge of metal' [B:216:12] = ***bloom***; *bruthdamnae* cpd. n. (*bruth* + *damnae*) 'specific charge or amount of metal (for melting or shaping)' [B:217.59], = ***raw bloom, product of a smelting operation***; *bruithenda* 'fiery, glowing' [B:215.79]; *bruithned* vn. 'smelting, refining' [B:216.20]; *bruithneóir* n. 'smelter' [B:216.24].

cf. *athbruithned* part. 'refined' [A:450.60] (c/w *cáer*).

Bruth would refer to impure metal, a raw bloom before or during consolidation (Scott 1981b, 153–154). The implication of the molten state is inappropriate when to ferrous metallurgical processes, except when referring to the melting of the *slag*. As with *bruinnid* and *legaid*, this is a part of the semantic field carried over from earlier non-ferrous metallurgical origins. The strong implication is of 'heat, glowing hot' from a furnace and reducing conditions. Thus, *bruth* should principally refer to the glowing bloom from the smelting furnace. Its use in connection with the term *íarn aithlegtha* is interesting when considered in connection with sword-blades, as in . . . *claideb do secht ṁbrattomon do iurn aithlegtha* '. . . a sword made from seven billets of carburised iron' (*Táin* 4486): this suggests that *bruth* also had proper usage referring to stock under carburisation in the reducing zone of the forge fire.

IEW (143–144) derives *berbaid* from a form **bh(e)reu-*, **bh(ĕ)ru-* 'Garen, Brauen, Kochen usw.', thence from **bher-* 'von quellendem oder siedendem Wasser'(*IEW* 132–133). From this latter comes O. Eng. *bræð* (Engl. 'breath') 'Dunst, Hauch, Wind' and O. Icel, *bráðna* 'schmelzen', and O. ir. *topur* 'well, spring' (*CCG* §512: < **brend-na-* >1 *bruinnid* 'springs forth, flows').

Cognates include Lat. *ferveō* 'boil, seethe etc.'.
See also ***aithlegad***, ***bruinnid***, ***cáer***, ***tanalaig***, ***tinne***, ***legaid***

boingid vb. 'breaks, smites, strikes, cuts, reaps etc.' [B:136.23ff], = ***quarries ore, crushes? ore (in metallurgical contexts)*** as in . . . *boingidh mein in uma* . . . ' . . . they quarry the copper ore . . . ' (*CIH* V:1571.19): see esp. *búain* vn. 'act of cutting, breaking' [B:225.35f], also of quarrying, crushing(?) or breaking ore, as in . . . *oc buain míanna* . . . (*Met. Dinds.* iv, 308.6).

In the passage from *Met. Dinds.*, *boingid* and *bruïd* refer to separate processes – quarrying and crushing (as part of the preparation of ore for the smelting furnace). It is tempting, however, to suggest that *boingid* might also include the breaking up of ore in its semantic field.
See also ***bruïd***, ***búaidrid***, ***méin***

bolg n. '(smith's) bellows' [B:138.80*f*].

Bolg 'bag' includes 'bellows' as an extension to the semantic field, covering 'containers made from animal skin'. In this meaning there are many cognates in the IE language group (e.g. *IEW* 125–126). For 'bellows', semantic parallels include Lat. *follis* 'skin, leather wine-bottle, bellows', Gk. ἀσκός 'skin container, bellows, Indo-Aryan *bhástrā* (sect. 7.1).

Cognates include Goth *blasebalg*, Engl. *belly*, *bellows*.

Mod. Ir. *boilg* (*gabha*) '(smith's) bellows' (*Ó Dónaill* 123).
See also ***airbuinne***, ***derc***, ***nes***(s), ***séitid***

bró n.'quern, millstone, grindstone, the act of grinding' [B: 194.78f].

cf. *airnem* n. 'whetstone' [A:233.32]; *lia* n. 'stone, whetstone' [L:142.35ff].

Also used of the upper millstone.

The cpd. *liacc-bron* appears in Mod. Ir, as *liabró* 'whetstone, grindstone' (*Dinn.* 660; *O Dónaill* 781).
See also ***áith***, ***límaid***

bruïd vb. 'breaks in pieces, smashes, crushes' [B:210.74f]: see esp. ? 4 *brú* = *brúd* vn. 'crushing of ore?' [B:207.82], = ***ore dressing (in metallurgical contexts)***.

The example of *brú* in *Met. Dinds.* iv, 308.47 appears to be an exceptional form in which the final *-d* (= *-dh*) is omitted (as in common speech in the 12th century) for the sake of the rhyme with Benntraigiu.
See also ***boingid***, ***búaidrid***, ***méin***

2 bruinnid vb. 'smelts, refines' [B:214.35]: see also *bruinnte* part. 'refined, welded(?)' [B:214.69], = *smelted, refined; airbruinniud* vn. 'act of smelting' [A:170.60].

Most commonly *bruinnid* is used of non-ferrous metals (conspicuously of gold – Scott 1981b) where operations in the molten state were commonplace. *Thurn. Gramm.* (§355) gives *bruinnid* as a weak *i*-verb as opposed to the strong BII verb *bruinnid* 'springs forth, flows', while *DIL* suggests *bruinnid* 'smelts' as a special use of *bruinnid* 'flows'. Examples of 2 *bruinnid* are all later than most of those of 1 *bruinnid*, and there was a general tendency for strong verbs in Middle Irish to adopt the present stem throughout the system, which seems to have happened here. Like *berbaid*, 1 *bruinnid* is ultimately derived from a form **bher-* 'springing, gushing, seething etc. water', and it is to this source that implications of liquation may be traced, through the period of non-ferrous metallurgical usage when melting of metals was an unremarkable achievement. As with *berbaid* and *legaid*, confusion may be introduced by the fact that it is also appropriate to the molten state of the slag during iron smelting. The suggested translation of *bruinnte* [B:214.72] as 'welded' is untenable.
See also ***aithlegad***, ***berbaid***, ***legaid***

búaidrid vb. 'disturbs, troubles, (of water) stirs up, muddies, makes turbid' [B:223.46f]: see esp. *búaidred* vn. 'act of troubling, stirring up, etc' [B:223.15f] = ? *buadradh* n. 'gangue, effluent' – Ó Corráin 1974, 68: *Contrib.* 281; *Vendr. Lex.* B-108.

Although the verbal noun occurs in a passage relating to the activity of miners (see sect. 7.2), we do not know if it was specifically iron ores that were involved. Concentrating of iron ore in this manner would, however, be quite appropriate.
See also ***boingid***, ***bruïd***, ***méin***

búailid, ***búailaid*** vb. 'beats, strikes' [B:224.49f]: see also *búalad* vn. 'attacking, beating, blow, stroke etc.' [B:226.73f].

Also 'forges', as in *Mad si a indeoin . . . is a ic d'fir bualta*. (*CIH* III:936.24) '. . . If it be the anvil . . . it is the man who strikes who pays it.

Mod. Ir. *iarann a bhualadh ar inneoin* 'to beat iron on an anvil' (*Ó Dónaill* 712).
See also ***benaid***, ***túarcon***, ***urlaide***

C

cáer n. 'a mass of cast metal (usually iron)' [C:19.19f], = ***a consolidated bloom, an ingot, a cast mass of metal (other than iron).***

The *DIL* rendering is unacceptable for ferrous metals. A satisfactory translation is 'metal ingot', referring to a piece of stock or a consolidated bloom, i.e. metal ready for working up. It could also refer to the raw bloom, since an alternative meaning of *cáer* is 'globular mass, berry', which in ferrous metallurgical contexts could nicely describe the botyroidal flow patterns of molten slag on the surface of the bloom (Scott 1981a). Knott draws attention to the further meaning of *cáer* as found in the expression *caer chomraicthe* 'focus, of objects grouped round a nucleus forming a more or less solid mass' (*BDD*[2] 105).
See also ***berbaid***, ***tanalaig***, ***tinne***

cepp 'tree-stump, log, block, anvil block' [C:136.16].

Used of the block on which the anvil was placed, as in . . . *in cepp i mbái ind inneóin* '. . . the block on which the anvil stood . . . ' (Joynt 1910, 110.71), . . . *t'indéin do béimim i cepp* ' . . . a striking of thy anvil into the block' (O'Neill, J. 1907 108.53).

Probably a borrowing from Lat. *cippus* 'tree-stump'.

Mod. Ir. *ceap* 'anvil base' (*Ó Dónaill* 712).
See also ***indeóin***, ***soc***

cerdchae n. (cpd. of *cerd* + *-cha* < *cáe*: *DIL* C:9.84ff) 'workshop, smithy' [C:140.64f].

From *cerd* 'artisan, (non-ferrous) metalworker' also 'handicraft, art, skill' [C:139.22f]. The Goidelic dialects differ markedly both from their Brittonic counterparts, and from other IE languages in that the words for 'blacksmith' and 'smithy' are unconnected (sect. 7.4).

Cognate with Welsh *cerdd* 'craft, song', Gk. κέρδος 'gain, profit, craft', Lat. *cerdō* 'labourer, craftsman'.

Mod. Ir. *ceárta* (*Ó Dónaill* 212).
See also ***gobae***

claidid vb. 'digs, digs up, excavates, mines' [C:214.19f], as in *Slan donti claides an mein . . .* (*CIH* I:270.15) 'I deem to be exempt the person who mines ore . . . ' : see also *claide* (? vn.) 'act of digging, digging out etc.' [C:212.72f].

cf. *tochailt* (vn. of **do-claid*) 'act of digging up, excavating, mining' [T:202.42f].

Claide is used interchangeably with *tochailt* in the versions of *Bla Miand Midchlais* (*claide* in *CIH* I:270.15: *tochailt* in *CIH* III:940.11: *tochailt* and *claide* together in *CIH* IV:1256.4). We find elsewhere that *claide mianna* 'mining ore' is glossed *.i. tochailt na miana* (*CIH* I:239.10).
See also ***clasaige***, ***méin***

clasaige n. (<*clas* 'ditch, trench, furrow' [C:222.39f] <*claidid*) 'ditcher, miner' [C:223.64].
See also ***claidid***, ***méin***

crithir n. 'spark, flash, spurt of flame, particle, drop' [C:535.51f].

Used to denote the sparks of hammer-scale from hot iron, as in . . . *crithri in iaraind 7 in tellaig . . .* (*CIH* I:267.30) . . . 'sparks from the (hot) iron and sparks from the forge-hearth . . .'.
See also ***drithle***, ***slaid***, ***spréd***, ***tuirenn***

crúaid as adj. 'hard, harsh etc.', as subst. 'steel' [C:556.29ff]: see also *crúad(aig)id* vb. 'hardens, stiffens, dries in a kiln' [C:555.57]; *crúadugud* vn. 'act of hardening, stiffening (also drying corn)' [C:556.15].

The translation as 'steel' before the early Modern

Irish period is tenuous, since most usage is metallurgically ambiguous. It might possibly have referred to steel in a few cases, but the best candidate for the early Irish term for steel is *íarn aithlegtha*. The *DIL* entry *crúadán* [text *chruadhan* – C:556.1] is a misreading of, or scribal variant for *crúan* 'red enamel. The verb *crúad(aig)id* (which, if *crúaid* meant 'steel', might reasonably be expected to include 'hardening of metals' in its semantic field) seemingly has no metallurgical connections, with the exception of the passage from *CCath* quoted in sect. 7.6, and refers consistently to the treatment of organic materials (corn, wood).

Mod. Ir. *cruach* 'steel' (*De Bhaldraithe* 704).
See also ***aithlegad***, ***íarn***

D

derc n. 'hole, aperture' [DII:33.45f] = ***bellows protector***.

Derc or *licc dercáin* refers to the perforated clay discs set up in front of the bellows in the forge hearth (Germ. *Essestein*: cf. Fig. 6.5.7h-j and Fig. 6.5.9). Stone examples of bellows protectors are well known from Europe and Scandinavia, but yet have to be identified here. Nevertheless, we may use this translation with confidence, for example, in the passage . . . *ba binde no chandais dord / fri derc a niss a da bolg* . . . '. . . sweetly his two bellows chanted a refrain against the bellows protector of the forge hearth . . .' (*Thr. Ir. Gl.* p. 32). *Derc* 'bellows protector' thus stands in contrast to *airbuinne* '(tubular) tuyère'.
See also ***airbuinne***, ***bolg***, ***nes(s)***, ***séitid***

derg adj. 'red, ruddy, red-hot, incandescent (of red-hot iron)' [D:35.77ff]: see also *dergaid* vb. (b) 'makes red-hot, heats, kindles' [D:37.40].

Also used in non-ferrous metallurgy in cpd. *dergór* to describe gold alloyed with copper (Scott 1981c).

Mod. Ir. *an t-iarann dearg* 'the red-hot iron' (*Ó Dónaill* 384).
See also ***berbaid***, ***bruinnid***

dlúthaid vb. 'makes close, compresses, condenses, joins, consolidates' [D:169.62f]: see also *dlúthad* vn. 'pressing together, condensing, confining, joining, connecting, welding, binding' [D:169.36f]; 1 *dlúth* adj. 'close, compact, dense' [D:168.32f]; *dlúthaigid* vb. 'brings together, presses together, makes compact' [D:170.7f].

cf. *táth* 'joining, welding, soldering, binding' [T:89.46f]; 1 *táthaid*, *táithid* vb. 'joins, welds, unites' [T:90.52f]; *táithe* part. 'welded' [T:90.83]; *táthaid* n. (*táth*) 'welder, smith, artificer' [T:91.4]; *táthaigid* vb. (*táth*) 'welds, cements, joins' [T:91.8]; *táthugad* vn. 'soldering' [T:91.30]; *táthaigtheóir*, *táthóir* n. 'welder, uniter, consolidator' [T:91.44]; *táthluib* 'cement' [T:93.10]; Mod. Ir. *táthaire* 'welder' (*De Bhaldraithe* 837), *táthaigh* 'to weld, solder, bind' (*Ó Dónaill* 1213); *úaigid* vb. 'stitches together, sews, unites, joins' [U:9.3f]; 1 *úaimm* vn. 'act of sewing, a seam, stitching, fastening, act of joining together, welding' [U:12.8f]; *úaigél* later vn. 'sewing, stitching, a seam, welding' [U:8.72ff]; *imm-úaimm* cpd. 'the act of joining or welding closely' [I:158.69].

This group of words seems to have little or no early connection with the fusion of hot iron or steel under the hammer. It appears likely that the various compilers of *DIL* have been influenced by modern usage of words such as *táthaím* 'welds' (*De Bhaldraithe* 837). Where citations refer to the joining of components, the sense seems to be invariably that of physical jointing (rivetting, stitching etc.), rather than fusion. While 'weld' in the figurative sense is acceptable, this group should really be excluded from early ferrous metallurgical terminology.
See also ***seim***

do-gní vb. 'does, 'makes, forms, constructs, practises' [D:291.8ff], as in *GOBAIND .i. doniat ind goibneacht*. . . (*CIH* I:268.10) '. . . blacksmiths, i.e. they practise the art of blacksmithing . . . ' : see also *dénum* vn. 'act of doing, performing etc.' [D:24.65f], as in *roboí Gobnenn goba isin cerdchai ag dénam calc 7 gaí 7 sleg* . . . 'Gobniu the smith was in the forge making swords and spears and javelins . . .' (*CMT*[2] 54.530); *dénta(e)* part. 'made (by an artificer), skilfully fashioned, elaborate' [D:24.35]; 1 *gním* vn. (of *gníid*) 'deed, action, work etc.' [G:121.54ff]. A general term denoting 'works, makes effort, fabricates etc.'.

Cognates include Lat. *gignō* 'beget, make', Mid. Welsh *gnif* 'labour, toil' (*CCG* §563).
See also ***imm-beir***

drithle n.'spark' [D:399.85f], = ***hammer-scale*** (***in ferrous metallurgical contexts***).

cf. *lad* n. 'sediment' [L:17.2f] = ***'hammer-scale'***, glossed *ferrarum* (leg. *ferrarium, ferraium*: *RMLW* 189) which refers to the hammer-scale that flakes off hot iron when thrust into water; 2 *oíbell* n. 'a spark or flame of fire' [N-O-P:115.3f].

The meaning 'hammer scale formed during forging' may be seen from . . . *caor ceardcha ag cur drithlionn* . . . [D:400.9] '. . . an ingot in the smith's forge giving off sparks . . . '.
See also ***crithir***, ***slaid***, ***spréd***, ***tuirenn***

F

felmac n. 'pupil, student, apprentice, including the pupil of a blacksmith' [F:73.76].

Well illustrated in *Blaí Ord Indeóin* (sect. 7.5).

cf. *foglamanta* 'apprentices' [F:234.5].
See also ***gobae***, ***ollam***

fo-bádi vb. 'submerges' [F:182.34]: see also *fobdud* vn. 'dipping, submerging, quenching' [F:183.18f], as in . . . *mád chían co tairi a fobdud*. (*LU* 7706) 'if its quenching be delayed'; *fobádud* late vn. 'submerging, immersion' [F:182.39]; *foibdide* part. 'dipped, immersed, (of metals) annealed' [F:240.85] = ***quenched*** (*in ferrous metallurgical contexts*).

Translations offered in *DIL* are inconsistent, suggesting lack of understanding of the terms 'quench', 'temper' and 'anneal' in the minds of the lexicographers. The suggestion of 'tempering by immersion', unless specifically referring to tempering in a heated

oil-bath (for which the only authority would appear to be *CCath* 5211f: sect. 7.6), seems improbable. No metallographic evidence for such a technique in use in early Ireland is as yet forthcoming. It may possibly stem from confusion with the modern term 'soaking'. There is also confusion particularly with forms from *fo-beir*. It seems most likely that forms from both *fo-bádi* and *fo-beir* refer to quenching, either the initial immersion from high temperature or that which would *follow* tempering rather than tempering (see sect. 7.6), and not at all to annealing, for which forms from *fo-geir* would seem better candidates. The *DIL* translations from metallurgical contexts must be treated with great caution, and really in many cases are best set aside.
See also ***fo-beir***, ***fo-geir***, ***fothraicid***

fo-beir vb. lit 'brings under, dips, submerges' [F:183.27], = ***quenched (in ferrous metallurgical contexts)***: see also *fobairt* vn. 'dipping, immersion, a tempering or heating in the fire, of tempering by immersion' [F:182.65]; *fobride* part. 'dipped, tempered, annealed' [F:184.62]; *foibred* (analogous vn.) 'dipping, immersing, annealing' [F:241.4]; *foibrigid* vb. 'anneals' [F:241.9]: see also *foibrigthe* vn. 'tempered' [F:241.11], illustrating the confusion in *DIL* over the nature of quenching, tempering and annealing.

As with *fo-bádi*, it certainly seems more prudent to translate *fo-beir* and forms from it as 'quenches' rather than 'tempers'. In the context of battle, the *DIL* example . . . *Rofoibred na fiarlanna isna lennaib fichecha a slessaib na saerclann*, is surely better translated '. . . The curved blades were *quenched* in boiling pools of blood in the thighs of warriors.' (*CMT* 44.17–18). Where it is possible to isolate a definite sense, it is of immersion in liquid from glowing heat, as opposed to the lower temperatures in tempering that precede the final quench, although, as with *fo-bádi*, this latter is also a possibility, as is that of tempering in a hot liquid. The *DIL* translations would be best set aside, or at least treated with extreme caution. It is fair to say also that there is much later confusion with forms from *fo-bádi* and *fo-geir*.
See also ***fo-bádi***, ***fo-geir***, ***fothraicid***

fo-geir vb. 'heats, inflames . . . tempers, seasons' [F:230.64f]: see also *fagairt* vn. 'tempering . . . hardening, seasoning' [F:18.63]; *fagairthe* part. 'burnished, polished' [F:18.67]; *foigred* late vn. 'heating, burnishing' [F:245.17].

cf. *guirid* vb. 'warms, burns' [G:174.79].

The strong implication of heating (<*gor-*, *guir-* vb. 'heats') suggests tempering. Given the use also for 'seasons' and 'hardens', annealing is a less likely interpretation.

Cognates of *guirid* include Lat. *formus* 'warm', *fornus* 'oven, furnace' (*IEW* 493–494).

Mod. Ir. *faghairt* 'temper' (*De Bhaldraithe* 745; *Ó Dónaill* 504).
See also ***fo-bádi***, ***fo-beir***, ***fothraicid***

fothraicid vb. 'bathes, washes, immerses' [F:395.64]: see esp. *fothrucud* vn. 'bathing, washing . . . annealing' [F:396.49f].

If this has a genuine metallurgical use, then either 'tempers' or 'quenches' would seem a more like meaning than annealing. The ferrous metallurgical connection however is exceedingly tenuous.
See also ***fo-bádi***, ***fo-beir***, ***fo-geir***

G

gobae n. 'a smith' [G:127.58f]; see also *goibne* 'smith's craft' [G:128.65]; *goibnecht* 'the craft or calling of a smith' [G:128.68]; *goibned* 'doing smith's work(?)' [G:128.80]; *goibnide* 'pertaining to a smith's work' [G:128.85].

For detailed discussion, see sect. 7.4.

Cognates include Scots. *gobhainn*, Welsh *gof*, Corn. *gof*, Bret. *gof*, *gov*, Gaul. **gobanno-*.

Mod. Ir. *gabha* 'blacksmith' (*Ó Dónaill* 598).
See also ***cerdchae, ollam***

gorm adj. 'blue (gen. deep blue, cærulean)' [G:137.73f].

cf. *glas* adj. 'shades of light green and blue, passing from grass-green to grey' [G:95.13ff], also 'metallic sheen, blued(?)', as in *glas-iarand*; *dub* 'black, swarthy' [D:425.38f], as in *dubga* 'black spear' (cf. Eng. *blacksmith*).

Also used of 'metallic sheen, steel blue', as in . . . *o chruaidh gurm...* 'steel blue(?)'. We may note in passing the relationship of *gorm* to the group of words which IEW (493–494) derives from *$g^{u}_{\circ}her$- 'heiss warm', in view of the semantic connection with 'steel'. During tempering, the heating causes oxidation of the surface of steel which undergoes a series of colour changes from light straw to a dark blue-purple.
See also ***aithlegad***, ***crúaid***, ***fo-geir***

gúal n. 'charcoal, coal [G:169.41f]; see also *gúalach* 'coal, charcoal' [G:170.33].

cf. *cúal* 'faggot, bundle of sticks (firewood)' [C:572.74].

Use in connection with ferrous metallurgy may be seen, for example, in the tract *Críth Gabl.* (§15.191 = *CIH* III:780.10) where, among the possessions of the *mruighfer* (a grade of farmer) is listed . . . *míach gúaile fri herna* . . . '. . . a sack of charcoal for [working] the irons . . .'.

Mod. Ir. *gual* 'coal', *gual gabha/ceárta* 'charcoal' (*Ó Dónaill* 677).

I

íarn n. 'iron, a weapon or implement made of iron' [I:28.69ff]; see also *íarnach* '(iron) fetters, ploughshare, file(?)' [I:29.49]; *íarnaide* 'iron, made of iron, an iron implement' [I:29.70f]; *íarndae* 'iron, made of iron' [I:30.15].

Not referring *sensu stricto* to pure iron, but rather to the relatively impure bloomery product which could be distinguished from its much harder alloy with carbon – 'steel' (i.e. carburised iron). In early periods, the word would certainly have referred to a material of highly variable quality and properties.

Cognates include Welsh *haearn*, Corn. *hoern*,

Bret. *houarn*, Celt. **īsarno-*, Goth. *eisarn*, O. Norse *járn*, Germ. **īsarna-*. The suggestion of a link between **īsarno-* and Skt. *ayas* through a form **ai̯os* seems unlikely.

Mod. Ir. *iarann* 'iron'; *iarann buailte* 'wrought iron' (*De Bhaldraithe* 377), *iarann saoirsithe* 'wrought iron' (*Ó Dónaill* 690).
See also ***aithlegad***, ***crúaid***

imm-beir vb.'plies, handles, wields, uses, works (in a metal)' [I:132.76ff]: see esp. *imbert* 'act of using, employing. handling a weapon or tool etc.' [I:69.15], as in . . . *Masa fis fir imerta 7 anfis a gobaind*. . . '. . . If it be with the knowledge of the striker [*i.e. the person wielding the hammer on the anvil*] and with ignorance on the part of the smith . . . ' (*CIH* III:936.21: sect. 7.5).

cf. 2 *gres(s)* n. 'an essay, bout, attempt, the amount of anything done at a time' [G:155.8ff], as in . . . *grod-greasa gaibnige le h-ordaibh* . . . '. . . keen blows of smithcraft with hammers . . .'.
See also ***do-gní***

indeóin n 'anvil' [I:224.45f].

Also the lower millstone of a rotary mill. There is an interesting parallel with Arm. *sal* 'flint, flagstone', modern meaning 'anvil'. It is also interesting to note that the word retained the subordinate meanings 'block of stone, navel' in modern Irish.

Cognate with Welsh *einnion* 'anvil'.

Mod. Ir. *inneoin* 'anvil, block of stone' (*O'Reilly* 305).
See also ***cepp***, ***soc***

L

legaid vb. '(trans.) melts, smelts, (intrans.) dissolves etc.' [L:72.34ff]: see also *legad* 'act of melting (away), dissolving, smelting' [L:71.74f]; see also *legtha* part 'molten, melted' [L:74.13].

DIL cites the *Aen.* (2500) translation of *secht-legtha* as 'seven times tempered' which is no better than those translations which include 'molten' in connection with iron, since *tempering* steel a number of times in succession is a quite useless process. As a verb meaning 'smelts (iron)', *legaid* is quite acceptable despite including 'melts' in its semantic field (cf. *berbaid* and *bruinnid*). *Legaid* and forms from it finds perfectly respectable use in connection with non-ferrous metals (Scott 1981b, 154–156). It is used in the same (ferrous metallurgical) sense as Lat. *liquo* (sect. 7.3).

Cognate with Welsh *dad-leithio* 'to smelt' (*IEW* 657).
See also ***aithlegad***, ***berbaid***, ***bruinnid***

límaid vb. 'sharpens, grinds, polishes' [L:154.27f]; see also *límad* vn. 'whetting, sharpening, polishing' [L:154.19]; *límugud* vn. 'act of sharpening' [L:154.70]; *límtha* part. 'sharpened, burnished (of weapons, edged tools)' [L:154.47].

cf. *grenaid* vb. 'carves, engraves' [G:153.57]; *grenta* 'carved, ground, polished (of weapons)' [G:154.77]: *slípaid* vb. (Germ. = O. Norse? borrowing) 'smoothes, polishes, whets' [S:276.1]; see also *slípad* vn. 'act of polishing, smoothing, whetting' [S:275.79].

It is interesting to note that the various words denoting 'removal of metal by abrasion' seem invariably to refer to grinding with siliceous materials. Now, while this is undoubtedly the best technique for producing a really sharp edge (cf. the use of oilstone to prepare the edges of wood chisels), the metal file (of which there are several cuts) is a vital tool of the metalworker. Given the paucity of metal files in the archaeological record, we could possibly see a concordance between the terminological bias to the grindstone, and the scarcity of references to the use of roughened metal surfaces. We may note in passing the section in *Lebor Aicle* entitled *Bla liac limaid* (*CIH* I:290.27f).
See also ***áith***, ***bró***

M

méin, also ***mían*** n.'mineral, ore, metal(?)' [M:83.17f]; see also *mían* 'mineral, ore' [M:127.37]; *mianach* 'ore, a vein of ore, a mine' [M:127.46]; *mianaige* 'a miner' [M:127.61]; *mianaigid* 'breaks ore' [M:127.72].

Like Gk. μέταλλον, *méin* was used to refer both to ore and to the metal smelted from it. The *DIL* quote, under *claide*, on mining repeats the rendering *rúd* as 'iron ore' (sect. 7.2: cf. note 6).

Cognates include Welsh *mwyn*, Eng. *mine* (*Vendr. Lex.* M-29).

Mod. Ir. *mianach* 'ore', *mianach iarainn* 'iron ore' (*De Bhaldraithe* 497).
See also ***búaidrid***, ***claidid***, ***clasaige***

meirg n. 'rust, corrosion' [M:87.5f].

Used of metal decay in the broadest sense.

Mod. Ir. *iarann ag tógáil meirge* 'iron rusting' (*Ó Dónaill* 851).
See also ***slaid***

N

nes(s) n. 'a wooden mould for shaping a smith's clay furnace, the clay furnace itself' [N-O-P:37.70f], ='***clay-lined furnace or forging hearth, a former for tuyères, possibly also a clay crucible (non-ferrous metals)***'.

cf. *sornn* n. 'a furnace, kiln, oven' [S:347.64*f*], a loanword from Lat. *furnus*, though with no immediate metallurgical connections; cognate with *guirid*, *fo-geir*.

The *DIL* offering of a 'mould for a furnace', i.e. a 'former' for a furnace (following Joyce 1903, II, 238) seems quite improbable. The meaning 'former' might refer to a stout stick used to form tuyères round (and by Gobniu to hit unwelcome visitors!). It might be possible to see a (late) use as a word for 'clay mould', since acceptable early terms for these are otherwise not easy to identify. But in the context of the proferred examples 'furnace, forging-hearth, hearth' seems more secure, particularly in the lines from the elegaic verse *ba binde no chandais dord / fri derc a niss a da*

bolg 'sweetly his two bellows chanted a refrain against the bellows protector of the hearth (*Thr. Ir. Gl.* 32). From other uses implying firing of clay vessels, it seems best to interpret *nes*(*s*) generally as referring to a clay lined (or built) heating chamber.
See also ***airbuinne***, ***derc***, ***teine***, ***tellach***

O

ollam 'the highest grade of *fili*, later an expert in any art of science' [N-O-P:138.45ff].
cf. *prím-* 'foremost, chief, principal . . . at the peak of one's profession' [N-O-P:203.30*f*], as in . . . *is hé fá primh-gobha* . . . 'he was chief smith . . .' , also *prímgoibnech* 'highest quality blacksmithing'.
The word *ollam* includes blacksmithing, as in *secht n-ollumain ghabhunn* 'seven master smiths' (*Lism. L.* 2931). The *DIL* comment that use of *ollam* for other than *fili* seems wrong, unless it is intended to refer merely to the dates of the texts in which this more general usage appears. Its use with *fili* is only one (the commonest) application: it occurs in 7th/8th century poetry for warriors [N-O-P:139.5ff].
See also ***gobae***

ord n. '(sledge-?)hammer' [N-O-P:155.45].
In early usage a generic term for hammers of all sizes.
Mod. Ir. *buille oird ar inneoin* 'the striking of a hammer on an anvil', *ord ceárta* 'smith's heavy hammer' (*Ó Dónaill* 712, 933).
See also ***búailid***, ***benaid***, ***indeóin***, ***túarcon***, ***urlaide***

S

seim n. 'a rivet' [S:145.83f], as in *Dugníth dano Crédne cerd na semonn* . . . 'Then Crédne Cerd would make the rivets . . . ' (*CMT*^{2} 54.536).
A term used to refer both to iron and non-ferrous rivets used in jointing, and to decorative studs (e.g. *scíath* . . . *semandaib findruine* 'a shield . . . with decorative studs of *findbruine*: *BDD*^{2} 967).
See also ***tairbir***

séitid vb. 'blows, blows across, wafts, of blowing bellows' [S:158.6f]; see also *séited* vn. 'act of blowing, a blast' [S:157.23f]; *séitire* n. 'a bellows, a blower' [S:159.5].
cf. *trefet* n.'act of farting, bellows(?), (bellows) blower' [T:289.22], cpd. of prep. *tre* + *séit* (<*séitid*: *Vendr. Lex.* T-32) = ***'tuyère'***.
This verb also finds use in non-ferrous contexts, e.g. in the cupel refining of gold (Scott 1981c).
See also ***airbuinne***, ***bolg***, ***derc***, ***nes(s)***, ***side***

side, ***sithe*** n. 'a blast or gust (of wind)' [S:219.67ff], used with *bolg* in cpd. *bolcsithe* glossing Lat. *follis* (*Thes.* II, 122.67a2); see also *sidemnaig* 'blowing (of bellows)' [S:221.74].
See also ***airbuinne***, ***bolg***, ***derc***, ***séitid***

slaid n. 'scrap metal' [S:258.7f], = ***hammer-scale***; see also *slaide* n. 'scrap metal' [S:258.34f], = ***hammer scale, rust***; *slaidred* 'metal dross' [S:258.85].
cf. *slaigthech* 'dross of metals' [S:259.3] = ***'hammer-scale, smithing slag, crucible slag (non-ferrous metals)'***.
The *DIL* translations of *slaid* and *slaide* are misleading, since the citation for *slaid* is . . . *slaid cacha cerdca, luaith cacha tellaig* . . . which should be rendered ' . . . the hammer-scale from every forge, the ashes from every hearth . . . '. For *slaide*, the correct translation of the cited gloss *in t-iarann 7 in slaidhi 7 in tuirenn* set against *ferrum ferrugo et scama ferri*, should be 'the iron, the rust/scale, and the hammer-scale'.
See also ***crithir***, ***drithle***, ***meirg***, ***spréd***, ***tuirenn***

soc n. 'a plough share, a snout' [S:316.76*f*], = ***bick-iron, anvil snout***.
Mod. Ir. *soc na hinneonach* 'anvil snout, i.e. bick iron' (*De Bhaldraithe* 27).
See also ***cepp***, ***indeóin***

spréd n. 'spark' [S:360.49*f*]; see also *sprédach* collect. 'fragments' [S:360.73].
Spréd occurs in metallurgical contexts as, for example, . . . *mas iad sbreide na teinid 'ga seidid* . . . (*CIH* III:936.35) ' . . . if it be sparks from the fire through blowing . . . '. It is, however, more likely to refer to 'sparks' in general, in contrast to *crithir* and *tuirenn* which appear to be more specific terms for hammer-scale or smithing slag.
See also ***crithir, drithle, slaid, tuirenn***

T

tairbir n. (cpd. with *bir* 'stake, spit, point, spear, spike' B:103.52f) 'spike, pointed weapon or implement' [T:33.63*f*] = ***punch, drift*** (*in blacksmithing*).
The term appears in *CMT*^{2} (54.536) in the sentence *Dugníth dano Crédne cerd na semonn . . . 7 níbo écen tairbir remib:* . . . which is best translated 'Crédne the brazier would make the rivets . . . and it was not necessary to punch holes for them: . . .'. Given that the second element is *bir*, the obvious translation is 'punch, drift', since this technique of making rivet-holes was commonplace in early Ireland, while the use of the drill is comparatively modern. In Triad 120, *bir* is better translated 'punch' rather than 'spit', to match the hearth and anvil which make up the trio of implements (*Triads* p.17).
See also ***seim***

tanalaig n. 'some stage in the working of metals, bar, mass(?), thin sheet(?)' [T:71.19f].
cf. *airlam* adj. 'ready, prepared (for use), finished product' [A:220.71ff].
DIL compares this rare term with *tanalach* (< *tana* 'thin, slender, narrow' T:68.3f) 'shallow water'. The term occurs in the passage in the laws concerned with 'claim jumping' (sect. 7.2) where it stands in contrast to *cáer* (for iron), *tinne* and *cló* (C:245.25ff: for copper) which are taken here to represent metal 'as smelted', and to *airlam* descibing a finished product. If the comparison with *tana* holds good, then connection with the verb *tanaigid* 'thins, thins out, dilutes'

(T:70.8f) might lead to *tanalaig* being seen as referring to metal with had been 'drawn down, beaten out etc', i.e. metal as 'semi-products'. The degrees of distraint against misappropriated materials are thus five days when in the unworked state, three days as semi-products, and one day when fully processed.
See also ***berbaid***, ***cáer***, ***tinne***

teine n. 'fire, a fireplace, a hearth' [T:115.79ff].
The term is also of a blacksmith's smithing hearth (cf. *tellach*), as in . . . *cinta in iaraind o theinid co hinneoin* . . . (*CIH* III:936.32) 'injuries done by the iron (*being carried*) from the fire to the anvil . . . '.
Cognates from **tep*- 'warmth, heat' are extensive (*IEW* 1069–1070).
See also ***nes(s)***, ***tellach***

tellach n. 'hearth, fireplace' [T:140.35ff].
This term, like *teine* can also mean 'smithing hearth' as in . . . *Mas iat spreda in tellaig rofogaill ann* . . . (*CIH* I:267.36) '. . . If it be the sparks from the hearth that have caused the injury . . .'. *DIL* derives *tellach* from *teine* through an earlier *tenlach*. In the context of the forge, *tellach* and *teine* appear to be used interchangeably to describe the site at which metal was heated during forging.
Mod. Ir. *teallach* 'a hearth or fireplace, a fire, often of a smith's fire', *uisce teallaigh* 'water used in a smithy for cooling hot irons' (*Dinn.* 1189), *teallach an gabha* 'a forge hearth' (*De Bhaldraithe* 272), *teallach ceárta* 'blacksmith's hearth' (*Ó Dónaill* 1217).
See also ***nes(s)***, ***teine***

tenchor n. 'tongs, pincers, forceps' [T:143.25].
Extensively used to denote a blacksmith's tongs, as in . . . *an gaba* . . . *7 bruth mor iairnn isan tenchair ina laim* . . . '. . . the smith . . . and a huge bloom in the tongs in his hands . . . ' (*ICM*[2] 142.7f): also . . . *teanncuir an ghobhann, 7 an t-íarann dearg 'na beal* 'smith's tongs with the red-hot iron in its jaws'.
Mod. Ir. *teanchair gabha* 'blacksmith's tongs' (*Ó Dónaill* 1217), *a gobaibh na teanchaire* 'out of the claw (mouth?) of the tongs' (*Dinn.* 557), *béal na teanchaire* 'the grip of the pincers' (*Dinn.* 1190).

tinne n. 'bar, rod of metal, ingot' [T:180.63].
Primarily a general term for metal stock, although in one instance possibly referring to 'steel' from a gloss of Lat. *calibs* (leg. *chalybs*: *Thes.* II, 227.23). It is worth noting the expression . . . *secht cendphartaib iarndaib a cind cacha slabraide trummither tinni deich bruthdamnae cach n-ae.* 'seven iron knobs at the end of each chain, each of them as heavy as an ingot made from the blooms of ten smelts.' which supports the suggestion that *tinne* refers to cold stock.
See also ***berbaid***, ***cáer***, ***tanalaig***

túarcon (vn. of *do-fúairc* 'grinds, crushes, beats, pounds etc' [D:274.85ff]) 'grinding, wearing away, hammering (in a forge), hammering, beating, smiting' [T:345.49ff]; see also *túaircnech* n. 'a beating, striking, hammering' [T:336.37]; *con-túairc* 'strikes, pounds' [C:467.1]; *comthúaircnide* part. 'firmly wrought' [C:421.18], = ***forged***; *comthúarcon* vn. 'act of (mutually) striking, crushing (together)' [C:421.22f].
A general term for the application of percussive force, used frequently of the forging of iron, as in . . . *comthuarccain na nord ag bualadh na nindéon* . . . '. . . the pounding of hammers forging (iron) on the anvils . . . ' (*BNnÉ* I, 71.124).
Mod. Ir. *ag tuargaint iarainn ar inneoin* 'hammering iron on an anvil' (*Ó Dónaill* 1281).
See also ***benaid***, ***búailid***, ***urlaide***

tuirenn n. 'dross, sparks of metal' [T:368.3f], = ***hammer-scale***.
It is used metallurgically (in connection with *slaide*) being glossed *scama ferri* ('iron scale': leg. *squama*: *RMLW* 449) as in . . . *geibidh tennchair caor nduibhdheirg / ima ndregaid alla uird / sceinnidh tuireann ar gach leith / ima seich seinnid na builg.* '. . . The tongs hold the black, red metal, about which hammers contend yonder. Sparks of hammer-scale flies on all sides, the bellows are blown alternately.' (*Lib. Hymn.* 90.6).
See also ***crithir***, ***drithle***, ***slaid***, ***spréd***

tulchube cpd. n. 'a large vessel for holding wine, vat, cup etc.' [T:379.8f].
Also used of a water trough in a forge, as in . . . *Ruc Fiachna in taulchuma thair* . . . '. . . Fiachna carried out (from the forge) the water trough on his back . . . ' (Joynt 1910, 110.70). The word is a compound of *tul* ('la face plane du tonneau': *Vendr. Lex.* T-181) and **cubae*, the latter a borrowing from Lat. *cupa*.
cf. *ommar* n. 'a trough for holding water' [N-O-P:141.30f].
See also ***cerdcha***, ***goba***, ***ord***

U

urlaide (vn. of *ar-slaid* < *slaidid* A:416.4) 'smiting, sledging' [U:92.21f]; see also *urlaidecht* 'striking, fighting, beating' [U:92.58].
Frequently used to signify the forging of iron on the anvil, as in . . . *is a ic d'aes urlaide* . . . (*CIH* I:267.39) '. . . it is the persons hammering [who are to pay]'.
See also ***benaid***, ***búailid***, ***túarcon***

* * *

7.8 – Summary

The Irish sources for the study of early ironworking present many problems, not the least being our modern inability to assess accurately the degrees of comprehension of writers of the significance of the terminology employed. In some instances, such as the legal stipulations on the operation of the forge, it is evident that the basics were laid down by one who had keenly observed, if not actually participated in, blacksmithing operations. From such texts we can construct, with reasonable confidence, the terminology (Table 7.7.1) for many aspects of smelting and smithing. Elsewhere, it is equally clear that writers were merely transmitting words. Frequently there are major difficulties in reconciling many statements in the texts with scientific fact, as with the term *íarn aithlegtha*, and those interpreted as referring to heat-treatments of steel. Undoubtedly, many of our problems stem from the very wide semantic fields of the words used by early writers to refer to metallurgical operations and materials. It is of no real consolation to note that the problems are as difficult with regard to the non-ferrous metals.

In section 7.1, I alluded to the problem of determining the stratigraphy of the various terms found from the Old Irish period through to early Modern Irish. Without a much more detailed survey, it is not possible to chart the changes – additions and subtractions – in vocabulary. When, for example does *íarn aithlegtha* cease to be used to denote 'carburised iron', and Mod. Ir. *cruach* become established to describe a regularly-produced material 'steel'? Similarly, although the verbs *fo-bádi*, *fo-beir*, and *fothraicid* appear to have been used to denote 'quenching' (possibly also, other heat-treatments), the Mod. Ir. term for quenching is *múch* (*miotal a mhúchadh* 'to quench metal' – *Ó Dónaill* 833: from an earlier *múchaid* 'covers over, stifles, extinguishes etc.': *DIL* M:179.61f), while it is from *fo-geir* that Mod. Ir. *faghairt* 'tempering' derives. The difficulties are particularly well illustrated by the absence of a clear term for welding in early texts. Mod. Ir. *táthaíd* 'welds (iron)' (*De Bhaldraithe* 837) certainly comes from an earlier *táth* 'joins etc.' (sect. 7.7 under *dlúthaid*), but good examples of *táth* = 'a weld between iron components' have eluded me at least. When did the

Table 7.7.1 Summary of the principal early metallurgical terms

ORE EXTRACTION AND PREPARATION

Ore	**Mine**	**Miner**	**Mining**	**Preparation**	**Gangue**
méin	*míanach*	*clasaige*	*claidid*	*búain*	*buadradh*
mian	*mianaige*		*tochailt*	*bruïd*	
míanach	*mianaigid*				

IRON PRODUCTION

Bellows	**Tuyère**	**Slag**	**Bloom**	**Smelting**
bolg	*airbuinne*	*slaidred*	*bruth*	*berbaid*
séitire	*trefet*	*slaigthech*	*cáer*	*bruinnid*
				legaid

Iron	**Steel**
íarn	*iarn aithlegtha*
	crúaid

THE FORGE AND ITS FURNITURE

Smithy	**Anvil**	**Hearth**	**Bellows**	**Tongs**	**Hammers**
cerdcha	*indeóin*	*teine*	*bolg*	*tennchor*	*ord*
	cepp	*tellach*	*séitire*		
		nes(s)			
		derc			

Drift	**Abraders**	**Water trough**
tairbir	*airnem*	*ommar*
	bró	*tulchuba*
	íarnach	
	lia	

FORGING AND FINISHING

Ingot	**Hammer**	**Joint**	**Carburise(d)**	**Quench**	**Temper**
cáer	*benaid*	*seim*	*aithlegad*	*fo-bádi*	*fo-geir*
tanalaig	*búailid*			*fo-beir*	*fothraicid*
tinne	*con-tuairc*			*fothraicid*	
	gres				
	imbert				
	túarcon				
	urlaide				

Scale	**Surface finish**	**Rust**
drithle	*aithigid*	*meirg*
lad	*glas(?)*	
slaid	*gorm(?)*	
spréd	*límaid*	
tuirenn	*slipaid*	

extension of the semantic field take place? The tracing of these developments as the complement of the study of ferrous metallurgical development would be of the greatest benefit to students of metalworking in early Ireland, hopefully also to linguists.

The make-up of early Irish metallurgical vocabulary itself presents a significant lack of loans, particularly from the Scandinavian world. Given the expertise of the intruders, and their impact on the vocabulary of their other sphere of exellence, shipping, this apparent lack is puzzling. Metallographic studies of but a few of the artefacts recovered from Viking Dublin have demonstrated just how expertly smiths there could shape and manipulate iron and steel. It might well be that a study of contemporary Scandinavian material would throw up expressions which could explain some of the Irish metallurgical terminology, for example with regard to heat-treatments and welding. Such a study is beyond my competence, as is the essential comparative study of Welsh, Breton and Cornish texts to fill out, and throw fresh light on the Irish evidence. Thus, while the Welsh sources are largely closed to me, even cursory scanning of the translations of *The Mabinogion* by Jones and Jones (1974) and by Ford (1977), and Jarman's (1988) edition of *Y Goddodin*, indicates rich potential sources of information yet to be tapped.

Another area requiring the attention of specialists is that of the symbolism of many of the passages dealing with metalworking. I have alluded briefly to the potential deep-seated significance of the forge episode in *Echtra mac Echdach Mugmedóin*, and the tale *Iarn Luchtna*. These, and many of the other passages referred to in this chapter seem to me to require thorough analysis by specialists. Valuable work has been done over the years in this field, though mainly on the principal themes of early Irish texts. There seems to be much of the pre-Christian stratum of the role of the metalworker still to be uncovered and elucidated.

Finally, as noted at the start of this chapter, it has not been my intention to try to produce a definitive study of the treatment of iron in early writings, but rather to demonstrate the potential of the field. Many problems remain to be tackled and solved, and I can but hope that the above synopsis of what I see to be the areas of prime importance, not just for early metallurgy but for the exploitation of natural resources in general, will perhaps stimulate the interest of others to engage in research. The era of the polymath should be past, and the future lies in co-operation between specialists in different disciplines which will allow whole suites of expertise to be brought to bear on problems on which, in recent years, of necessity have been tackled by non-specialists.

* * *

Notes to Chapter 7

1 Extracts from the *Zoilomastix* of Philip O'Sullivan Beare have been edited by O'Donnell (T.J. O'Donnell *Selections from the Zoilomastix of Philip O'Sullivan Beare*, Dublin 1960). The Irish version of Albertus Magnus' *De mineralia* is unpublished, and at present under study by the writer.

2 Mac Neill's translation of *caladcholg* (*Duan. Finn* I, e.g. 55.90) as 'tempered blade' (*calad* 'hard' – *DIL* C:58.35ff) is unjustified.

3 It is to be noted that many of the words listed in the Glossary also occur in connection with crafts other than ironworking.

4 Magrill (1980) estimated that there then were some 4.5 million substances listed in chemical literature, with information on some 350,000 further substances added each year. Most of these come under a variety of names, depending on the system of nomenclature used. He cites the example of the artificial sweetener *saccharin* (trivial name), which can also be found in the literature as *1,2-benzisothiazol-3(2H)-one,1,1-dioxide* (systematic name), *1–Sulphobenzamide, Benzoic sulphamide, 1,2-dihydro-2-ketobenzziso-sulphonazole*, and *benzosulphamide*. While the organic chemist should not find too much difficulty with this proliferation, the layman surely would! In another paper (Scott 1983a) I have discussed some aspects of the development of early metallurgical nomenclature (not treated in this work), suggesting that names for artefacts or for substances which evolved from terms for the method of manufacture, or for shape, or for form, would in every way have been as readily comprehensible to early technologists as our nomenclature is to those brought up under modern scientific discipline. A good early example of this is O Ir. *murluaith* which literally translates as 'sea ashes' – i.e. the product of evaporating seawater over a fire (Scott 1982)!

5 In *BR*[2], the king of the Cuirc is described as *rí na Corc ón choill* 'The king of the Cuirc from the forest' (*BR*[2] 58.861). In view of the quantity of timber required to produce the tribute in iron supposedly due from the Cuirc, close proximity to a good supply of wood for charcoal would have been essential.

6 The juxtaposition of *rú*, which they presumably took as a form of *rúad* 'brownish red, dark red' (*DIL* R:106.44f), with iron, led the editors of *Laws* to translate it as 'iron ore', a reasonable explanation, albeit one which raises problems. This difficult passage refers both to the collection of *rú* and to the mining of ore. The term *trilsén* (dim. of *trilis* 'tress, plait of hair': *DIL* T:311.34f) refers to a bundle or sheaf (*DIL* T:312.4), association of *rú* and *trilsén* also occurs in the 'appendix' to *Cáin Lánamna* (*CIH* I:177.12f) where as part of a 'divorce settlement', we are told that a woman is entitled to:

> *Nomadh do mnaoi a roi(n)d . . . nomhadh na roidi di ara buain . . . .ui. sedh ara trillsinaib du denumh de . . .*
>
> A ninth of the *rú* to the woman . . . a ninth of the *rú* when it is reaped . . . a sixth when it is made into sheaves . . .

In *Tír Cumaile*, one division of land is 'first class

cultivable' on which may be grown, amongst other things *rú*, translated by Mac Niocaill as 'madder' (1971, 84). Earlier identification as a plant from the family *Rubiaceae* ('Bedstraw'), specifically lady's bedstraw (*G. verum* L.) was also favoured by Sullivan in his introduction to O'Curry's *Manners and Customs*, I, ccccii), although field madder (*S. arvensis* L.) is another possibility: both of these favour the sort of dry, grassy habitats which would fall into the category of 'first class cultivable land'. The comparatively rare wild madder (*R. peregrina* L.) is to be found locally on cliffs around the south and west coasts of Ireland, and is only very rarely found inland. If we translate *all* (*DIL* A:285.39f) in this context as 'a rocky exposure, steep outcrop', then we might draw the two, otherwise seemingly unrelated, strands together. Red/red-brown dyestuffs were made in antiquity from both from plants, and from lichens which are found on exposed rocky surfaces. Both source and product were considered to be of some worth, and the fruits of communal land presumably were to be protected. If *rú* can be taken as applying not specifically to one of the *Rubiaceae*, but more generally to plants from which a red dye could be prepared, then the otherwise confused references to iron and copper may also be explained by translating *méin* in its more general sense as 'ore, metal', so that the passage then reads as a stipulation concerning the length of the distraint to be placed on materials removed from communal land depending on the degree of processing it has received when the 'crime' is detected.

7 I am also indebted to Ms. E. Bhreatnach for her kind assistance in grappling with this difficult text.

8 In their edition of *De Re Metallica*, Hoover and Hoover (1912, 86, note 6) give a useful summary of the development of European mining law as regards mineral rights (cf. Davies 1935, 2f; Healy 1978, 104f). The question of the ownership of minerals as distinct from the land under which they occur is one which might perhaps repay the attention of students of early Irish legal institutions.

9 Piaskowski (1982a) suggests that the Chalybes smelted the iron-nickel ore, chlöanthite, to produce their famed metal (cf. Photos 1989b). However, one objection which has been raised to this is the limited distribution of that mineral.

10 It is not unusual to find an orthographic lenited *-d-* in words which originally contained two vowels in hiatus, as in *luäm* > *luadam* 'steersman'. The *DIL* entry [C:556.1] gives only one example for *crúadán: sgíath . . . ar na choimhegair do chruadhan* [leg.' *chrúan*] which can be translated as 'a knife . . . inlaid with *red enamel*', such inlaying being well known on Iron Age ironwork both in Ireland (e.g. the Moyfin, Co. Meath spearhead – Fig. 4.1.1g) and elsewhere.

11 As explained in Chapter 1.4, smelting involves the *reduction* of iron oxide to iron. Since iron is an element, it is incapable of further reduction, and hence the term 'resmelting' (unless referring to iron that has wholly rusted to iron oxides, hardly a plausible proposition) is a chemical impossibility! Only if the term referred to the resmelting of hammer scale, which, as iron oxide, could be treated as a very pure ore, would such a translation be tenable. However, there is no evidence for the use of such a process in Ireland.

12 The term *ór forloiscthe* denotes fire-assayed gold (Scott 1981c, 248), from *for-loisci* 'burns, sets on fire'. It is mildly surprising that the term *íarn aithlegtha* is not, in fact, *íarn forloiscthe* which would be perhaps more appropriate, but I know of no such occurrence.

13 There is a semantic bond between Germ. *smid* and Slav. *kovar*, in that each is traced back to a form whose primary sense includes 'strike, hew' (**smēi-*, **smǝi-*, **smi IEW 968: *kāu-*, **kǝu-* IEW 535). Is it possible that Celt. **gobann-* might also go back to a form with a similar semantic field?

14 The inscription on an Ogam stone at Ardmore, Co. Wicklow was read by Macalister (1907, III, 172–178: cat. no. 263) as *DOLATIBIGAIS GOB(ANN) LUGUDECCAS MAQI MUCOI NETASEGAMONAS* which he translated as 'Of Dolativix the Smith, Lugud's son, Tribesman of Nia Segamain'. Drs. P. Ní Chatháin and A. Hamlin, however, have kindly informed me that this reading and translation are quite incorrect, and may be discounted as an early written example of *gobae* in Ireland.

15 The prime example of *cerd* being translated consistently as 'smith' is in the name *Culand Cerd* from whom the infant *Sétanta* took his better-known name *Cú Chulainn* ('Hound of Culand'). That Culand was indeed a metalworker as may be seen from a passage in *Táin* 23.822f which tells of a visit to him by Conchobor:

> *Culand Cerd buí i crích Ulad. Ro urgnastar fleid do Chonchobur 7 dochúaid dá thoc[h]uriud co Emain. Rádis friss ara tísed úathad leis meni thucad fíraígid leiss ár nach crích nó ferand baí aice acht a uird 7 a indeóna 7 a duirn 7 a thendchore.*

> Culand the *cerd* dwelt in Ulster. He prepared a feast for Conchobor and went to Emain to invite him. He told him to come with a small number unless he could bring a few genuine guests, for neither land nor domain had he but only his hammers and his anvils, his fists and his tongs.

The problem of unspecifically rendering *cerd* as 'smith' is clear, since the anvils and tongs claimed to be the only visible means of support would (in general terms at least) be equally at home in the forge of a blacksmith or the workshop of a cerd. It seems equally likely that a famed jeweller would entertain a king as a master blacksmith.

16 A not dissimilar motif occurs in a Welsh poem from the *Black Book of Carmarthen* in which Arthur is challenged to enumerate his followers by the doorkeeper of a house (Jones and Jones 1974, xxiii).

17 Unfortunately we do not appear to have any literary references to the making of charcoal from

different tree species. To date, little or no work has been done specifically on the identification of charcoal associated with smelting sites (cf. Chapter 6.5).

18 Incidentally, one cannot agree with Charles (1980, 177f) that Hephaistos/Vulcan was said to be lame on account of a tradition in which recognition of the health hazard of working with arsenical copper under oxidising conditions was enshrined. His lameness specifically was attributed to grievous bodily harm inflicted by an angry Zeus/Jove, seemingly with no retained element attributing it to a work-related illness.

19 O'Curry quotes from 'a copy of an ancient tract, prose and verse', which sited Gobniu's forge, called *Cerdcha Ghaibhinn*, in *Glenn Treithim*, near to Mullaghmast in Co. Kildare. He states 'The name and situation of this primitive Irish forge were well known at the time of compiling the tract . . . the place of the furnace and the debris of ore and coals then still remained.' (1873, II, 246–247). *Hog. Onom.* 446 cites a reference in the Book of Lismore to *Glenn Treicím* at the base of *Mullach Maisten*.

20 In this test of rescuing of articles from a burning forge we may detect a common element from European traditions surrounding the power of the blacksmith in his role as 'master of fire'. The forge has been taken in other traditions as symbolic of the womb (cf. Rees and Rees 1961, 252f), and in this respect, the smith Sithcenn may be viewed as bringing about the rebirth of Eochaid's sons through the medium of fire, and thus, in fact, performing an initiation rite for the king's sons.

21 The term *bruithneóir* appears late, being a cpd. of *bruithne* + *óir*. The second element derives from Lat. *-arius* (*Thurn. Gramm.* §269).

22 We may compare these descriptions (also *Blaí Ord Indeóin* cited later) with the late (early Modern Irish) version of *Cath Maige Mucrama*. Here we are given another vision of frantic activity in a forge, this time a field armoury engaged in preparing weaponry for the forthcoming battle. One text was edited by Ó Dúnlainge (1907, 436), while a translation of another (by O'Looney) is given by O'Sullivan in the Introduction to O'Curry's *Manners and Customs of the Ancient Irish* (Vol. I, ccccxxxiv, note 767). The portrayal of movement and noise is typified by the following extract from the edition by Ó Dúnlainge for whose translation I am grateful to Professor G. Mac Eoin:

> *Acht cheana, níor chonar chómráidh an t-árdárus sin le foghar-ghréasacht na mbolg agus na n-órd fór na h-inneónaibh, le fíor-thiomáin na lann for na ligh-bhróintibh, le fiucha na bhogha bhfaobhrach, agus le cas-chuimilt na gcead chloidheamh ris na h-iarnaighibh;*
>
> However, that high dwelling [the forge] was no place for conversation, with the noise of the bellows and of the sledges on the anvils. with the constant rubbing of the blades on the whetstones, with the seething of the sharp javelins, and with the backwards and forwards rubbing of the hundreds of swords with the files.

Another early Modern Irish description of a forge may be found in the poem *The Lay of the Smithy* (*Duan. Finn* II, 2ff). In it, Finn and his companions race a giant, *Lon mac Líomtha*, who professes to have mastered every craft, and to be *ollamh gabhonn ag righ Lochlann sa Bheirbhe* 'teacher of smiths to the king of Lochlann (Norway) in Bergen'. Having heroically traversed a good part of Ireland, Finn's band arrive at Lon's smithy, where one of the group, Daolghus, assists in the forging out of a sword blade:

> [28] *Tíaghad ar áon san úamhaidh Fionn ga ffúabhairt go fearrdha fúarattar builc da séideadh fuarattar cré is ceardcha*
>
> [29] *Inneóin is uird ga mbúaladh lé trénbhuinne lúath lasrach seacht ccloidhme leó arna síneadh 'na mbróinibh dirge dhata . . .*
>
> [31] *Cuiris teanchoir fon tellaigh togbhais lía ceithre ndromann do-rinne in gabha is Dáolghus urlaighi fháobrach udmhall*
>
> [32] *Da ord mora ag in ghabha 7 teanchair thailc tháobhglas tri lamha leis ga ffreastal is maith do freagair Dáolghus*
>
> [33] *Do-níd lann crúaidghér crithreach fa maith biseach ré búaladh feadh duirn dhi gan da fháobhar 'na buinne caomhghlas crúadha*
>
> [34] *Teidhis re cneas na ceardcha Dáolghus ba teann a throda ba deirge na gúal gartha a shnúadh d'aithle na hopra*
>
> [35] *Do ráidheastar na gaibhne ba lór gairbhe is grúamdhacht cía in fear caílti gan time shineas in tinne crúadha*
>
> [28] They go together into the cavern, Fionn pressing upon them manfully; they found bellows a-blowing: they found earth and a forge,
>
> [29] Anvils and sledgehammers being smitten and a swift strong blast of flame, seven swords too stretched out in straight and lovely masses(?).
>
> [31] He (*Lon*) put tongs into the hearth and lifted a stone with four ridges: the smith and Daolghus carried out some keen and nimble hammering.
>
> [32] The smith had two great sledgehammers and strong grey-sided tongs: he had three hands attending to his tools: Daolghus responded well.
>
> [33] They make a hard sharp sparkling blade of good success(?) in striking: for a hilt's length it lacked two edges, a fair grey shoot of steel.
>
> [34] Daolghus grew warm beside the forge

(stout were his combats): redder than glowing coal was his complexion after the work.

[35] The smiths, who were full rough and surly, said "Who is the slender warm man without weakness who is stretching the bar of steel?"

Líomtha is the verbal noun of *límaid* 'smooths, polishes etc.'. Of interest here is the description *lía ceithre ndromann* '(lit.) a stone with four ridges'. It is not impossible that this refers to a bottom swage (Fig. 1.5.1g) which would have been used in forging out a blade with a complex section. The common adjective for a grooved or fullered blade, *clas* (Mallory 1982, 101), does not occur. We might also note the use of *síneadh* 'stretching, extension etc.' (*Ó Dónaill* 1095), possibly as a term for drawing down (Chapter 1.5).

23 I am most grateful to the Dr. B. Meehan, Keeper of Manuscripts, Trinity College Dublin, for supplying this plate, and to the College authorities for their permission to reproduce it here.

24 The use in smelting by ironworkers of bellows with two or more chambers is known modern ethnographic parallels (e.g. Wulff 1966, 50). I am most grateful to Professor G. Mac Eoin for assistance with this passage.

25 This episode in *Orgain Denna Ríg* has been borrowed as the story of the iron house in *Branwen Daughter of Llŷr* (Ford 1977, 64):

> . . . they had a chamber made entirely of iron; when it was ready, they summoned all there were of blacksmiths in Ireland there, and all the owners of tongs and hammers and had coal piled as high as the roof of the chamber. . . . they kindled the fire around the chamber and blew the bellows that had been placed around the house – one man to each two bellows – and began to blow the bellows until the house was white-hot . . .

I am unclear as to whether the text means that one man worked two sets of bellows, or that one man worked a two-chambered bellows (the latter seems more likely).

26 In *De ceithri slichtaib Athgabála*, one of the articles to which a stay of one day attached in cases of distraint was:

> . . . *trefet tigi srotha*

to which attached the gloss on *trefet*:

> *.i. treithi seter in teni tresa an leidb, 7 imin feith trein tresa seiter teni i tig cach uasail .i. in seitire* *CIH* II:376.13

> The *trefet* of the house of an elder i.e. through it the fire is blown through the leather, or it refers to the strong blast with which the fire is blown in the house of every noble: i.e. a blower.

CIH III:889.5 has:

> *Im trefet tigi .s. .i. eolus seidi tenid.*

> The *trefet* of a house, that is, the opening through which the fire is blown.

Joyce (1903, II, 310) suggested that the term *trefet* referred to a different type of bellows than that called *bolg*. However, the first explanation of *trefet* (i.e. *tre* + *séit*) given in this gloss is correct (cf. *Vendr. Lex.* T.132). The insertion of *trein* after *feit* is merely an attempt to explain the first syllable.

27 *colgdirech* = 'straight as a *colg*' – Mallory 1982, 100f.

28 In the Germanic tales of Weland the Smith, there is a passage describing how he improved iron by filing it down and feeding the filings to tame birds. Experimental work (Willen et al. 1976; Martin and Foley 1979) have shown that in terms either of introducing nitrogen or reducing phosphorous, all that this achieves is magnetic hens! However, heating iron filings in contact with chicken dung does increase the nitrogen content. This suggests that as with the technological core of quenching in *BDD*, Germanic storytellers dressed up a simple process into something magical.

29 The term *airtem* refers to a measure of the hand with extended thumb.

30 The colour blue is attributed to spears in a Welsh poem from the *Black Book of Carmarthen* (Jones and Jones 1974, xxiv), where we find:

> Cei Wynn and Llacheu
> They used to make battles
> Before the pangs of the blue spears . . .

In *Y Gododdin* we find reference to . . . *Cleddyfawr glas glân* . . . '. . . Bright blue swords . . .' (Jarman 1988, 2–3).

Note added in press

One other possible interpretation of *íarn aithlegtha* is that it refers to iron produced by the smelting of the sintered mass of unreduced ore, slag and iron particles that can result from what researchers usually regard as a failed smelt. Preliminary experimental evidence (see also p. 170 above) indicates that in a bowl furnace at least, such a product can be successfully processed to useable iron in a second smelt. There appears also to be a degree of support for this hypothesis from ethnographic studies.

CHAPTER 8

Summary and the Course of Future Research

8.1 – Summary of conclusions

There seems good evidence to indicate that knowledge of iron and its working reached Ireland sometime before the middle of the 1st millennium BC. At present, however, only a handful of artefacts which can be ascribed with any confidence to the transition period have been identified. While the Aughinish horsebit and the Rathtinaun material come from excavations, the majority of this small group have no reliable contexts. Their ascription is open to debate, and the conclusions reached here will not be universally accepted. There is certainly reason to be cautious, and the matter will not be settled until compatible material is found in secure contexts. It is little short of a disaster, therefore, that the application of high-precision C-14 dating of the carbon contained by these artefacts would be of no use because of the impossibility of using C-14 for most of the last millennium BC. But without these objects, and the clues which they provide to events in the latter stages of the Bronze Age, interpretation is well-nigh impossible.

The alternative to accepting an early date for these artefacts, and hence for the introduction of iron, is to propose that ironwork belonging to what we may describe as the Lisnacrogher horizon, dating no earlier than the 3rd century BC, is the product of an industry of no obvious Irish ancestry. There seems little disagreement over the interpretation of the archaeological evidence for a quite sudden and dramatic decline of the vigorous and productive non-ferrous metalworking industries which flourished in the earlier 1st millennium BC. Their disappearance leaves us with a 'dark age' between the 7th and 3rd centuries BC. Dismissal of what is proposed here as the earliest material thus leaves Ireland lagging some four centuries behind Britain and Atlantic Europe in technological development – hence in the industrial and economic developments which accompanied the adoption of iron technology elsewhere. The existence of iron artefacts dating to the 7th/6th centuries BC provides insights into the process of decline of the Later Bronze Age industries of which they are products.

Technological development elsewhere in barbarian Europe went together with major realignments in industrial organisation which saw the rise of high-output mass production iron industries. The emergence of these industries was coupled with changes in social and economic organisation which provided the markets for their output. The metallographic evidence demonstrates that in the Irish transition period, those experimenting with the new metal were acquainted with virtually all of the major techniques of the blacksmith, even if they were not not wholly proficient in their application. This is in accord with evidence from other regions where the earliest iron artefacts show a comparatively high degree of technical skill in contrast with the majority of those from the subsequent period when iron has ceased to be a rarity and is in common use. The inference which may be drawn then is that the conditions which allowed experimentation with iron here, in a Later Bronze Age industrial setting, were broadly comparable with those in other regions of barbarian Europe. That we then face an dark age indicates that, for what ever set of causes, Ireland deviated from the general trend of industrial development to follow its own, distinctly insular course towards dispersed and small-scale production geared to serve a fragmented society.

The appearance of finely-shaped iron artefacts around the 3rd century BC marks the real establishment of iron technology here. But although artefact design is pleasing, the quality of the material is not. Even such an attractive weapon as the triple-ribbed Lisnacrogher sword is made from iron of indifferent quality. None of the edged tools and weapons examined so far really shows decisive use of heat-treatments, although a very high proportion have fold-welded structures which indicate an awareness of the need to improve quality. Throughout EIA B, the amount of iron in use is small compared to that of other metals, and again we may interpret this, and the lack of really high-quality technology, as resulting from the small-scale organisation of production. Although there was ongoing contact of one sort or another with Roman-dominated Britain, studies there indicate that the majority of artefacts were of indifferent quality also. But a major difference between the two islands was in smelting technology, with the shaft-furnace coming into Britain at a comparatively early date, possibly before the Roman conquest, while Irish smelters stayed (or were stuck) with the bowl furnace. The implications of this are discussed in Chapters 5 and 6, but it is worth restating that there is a significant difference in the productivity which can be achieved from each type, with shaft furnaces giving by far the higher yield. There is no reason to suggest that Irish smelters (at any stage) could not gather together enough stocks of charcoal and ore to feed the more demanding shaft furnace. The primary reason would seem to be the barrier to the introduction and spread of the shaft furnace from across the Irish sea posed by fragmentary production and poor communication between craftsmen.

Although in LIA A, iron was commonplace, the overall level of technology revealed by metallographic examination is still low, with only a comparatively small proportion of artefacts showing heat-treatments. In contrast, LIA A is a period of excellence in the production of fine non-ferrous metalwork. Outside ireland, the levels of expertise in manipulating the properties of iron are much greater. Again, the poor quality of LIA A ironwork must be put down to the lack of communication between skilled workers, even those who had possibly learned advanced techniques outside Ireland (as, for example, would have been possible with Irish monks working on the Continent). To use the contemporary existence of fine non-ferrous metalwork as evidence against this hypothesis is to fail to appreciate the significant differences between ferrous and non-ferrous metalworking, and, in particular, to confuse to some degree technological and stylistic considerations.

There is a strong uniformity of shape and decoration on non-ferrous metalwork throughout LIA A Ireland, but this can be explained simply. A skilled craftsman could take a highly decorated brooch and produce a copy in the same style, drawing on his skills, and the fact that the shape and patterns which he wished to reproduce were evident. If there was a feature with whose fabrication be was unfamiliar, the form would be plainly visible and could be replicated by experiment or duplicated by use of a different material. A smith, given an iron artefact, again could duplicate visible shape, but unless he was familiar with the properties of his material and how they could be altered, would not have been able to produce a hardened steel cutting edge using soft uncarburised bloomery iron. The point is that the craftsman producing finely decorated non-ferrous metalwork was manipulating shape and not the properties of his materials. Shape is visible and easily transmissible without explanation. Understanding of invisible technology is not. And we might look also at non-metal technology, specifically to the absence of native Irish wheel-thrown pottery. We know that the Irish became acquainted with wheel-thrown pottery which was being imported to Ireland in significant amounts in LIA A. But familiarity with shape does not mean familiarity with the technology that produced it. Very likely, such shapes could have been produced by hand, but without first-hand knowledge of how the originals were produced, and practice, wheel-throwing was a technique not adopted before the mediaeval period. In short, for a technique to spread and be widely adopted, without the intervention of immigrants, requires that there are communicators, lines of communication, and communicants. There was undoubtedly a learning from outside of millifiori and enamel working, but the basics of these techniques were well established by EIA B. But even though by LIA A times such techniques as shaft furnace smelting and pattern welding were in common practice in Britain and Atlantic Europe, and even though there were strong contacts between Ireland and these regions, neither technology reached here.

The present state of our evidence suggests that following a promising start in line with neighbouring regions, the growth of iron technology in Ireland was stunted by socio-political developments which led to the rise of dispersed, small-scale production. The lack of major production centres and of communications between skilled workers both inside and outside the country prevented the introduction and adoption of the shaft furnace, definitely a non-portable idea, and hence limited production potential. However, since technological innovations are adopted because they are deemed to be of real benefit, it is likely that no one group actually needed the enhanced output which the shaft furnace would have provided to serve its market. Poor communication also stifled the development of regular high quality artefact production, and those best-made artefacts that we can identify at present seem to be the products of a small number of blacksmiths scattered in space and time, representing the skills of individuals rather than groups. Despite the promising indications of literary and linguistic evidence, dating from LIA A times onwards, we have yet to identify positively any strong indications of European state-of-the-art ferrous technology here.

All of this might seem to characterise Ireland as a backwater in terms of iron, this depiction seeming all the more extreme when we consider the ranges and depths of skills shown, particularly in LIA A, by craftsmen in other media. One must be cautious however, since, as I have tried to stress throughout this work, these interpretations are based on a preliminary survey and may require major revision in the light of further studies. It is to potentially rewarding lines of future research that we must now look to provide clarification and development of our understanding of early Irish ironworking.

* * *

8.2 – Areas for future research

The main stages in the production of iron from its ores were outlined in Chapter 6.1 as:

prospecting —— locating suitable ores

mining —— retrieving ore

ore processing —— concentrating, roasting

smelting —— reduction of ore to metal

bloomsmithing —— consolidation of iron/slag sponge

forging —— shaping of artefacts from the consolidated bloom

Each stage produced residues that we might expect to survive on sites:

operation	**archaeological survivals**
mining	mine workings, tools, containers
ore processing	ore roasting pits, charcoal, ore fines, gangue, roasted ore
smelting	furnace remains, tuyères, charcoal, unreduced ore, 'slags', raw bloom
bloomsmithing	hearth remains, tuyères, charcoal, anvil, hammers, 'slags', spheroidal gromps, hammer-scale consolidated blooms
forging	hearth remains, tuyères, charcoal, tools, 'slags', hammer-scale, semi-products

Comparing this list with the archaeological record, it will be evident that there are significant gaps in our knowledge at every stage of the process. As but one example, the assemblage of early ironwork from the period under review, and beyond, must now run into many thousands of artefacts, the majority of which have some form of context or association. Apart from *CIIAA* which, arguably excludes a number of EIA A and B specimens, there is no *corpus*, descriptive or analytical, published of such a large and important body of material. And from this assemblage, less than one hundred have been subjected to detailed metallographic examination and chemical analysis. The paucity of typological and technological information on the final stage of production is matched for every other stage of the process.

We have little knowledge of the distribution and composition of useable ores that would have been accessible to early ironworkers and less of the actual deposits which were mined. The long history of cutting peat for fuel, coupled with modern peat cutting practice, and the industrial mining of bog ores in the 19th and earlier 20th centuries, may well have altered drastically one source which is often cited as the staple of early ironworking. Similarly, the comparatively short-lived but viciously exploitive Elizabethan blast-furnace industry here will have also probably obliterated many bedded ore sources, and also traces of earlier workings. Surveys of ore deposits in recent times have concentrated on what have been considered commercially viable deposits. There is an obvious need for detailed field survey and sample collection and analysis. Thus, the occurrence of artefacts with high arsenic and phosphorous contents indicates ores possibly quite limited in their occurrence and distribution. Where are they? Similarly, the Antrim laterites have distinctively high contents of Al, Ti and V, and analyses of some of the EIA A and B material from that county indicate links with these deposits. There are obvious problems concerned with the partitioning of elements during the smelting process, and these require to be addressed as part of the study of the possibilities of provenance analysis for Irish iron and ironworking debris.

One obvious starting point with such an investigation is with ores found in the course of excavation. Of the sites which have produced evidence for ironworking, finds of ore have been reported only rarely. Just how much of the ore recovered from sites in previous years survives in our museum collections? Allowing for the problems of distinguishing between smelting and smithing activities, hence of the liklihood that ore would have been present on a particular site, does this mean that ores were not present on those from which it was not recorded, or that on some at least that it went unrecognised? How does each find relate to the local lithology of the site where it was found? At Ballyvourney, Co. Cork, O'Kelly (1952, 35–36) found '. . . an indurated iron pan almost black in colour and 10" to 14" thick . . .'. This extensive layer of readily available 'limonite' was probably exploited as an iron ore . . .'. But no analysis of this material is given. An excavator working in 1952 can be excused for not having realised the need to provide such data. But no published excavation report from Ireland of which I am aware has provided adequately detailed analytical data on ores found on site.

Two of the questions posed by the finding of ore on a site are 'which are the nearest sources of this type?', and 'where were the workings from which it was extracted?'. The former is one which is best addressed to the geologists, who may also be able to pinpoint the most likely locations where traces of mining might be located through field survey. Even when a source for ore cannot be identified, the recording and reporting of such a find, along with detailed compositional data, puts down a marker for future work. Another question to be addressed is that of the condition of the ore – 'is it freshly mined, or has it been processed by concentrating (to remove gangue) and/or roasting?'. Again, the geologist can make an important contribution here, although the simple but informative roasting experiment conducted by O'Kelly (1964, 103) on material from Garryduff I shows what can be done. Bearing in mind that ore roasting will inevitably produce fines, the geologist again should be consulted over analysis of soil samples from areas where there is evidence of metallurgical activity.

Among the sites which are reported to have produced the debris of ironworking, only about half provide what I can definitely describe as structures which have been used for iron smelting. Those sites where I would agree with the identification of furnace remains are mostly those which I have seen for myself during excavation. If this seems to be an arrogant dismissal of the work of others, I would admit my own fallibility in compiling this list, but submit also that standards of reporting have been so poor on most of the other sites that it is impossible for me (or anyone else, for that matter) now to adjudicate. In none of the reports in which the existence of a smelting furnace or furnaces is claimed, and which I relegate to the 'dubious' list, are there published informative photographs, let alone decent sections of the structure(s) in question. The unillustrated description of a feature as 'a depression *x* cm in diameter by *y* cm deep', followed by the claim that it was a smelting furnace on the basis of association with 'slag' and 'burnt' or 'vitrified' clay, cannot be entertained. All excavators are aware of just how far 'slags' could be scattered throughout occupation layers, and, apart from Lisleagh, Co. Cork, there is little to suggest that early ironworkers did much other than rake hot debris away from the immediate site of operations. Such hot material raked into a shallow natural depression and left there is likely to have left some evidence of burning and to give rise to uncritical interpretation as 'a bowl furnace'. Criteria for identification are difficult to establish, but a primary requirement is the provision of detailed records of the structure and stratigraphy of any feature suspected of being the remains of a bowl smelting furnace. Once excavators begin to record and to report in full structures suspected of being smelting furnaces (or smithing hearths), progress can be made.

Of the structures which I would admit as being bowl smelting furnaces, none has yet provided evidence either for the positioning or type of tuyère used, or for the existence (or non-existence) of any superstructure. All of the bowl furnaces identified have not survived to what would have been the original level of the ground into which they were dug. And it would have been here that the tuyère would have been inserted. Future excavation of such features should take account of the possibility of identifying tuyère position. It is possible that even a superstructure of the thickness required for good insulation against heat loss, and one whose inner surface showed traces of vitrification, once discarded could have been reduced to a large number of small fragments, distributed over quite a few square metres. Reports not infrequently refer to finds of such material as representing 'furnace lining', implying the material in the bowl of the furnace. Careful sorting of such material in future might identify not only the lower lining of the bowl, but also fragments of any superstructure which might have been in place over it.

Another material associated with ironworking activity is charcoal. The exploitation of woodland and forest to provide fuel for domestic and industrial purposes was a very significant activity. But again, remarkably few excavators have attempted to have identified the variety of species represented. A notable recent exception is the report of the Reask, Co. Kerry excavations by Fanning (1981), which can serve as a model for future reports. In theory at least, it should be possible to see whether particular charcoals associate with industrial rather than domestic activity, and to compare the use of different species with their potential availability as seen in the pollen record. It should be remembered that local declines in tree pollens following the introductions of non-ferrous and ferrous metallurgies need not necessarily indicate clearance of land for agricultural purposes, but could result from felling of timber for charcoal. The obvious example, albeit an extreme one, is the post-mediaeval decline attributable to the activities of the 17th century ironmasters and the makers of barrel staves (e.g. Mitchell 1976, 195ff).

Turning now to the vexed questions of the interpretation and identification of slags, it has to be repeated (cf. Chapter 6.4) that these admit of no easy answers. But as with other areas of this research, simple observation and recording of basic facts is often omitted from excavation reports. Thus, very few excavators actually record the total weight of slags found on their sites, and even the same excavator can give total weight in one report, and omit this information in the next. While such information is of limited use, it does at least allow some measure of activity, and can allow some crude observations to be made. Thus, one tonne of smelting slag might be calculated to represent the processing of 1.5–2 tonnes of ore, the consumption (assuming an ore to fuel ratio of 1:3) of 4.5–6 tonnes of charcoal, the felling of 30–35 tonnes of wood, and the production (at an efficiency of 25% recovery of iron from processed ore) of something in excess of 250 kilos of iron. While the figures may be quite imprecise, they nevertheless allow some broad ideas to be obtained about production and consumption, particularly if activity can be assigned a date range with any reasonable degree of precision. Such calculations are more difficult, and subject to a greater degree of uncertainty when we are dealing with smithing residues, Nevertheless they can still provide simple guidelines for interpretation.

Although much detailed mineralogical and chemical research has been carried out in Europe and Scandinavia, there is still no consensus on an overall scheme for the differentiation between smelting and smithing residues. Simple visual identification is uncertain, although the slags produced by the more advanced types of furnace with provision for tapping are, by and large, clearly identifiable. But this is (to date) irrelevant to Ireland, leaving a very large body of material which is very difficult to characterise. At the most basic level, there is a crucial distinction to be made between smelting and smithing residues. In addition to visual examination, attempts at characterisation must involve detailed chemical analysis and mineralogical examination of not one, but many samples from an assemblage. A small amount of such work has been done, and is included in Chapter 6.4, but what is necessary is the inclusion of detailed slag studies for all sites on which they occur. If this seems excessive, it must be remembered that there is a pressing need to determine, even roughly, the proportion of each activity represented in a 'slag' assemblage. In the light of our present knowledge, the reader will appreciate that I would not be too surprised were it to be shown that activity identified tentatively here as predominantly smelting proved, instead to be predominantly smithing. In this case, the overall model proposed for the organisation of production in Later Iron Age Ireland would have to be scrapped. While the construction, demolition and reconstruction of models is a necessary part of the archaeological process, it would be made less fraught for the study of early Irish ironworking if more data were available.

As Polish workers have demonstrated in recent years (e.g. Nosek 1975; Bielenin 1987), a clear indication of bloomsmithing is provided by the highly magnetic blobs of slag and iron, produced during consolidation of the bloom, and christened *gromps* by Nosek. They can be readily found where they occur by passing a magnet over soil samples. These have been identified at Lisleagh, but have not been reported from any other site. The forging of iron at heat causes loss of metal due to oxidation, and this falls away as hammer-scale. Although it has been reported from British and European sites, I know of no certain occurrence here. It is very difficult to identify unless the excavator knows precisely what to look for.

Finds of tubular tuyères, ore, slags and unconsolidated blooms will be a good indicator of smelting activity. It is a subjective judgement, but a large amount (by Irish standards – say in excess of 100 kg) of slag, including large plano-convex lumps is likely to have come from predominantly smelting activity. Unconsolidated blooms, such as the example from Ballyhenry, can be identified by X-radiography screening of slag assemblages.

The occurrence of bellows-protectors, gromps, hammer-scale and small slag cakes (so-called smithing hearth bottoms) indicate bloom- and artefact-smithing. Another find indicative of artefact smithing is the consolidated bloom (providing basic stock for working up). So far the only examples come from Lissue (unstratified) and Garannes, although there was reported a second find from Lough Faughan. These are distinctive in the hand, both from their weight and magnetism, although it must be admitted that visual identification will not always be easy. X-radiographic screening of slag assemblages is a simple way to find blooms or fragments of bloom amongst a mass of slag lumps. A further characteristic find to be associated with smithing activity is the unfinished artefact or 'semi-product'. As with other indicators, none has so far been reported from a site of the period under review, but it is quite possible that some lurk unrecognised due to poor preservation in site assemblages.

There is a suitable irony in the fact that what may be reasonably identifiable as a 'bowl furnace' is not a certain indicator of smelting activity alone. As we have seen earlier, a hemispherical depression in the ground could function as a smelting furnace or as a hearth for bloomsmithing and forging. Iron tongs are one of the primary blacksmithing tools, used to grip hot objects. But heating is common to ferrous and non-ferrous metalworking, and tongs of the type found at Garannes could as easily have been used to lift a crucible of molten copper from the fire as hold red-hot iron. Finds of large amounts of charcoal could represent smelting or smithing of iron, but also smelting and working of non-ferrous metals and other materials such as glass and enamel. We lack a comprehensive survey of the iron artefacts in our national collections, even in the form of simple descriptive museum catalogues. Many are illustrated in excavation reports and in less accessible publications such as student dissertations and theses. In addition from the obvious typological seriations that such a study would yield, the reader will appreciate, hopefully, from the brief description in Chapter 1.5, that even from a basic drawing it is possible to make important deductions about forging tools and techniques. But unless a *corpus* of ironwork is illustrated to acceptable standards, unlike some recent excavation reports, its value will be seriously diminished. This would be a sizeable undertaking, beyond the limits to be expected even from the most assiduous doctoral candidate. Rather it requires input of time and resources by our national museums and archaeological surveys.

In conjunction with the compilation of a record of ironwork should be a comprehensive program of metallographic examination and chemical analysis. The artefacts examined here were selected with a reasonable degree of randomness but do not constitute a statistically valid sample of iron artefacts of the period ca 7th century BC to 10th century AD. Several assumptions were made as to the information likely to derive from their study, and the body of data presented here is intended to be indicative, not definitive. The studies of scholars such as Piaskowski and Pleiner on statistically useful population samples demonstrate clearly what can be done in the study of iron technology over territories far larger than Ireland. We need perhaps 500 additional examinations and analyses before fully reliable models of ironworking technology from EIA A to LIA A can be constructed. And there are still problems of the metallurgy of bloomery iron to be addressed. For example, what range of compositions do 'acicular precipitates' represent, and is there any real basis to their apparent association with enlarged ferrite grains? Also, what is the nature of the correlation between P and As, and exactly how do their distributions affect carbon distribution and structure? The process of taking, preparing, analysing and recording a single sample can take up to a week, while restoring the sample site to leave the artefact in 'displayable' condition may take several hours. To produce the raw data could thus take a minimum of five years, discounting additional time necessary for the synthesis and publication of results. In addition, the metallographer needs analytical services, and so such a program would not come cheap when we include salary costs.

One task of major importance to future studies of early Irish iron technology (and for other technologies also) is a joint 'excavation', by a team of scientists, linguists, archaeologists, and ethnographers of Binchy's monumental *Corpus Iuris Hibernici*. With more than a tinge of sadness, Binchy himself (*CIH* I.xviii) expressed concern over its limited 'readership'. Yet I have found in those few sections accessible to me indications that, whatever their value to historians of early Irish legal institutions, the contents of *CIH* are a potential goldmine of technical information and observation, even if much of the significance is not immediately clear. Added to this is the mass of early textual material into which I have dipped here and there, again finding indications of great potential value if only the interests of Celtic scholars could be aroused and harnessed to that of other specialists. There is surely the strongest possible case for a concerted study program, and here we might look to bodies like the *Royal Irish Academy* and the *School of Celtic Studies* to act as sponsors of long-term research. It could well take the same length of time as was needed to complete *DIL* and *CIH*, but the return of information would be just as great.

To summarise, the areas which require detailed attention are ore distribution and composition, mining remains, charcoal manufacture and use, furnace and hearth structures, slag characterisation, quantification, and occurrence (i.e. levels of activity and the proportions of smelting to smithing remains), artefact listings and typologies, metallographic and chemical studies of artefacts, and the literary and linguistic evidence. Problems of the theories of smelting and bloomery iron metallurgy will require collaboration in the programs of research continuing outside Ireland.

All of this work will provide a vastly improved picture of the establishment and development of native Irish ironworking than can be presented here. It will also provide the platform for the next three areas of work, namely the production of iron by the Vikings in their colonial settlements in Ireland, definition of the interface between native and intrusive technologies, and the relationship of ironworking with other pyrotechnologies. Preliminary work indicates that the Scandinavians who established the first truly urban centres in Ireland were working in the mainstream of European iron technology. Metallographic examination of a small number of tools from Viking Dublin confirms the presence of virtually all of technical sophistication of ironwork used at major centres like York and Haithabu. This is a tantalising glimpse of what a full-scale research program will uncover. There is ample evidence for interaction and exchange between native and intruder in other fields, and the development of our knowledge in this area will enhance also our understanding of how life was lived in LIA B Ireland.

This book represents nearly twenty years of involvement in the study of Irish ironworking. I find it fascinating to observe just how little I have managed to do in comparison with the work that remains to be done before archaeometallurgical studies of iron in Ireland give as clear an understanding as in other regions. To achieve this will require time, money and interest. Unfortunately there are major questions to be addressed before Irish archaeology permits the serious deployment of resources to the study of 'any old iron'.

* * *

8.3 – Barriers to future research

If one accepts that the adoption and exploitation of iron technology was a major step in the progress of early society, then arguably there is an need to devote to its study and description the same dedication regularly expended on the study, say, of art styles. While I have an obvious interest in promoting Irish archaeometallurgy (with particular reference to iron!), the following remarks are equally applicable to other areas of 'archaeological science' here.

Hopefully it will be clear to the reader that the prerequisite for future research is the collection and synthesis of information and I have attempted throughout this book to highlight areas where data are inadequate. But it is one thing to define lines of future research and quite another to initiate and implement it. There is an obvious need for funding, both to provide the salaries of researchers and teachers, and to provide facilities. But before funds can be deployed effectively, there is a need to have those trained researchers and teachers in place.

Archaeometallurgy is not a single-discipline field, whichever metal is under study, and progress requires question formulation and data interpretation by archaeologists, linguists, historians, metallurgists, chemists and geologists. Thus, a fundamental requirement for any future archaeometallurgical research is the development of regular co-operative links between specialists in these fields. While this exists on an international basis for iron metallurgy under the auspices of the *Comité pour la Sidérurgie Ancienne de l'UISPP*, which provides a forum for the discussion and debate of fundamental problems of theory and explanation, it is less common at regional levels. In Britain, the *Archaeology Committee* of the *Historical Metallurgy Society* fulfills a similar co-ordinating role for all aspects of early metallurgy. For Ireland, there is no such body, the primary reason being that there are, at most, two or three people actively interested in the study of early metal use.

The lack of archaeometallurgists in Ireland stems from the fact that while there is good training in 'traditional' archaeology in Irish universities, the approach is overwhelmingly arts-based. Despite the long tradition of palaeoecological studies here, and the more recent development of world-class research in dating techniques, most archaeology students, and most professional archaeologists, are still woefully ill-equipped to deal with the applications of science in archaeology. This has led, and continues to lead to uncritical acceptance or rejection of ideas and interpretation, depending on individual enthusiams. There are numerous areas in which techniques borrowed from the sciences could advance our knowledge and understanding of the life of man in early Ireland. This requires that Irish archaeologists know what questions to frame, and how to frame them and that they are able to understand and to use the answers! Reading through archaeological literature of even the last ten years shows how few archaeologists here can do either. But despite regular demonstrations of the value of such research, its practitioners are a small band, usually without full-time employment or else employed in another field.

What might be regarded as the first step in the integration of science in Irish archaeology, the gearing of university training towards providing the basic science of early materials as an integral part of archaeology courses, has been taken only in Queen's University, Belfast. It should not have proved impossible to open up regular channels of communication to allow specialist practitioners of chemistry, metallurgy and geology, for example, with an interest in archaeology, formally to divert some of their time away from their main employment into teaching and research. But to date, such collaboration is nearly always on an *ad hoc* basis. This highlights another unfortunate aspect of much of the teaching of archaeology here and that is the insulation of the subject from the benefits of interaction with other disciplines, not only the sciences, but also language and history. While the era of the polymath should have given way here (as it has in other regions) to the era of multidisciplinary collaboration, there are few signs of changes in practice. Until the teaching of archaeology changes, the progress of the subject will be, at best, halting.

Even allowing for the fact that all aspects of archaeology in Ireland are underfunded disgracefully, the lack of understanding of the potential benefits of science to archaeology amongst the profession is reflected both by the tiny number of specialists employed in archaeological bodies and by the meagre allocations of funds to such research. While most professional Irish archaeologists will pay lip-service to 'science', there are no obvious moves to divert a proportion of funds towards its establishment in the archaeological mainstream. Thus, neither the two state archaeological surveys, nor the two national museum Antiquities departments, employ a scientific specialist, and only two of the four university departments of Archaeology employ scientists, these being in dating and environment. There are no specialists in the science of early materials of any kind employed on a full-time basis by any of the bodies charged with teaching and practising archaeology in Ireland. What is equally deplorable is that even were posts to be established immediately, the field of Irish applicants with a scientific specialism and an understanding of any period of Irish archaeology, would be severely limited.

The long-term ideal would be for the establishment of an institute of archaeological research which would bring together archaeologists trained in traditional methods, specialists in the various scientific fields which can contribute to archaeology, linguists and historians. Such a proposition was discussed as long ago as 1972 at a public meeting in Dublin, sponsored by the *Association of Young Irish Archaeologists*, and attended by many senior professional archaeologists, and by a significant number of students who were to become today's senior professionals. In the intervening 18 years, the degree of progress towards achieving what was then generally agreed to be a good and desirable goal has been precisely nil!

All of this has implications for a whole variety of fields, including studies of the early technologies of metals, ceramics, glass, lithics, wood, textiles, leather, and the studies of animal and human bone, all of which could make major contributions to our knowledge of how people lived here in the past. I trust, however, that I may not be criticised too harshly for regretting most that at present, I see no prospects of my own work being taken up and developed by the product of an Irish Department of Archaeology.

Appendices

This listing of sites which have produced evidence of ironworking was compiled from a literature search of the major Irish archaeological journals, but is not claimed as exhaustive. A major problem for any researcher is the great variability in the quality of reporting over the years. In recent years also, there has been a significant increase in excavation activity, particular rescue excavation,throughout the island. Pressure on archaeologists has meant that a significant backlog of publication has built up. Although summary reports for many sites are available in the publication *Excavations*, these obviously are intended as indicative only, and not definitive. For these reasons, the listing below will require expansion, updating and emendment in the coming years.

Appendix 1

Sites producing evidence of ironworking in the period EIA A to LIA

No.	Site name	site type	date	O	F	TY	B	S	A	T	H	BP	I	comments	report
	Co. Antrim														
1	Ballymacash	ringfort	LIA A?	–	–	–	–	+	–	–	–	–	–		summary in Proudfoot 1961
2	Ballymacrea	sandhills		–	–	–	–	+	–	–	–	–	–		Flanagan 1966
3	Ballyhenry	ringfort		–	–	–	1	+	–	–	–	–	–		Lynn 1983b
4	Ballyvollen	industrial site	LIA A	–	–	3	–	178	–	–	–	–	–	production centre?	Williams 1985a
5	Ballywee	ringfort	LIA A	–	–	–	–	+	–	–	–	–	+		Hamlin and Lynn 1988
6	Cleggan	stone hut circle	LIA A?	–	1?	–	–	+	–	–	–	–	–		Mogey 1946
7	Craigywarren	crannog	LIA A	–	–	–	–	+	–	–	–	–	–		Knowles and Coffey 1907
8	Deer Park Farms	raised rath	LIA A	–	–	–	–	+	–	–	–	–	+		Lynn 1987
9	Downview Pk. Belfast	souterrains	LIA A	–	–	–	–	+	–	–	–	–	–		Davison 1964
10	Dromore	earthwork	LIA A	–	–	–	–	+	–	–	–	–	–		Collins 1968
11	Drumnakill	souterrain	LIA A	–	–	–	–	+	–	–	–	–	–		Evans, E.E. 1945
12	Galgorm	souterrain	LIA A	–	–	–	–	+	–	–	–	–	–		Evans, E.E. 1946
13	Larrybane	promontory fort	LIA A	–	–	–	–	+	–	–	–	–	–		Childe 1936
14	Lissanoure	crannog	LIA A	–	–	–	–	+	–	–	–	–	–		Evans 1945
15	Lissue	ringfort	LIA A	–	–	?	1	+	–	–	–	–	?		Bersu 1947
16	Murlough Bay	sandhills	LIA A>	–	–	–	–	+	–	–	–	–	–		Evans 1945
17	Shaneen Pk, Belfast	ringfort	LIA A	–	–	–	–	+	–	–	–	–	–		Proudfoot 1958
18	Shane's Castle	ringfort	LIA A	–	–	–	–	+	–	–	–	–	–	small amount of slag	Warhurst 1971
	Co. Armagh														
19	Cathedral Hill Armagh	hill settlement	EIA B>	–	–	7	–	+	–	–	–	3+	+		Brown and Harper 1984
20	Dressogagh	ringfort	LIA A	–	1	–	–	+	–	–	–	–	–		Collins 1966
	Co. Clare														
21	Cahercommaun	stone ringfort	LIA A>	–	–	–	–	+	–	–	–	–	?	slag throughout	Hencken 1938
22	Inishcealtra	monastic	LIA A>	–	?	–	–	+	–	–	–	–	–	'iron reduction pits'	de Paor 1970
	Co. Cork														
23	Ballyara	souterrain	LIA A	–	–	–	–	+	–	–	–	–	–		Fahy 1953
24	Ballycatteen	ringfort	LIA A	–	–	–	–	15	–	–	–	–	+		O'Riordáin and Hartnett 1943
25	Ballyvourney	ringfort	LIA A	+	9?	13	–	+	–	–	–	+	–		O'Kelly 1952
26	Garannes	ringfort	LIA A	–	–	1	–	+	–	2+	–	–	+	royal site	O'Riordain 1942
27	Garryduff I	ringfort	LIA A	+	??	>8	–	+	?	–	+	+	+	'much slag throughout'	O'Kelly 1964
28	Glannafeen	stone hut circle	LIA A?	–	1?	+	–	+	–	–	–	–	–		O'Cuileanáin 1955
29	Lisleagh	ringfort	LIA A	–	1	2	–	800	–	–	–	–	–	slag dump in ditch	Monk 1988
30	Lisnagun	ringfort	LIA A	–	–	–	–	+	–	–	–	–	–		O'Sullivan 1988
31	Old Court	ringfort	LIA A	+	?	?	–	+	–	–	–	–	–	'roasted' local limonite	Murphy 1961

'date' refers to the earliest evidence for ironworking, not to the overall dating of the site. Suffix '>' indicates continuity with subsequent period(s).
O = ore F = furnace TY = tuyère B = bloom S = slag A = anvil T = tongs H = hammer BP = bellows protector
I = other industrial activity – = absent or not detected + = present but unquantified ? = questionable occurrence
For ore and slag, a number gives the quantity found in kg. For furnaces, tuyères, blooms and tools, a number is the number found.

Appendix 1 ctd.

No.	Site name	site type	date	O	F	TY	B	S	A	T	H	BP	I	comments	report
	Co. Derry														
32	Big Glebe	mound	LIA A	–	–	–	–	+	–	–	–	–	?		Hamlin and Lynn 1988
33	Largantea	megalith	LIA A	–	–	–	–	+	–	–	–	–	–	re-used in LIA A	Herring 1938
34	Lismurphy	ringfort	LIA A?	–	–	–	–	+	–	–	–	–	–		Lynn 1980
35	Lough Enagh	crannog	LBA>	–	–	–	–	+	–	–	–	–	–		Davies 1941
	Co. Donegal														
36	Magheragallen	sandhills	EIA B?	?	?	–	–	+	–	–	–	–	–	EBA ceramics	Evans, E.E. 1941
37	Moylederg	crannog	LIA A	–	–	?	–	+	–	–	–	–	–		Davies 1946
38	Rinarraw	cashel, house site	LIA A?	–	–	–	–	+	–	–	–	–	–		Fanning 1987
	Co. Down														
39	Ballyfounder	ringfort	LIA A	–	–	–	–	+	–	–	–	–	–		Waterman 1958
40	Ballynarry	raised ringfort	LIA A>	–	–	–	–	+	–	–	–	–	–		Davison 1962
41	Castleskreen 2	ringfort	LIA A>	–	–	–	–	+	–	–	–	–	+		Dickinson and Waterman 1959
42	Downpatrick	monastic	?LIA A>	–	–	–	–	+	–	–	–	–	+	on Cathedral Hill	Hamlin and Lynn 1988
43	Lisnagade 2	ringfort	LIA A	–	–	–	–	+	–	–	–	–			summarised in Proudfoot 1961
44	Lough Faughan	crannog	LIA A	+	–	–	2?	+	–	–	1	–	+	'bog ore' and blooms?	Collins 1955
45	Lough Island Reevy	island habitation	LIA A	–	–	–	–	+	–	–	–	–	–		Davies and Gaffikin 1938
46	Movilla Abbey	monastic	LIA A>	–	–	–	–	+	–	–	–	+	+		Ivens 1984a
47	Nendrum	monastic	LIA A	?	–	?	–	+	–	+	–	–	+	'ironstone'?	Lawlor 1925
48	Rathmullan	raised ringfort	LIA A	–	–	–	–	+	–	–	–	–	+		Lynn 1982b
49	Tullylish	monastic	LIA A>	–	–	+	–	+	–	–	–	–	+		Ivens 1987
	Co. Dublin														
50	Ballyman	ecclesiastical	LIA A>	–	–	–	–	+	+	–	–	–	–		O'Brien, E. 1986
51	Feltrim Hill	hilltop settlement	LIA A	–	–	–	–	+	–	–	–	–	–		Hartnett and Eogan 1964
52	Taylorsgrange	open settlement	LIA A	–	1	–	–	+	–	–	–	–	–		Keeley 1987
	Co. Fermanagh														
53	Ballydoolough	crannog	LIA A?	–	–	–	–	+	–	–	–	–	–		Wakeman 1871
54	Boho	ringfort	LIA A	–	1?	–	–	+	–	–	–	–	–		Proudfoot 1953
55	Carn	stone ringfort	LIA A	–	–	–	–	+	–	–	–	–	–		Brannon 1981
56	Coolcran	ringfort	LIA A	–	1+?	1	–	+	–	–	–	–	–	smelting/smithing?	Williams 1985b
57	Lough Carrick	shore finds	LIA A?	–	–	–	–	+	–	–	–	–	–	'derived from crannog'?	Gaffikin and Davies 1938
	Co. Galway														
58	Feerwore	ringfort	EIA B	–	–	–	–	+	–	–	–	–	+		Raftery, J. 1944
59	Kiltiernan East	monastic	LIA A>	–	–	–	–	+	–	–	–	–	–		Duignan 1951
60	Rathgurreen	ringfort	LIA A	–	?	–	–	+	–	–	–	–	–		unpublished – info from UCG
	Co. Kerry														
61	Beginish	monastic	LIA A	–	–	+	–	+	–	–	–	–	+		O'Kelly 1956
62	Church Island	monastic	LIA A	–	–	–	–	+	–	–	–	–	+		O'Kelly 1959
63	Leacanabuaile	ringfort	LIA A	–	–	–	–	+	–	–	–	–	+		O'Riordáin and Foy 1941
64	Reask	monastic	LIA A	–	3	8	–	+	–	–	–	+	+	smelting and smithing	Fanning 1981
65	Sceilg Mhiceal	monastic	LIA A	–	–	–	–	+	–	–	–	–	+	smithing?	O'Kelly 1973
	Co. Kilkenny														
66	Freestone Hill	hillfort	EIA B	?	–	–	–	+	–	–	–	–	+	ironstone nodules	Raftery, B. 1969
	Co. Limerick														
67	Carraig Aille I & II	ringforts	LIA A	–	–	–	–	+	–	–	–	–	–		O'Riordain 1949a
68	Croom East	ringfort	?	–	–	–	–	+	–	–	–	–	–		Shee 1974a
69	Cush sites 3,4, 6,7,10	earthworks	EIA B>	?	–	–	–	+	–	–	–	–	–		O'Riordain 1940
70	Grange	ringforts	LIA A	–	–	?	–	+	–	–	–	–	–		O'Riordain 1949b
71	Knockea	ringfort	LIA A	–	–	1	–	+	–	–	–	–	–		O'Riordain 1949b
72	Sluggary	ringfort	LIA A	–	?	–	–	?	–	–	–	–	–		Shee 1974b

'date' refers to the earliest evidence for ironworking, not to the overall dating of the site. Suffix '>' indicates continuity with subsequent period(s).
O = ore F = furnace TY = tuyère B = bloom S = slag A = anvil T = tongs H = hammer BP = bellows protector
I = other industrial activity – = absent or not detected + = present but unquantified ? = questionable occurrence
For ore and slag, a number gives the quantity found in kg. For furnaces, tuyères, blooms and tools, a number is the number found.

Appendix 1 ctd.

No.	Site name	site type	date	O	F	TY	B	S	A	T	H	BP	I	comments	report
	Co. Louth														
73	Aghnaskea A	megalith	LIA A	+	1	?	–	+	–	–	–	–	–	re-used in LIA A	Evans, E.E. 1935
74	Aghnaskea B	megalith	LIA A	–	1	–	–	+	–	–	–	–	–	re-used in LIA A	Evans, E.E. 1940
75	Ballymakellett	ringfort	LIA A?	–	+	+	–	+	–	–	–	–	–	field survey only	McCabe 1968
76	Faughart	ringfort?	LIA A	–	–	–	5?	+	–	–	–	–	–		O'Connor 1966
	Co. Mayo														
77	Ardcloon	ringfort	LIA A	–	–	–	–	3	–	–	–	–	+		Rynne 1955
78	Inishkea North	monastic?	LIA A	–	?	–	–	+	–	–	–	–	–	smithing?	Henry 1945
79	Letterkeen	ringfort?	LIA A	–	–	–	–	+	–	–	–	–	–		O'Riordain and McDermott 1952
	Co. Meath														
80	Kells, Townparks	by monastic site	LIA A	–	–	–	–	+	–	–	–	–	–		Byrne, G. 1987
81	Knowth	megalith	EIA B>	–	3?	?	–	+	–	–	–	–	+	occupied from EIA B on	Eogan 1972
82	Moynagh Lough	crannog	LIA A>	–	3?	?	–	+	–	–	–	–	+		Bradley 1986
83	Simonstown	ringfort	LIA A	–	–	–	–	+	–	–	–	–	–		Kelly 1976
	Co. Offaly														
84	Ballinderry II	crannog	LIA A>	–	–	–	–	+	–	–	–	+	+		Hencken 1942
85	Gallen Priory	monastic	LIA A	–	?	–	–	+	–	–	–	–	–	'ore crushing basin'?	Kendrick 1939
	Co. Sligo														
86	Achonry	early monastic	LIA A	–	–	–	–	+	–	–	–	–	–		OS 6" sheet 32 – unpublished
87	Creevykeel	court-grave	LIA A	–	1	–	–	+	–	–	–		–	re-used in LIA A	Hencken 1939
88	Carrowntemple	monastic enclosure	LIA A	–		–	–	+	–	–	–	–	–		Wallace and Timoney 1987
89	Derryleghan	monastic enclosure	LIA A	–	–	–	–	+	–	–	–	–	–		OS 6" sheet 5 – unpublished
90	Drumcliffe South	monastic site	LIA A	–	–	–	–	+	–	–	–	–	+		Enright 1986
	Co. Tipperary														
91	Liathmore	monastic	LIA A	–	–	–	–	+	–	–	–	–	?		Leask and Macalister 1945
92	Poulacapple	ringfort	LIA A	–	–	–	–	+	–	–	–	–	–		Reynolds 1972
	Co. Tyrone														
93	Altanagh	ringfort	EIA B>	–	3?	–	–	+	–	–	–	–	–	smelting, smithing?	Williams 1986
94	Clogher	hillfort	EIA B>	–	3?	–	–	+	1	–	–	–	+	royal site	Warner 1972
95	Doras	ecclesiastical	LIA A>	–	–	–	–	8	–	–	–	–	–		McDowell 1987
96	Dunmisk	monastic enclosure	LIA A	–	–	–	–	+	–	–	–	–	+		Ivens 1986
97	Lough Enagh	crannog	LBA>?	–	–	1?	–	+	–	–	–	–	–		Davies 1941
98	Mullaghbane	ringfort	LIA A	–	–	–	–	+	–	–	–	–	–	'large amount of slag'	Harper 1972
99	Urney Park	ringfort	LIA A	–	–	–	–	+	–	–	–	–	–		Scott 1970
	Co. Waterford														
100	Brothers' Cave	cave	LIA A?	–	–	–	+	–	–	–	–	–	–		unpublished – info from UCC
101	Carrigmuirish	ringfort	LIA A?	–	–	1	1	–	–	–	–	–	–		in O'Kelly 1964
102	Kiltera	monastic	LIA A	–	?	–	–	+	+	–	–	–	+	stone anvil and slag	Macalister 1935
	Co. Westmeath														
103	Kilpatrick	monastic	LIA A	–	1	?	–	+	–	–	–	–	+		Swann 1973
104	Lagore	crannog	LIA A>	–	–	–	–	+	–	–	1	–	+	royal site	Hencken 1950
105	Newtonlow	crannog	LIA A>	–	–	–	–	+	–	–	–	–	?		Bourke 1986
	Co. Wicklow														
106	Rathgall	hillfort	EIA B	–	1	–	–	+	–	–	–	–	+		Raftery, B. 1976b

'date' refers to the earliest evidence for ironworking, not to the overall dating of the site. Suffix '>' indicates continuity with subsequent period(s).
O = ore F = furnace TY = tuyère B = bloom S = slag A = anvil T = tongs H = hammer BP = bellows protector
I = other industrial activity – = absent or not detected + = present but unquantified ? = questionable occurrence
For ore and slag, a number gives the quantity found in kg. For furnaces, tuyères, blooms and tools, a number is the number found.

Appendix 2

Samples for metallographic examination were removed by hand, and prepared by standard techniques. Preliminary inspection followed etching in 2% Nital, and specimens were then repolished and etched for further examination and macro- and microphotography. Etchants used were 4% Nital, Picral or Bisulphite. Indications of segregations other than of carbon were sought after etching with Oberhoffer's reagent. Measurements of macro- (H_v) and microhardness (mH_v) were made on the Vicker's scale using loads of 1kg and 50gm respectively onto a diamond pyramid indenter. The hardness profiles presented with the reports of the examinations show the macrohardnesses at points along lines as close as possible to the central long axes of the specimens, and in areas where there was sufficient thickness of metal to allow hardness measurements to be made. Those reports lacking macrohardness profiles are of specimens where there was not sufficient thickness of metal. Grain sizes were measured using the ASTM standard, and carbon content was measured from the amount of pearlite present in the section. Portions of each sample were kept for bulk chemical analysis which was conducted by absorption spectrophotometry. Results are presented both in text descriptions and in summary tables set out in accordance with the guidelines laid down by J. Piaskowski (1987a).

References

A

Aen. = Calder, G. (1907) *The Irish Aeneid*, Dublin (*Irish Texts Soc.* VI).

Alcock, L. (1972) 'The Irish Sea zone in the Pre-Roman Iron Age', in ed. C. Thomas, London, 99–112. Alexander, J. (1981) 'The coming of iron to Britain', in ed. R. Pleiner, Schaffhausen, 57–68.

Allen, I.M., Britton, D. and Coghlan, H. (1970) *Metallurgical Reports on British and Irish Bronze Age Weapons and Implements in the Pitt Rivers Museum*, Oxford (*Pitt Rivers Occasional Papers on Technology, 10)*.

Anderson, A.O. and Anderson, M.O. (1961) *Adomnan's Life of Columba*, Edinburgh.

Anecd. = Bergin, O., Best, R.I., Meyer, K. and O'Keeffe, J.G. (1913) *Anecdota from Irish Manuscripts*, 5 Vols., Dublin, 1907–1913.

Andrieux, P. (1989) 'Dix ans d'experimentation: Le feu entre terre et metal', in ed. R. Pleiner, Prague, 77–92.

Angus, N.S., Brown, G.T. and Cleere, H.F. (1962) 'The iron nails from the Roman legionary fortress at Inchtuthill, Perthshire', *J. Iron and Steel Inst.* 200(11), 956–968.

Arrhennius, O. (1967) 'Ore, iron, artifacts and corrosion', *Sveriges Geol. Unders.* 61(11), 1–39.

Armstrong, E.C.R. (1923) 'The La Tène period in Ireland', *J. Roy. Soc. Antiq. Ireland* 53, 1–33.

----- (1924) 'The Early Iron Age, or Hallstatt period in Ireland', *J. Roy. Soc. Antiq. Ireland* 54, 1–14.

ASCD = *An Archaeological Survey of Co. Down*, Belfast, 1966.

* * *

B

Bachmann, H.G. (1982) *The Identification of Slags from Archaeological Sites* (*Institute of Archaeology Occasional Papers,* 6), London.

Bailey, H.W. (1977) 'Khotan Saka Metal and Mineral Names', *Stud. Orientalia* 47, 19–23.

Baillie, M.G.L. (1985) 'Irish dendrochronology and radiocarbon calibration', *Ulster J. Archaeol.* 48, 11–23.

----- and Pilcher, J. (1983) 'Some observations on the high-precision calibration of routine dates', in ed. B.S. Ottaway *Archaeology, Dendrochronology and the Radiocarbon Calibration Curve*, Edinburgh, 51–62.

Baldinger, K (1980) *Semantic Theory*, Oxford.

BDD = Stokes, W. (1902) 'Togail Bruidne Dá Derga', *Rev. Celtique* XXII (separate reprint).

BDD^2 = Knott, E. (1963) *Togail Bruidne Dá Derga* (*Med. and Mod. Irish Ser.*, VIII) Dublin.

Benveniste, E. (1956) 'Le nom celtique du fer', *Celtica* 3, 279–283.

Bergin, O. (1910) 'The Harrowing of Hell', *Ériu* IV, 112–119.

Bersu, G. (1947) 'The rath in Lissue Townland, Co. Antrim', *Ulster J. Archaeol.* 10, 30–58.

de Bhaldraithe = de Bhaldraithe, T. (1959) *English-Irish Dictionary*, Dublin.

Bielenin, K. (1973) 'Schmelzversuche in halbeingetieften Rennöfen in Polen', in eds W.U. Guyan, et al., Schaffhausen, 62–71.

----- (1974) *Starożytne Górnictwo i Hutnictwo Zelaza w Górach Swiętokrzyskich*, Kraków.

----- (1987) 'Bloomsmithies on early historic sites in the Holy Cross Mountains', in eds. B.G. Scott and H. Cleere, Belfast, 35–46.

Biborski, M. and Kaczanowski, P. (1989) 'Zur Identifizierung römischer Schwertimporte im Lichte metallographischer Untersuchungen', in ed. R. Pleiner, Prague, 379–386.

Binchy, D.A. (1955) 'Bretha Nemed', *Ériu* 17, 4–6.

----- (1970) *Celtic and Anglo-Saxon Kingship*, Oxford.

----- (1973a) 'Distraint in Irish law', *Celtica* 10, 22–71.

----- (1973b) 'A text on the forms of distraint', *Celtica* 10, 72–86.

----- (1982) 'Brewing in eighth-century Ireland', in ed. B.G. Scott, Belfast, 3–6.

Birkhan, H. (1970) *Germanen und Kelten bis zum Ausgang der Römerzeit*, Wien.

BNnÉ = Plummer, C. (1922) *Bethada Náem nÉrenn*, 2 Vols., Oxford (1968 reprint).

Boate, G. (1652) *Natural History of Ireland*, London.

Boll, P.O., Erisman, T.H. and Muster, W.J. (1981) 'Metallkundliche Untersuchung eines frühen mitteleuropäischen Eisenschwertes', in ed. R. Pleiner, Schaffhausen, 45–52.

Bourke, C. (1986) 'Newtownlow, Co. Westmeath', *Excavations 1986*, 37.

BR^2 = Dillon, M. (1962) *Lebor Na Cert* (*Irish Texts Soc.*, XLVI), Dublin.

Bradley, J. (1986) 'Excavations at Moynagh Lough 1984', *Ríocht na Midhe* VII(4), 79–92.

Brannon, N. (1981) 'Excavation of a cashel in Carn townland, Co. Fermanagh', *Ulster J. Archaeol.* 44, 60–64.

Bratt, A. (1976) 'Deerfin Lower', *Excavations 1975–1976*, 5–6.

Bretz-Mahler, D. (1971) *La Civilisation de La Tène I en Champagne*, Paris.

Briggs, C.S. (1976) 'The indigenous minerals and metallurgy of the earliest Irish Bronze Age', *Irish Archaeol. Res. Forum* 3(1), 9–15.

----- (1984) 'The discovery and description of trench mines at Derricarhoon td., Co. Cork, in 1846', *J. Irish Archaeol.* II, 33–40.

----- (1987) 'Buckets and Cauldrons in the Late Bronze Age of North-West Europe; a review' in *Actes du Colloque de Bronze de Lille*, 161–187.

Brown, C. and Harper, A.E.T. (1984) 'Excavations on Cathedral Hill, Armagh 1968', *Ulster J. Archaeol.* 47, 109–161.

Bruce-Mitford, R. (1972) *The Sutton Hoo Ship Burial*, London (British Museum).

Buchwald, F. and Mosdal, G. (1985) 'Meteoric iron, telluric iron and wrought iron in Greenland', *Meddelelser om Grønland* (*Man and Society*) 9, 3–49.

Burgess, C. (1974) 'The Bronze Age' in ed. C. Renfrew *British Prehistory*, London, 165–232.

Byrne, F.J. (1971) 'Tribes and tribalism in early Ireland', *Ériu* XXII, 128–165.

----- (1973) *Irish Kings and High Kings*, London.

Byrne, G. (1987) 'Kells, Townparks, Co. Meath', *Excavations 1987*, 23–24.

* * *

C

Calissendorf, K. (1979) 'Linguistic evidence for early iron production' in ed. H. Clarke, Stockholm, 157–175.
Carney, J. (1958) 'Two Old Irish poems, *Ériu* XVIII, 1–43.
Caulfield, S. (1977) 'The beehive quern in Ireland', *J. Roy. Soc. Antiq. Ireland* 107, 104–138.
----- (1981) 'Some Celtic problems in the Irish Iron Age', in ed. D. Ó Corráin, Cork, 205–215.
CCath. = Stokes, W. (1909) *In Cath Catharda* (*Irische Texte*, IV) Leipzig.
CCG = Lewis, H. and Pedersen, H. (1961) *A Concise Comparative Celtic Grammar*, Göttingen.
Chadwick, G.A. (1972) *Metallography of Phase Transformations*, London.
Champion, T.C. (1971) 'The end of the Irish Bronze Age', *N. Munster Antiq. J.* 14, 17–24.
----- (1982) 'The myth of Iron Age invasions in Ireland' in ed. B.G. Scott, Belfast, 39–44.
Charles, J.A. (1980) 'The coming of copper and copper-base alloys and iron: a metallurgical sequence', in eds. T.A. Wertime and J.D. Muhly, Yale, 151–182.
----- (1984) 'The Middle Bronze Age iron punch of southeast Drenthe', *Palaeohistoria* 26, 95–99.
Chikashige, M. (1936) *Alchemy and Other Chemical Achievements of the Ancient Orient*, Tokyo.
Childe, V.G. (1936) 'A promontory fort on the Antrim coast', *Antiq. J.* XV1, 179–198
----- (1950) *Prehistoric Migrations in Europe*, Oosterhout (1969 reprint).
Chilton, J.P. and Evans, U.R. (1955) 'The corrosion resistance of wrought iron', *J. Iron and Steel Inst.* 181, 113–122.
CIH = Binchy, D.A. (1978) *Corpus Iuris Hibernici*, 5 Vols., Dublin.
CIIAA = Raftery, B. (1983) *A Catalogue of Irish Iron Age Antiquities*, 2 Vols., Marburg.
Clarke, H. ed. (1979) *Iron and Man in Prehistoric Sweden*, Stockholm.
Cleere, H.F. (1963) 'Primitive Indian iron making furnaces', *British Steelmaker* April 1963, 154–158.
----- (1972) 'The classification of early iron-smelting furnaces', *Antiq. J.* 52(1), 8–23.
----- (1987) 'Ironmaking in the economy of the ancient world: the potential of archaeometallurgy', in eds. B.G. Scott and H. Cleere, Belfast, 1–6.
Clough, R.E. (1987) 'The bloomery process – observations on the use of rich ores and the production of natural steel' in eds. B.G. Scott and H. Cleere, Belfast, 19–27.
CMM = O Daly, M. (1975) *Cath Maige Mucrama* (*Irish Texts Soc.*, 50), Dublin.
CMT = Fraser, J. (1915) 'The first Battle of Moytura', *Ériu* 8, 1–63.
CMT^2 = Gray, E.M. (1982) *Cath Maige Tuired* (*Irish Texts Soc.*, 52), Dublin.
Coffey, G. (1907) 'Early iron sword found in Ireland', *Proc. Roy. Irish Acad.* 34(C), 42–43.
----- (1910) 'Archaeological evidence for the intercourse of Gaul with Ireland before the first century', *Proc. Roy. Irish Acad.* XXVIII(C), 96–106.
----- and Armstrong, E.C.R. (1910) 'Scandinavian objects found at Island-bridge and Kilmainham', *Proc. Roy. Irish Acad.* XXVIII(C), 107–122.
Cole, G.A.J. (1922) *Memoir and Maps of Localities of Minerals of Economic Importance and Metalliferous Mines in Ireland,* Dublin.
----- and Halissy, T. (1921) *Handbook of the Geology of Ireland,* London.
Collins, A.E.P. (1955) 'Excavations in Lough Faughan crannog, Co. Down', *Ulster J. Archaeol.* 18, 45–80.
----- (1966) 'Excavations at Dressogagh rath, Co. Armagh', *Ulster J. Archaeol.* 29, 117–129.
----- (1968) 'Excavations at Dromore ringwork, Co. Antrim', *Ulster J. Archaeol.* 31, 59–66.
----- and Seaby, W.A. (1960) 'Structures and small finds discovered at Lough Eskragh, Co. Tyrone', *Ulster J. Archaeol.* 23, 25–37.
Collis, J. (1976) 'Town and Market in Iron Age Europe' in eds. B. Cunliffe and T. Rowley, Oxford, 3–24.
Contrib. = Meyer, K. (1906) *Contributions to Irish Lexicography*, Halle.
Cooke, S.R.B. and Aschbrenner, S. (1975) 'The occurrence of metallic iron in ancient copper', *J. Field Archaeol.* 2, 251–266.
Cowen, J.D. (1967) 'The Hallstatt sword of bronze: on the Continent and in Britain', *Proc. Prehist. Soc.* 33, 377–454.
Craddock, P.T. and Meeks, N.D. (1987) 'Iron in ancient copper' *Archaeometry* 29(2), 187–204.
Crew, P. (1987) 'Bryn y Castell hillfort – a late prehistoric ironworking settlement in north-west Wales' in eds. B.G. Scott and H. Cleere, Belfast, 91–100.
----- (1988) 'More experimental smelting of bog iron ores', *Hist. Met. Soc. News* 11, 3.
Críth Gabl. = Binchy, D.A. (1970) *Críth Gablach* (*Med. and Mod. Irish Ser.*, XI), Dublin.
CRR = Hogan, E (1892) *Cath Rois na Riogh*, Dublin.
Cunliffe, B. (1974) *Iron Age Communities in Britain*, London.
----- (1976) 'The origins of urbanisation in Britain', in eds. B. Cunliffe and T. Rowley, Oxford, 135–162.
----- and Rowley, T. eds. (1976) *Oppida in Barbarian Europe* (*British Archaeol. Rep. Supp. Ser.*, 11), Oxford.
Curwen, E.C. (1948) 'A bronze cauldron from Sompting, Sussex', *Antiq. J.* 28, 157–163.

* * *

D

Danaher, K. (1962) *In Ireland Long Ago*, Cork.
Davies, O. (1935) *Roman Mines in Europe*, London.
----- (1941) Trial excavation at Lough Enagh, Co. Tyrone', *Ulster J. Archaeol.* 4, 88–101.
----- (1946) 'Excavations around Lough Eske, Donegal', *Ulster J.. Archaeol.* 9, 91–99.
----- (1950) *Excavations at Island McHugh*, Belfast.
Davis, D., Maddin, R., Muhly, J.D. and Stech, T. (1985) 'A steel pick from Mount Adir in northern Galillee', *J. Near Eastern Stud.* 44(1)), 41–51.
Davison, B.K. (1962) 'Excavations at Ballynarry rath, Co. Down', *Ulster J. Archaeol.* 24–25, 1961–1962, 39–87.
----- (1964) 'Three souterrains at Downview Park West, Belfast', *Ulster J. Archaeol.* 27, 124–128.
Dickinson, C.W. and Waterman, D.M. 'Excavations of a rath with motte at Castleskreen, Co. Down', *Ulster. J. Archaeol.* 22 1959, 67–82
DIL = *A Dictionary of the Irish Language* (*Royal Irish Academy*), Dublin, 1913–1975.
Dillon, M. and Chadwick, N. (1972) *The Celtic Realms*, London.
Dinn. = Dinneen, P.S. (1927) *Foclóir Gaedhilge-Béarla*, Dublin (1970 reprint).
Dodwell, C.R. (1961) *Theophilus De Diversis Artibus*, London.
Duan. Finn = MacNeill, E. and Murphy, G. (1953) *Duanaire Finn* (*Irish Texts Soc.* VII, XXVIII, XLIII), Dublin, 1908–1953.
Dubordieu, J. (1812) *A Statistical Survey of Co. Antrim*, 2 Vols. London.

Dugan, C.W. (1897) 'Interesting "find" in the Montiaghs, Co. Armagh', *J. Roy. Soc. Antiq. Ireland* 27, 437–438.

Duignan, M.V. (1944) 'Irish agriculture in Early Historic times', *J. Roy. Soc. Antiq. Ireland* 64, 124–145.

----- (1951) 'Early monastic site at Kiltiernan East, Co. Galway', *J. Roy. Soc. Antiq. Ireland* 81, 73–75.

----- (1976) 'The Turoe stone: its place in insular La Tène art' in eds. Duval, P-M. and Hawkes, C.F.C *Celtic Art in Ancient Europe: Five Protohistoric Centuries*, Oxford, 1976, 201–218.

Dunning, G.C. (1934) 'The swan's neck and ring-headed pins of the Early Iron Age in Britain', *Archaeol. J.* 91, 269–295.

Dyer, J. (1976) 'Ravensburgh Castle, Hertfordshire', in ed. D.W. Harding, London, 153–161.

* * *

E

Ehrenreich, R. (1985) *Trade, Technology and the Ironworking Community in the Iron Age of Southern Britain* (*British Archaeol. Rep.* 144), Oxford.

Eichhorn, P., Rollig, H., Schwarz, U, Urbon, B. and Zwicker, U. (1974) 'Untersuchungen über die hallstattzeitliche Technik für Bronzeeinlagen in Eisen', *Fundber. aus Baden-Württemburg*, 1, 243–398.

Eliade, M. (1978) *The Forge and the Crucible*, London (2nd. edn.).

Enright, J. (1986) 'Drumcliffe South, Co. Sligo', *Excavations 1986*, 32.

Eogan, G. (1964) 'The Later Bronze Age in Ireland in the light of recent research', *Proc. Prehist. Soc.* 30, 268–351.

----- (1965) *The Bronze Sword in Ireland*, Dublin.

----- (1972) 'Knowth, Co. Meath', *Excavations 1972*, 23–24.

----- (1974a) 'Regionale Grüppierung in der Spätbronzezeit Irlands', *Archäol. Korresp.* 4, 319–327.

----- (1974b) 'Pins of the Irish Later Bronze Age', *J. Roy. Soc. Antiq. Ireland* 104, 74–119.

----- (1981) 'The gold vessels of the Irish Bronze Age and beyond', *Proc. Roy. Irish Acad.* 81C(14), 345–382.

----- (1983) *Hoards of the Irish Later Bronze Age*, University College, Dublin.

Evans, D.E. (1967) *Gaulish Personal Names*, Oxford.

Evans, E.E. (1936) 'Excavations at Aghnaskea, Cairn A', *J. Co. Louth Archaeol. Soc.* VIII(3), 1933–1936, 235–256.

----- (1940) 'Excavations at Aghnaskea Cairn B, Co. Louth', *J. Co. Louth Archaeol. Soc.* IX, 1937–1940, 1–18.

----- (1941) 'A Sandhills site in Co. Donegall', *Ulster J. Archaeol.* 4, 71–75.

----- (1945) 'Field archaeology in the Ballycastle District', *Ulster J. Archaeol.* 8, 14–32.

----- (1946) 'Newly discovered souterrains in Co. Antrim', *Ulster J. Archaeol.* 9, 79–83.

----- (1948) 'Strange iron objects from Co. Fermanagh, Ireland', *Ulster J. Archaeol.* 11, 58–64.

Eyles, V.A. (1952) *The Composition and Origin of the Antrim Laterites and Bauxites*, Belfast (HMSO).

* * *

F

Fahy, E.M. (1953) 'A souterrain at Ballyara', *J. Cork Archaeol. Hist. Soc.* 58, 55–59.

Fanning, T. (1981) 'Excavation of an Early Christian cemetery and settlement at Reask, Co. Kerry', *Proc. Roy. Irish. Acad.* 81C, 3–172.

----- (1987) 'Rinnaraw, Co. Donegal' *Excavations 1987*, 12.

Fianaig. = Meyer, K. (1910) *Fianagecht* (*Roy. Irish Acad. Todd Lecture Series*, XVI), Dublin.

Flanagan, L.N.W. (1966) 'Dark Age sites at Ballymacrea Lower, Co. Antrim', *Ulster J. Archaeol.* 29, 115–116.

----- (1978) 'Industrial resources, production and distribution in earlier Bronze Age Ireland', in ed. M.F. Ryan, Dublin, 145–163.

Ford, P.K. (1977) *The Mabinogi and other Medieval Welsh Tales*, Berkeley.

Forbes, R. (1964) *Studies in Ancient Technology*, IX, Leiden.

Fowler, A. and Robbie, J.A. (1961) *Geology of the Country around Dungannon, Co. Tyrone*, Belfast (HMSO).

Fox, C. (1939a) 'A second cauldron and an iron sword from the Llyn Fawr Hoard, Rhigos, Glamorganshire', *Antiq. J.* 19, 369–404.

----- (1939b) 'The socketed bronze sickles of the British Isles', *Proc. Prehist. Soc.* 5, 222–248 .

----- (1946) *A Find of the Early Iron Age From Llyn Cerrig Bach, Anglesey*, Cardiff (Nat. Mus. Wales).

* * *

G

Gaffikin, M. and Davies, O. (1938) 'Lough Island Reevy, Co. Down', *Ulster J. Archaeol.* 1, 201–204.

Gilles, J.W. (1958) 'Versuchschmelze in einem vorgeschichtlichen Rennofen', *Stahl u. Eisen* 78, 1690–1695.

Greene, D. (1954) 'Early Irish Literature' in ed. M. Dillon, *Early Irish Society*, Dublin, 22–35.

----- (1955) 'Orgain Denna Rig' in *Fingal Rónáin and Other Stories* (*Med. and Mod. Irish Ser.*, XVI), Dublin, 16–26.

Gregory, A. (1920) *Visions and Beliefs in the West of Ireland*, London (1970 reprint).

Groves, A.W. (1952) *Wartime Investigations into the Haematite and Manganese Ore Resources of Great Britain and Northern Ireland*, London (*Ministry of Supply*).

Guyan, W.U., Pleiner, R. and Fabešová, R. eds. (1973) *Die Versuchschmelzen und Ihre Bedeutung für die Metallurgie des Eisens und dessen Geschichte*, Schaffhausen.

Gwynne, E.J. (1942) 'An Old-Irish tract on the priveleges and responsibilities of poets', *Eriu* XIII, 1–60 and 220–236.

* * *

H

Halleux, R. (1974) *Le Problème des Métaux dans la Science Antique*, Liège.

Hamlin, A. and Lynn, C. (1988) *Pieces of the Past*, Belfast (HMSO).

Harding, D.W. (1974) *The Iron Age in Lowland Britain*, London.

----- ed. (1976) *Hillforts*, London.

Härke, H.G.H. (1979) *Settlement Types and Patterns in the West Hallstatt Province*, Oxford (*British Archaeol. Rep. Supp. Ser.*, 57).

Harper, A.E.T. (1972) 'Excavation of a rath in Mullaghbane townland, Co. Tyrone', *Ulster J. Archaeol.* 35, 37–44.

Hartnett, P.J. and Eogan, G. (1964) 'Feltrim Hill, Co. Dublin: a Neolithic and Early Christian site', *J. Roy. Soc. Antiq. Ireland* 94, 1–38.

Hawkes, C.F.C and Smith, M.A. (1957) 'On some buckets and cauldrons of the Bronze and Early Iron Ages', *Ant. J.* 37, 131–198.

Healy, J.N. (1978) *Mining and Metallurgy in the Greek and Roman World*, London.

Hedges, R.E.M. and Salter, C.J. (1979) 'Source determination of iron currency bars through the analysis of slag inclusions', *Archaeometry* 21(2), 161–175.

Hencken, H. O'N (1938) *Cahercommaun: A Stone Fort in Co. Clare, Roy. Soc. Antiq. Ireland* (extra volume).

----- (1939) 'A long cairn at Creevykeel, Co. Sligo', *J. Roy. Soc. Antiq. Ireland* 49, 53–98.

----- (1942) 'Ballinderry crannog no. 2', *Proc. Roy. Irish Acad.* 47C, 1–76.

----- (1950) 'Lagore crannog: an Irish royal residence of the 7th to 10th centuries A.D.', *Proc. Roy. Irish Acad.* 53C 1–247.

----- (1968) *Tarquinia, Villanovans and Early Etruscans,* 3 Vols., Harvard (*Peabody Museum Bulletin*, 23).

Henry, F. (1945) 'Remains of the Early Christian Period at Iniskea North, Co. Mayo', *J. Roy. Soc. Antiq. Ireland* LXXV, 127–155.

Herity, M. and Eogan, G. (1977) *Ireland in Prehistory*, London.

Herring, I. (1938) 'Cairn excavation at Well Glass Spring, Largantea', *Ulster J. Archaeol.* 1, 164–188.

Hingst, H. (1981) 'Die Eisenverhüttungsplätzen in Schleswig-Holstein, in ed. R. Pleiner, Schaffhausen, 79–88.

Hog. Onom. = Hogan, E.I. (1910) *Onomasticon Goedelicum*, Dublin.

Holder, A. (1896) *Alt-Celtischer Sprachschatz*, 3 Vols., Leipzig.

Holdsworth, J. (1858) *Geology, Minerals and Mines of Ireland*, London.

Holmes, P. (1978) 'The manufacturing technology of Irish Bronze Age horns', in ed. M.F. Ryan, Dublin, 165–288.

Hoover, H.C. and Hoover, L.H. (1912) *Georgius Agricola, De Re Metallica*, New York (1950 reprint).

Howarth, R.G. (1971) 'The horse-harness of the Irish Early Iron Age', *Ulster J. Archaeol.* 34, 26–49.

Hume-Rothery, W. (1966) *The Structures of Alloys of Iron*, Oxford.

Hurley, M. (1985) 'Castletown, Co. Kilkenny', *Excavations 1985*, 27.

* * *

I

ICM² = Oskamp, H.P.A. (1970) *The Voyage of Máel Dúin*, Groningen.

IEW = Pokorny, J. (1959) *Indogermanische Etymologisches Wörterbuch*, Berne.

IGT Decl. = Bergin, O. (1928) 'Irish Grammatical Tracts, Declensions', *Ériu* VIII(1916)-X(1926–1928).

Ivanov, I. (1978) *Treasures of the Varna Chalcolithic Necropolis*, Varna.

Ivens, R. (1984a) 'Movilla Abbey, Newtownards, Co. Down: Excavations 1981', *Ulster J. Archaeol.* 47, 71–108.

----- (1984b) 'Killyliss rath, Co. Tyrone' *Ulster J. Archaeol.*, 47, 9–35.

----- (1986) 'Dunmisk Fort, Co. Tyrone', *Excavations 1986*, 33–34.

----- (1987) 'The Early Christian monastic enclosure at Tullylish, Co. Down', *Ulster J. Archaeol.* 50, 55–121.

* * *

J

Jackson, J. (1978) 'Metallic ores in Irish prehistory: copper and tin', in ed. M.F. Ryan, Dublin, 107–125.

Jackson, K. (1953) *Language and History in Early Britain*, Edinburgh.

----- (1964) *The Oldest Irish Tradition: A Window on the Iron Age*, Oxford.

Jacobi, G. (1974) *Werkzeug und Gerät aus dem Oppidum von Manching*, Wiesbaden.

Jarman, A.O.H. (1988) *Aneirin: Y Gododdin*, Llandysul.

Jensen, (1968) 'Ole Evenstadt's *Treatise on iron ore and the process of turning it into iron and steel*', *Bull. Hist. Met. Gr.* 2(2), 61–65.

Jessen, K. (1949) 'Studies in Late Quaternary deposits and flora-history in Ireland' *Proc. Roy. Irish Acad.* 52B, 85–290.

Jones, G. and Jones T. (1974) *The Mabinogion*, London.

Jope, E.M. (1954) 'An Iron Age sword-scabbard from the River Bann at Toome', *Ulster J. Archaeol.* 17, 81–91.

----- (1955) 'An iron spearhead of Germanic type from Co. Fermanagh', *Ulster J. Archaeol.* 18, 16–17.

----- (1961) 'Daggers of the Early Iron Age in Britain', *Proc. Prehist. Soc.* 27, 307–343.

----- (1962) 'Iron Age brooches in Ireland: a summary', *Ulster J. Archaeol.* 24–25 (1961–1962), 25–38.

----- (1966) 'Man's use of natural resources', *Adv. Sci.* 22, 1965–1966, 1–14.

----- and Wilson, B.C.S. (1957) 'The decorated cast bronze disc from the River Bann near Coleraine', *Ulster J. Archaeol.* 20, 95–102.

Joyce, P.W. (1903) *A Social History of Ancient Ireland*, London.

Joynt, M. (1910) 'Echtra mac Echdach Mugmedóin', *Ériu* IV, 94–111.

* * *

K

Kane, R. (1844) *Industrial Resources of Ireland*, Dublin.

Kędzierski, Z. and Stępiński, J. (1989) 'Classification of the manufacturing technology of double-edged swords of the 1st to 4th centuries AD from Europe', in ed. R. Pleiner, Prague, 387–392.

Keeley, V. (1987) 'Brehon's Chair, Taylorsgrange, Co. Dublin', *Excavations 1987*, 14–15.

Keesman, I. (1989) 'Chemische und mineralogische Detailuntersuchungen zur Interpretation eisenreicher Schlacken' in ed. R. Pleiner, Prague, 17–34.

-----, Bachmann, H.G. and Hauptmann, A. (1982) 'Norm-Berechnungsverfahren für eisenreiche Schlacken', *Forschritte der Mineralogie* 60(1), 110–111.

Kellcher, J.V. (1971) 'The *Táin* and the Annals', *Ériu* XXII, 107–127.

Kelly, E.P. (1974) 'Aughinish Island, sites 1 and 2', *Excavations 1974*, 21.

----- (1976) 'Simonstown, Co. Meath', *Excavations 1975–1976*, 17–18.

Kelly, F. (1976) *Audacht Morainn*, Dublin.

----- (1988) *A Guide to Early Irish Law*, Dublin.
Kendrick, T.D. (1939) 'Excavations at Gallen Priory, Co. Offaly, 1934–1935', *J. Roy. Soc. Antiq. Ireland* 49, 1–20.
Kennon, N. (1980) 'Nomenclature of phases and constituents' in L.E. Samuels, 1980, 5–39.
Kilroe, J.R. (1907) *A Description of the Soil Geology of Ireland*, Dublin.
Kimmig, W. (1964) 'Seevölkerbewegung und Urnenfelderkultur', *Studien aus Alteuropa* I, Köln, 220–284.
----- (1981) 'Ein Grabfund der jungeren Urnenfelderzeit mit Eisenschwert von Singen am Hohentweil' in ed. R. Pleiner, Schaffhausen, 37–44.
Kinahan, G. (1878) *Geology of Ireland*, London.
Knowles, W.J. and Coffey, G. (1907) 'Craigywarren crannog', *Proc. Roy. Irish Acad.* XXVI, 1906–1907, 109–118.
Kolčin, B.A. (1959) *Zhelezoobrabatvaguscheye remeslo Novgoroda Velikogo*, in *Trudy Novgorodskoy Ekspediciyi*, II, Moscow.
Krawczuk, A. and Piaskowski, J. (1958) 'Metalurgia w pismach Aristotelesa', *Kwart. Hist. Naukii Mat.* VI, 232–342.
Kresten, P. (1984) *The mineralogy and chemistry of selected iron slags from Dalarna, Sweden*, Stockholm (*Jernkontoret*).
Kromer, K. (1959) *Das Gräberfeld von Hallstatt*, 2 Vols., Firenze.

* * *

L

Landgren, S. (1948) 'On the geochemistry of Swedish iron ores and associated rocks', *Sveriges. Geol Unders.* 496C, 5–179.
Lang, J. (1987) 'The technology of Celtic iron swords', in eds. B.G. Scott and H.F. Cleere, Belfast, 61–72.
Lawlor, H.C. (1925) *The Monastery of St. Mochaoi of Nendrum*, Belfast.
Laws = O'Donovan, J. and O'Curry, E. (1901) *The Ancient Laws of Ireland*, 5 Vols., London (1865–1901).
Leask, H.G. and Macalister, R.A.S. (1945) 'The partial excavation of a site at Liathmore or Leigh, Co. Tipperary', *Proc. Roy. Irish Acad.* 47C, 77–150.
Leb. Gab. = Macalister, R.A.S. (1956) *Lebor Gabála Érenn* (*Irish Texts Soc.*, 34, 35, 39, 41, 44), Dublin, 1938–1956.
Lee, H.D.P. (1962) *Aristotle: Meteorologica*, London.
Lib. Hymn. = Todd, J.H. (1897) *Liber Hymnorum*, London.
Lism. L. = Stokes, W. (1890) *Lives of the Saints from the Book of Lismore*, Oxford.
LL = Best, R.I. and O'Brien, M. (1967) *The Book of Leinster*, I - V, Dublin 1954–1967.
Lowery, R.P., Savage, R.D.A. and Wilkins, R.L. (1971) 'Scriber, graver, scorper, tracer: notes on experiments in bronzeworking technique', *Proc. Prehist. Soc.* 37,167–182.
LS = Liddell, H.G. and Scott, R. (1968) *A Greek-English Lexicon*, Oxford.
LU = Best, R.I. and Bergin, O. (1929) *Lebor Na Huidre*, Dublin.
Lynn, C.J. (1980) 'A destroyed rath at Lismurphy td., Co. Derry', *Ulster J. Archaeol.* 43, 75–78.
----- (1982a) 'The Dorsey and other linear earthworks', in ed. B.G. Scott, Belfast, 121–128.
----- (1982b) 'Excavation of Rathmullan, a raised rath and motte, Co. Down', *Ulster J. Archaeol.* 45, 65–172.
----- (1983a) 'Some "early" ringforts and crannogs', *J. Irish Archaeol.* I, 47–58.
----- (1983b) 'Two raths at Ballyhenry, Co. Antrim', *Ulster J. Archaeol.* 46, 67–91.
----- (1987) 'Deerpark Farms', *Archaeol. Ireland* 1, 11–15.

* * *

M

McCabe, J.F. *An Archaeological Survey of Co. Louth*, unpublished M.A. Thesis (University College Galway).
Macalister, R.A.S. (1907) *Studies on Irish Epigraphy*, 3 Vols., London 1897–1907.
----- (1938) 'The excavation of Kiltera, County Waterford', *Proc. Roy. Irish Acad.* 43C, 1–16.
McDonnell, J.G. (1986) *The Classification of Early Ironworking Slags*, unpublished Ph.D. Thesis (University of Astin in Birmingham).
----- (1987) 'The metallurgy of Anglo-Scandinavian knives', in eds. B.G. Scott and H. Cleere, Belfast, 87–89.
----- (1989) 'Iron and its alloys in the fifth to eleventh centuries in Britain', *World Archaeol.* 20(3), 373–382.
McDowell, J. (1987) 'Excavation of an ecclesiastical enclosure at Doras, Co. Tyrone', *Ulster J. Archaeol.* 50, 137–154.
Mac Eoin, G. (1982) 'The early Irish vocabulary of mills and milling', in ed. B.G. Scott, Belfast, 13–19.
----- (1986) 'The Celticity of Celtic Ireland', in ed. K.H. Schmidt *Geschichte und Kultur der Kelten*, Heidelberg, 161–174.
McGrath, J.N. (1968) 'Preliminary report on the metallographic examination of sword blades from Llyn Cerrig Bach, Anglesey', *Bull. Hist. Met. Gr.* 2(2), 78–80.
Mac Niocaill, G. (1971) 'Tír Cumaile', *Ériu* XXII, 81–86.
Maddin, R. (1982) 'The beginning of the Iron Age in the Eastern Mediterranean', *Trans. Indian Inst. Met.* 35(1), 14–24.
----- (1987) 'The early blacksmith', in eds. B.G. Scott and H. Cleere, Belfast, 7–17.
----- Muhly, J.D. and Wheeler, T.S. (1977) 'How the Iron Age began', *Sci. American* 237(4), 122–131.
Magnusson, G. (1986) *Lagteknisk Järnhantering i Jämtlands Län*, Stockholm.
----- (1987) 'Who were the blacksmiths in Jämtland and Härjedalen?', in eds. B.G. Scott and H. Cleere, Belfast, 121–130.
Magrill, D. (1980) 'A name for every chemical', *New Scientist* 87 (1208), 16–18.
Mahr, A. (1934) 'A wooden cauldron from Altartate, Co. Monaghan', *Proc. Roy. Irish Acad.* 43(C), 11–29.
Mallett, W. (1853) 'Report on the chemical examination of antiquities from the museum of the Royal Irish Academy', *Trans. Roy. Irish Acad.* XXII, 313–342.
Mallory, J.P. (1982) 'The sword of the Ulster cycle', in ed. B.G. Scott, Belfast, 99–114.
----- (1984) 'The first Irishmen: the state of the Goidelic question', *J. Irish Archaeol.* II, 65–69.
----- (1987) 'Haughey's Fort, Tray, Co. Armagh', *Excavations 1987*, 10.
Manning, W. and Saunders, C. (1972) 'A socketed axe from Maids Moreton, Bucks., with a note on the type', *Antiq. J.* 52(2), 276–292.
Maréchal, J.R. (1958) 'La nituration du fer était utilisée par les Anciennes', *Met. Corr. Ind.* 391, 133–137.
Martens, I. (1987) 'Iron extraction, settlement and trade in the Viking and early Middle Ages in south Norway', *Univ. Oldsaksamling Skrifter* 9, Oslo, 69–80.
Martin, G. and Foley, V. (1979) 'Weland the Smith: some findings', *J. Hist. Met. Soc.* 13(1), 38–39.

Maryon, H. (1960) 'Pattern welding and damascening of sword blades', *Stud. Cons.* 5, 25–37 and 52–60.
van der Merwe, N. (1969) *The Carbon-14 Dating of Iron*, Chicago.
----- (1980) 'The advent of iron in Africa' in eds. T.A. Wertime and J.D. Muhly, Yale, 463–506.
Met Dinds. = Gwynne, E. (1935) *Metrical Dindsenchas*, I-V (*Roy. Irish Acad. Todd Lecture Series*, VIII-XII), Dublin 1903–1935.
Meyer, K. (1890) 'The oldest version of Tochmarc Emire' *Rev. Celtique* XI, 433–457.
Milligan, S.F. (1886) 'Crannogs in Co. Cavan', *Trans. Kilkenny Archaeol. Soc.* VII (4th. ser.), 148–152.
Mitchell, G.F. (1976) *The Irish Landscape*, London.
ML = O'Curry, E. (1855) *The Battle of Magh Leana*, Dublin (*Celtic Society*).
Mogey, J.M. (1946) 'Notes on some field monuments in the Braid, Co. Antrim', *Ulster J. Archaeol.* 9, 130–138.
Monk, M. (1988) 'Excavations at Lisleagh ringfort, north Cork', *Archaeol. Ireland* 2(2), 57–60.
Morton, G.R. and Wingrove, R. (1969) 'The constitution of bloomery slags: Part 1 – Roman', *J. Iron and Steel Inst.* 207(2), 1556–1664.
----- (1971) 'The constitution of bloomery slags: Part 2 – Mediaeval', *J. Iron and Steel Inst.* 210, 478–488.
MR = O'Donovan, J. (1840) *Cath Maige Rath*, Dublin.
Muhly, J.D. (1982) 'The nature of trade in the LBA Eastern Mediterranean: the organisation of the metals' trade and the role of Cyprus' in eds. J.D. Muhly, R. Maddin and V. Karagheorghis 1982, 251–270.
----- Maddin, R. and Karagheorghis, V. eds. (1982) *Early Metallurgy in Cyprus 4000–500 B.C.*, Nicosea.
Murphy, T.F. (1961) 'A ringfort at Old Court, Co. Cork', *J. Cork Archaeol Hist. Soc.* 66, 79–92.

* * *

N

Needham, J. (1970) *Clerks and Craftsmen in China and the West*, Cambridge.
Northover, P. (1982) 'The exploration of the long-distance movement of bronze in Bronze and Early Iron Age Europe', *Bull. Inst. Archaeol.* 19, 45–72.
----- (1985) 'The complete examination of archaeological metalwork', in ed. P. Phillips *The Archaeologist and the Laboratory*, Council for British Archaeology Research Report no. 58.
Nosek, E. (1975) 'The investigation of the iron-sponge fragments from Burgenland', in *Archäologische Eisenforschung in Europa*, Eisenstadt (*Wissenschaftliche Arbeiten aus dem Burgenland* 59), 71–82.
----- and Mazur, A. (1987) 'The oxidation of iron-carbon alloy at low temperature', in eds. B.G. Scott and H. Cleere, Belfast, 53–59.

* * *

O

O'Brien, E. (1986) '"Glen Munire", Ballyman, Co. Dublin', *Excavations 1986*, 16.
O'Brien, M.A. (1932) 'Old Irish *findruine*', *Ériu* XI, 170.
Ó Cathasaigh, T. (1977) *The Heroic Biography of Cormac Mac Airt*, Dublin.
O'Connor, A. and Clarke, D.V. eds. (1983) *From the Stone Age to the 'Forty-Five*, Edinburgh.
O'Connor, L. (1966) 'Faughart investigations 1966', *Co. Lough Archaeol. J.* 16(2), 125–129.
Ó Corráin, D. (1971) 'Irish regnal succession', *Stud. Hibernica* 11, 7–39.
----- (1974) 'Aspects of Early Irish History', in ed. B.G. Scott Belfast, 65–74.
-----, Breatnach, L. and Breen, A. (1984) 'The laws of the Irish', *Peritia* 3, 382–438.
Ó Corráin, D. ed. (1981) *Irish Antiquity*, Cork.
Ó Cuileanáin, C. (1955) 'Excavation of a circular stone house at Glannafeen, Co. Cork', *J. Roy. Soc. Antiq. Ireland* 85, 94–99.
O'Curry, E. (1875) *Manners and Customs of the Ancient Irish*, III Vols., Dublin.
Ó Dónaill = Ó Dónaill, N. (1977) *Foclóir Gaeilge-Béarla*, Dublin.
Ó Dúnlainge, M. (1907) 'Cath Mhuighe Mocruimhe IV', *Gaelic J.* 17(28), 434–439.
O'Keeffe, J.G. (1905) 'Cáin Domnaig', *Ériu* II, 189–214.
O'Kelly, M.J. (1952) 'St. Gobnet's House, Ballyvourney, Co. Cork', *J. Cork Archaeol. Hist. Soc.* 57, 18–40.
----- (1956) 'An island settlement at Beginish, Co. Kerry', *Proc. Roy. Irish Acad.* 57(C), 159–194.
----- (1959) 'Church Island, near Valencia, Co. Kerry', *Proc. Roy. Irish Acad.* 59(C), 1957–1959, 57–136.
----- (1961a) 'The ancient Irish method of smelting iron', *V Internationaler Kongress fur Vor- und Frühgeschichte*, Hamburg, 459–461.
----- (1961b) 'The Cork Horns, the Petrie Crown and the Bann Disc', *J. Cork. Hist. Archaeol. Soc.* 66, 1–12.
----- (1964) 'Two ringforts at Garryduff, Co. Cork', *Proc. Roy. Irish Acad.* 63(C), 1962–1964, 17–125.
----- (1973) 'Monastic sites in the west of Ireland', *Scottish Archaeol. Forum* 5, 1–16.
----- (1989) *Early Ireland*, Oxford.
Olshausen, O. (1915) 'Über Eisen in Altertum', *Prähist. Zeit.*, VII, 1–45.
Ó Máille, T. (1952) *Sean-fhocla Chonnacht*, 2 Vols., Dublin 1948–1952.
O'Neill, B.H. (1942) 'Excavations at Ffryd Faldwyn Camp, Montgomeryshire, 1937–1939', *Archaeol. Cambrensis* 97, 1–57.
O'Neill, H. (1967) 'Monastic mining and metallurgy', *Met. and Mat.* 1(6), 182–190.
O'Neill, J. (1907) 'The Rule of Ailbe of Emly', *Ériu* III, 92–115.
O'Rahilly, T.F. (1932) *Irish Dialects Past and Present*, Dublin (1972 reprint).
----- (1942) '*Iarann, Láraig* etc.', *Ériu* XIII, 119–127.
----- (1946) *Early Irish History and Mythology*, Dublin (1971 reprint).
O'Reilly = O'Reilly, E. (1864) *An Irish-English Dictionary*, Dublin.
Ó Ríordáin, S.P. (1940) 'Excavations at Cush, Co. Limerick', *Proc. Roy. Irish Acad.* 45C, 83–181.
----- (1942) 'The excavation of a large earthern ringfort at Garranes, Co. Cork' *Proc. Roy. Irish Acad.* 47C, 77–150.
----- (1946) 'Prehistory in Ireland, 1937–1946', *Proc. Prehist. Soc.* 12, 142–171.
----- (1949a) 'Lough Gur excavations: Carraig Aille and the "Spectacles"', *Proc. Roy. Irish Acad.* 52C, 39–111.
----- (1949b) 'Lough Gur excavations: 3 marshland habitation sites', *J. Roy. Soc, Antiq. Ireland* 79, 126–134.
----- and Foy, S. (1941) 'The excavation of Leacanabuaile stone fort, near Cahirciveen, Co. Kerry', *J. Cork Archaeol. and Hist. Soc.* 46, 85–99.
----- and Hartnett, P. (1943) 'The excavation of Ballycatteen fort, Co. Cork', *Proc. Roy. Irish Acad.* 49C, 1–43.
O'Sullivan, J. (1987) 'Lisnagun, Darrara' *Excavations 1988*, 11–12.

Ottaway, P. (1987) 'Anglo-Scandinavian knives from 16–22 Coppergate, York', in eds. B.G. Scott and H. Cleere, Belfast, 83–86.

* * *

P

de Paor, L. (1970) 'Iniscealtra', *Excavations 1970*.

PH = Atkinson, R. (1887) *The Passions and Homilies from Leabhar Breac*, (*Roy. Irish Acad. Todd Lecture Series*, II), Dublin.

Photos, E. (1988) *Early Extractive Metallurgy in Northern Greece: a unified approach to regional archaeometallurgy*, unpublished Ph.D. Thesis (Institute of Archaeology, London).

----- (1989a) 'The furnace illustrations on Greek vases: a new look at an old problem', in ed. R. Pleiner, Prague, 287–306.

----- (1989b) 'The question of meteoritic versus smelted nickel-rich iron: archaeological evidence and experimental results', *World Archaeol.* 20(3), 403–421.

Piaskowski, J. (1960) 'A socketed axe from Wietrzno-Bóbrka in the Carpathians', *J. Iron and Steel Inst.* 194, 336–340.

----- (1964) 'The method of determination of the characteristics and origins of iron objects on the basis of metallographic examinations', *Archaeol. Polonia* 6, 124–160.

----- (1965) 'Correlation between the phosphorous content of ore and slag and that in bloomery iron', *Archeol. Polonia* 7, 83–103.

----- (1969) 'Metallographische Untersuchungen der eisernern Erzeugnisse in der Hallstattzeit im Gebeit zwischen Oder und Weichsel', *Arbeits- u. Fundber. Sächsichen Bodendenkmalpflege* 7, 179–210.

----- (1971) 'Über die Kennzeichen der Zementation und des Schweissens an der frümittelalterlichen eisernen Gegenständen', *Acta Archaeol. Carpathica* 12(2), 127–159.

----- (1972) 'Kryteria okreslania technologii wyrobów z żelaza dymarskiego', *Archaeol. Polski* 17(1), 7–45.

----- (1974) 'Metallkundliche Untersuchungen an Eisengegenstaden aus dem Gräberfeld von Kornye', *Acta Archaeologica Hungaricae* 26, 117–130.

----- (1976) 'Classification of the structures of slag inclusions in early objects made of bloomery iron', *Archaeol. Polonia* XXVII, 139–149.

----- (1981) 'Zur Lokalisierung der antiken Kotiner', *Beiträge Ur- u. Frühgesch.* I, 675–686.

----- (1982a) 'A study of the origin of the ancient high-nickel iron generally regarded as meteoric' in eds. T.A. Wertime and S.F. Wertime *Early Pyrotechnology: The evolution of the First Iron-using Industry*, Washington (Smithsonian Institute), 237–243.

----- (1982b) 'Występowanie arsenu w starożytnych i wczeczesnośredniowieczbych przedmiotach z żelaza dymarskiego', *Kwart. Hist. Nauki i Tech.* 27(2), 397–409.

----- (1984) 'Das Vorkommen von Arsen in antiken und frühmittelalterlichen Geganständen aus Renneisen', *Zeit. Archaeol.* 18, 213–226.

----- (1985) 'Bemerkungen zu den Eisenverhüttungszentren auf polnischen Gebeit in ur- und frühgeschichtlicher Zeit', in *Produktionverhältnisse in ur- und frühgeschichtlicher Zeit*, Berlin, 231–243.

----- (1987a) 'A standardisation procedure for the presentation of metallographic examinations of early iron implements', in eds. B.G. Scott and H. Cleere, Belfast, 157–168.

----- (1987b) 'Proposals for a standardisation of the criteria for determining technological processes in early iron and steel', in eds. B.G. Scott and H. Cleere, Belfast, 169–178.

Píč, J. L. (1899) *Starožitnosti Země České,* Vol. I, Prague,

Piggott, V. (1982) 'The innovation of iron', *Expedition* Fall 1982, 20–25.

----- and McGovern, P.E. (1982) 'The earliest steel from Transjordan', *MASCA J.* 2(2), 35–39.

Pleiner, R. (1958) *Základy Slovanského Zelezářského hutnictví Českých Zemích,* Prague.

----- (1962) *Staré Evropské Kovářstvi,* Prague.

----- (1964) 'Die Eisenverhüttung in der "Germania Magna" zur römischen Kaiserzeit', *Ber. Röm-Germ. Komm.* 45, 11–86.

----- (1967) 'Die Techologie des Schmiedes in der grossmährischen Kultur', *Slovenská Archeol.* XV(1), 77–188.

----- (1969) 'Experimental smelting of steel in early Mediaeval furnaces', *Pam. Archeol.* LX, 458–487.

----- (1969b) *Iron Working in Ancient Greece*, Prague.

----- (1970) 'Zur Schmiedetechnik im römerzeitliche Bayern' *Bayerische Vorgeschichtsbl.* 35(1/2), 113–141.

----- (1971a) 'Zur Schmiedetechnologie des Grossmärischen Reiches (Jerzy Piaskowski und die Deutung der metallographischen Befunde von Eisenfundstücken)', *Acta Archaeol. Carpathica* 12, 107–223.

----- (1971b) 'The problem of the beginning of the Iron Age in India', *Acta Praehist. Archaeol.* 2, 5–36.

----- (1973) 'Metallography of early artifacts: the problem of welding together iron and steel', *Antikvarist Arkiv* 53, 17–28.

----- (1974) 'Nochmals zu J. Piaskowski's Erkennungsmerkmalen der alten Schmiedetechnik', *Acta Archaeol. Carpathica* XIV, 151–153.

----- (1979) *Otázka Státu Ve Staré Galii*, Prague.

----- (1980) 'Early iron metallurgy in Europe', in eds. T.A. Wertime and J.D. Muhly, Yale, 375–416.

----- (1981) 'Die Wege des Eisens nach Europa', in ed. R. Pleiner, Schaffhausen, 115–128.

----- (1982) 'Untersuchungen zur Schmiedetechnik auf den keltischen Oppida', *Pam. Archeol.* LXXIII, 86–173.

----- (1986) 'Über das Eisen der Bronzezeit', *Veröffentlichungen des Museums für Ur- und Frühgeschichte Potsdam* 20, 237–240.

----- (1987) 'Examination of some early La Tène period knives from Bohemia', in eds. B.G. Scott and H. Cleere, Belfast, 73–82.

----- (1988) 'Investigation into the quality of the earliest iron in Europe', *PACT* 21, 33–36.

Pleiner, R. ed. (1981) *Frühes Eisen in Europa, Festschrift W. U. Guyan*, Schaffhausen.

Pleiner, R. ed. (1989) *Archaeometallurgy of Iron 1967–1987*, Prague.

Podborský, V. (1970) *Mähren in der Spätbronzezeit und an der Schwelle der Eisenzeit*, Brno.

Pokorny, J. (1914) 'Herkunft und etymologie des Wortes *Eisen*', *Zeit. f. vergleichende Sprach.* 46, 292–294.

----- (1920) '*Aes* und *isarnon*', *Zeit. f. vergleichende Sprachforschung* 49, 126–128.

Portlock, J.E. (1845) *Report on the Geology of the County of Londonderry, and of Parts of Tyrone and Fermanagh*, Dublin.

Powell, T.G.E. (1976) 'The inception of the Iron Age in temperate Europe', *Proc. Prehist. Soc.* 42, 1–14.

Proudfoot, B. (1953) 'Excavation of a rath at Boho, Co. Fermanagh' *Ulster J. Archaeol.* 16, 41–57.

----- (1954) 'Excavations at the Cathedral Hill, Downpatrick, Co. Down, 1953', *Ulster J. Archaeol.* 17, 97–102.

----- (1958) 'Further excavations at Shaneen Park, Belfast, Ballyaghagan townland', *Ulster J. Archaeol.* 21, 18–38.

----- (1961) 'The economy of the Irish rath', *Med. Archaeol.* 5, 94–122.

----- (1970) 'Irish raths and cashels: some notes on chronology, origins and survivals', *Ulster J. Archaeol.* 33, 37–48.

R

Rackham, H. (1952) *Pliny, Natural History, Books XXXIII-XXXV*, London.

Raftery, B. (1969) 'Freestone Hill, Co. Kilkenny: an Iron Age hillfort and Bronze Age cairn', *Proc. Roy. Irish Acad.* 68(C), 1–108.

----- (1972) 'Irish Hillforts', in ed. C. Thomas, London, 37–58.

----- (1976a) 'Dowris, Hallstatt and La Tène in Ireland: problems of the transition from bronze to iron', in ed. S.J. de Laet *Acculturation and Continuity in Atlantic Europe*, Brugge, 189–197.

----- (1976b) 'Rathgall and Irish hillfort problems', in ed. D.W. Harding, London, 1976, 339–358.

----- (1981) 'Iron Age burials in Ireland', in ed. D. Ó Corráin, Cork, 173–204.

----- (1984) *La Tène in Ireland*, Marburg.

----- (1987) 'Ancient trackways in Corlea Bog, Co. Longford', *Archaeol. Ireland* 1(2), 60–64.

Raftery, J. (1940) 'A suggested chronology for the Irish Iron Age', in ed. J. Ryan *Féilsgríbhinn Eóin Mhic Néill*, Dublin, 272- 281.

----- (1943) 'Knocknalappa crannog, Co. Clare', *N. Munster Antiq. J.* 3, 1942–1943, 53–74.

----- (1944) 'The Turoe stone and the rath of Feerwore', *J. Roy. Soc. Antiq. Ireland* 74, 23–52.

----- (1963) 'A matter of time', *J. Roy. Soc. Antiq. Ireland* 93, 101–114.

----- (1972) 'Iron Age and Irish Sea: problems for research', in ed. C. Thomas, London, 1–10.

----- (1981) 'Concerning chronology', in ed. D. Ó Corráin, Cork, 82–92.

Rainbow, R. (1928) 'Socketed and looped axes from the British Isles', *Archaeol. J.* 85, 170–175.

Randsborg, K. (1972) *From Period III to Period IV*, Copenhagen.

Reed-Hill, R.E. (1973) *Physical Metallurgy Principles*, London (2nd edn.)

Rees, A. and Rees, B. (1961) *Celtic Heritage*, London (1973 reprint).

Rehder, J.E. (1989) 'Ancient carburisation of iron to steel', *Archaeomaterials* 3, 27–37.

Reith, A. (1942) *Der Eisentechnik der Hallstattzeit*, Leipzig (*Mannus Bucherei*).

RMLW = Latham, R.E. (1965) *Revised Medieval Latin Word-List*, London.

Reynolds, P.J. (1972) 'Poulacapple ringfort, Co. Tipperary', *Excavations 1972*, 27.

Rollason, E.C. (1973) *Metallurgy for Engineers*, London.

Rosenqvist, A. (1954) 'Studier av bronseteknikken i Vestby-Funnet', *Viking* (special edn. 1954), 125–155.

Ryan, M.F. (1982) 'Some archaeological comments on the occurrence and use of silver in pre-Viking Ireland' in ed. B.G. Scott, Belfast, 45–50.

Ryan, M.F. ed. (1978) *The Origins of Metallurgy in Atlantic Europe: Proceedings of the Vth Atlantic Colloquium*, Dublin.

Ryan, M.F. ed. (1987) *Ireland and Insular Art A.D. 500–1200*, Dublin (Royal Irish Academy).

Rynne, E. (1956) 'Excavation of a ringfort at Ardcloon, Co. Mayo', *J. Roy. Soc. Antiq. Ireland* 86, 203–212.

----- (1958) 'Iron axehead of La Tène date', *J. Roy. Soc. Antiq. Ireland* 88, 149–150.

----- (1961) 'The introduction of La Tène in Ireland', *Bericht über den V Internationaler Kongress für Vor- und Frühgeschicht Hamburg 1958*, Hamburg, 705–709.

----- (1979) 'An early Celtic Spanish-North Munster connection', *N. Munster Ant. J.* XXI, 7–10.

----- (1982) 'A classification of pre-Viking Irish iron swords', in ed. B.G. Scott, Belfast, 93–97.

----- (1983) 'Some Early Iron Age sword-hilts from Ireland and Scotland', in eds. A. O'Connor and D.V. Clarke, Edinburgh, 188–196.

----- (1984) 'Military and civilian swords from the River Corrib', *J. Galway Archaeol. Hist. Soc.* 39, 5–26.

----- (1985) 'An important seventh century sword from Co. Roscommon', *J. Old Athlone Soc.* II(6), 148–151.

----- (1987) 'The date of the Ardagh Chalice', in ed. M.F. Ryan, Dublin, 85–89.

* * *

S

Salter, C.J. (1982) 'The relevance of chemical provenance studies to Celtic ironwork in Britain', *Bull. Inst. Archaeol. Univ. London* 19, 73–81.

Samuels, L.E. (1980) *Optical Microscopy of Carbon Steels*, Ohio (American Society for Metals).

Savory, H.N. (1976) 'Welsh hillforts: a reappraisal of recent work' in ed. D.W. Harding, London, 237–292.

Scott, B.G. (1970) 'Excavation of a ringfort at Urney Park, Strabane, Co. Tyrone', *Excavations 1970*.

----- (1971) 'Applications of metallographic examination of iron artifacts to Irish archaeology', *Ulster J. Archaeol.* 34, 87–95.

----- (1973) 'Early Irish "tribes" and "tribalism"', *Ogam* XXII-XXV, 1970–1973, 197–208.

----- (1974a) 'Notes on the development of metallurgical studies on ancient iron', *J. Hist. Met. Soc.* 8(2), 88–91.

----- (1974b) 'Notes on the transition from bronze to iron in Ireland', *Irish Archaeol. Res. Forum* I(1), 9–24.

----- (1974c) 'Reply to R.B. Warner', *Irish Archaeol. Res. Forum* I(2), 49–52.

----- (1974d) 'Applications of techniques of physical examination to archaeological research', in ed. B.G. Scott, Belfast, 107–119.

----- (1976a) 'A note on the application of X-radiography in the conservation and study of archaeological ironwork', *Irish Archaeol Res. Forum* III(1), 1–8.

----- (1976b) 'Remanent structures in corrosion products from an early mediaeval iron knife', *Irish Archaeol. Res. Forum* III(2), 61–64.

----- (1977a) 'Notes on the introduction of non-ferrous metal technology to Ireland, and the transition from stone-use to indigenous non-ferrous metal-use', *Irish Archaeol. Res. Forum* IV(2), 7–15.

----- (1977b) 'Metallographic examination of some early iron tools and weapons from Ireland', *Proc. Roy. Irish Acad.* 77(C)12, 301–317.

----- (1978a) 'The introductions of non-ferrous and ferrous metallurgies to Ireland: motives and mechanisms', in ed. M.F. Ryan, Dublin, 189–204.

----- (1978b) 'Iron 'slave collars' from Lagore crannog, Co. Westmeath', *Proc. Roy. Irish Acad.* 78C, 213–230.

----- (1981a) 'The origins and early development of iron use in Ireland', in ed. R. Pleiner, Schaffhausen, 101–108.

----- (1981b) 'Early Irish *cáer, íarn aithlegtha, créduma*', *Ériu* XXXII, 153–157.

----- (1981c) 'Goldworking terms in early Irish writings', *Zeit. f. Celtische Philologie* 38, 242–254.

----- (1982) 'Some conflicts and correspondences of evidence in the study of Irish archaeology and language', in ed. B.G. Scott, Belfast, 115–119.

----- (1983a) 'Some problems in the reconstruction of Indo-European metalworking vocabularies', *Offa* 40, 19–28.

----- (1983b) 'An early law tract on the blacksmith's forge', *J. Irish Archaeol.* I, 59–62.

----- (1985) 'The blast-furnace in Ireland: a failed industry', in *Medieval Iron in Society, Vol. I*, Stockholm (*Jernkontorets Forskning* H34), 286–298.

----- (1987) 'The status of the smith in early Irish literature', in eds. B.G. Scott and H. Cleere, Belfast, 153–156.

----- (1988) 'Some references to the mining of iron ores in early Irish texts', *PACT* 21, 111–116.

----- (1989a) 'The retrieval of information from ferrous corrosion products', in eds. R.C. Janaway and B.G. Scott *Evidence Preserved in Corrosion Products: New Fields in Artifact Studies*, London (*UKIC Occasional Papers* no. 8), 1988, 8–14.

----- (1989b) 'Analysis of texts and metal artefacts: problems and potential' in ed. J. Henderson *Scientific Analysis in Archaeology, and its Interpretation*, Oxford (*Oxford University Committee for Archaeology Monograph 19*), 237–249.

----- and Francis, P.J. (1981) 'Native copper in north-east Ulster', in ed. D. Ó Corráin, Cork, 28–42.

Scott, B.G. ed. (1974) *Perspectives in Irish Archaeology*, Belfast.

----- ed. (1982) *Studies on Early Ireland: Essays in Honour of M.V. Duignan*, Belfast.

----- and Cleere, H.F. eds. (1987) *The Crafts of the Ancient Blacksmith*, Belfast.

Scott, J.G. and Powell, T.G.E. (1969) 'A bronze horse figurine found near Birkwood, Lesmahgow, Lanarkshire', *Ant. J.* 49, 118–126.

Selkirk, A. and Waterman, D.M.W. (1970) 'Navan Fort', *Current Archaeol.* 22, 304–308.

Serning, I. (1984) 'The dawn of Swedish iron metallurgy', *Bull. Met. Mus. Japan Inst. Met.* 9, 3–19.

Service, E. (1962) *Primitive Social Organisation*, New York.

SG = O'Grady, S.H. (1892) *Silva Gadelica*, 2 Vols., New York (1970 reprint).

Shee, E. (1974a) 'Croom East, Co. Limerick', *Excavations 1974*, 23–24.

----- (1974b) 'Sluggary, Co. Limerick', *Excavations 1974*, 22–23.

Sheldon-Williams, I.P. (1968) *Iohannes Scotti Eriugenae, Periphyseon* (*De Diuisione Naturae*), 2 Vols., Dublin (*Scriptores Latini Hiberniae* 7).

Smith, C.S. (1960) *The Coming of the Age of Steel*, Harvard.

----- (1967) 'The interpretation of microstructures of metallic artifacts' in ed. W.J. Young *Application of Science in Examination of Works of Art*, Boston, 20–52.

----- (1973) 'An examination of the arsenic-rich coating on a bronze bull from Horoztepe' in ed. W.J. Young *Application of Science in Examination of Works of Art*, Boston, 96–102.

Smith, R.H., Maddin, R., Muhly, J.D. and Stech, T. (1984) 'Bronze Age Steel from Pella, Jordan', *Current Anthrop.* 25(2), 234–236.

Snodgrass, A. (1980) 'Iron and early metallurgy in the Mediterranean' in ed. T.A. Wertime and J.D. Muhly, Yale 1980, 335–374.

Sommerfelt, A. (1975) 'The Norse influence on Irish and Scottish Gaelic', in ed. B. Ó Cuív *The Impact of the Scandinavian Invasions on the Celtic-speaking Peoples*, Dublin, 73–77.

Šorm, F. and Schwarz, Š. eds. (1969) *Hallstatt a Býcí Skála*, Prague.

Sperl, G. (1980) *Über die Typologie urzeitlicher frügeschichtlicher und mittelaltlicher Eisenhüttenschlacken*, Wien (*Österreichischen Akademie der Wissenschaften*).

Spratt, D.A. (1982) 'The analysis of innovation processes', *J. Archaeol. Sci.* 9, 79–94.

Sprockhoff, E. (1931) *Die Germanischen Griffzungenschwerter*, Berlin.

St. Ercuil = Quin, E.G. (1939) *Stair Ercuil agus a Bás* (*Irish Texts Soc.*, XXXVIII), Dublin.

Stech-Wheeler, T. and Maddin, R. (1980) 'Metallurgy and ancient man', in eds. T.A. Wertime and J.D. Muhly, Yale, 99–126.

----- Muhly, J.D., Maxwell-Hyslop, K.R., and Maddin, R. (1981) 'Iron at Ta'anach and early iron metallurgy in the Eastern Mediterranean', *American J. Archaeol.* 85, 246–268.

Sternqvist, B. (1961) *Simris II, Bronze Age Problems in the Light of the Simris Excavations*, Bonn.

Stokes, W. (1891) 'Scél na Fír Flatha, Echtra Cormaic i Tir Tairngiri ocus Ceart Claidib Cormaic', *Irische Texte*, 3rd. ser., I, Leipzig, 185–229.

----- (1901) 'Orgain Dind Ríg', *Zeit. f. Celtische Phil.* III, 3–14.

----- (1903) 'The death of Crimthann, Son of Fidach, and the adventures of the Sons of Eochaid Muigmedon', *Rev. Celtique* XXIV, 172–207.

Straker, E. (1931) *Wealden Iron*, London (1969 reprint).

Stránský, K. (1984) 'Novodobá historie železného prstenu z Býči Skály', in ed. C. Němec, *Wankeluv Nález v Býči Skále,* Blansko (Okresní Muzeum), 45–54.

Swann, L. (1973) 'Kilpatrick, Co. Westmeath', *Excavations 1973*, 26–27.

Swanton, M.J. (1974) *A Corpus of Pagan Anglo-Saxon Spear-Types* (*British Archaeol. Rep.*, 7), Oxford.

* * *

T

Táin = O'Rahilly, C. (1970) *Táin Bó Cúalnge from the Book of Leinster*, Dublin.

TBC² = Strachan, J. and O'Keeffe, J.G. (1967) *Tain Bo Cualnge from the Yellow Book of Lecan*, Dublin (reprinted from *Ériu*).

Thålin, L. (1967) 'Metallografisk undersökning av ett vendeltidska praktsvärd', *Fornvännen* 62, 225–240.

Thålin-Bergmann, L. (1979) 'Blacksmithing in prehistoric Sweden', in ed. H. Clarke, Stockholm, 99–133.

Thes. = Stokes, W. and Strachan, J. (1901) *Thesaurus Palaeohibernicus*, Dublin (1975 reprint).

Tholander, E. (1971) 'Evidence of the use of carburised steel and quench hardening in Late Bronze Age Cyprus', *Opuscula Atheniensia* 10, 15–22.

----- (1978) 'A study of the technology behind nickel-alloyed prehistoric steel having a laminated structure', in ed. M.F. Ryan, Dublin, 319–334.

----- (1987) *Experimental Studies on Early Iron-Making*, Stockholm.

Thomas, C. ed. (1972) *The Iron Age in the Irish Sea Province*, London (*Council British Archaeol. Res. Rep.*, 9).

Thomsen, R. (1964) 'Trial reconstruction of an early form of iron production', *Kuml* 1964, 60–74.

----- (1971) 'Essestein und Ausheizschlacken – Zur Technik des wikingzeitlichen Schmeides', *Ausgrabungen in Haithabu* 5, 100–109.

----- (1989) 'Pattern-welded swords from Illerup and Nydam', in ed. R. Pleiner, Prague, 371–378.

Thr. Ir. Gl. = Stokes, W. (1862) *Three Irish Glossaries*, London.

Thurn. Gramm. = Thurneysen, R. (1946) *A Grammar of Old Irish*, Dublin (1970 reprint).

Thurneysen, R. (1925) 'Aus dem Irischen Recht II' *Zeit. f. Celtische Phil.* XV, 238–296.

Toffler, A. (1981) *The Third Wave*, London.

Tomtlund, J-E. (1973) 'Metallographic investigation of 13 iron knives from Helgö', *Early Med. Stud.* 5, 42–63.

Triads = Meyer, K. (1906) *The Triads of Ireland* (*Roy. Irish Acad. Todd Lecture Series*, XIII), Dublin.

Trip. = Stokes, W. (1887) *Tripartite Life of St. Patrick*, 2 Vols., London.

Tylecote, R.F. (1962) *Metallurgy in Archaeology*, Edinburgh.
----- (1965) 'Iron smelting in pre-industrial communities', *J. Iron Steel Inst.* 203, 340–348.
----- (1976) *A History of Metallurgy*, London.
----- (1986) *The Prehistory of Metallurgy in the British Isles*, London (Institute of Metals).
----- (1987) *The Early History of Metallurgy in Europe*, London.
----- Austin, J.N. and Wraith, A.E. (1971) 'The mechanism of the bloomery process in shaft furnaces', *J. Iron and Steel Inst.* 342–363.
----- and Boydell, P.J. (1978) 'Experiments on copper smelting based on early furnaces found at Timna', *Archaeo-metallurgy* 1, 27–51.
----- and Gilmour, B.J.J. (1986) *The Metallography of Early Ferrous Edge Tools and Edged Weapons* (*British Archaeol. Rep.*, 155), Oxford.
----- and Thomsen, R. (1973) 'The segregation and surface enrichment of arsenic and phosphorous in early iron artefacts', *Archaeometry* 15(2), 193–198.
Tyrkiel E.F. (1977) *Dictionary of Physical Metallurgy*, Warsaw.

* * *

V

Varoufakis, G. (1981) 'Investigation of some Minoan and Mycenean iron objects', in ed. R. Pleiner, Schaffhausen, 25–32.
Vendr. Lex. = Vendryes, J. *Lexique Etymologique de l'Irlandais Ancien*, Dublin and Paris, 1959–present.
van Vlack, L. H. (1968) *Elements of Materials Science*, Massachussets.
de Vries, J. (1962) *Altnordisches Etymologisches Wörterbuch*, Leiden.
Vouga, P. (1923) *La Tène: monographie de la station*, Leipzig.

* * *

W

Waddell, J. (1978) 'The invasion hypothesis in Irish prehistory', *Antiquity* 52, 121–128.
Wagner, D.B. (1990) 'Ancient carburisation of iron to steel: a comment', *Archeomaterials* 4, 111–117.
Wailes, B. (1976) 'Dún Ailinne: an interim report' in ed. D.W. Harding, London 1976, 319–338.
Wakeman, W.F. (1871) 'Remarks on the crannog at Ballydoolough, Co. Fermanagh', *J. Roy. Soc. Antiq. Ireland* XI, 1870–1871, 360–371.
----- (1898) 'On the antiquity of iron as used in the manufacture of certain weapons, implements and ornaments found in Ireland', *J. Roy. Soc. Antiq. Ireland* 8, 237–244.
Waldbaum, J.C. (1978) *From Bronze to Iron*, Göteborg, (*Studies in Mediterranean Archaeology*, 56).
----- (1982) 'Bimetallic objects from the Eastern Mediterranean and the question of the dissemination of iron' in eds. J.D. Muhly, R. Maddin and V. Karagheorghis, Nicosea, 325–350.
Wallace, P. and Timoney, M. 'Carrowntemple, Co. Sligo and its inscribed slabs' in ed. E. Rynne *Figures from the Past*, Dublin, 43–61.
Walshe, A. (1987) 'Excavating the Black Pig's Dyke' *Emania* 3, 4–11.
Warden, P.G., Maddin, R., Stech, T. and Muhly, J.D. (1982) 'Copper and iron production at Poggio Civitate', *Expedition* Fall 1982, 26–35.
Warhurst, C. (1971) 'Excavation of a rath at Shanes Castle, Co. Antrim', *Ulster J. Archaeol.* 34, 58–64.
Warner, R.B. (1972) 'The excavations at Clogher and their context', *Clogher Record* 8, 5–12.
----- (1974) 'The bronze/iron transition: a pessimistic view', *Irish Archaeol. Res. Forum* I(2), 41–42.
----- (1976) 'Some observations on the context and importation of exotic material in Ireland', *Proc. Roy. Irish Acad.* 76C, 267–292.
----- (1979) 'The archaeology of historic sites: some observations', *Bull. Ulster Placenames Soc.* 2, 43–45.
----- (1982) 'The Broighter hoard: a re-appraisal, and the iconography of the collar', in ed. B.G. Scott, Belfast, 29–38.
----- (1983) 'Ireland, Ulster and Scotland in the Earlier Iron Age', in eds. A. O'Connor and D.V. Clarke, Edinburgh, 160–187.
----- (1986) 'The date of the start of Lagore', *J. Irish Archaeol.* III, 1985–1986, 75–77.
----- (1987) 'Ireland and the origins of escutcheon art' in ed. M.F. Tyan, Dublin, 19–22.
----- (1988) 'The archaeology of early historic Irish kingship' in eds. S.T. Driscoll and M.R. Nieke *Power and Politics in Early Mediaeval Britain and Ireland*, Edinburgh, 47–68.
Waterman, D.M. (1958) 'Excavations at Ballyfounder rath, Co. Down', *Ulster J. Archaeol.* 21, 39–61.
Weatherup, D.R.M (1982) 'Armagh County Museum Archaeological Acquisitions: The Collection of the Armagh Natural History and Philosophical Society', *J. Roy. Soc. Antiq. Ireland* 112, 51–71.
Wertime, T.A. (1980) 'The pyrotechnological background' in eds. T.A. Wertime and J.D. Muhly, Yale, 1–24.
Wertime, T.A. and Muhly, J.D. eds. (1980) *The Coming of the Age of Iron*, Yale.
West, M.L. (1966) *Hesiod: Theogony*, Oxford.
----- (1988) *Hesiod: Theogony and Works and Days*, Oxford.
Wheeler, R.E.M. (1932) *Excavation of the prehistoric, Roman and post-Roman site in Lydney Park, Gloucs.*, London (*Rep. Soc. Antiq. London*, no. IX).
Willen, D., Soedel, W. and Foley, V. (1976) 'The story of Weland the Smith', *J. Hist. Met. Soc.* 10(2), 84–86.
Williams, B.B. (1978) 'Excavations at Lough Eskragh, Co. Tyrone', *Ulster J. Archaeol.* 41, 37–48.
----- (1985a) 'Excavations at Ballyvollen townland, Co. Antrim', *Ulster J. Archaeol.* 48, 91–102.
----- (1985b) 'Excavation of a rath at Coolcran, Co. Fermanagh', *Ulster J. Archaeol.* 48, 69–80.
----- (1986) 'Excavations at Altanagh, County Tyrone', *Ulster J. Archaeol.* 49, 33–88.
Wilson, H.E. (1964) 'The rise and decline of the iron ore and bauxite industry of Co. Antrim', *Proc. Belfast Nat. Hist. Phil. Soc.* 7 (2nd ser.), 14–23.
----- and Robbie, J.E. (1966) *Geology of the Country around Ballycastle*, Belfast (HMSO).
Wood-Martin, W.G. (1886) *The Lake Dwellings of Ireland*, London (1983 reprint).
Wulff, H.E. (1966) *The Traditional Crafts of Persia*, London.
Wynne, E.J. and Tylecote, R.F. (1958) 'An experimental investigation into primitive iron smelting techniques', *J. Iron and Steel Inst.* 190, 339–348.

Topographical Index

Emboldened numbers refer to illustrations: suffix n *refers to footnotes*

Subject Index

Emboldened numbers refer to illustrations: suffix n *refers to footnotes*